The Physics and Mathematics of Electroencephalogram

This book focuses on a systematic introduction to the knowledge of mathematics and physics of electroencephalogram (EEG) and discusses an in-depth application of EEG and the development of new methods and technologies for mining and analyzing EEG.

The Physics and Mathematics of Electroencephalogram offers a systematic overview of the technology for brain function and disease. It covers six parts: background knowledge of EEG, EEG forward problems, high-resolution EEG imaging, EEG inverse problems, EEG reference electrode, and EEG cloud platform. The author reviews the critical technologies in brain function and disease, such as EEG sourcing, EEG imaging, and EEG reference electrode standardization technique. The book's aim is to clarify the mechanism of EEG from the perspective of physics, mathematics, and engineering science to help multidisciplinary readers better understand and use EEG information more effectively.

This book can be used as reference for researchers in the fields of neuroengineering, cognitive neuroscience, neurology, psychiatry, applied mathematics, and brain-like intelligence.

"I strongly recommend *The Physics and Mathematics of Electroencephalogram* by Dezhong Yao. This comprehensive text provides scientists with valuable source material explaining the theoretical bases for the brain's large scale electric fields plus practical implementations of this technology. Yao's research group has developed new ideas as well as providing complementary treatments of existing EEG science."

Paul L. Nunez

Author (with Ramesh Srinivasan) of Electric Fields of the Brain: The Neurophysics of EEG, 2nd edition *(2006).*

"This book provides a captivating exploration of the electromagnetics and biophysics that underlie the enigmatic realm of brain electric fields. It offers a thorough examination of the intricacies of EEG signals, precisely explaining how these electric fields manifest on the scalp and how this knowledge enables the solution of the EEG inverse problem. The book is an indispensable resource and invaluable guide for both novices and seasoned professionals."

Christoph M. Michel

Professor at the University of Geneva, Co-Editor of the book Electrical Neuroimaging, *Co-Editor-in-chief of the journal* Brain Topography

"In this seminal work, the intricate science of scalp electroencephalography (EEG) is demystified, blending physics and mathematics to shed light on previously opaque areas. The text delves into fundamental and complex topics, offering unprecedented solutions to the longstanding conundrum of EEG zero reference and the multifaceted nature of source localization methods. Authored by a central figure in China's burgeoning EEG research landscape, this book has propelled theoretical and technical strides in the field. Now accessible in English, it extends its rich insights to an international cadre of scholars and practitioners."

Pedro A. Valdes-Sosa

M.D., Ph.D., D.Sc, Co-Director of the Global Brain Consortium; Member Emeritus of the Cuban Academy of Science; Fellow of the Organization for Human Brain Mapping

The Physics and Mathematics of Electroencephalogram

Dezhong Yao

CRC Press is an imprint of the
Taylor & Francis Group, an **informa** business

First edition published 2024
by CRC Press
2385 NW Executive Center Drive, Suite 320, Boca Raton FL 33431

and by CRC Press
4 Park Square, Milton Park, Abingdon, Oxon, OX14 4RN

CRC Press is an imprint of Taylor & Francis Group, LLC

ISBN: 978-1-032-62247-7 (hbk)
ISBN: 978-1-032-63924-6 (pbk)
ISBN: 978-1-032-63926-0 (ebk)

DOI: 10.1201/9781032639260

Typeset in Times
by codeMantra

Contents

Author

Dezhong Yao received the first PhD in Applied Geophysics from the Chengdu University of Technology (1991) and the second PhD in Biomedical Science and Engineering from Aalborg University (2005). He has been a faculty member with the University of Electronic Science and Technology of China(UESTC) since 1993, a full professor since 1995. He has been a one-year visiting scholar at the University of Illinois at Chicago from September 1997 and a half-year visiting professor at McMaster University from November 2000. He won the "Outstanding Youth Research Fund" of National Natural Science Foundation of China in 2005 and selected as the Yangtze River Scholar Professor in 2006 and National Excellent Teacher in 2010. He innovated REST for zero electroencephalogram (EEG) reference at infinity and proposed the concept of "Brain-Apparatus Communication" (BAC) to integrate brain–computer interface and psychosomatics. He set up the School of Life Science and Technology at the UESTC in 2001 and founded the "Sichuan Institute for Brain Science and Brain-Inspired Intelligence" in 2018. He has cultivated 60+ PhDs, published 5 books and 300+ journal papers, hosted 20+ conferences, and presented 100+ talks on zero reference of EEG, force of music, simultaneous EEG-functional magnetic resonance, BAC-based brain modulation, cloud platform of EEG, etc. He is the fellow of the American Institute for Medical and Biological Engineering, the winner of the Roy John Award of the EEG and Clinical Neuroscience Society, the chief editor of the BAC journal, and the associate editor of IEEE T-NSRE. He is the chairman of the Chinese EEG Consortium (2019–now), the vice-chairman of the Chinese Society of Biomedical Engineering (2015–2023), and the steering committee member of the Global Brain Consortium (2018–now).

Preface

During the last century, especially in more recent decades, substantial progresses have been made at the microscopic scale of neurochemistry and neurophysiology in brain research. For example, the important roles of synapses and neurotransmitters in control synaptic transmission are disclosed. In membrane biophysics, the basic laws of action potential propagation along axons are formed. These knowledges are critical for us to understand how sensory signals reach the central nervous system (input) and how the signals generated by the brain are transmitted to body organs (output). However, little is known about the higher brain function and diseases of the brain. Actually, look at the vivid brains, we can only use very few non-invasive means to understand them with limited spatial resolution. And electroencephalogram (EEG) may be the most commonly used one.

The higher brain function, such as thought and emotion that determine our behavior, is definitely supported by cell populations in the brain, which needs more in-depth studies. Although we know that some cell populations are essential for the normal function of the brain, such as the hippocampus for memory and thalamus in generalized epilepsy. However, memory is not limited to the hippocampus, or in other words, the hippocampus is just a part of a larger network, which means that anatomy cannot give us all we need to explore the whole brain network. EEG is one of the most useful non-invasive technology for brain function and functional network.

EEG mainly is the sum of the post-synaptic potentials of a large neuron population. It is the overall representation of the electrophysiological activities of the all cells in the brain. The magnetoencephalogram (MEG), closely related to EEG, is the sum of the extracranial magnetic field generated by the weak current from the migration of charged ions related to neurons. EEG and MEG are from the same neural source but sensitive to different aspects of the source; no one is absolutely superior to the other. For example, EEG is more sensitive to radial current, while MEG is more sensitive to tangential current. EEG equipment is inexpensive and widely used. MEG equipment and maintenance are expensive, although MEG is less affected by the non-uniformity of the brain structure and is free of the reference electrode problem. In general, EEG and MEG are not a simple distinction between good and bad but two important complementary technologies. In this book, presented is EEG only for its greater acceptance in practice.

Two other non-invasive techniques commonly used in brain function research in recent years are functional near-infrared spectroscopy (fNIRS) and functional magnetic resonance (fMRI), which are blood oxygen level-dependent signals; thus, they are indirect reflections of neural activities, with a higher spatial resolution but a lower time resolution. As shown in Figure A, based on "Web of Science," the number of journal papers on EEG, MEG, fNIRS, and fMRI from 2001 to 2021 implicated an increasing trend of the all four techniques year by year. Among them, EEG is almost always at the leading position, except being shortly surpassed by fMRI from 2012 to 2014. It means an unshakable and unique position of EEG in brain function and disease studies.

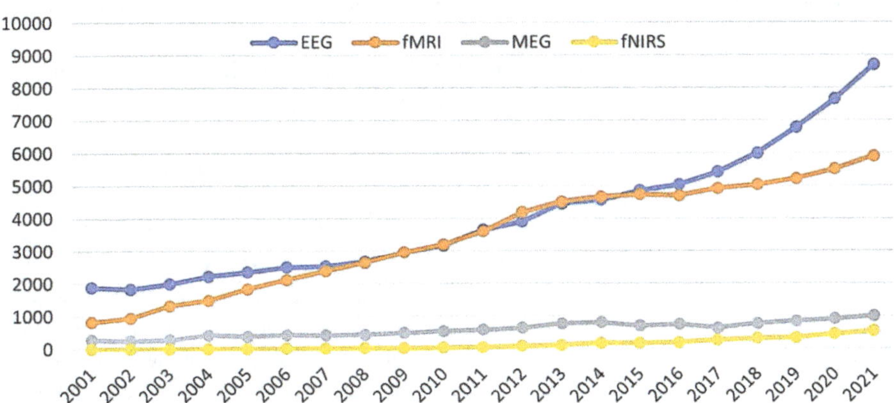

FIGURE A Journal article number of typical non-invasive brain imaging technologies from 2001 to 2021. EEG, electroencephalogram; MEG, magnetoencephalography; fNIRS, functional near-infrared spectroscopy; fMRI, functional magnetic resonance imaging.

EEG problems can be simply divided into two categories. One is to invert the source position information in the brain from the spatial potential distribution on the scalp surface, that is, the EEG inverse problem, which includes source localization techniques and high-resolution EEG mapping (compared to the resolution of the scalp potential distribution). The other is to mine the time-domain features from the temporal series in EEG records, including identification of characteristic waves in an EEG trace, linear, and non-linear measures of the trace, and the correlation between the activities of different brain regions (traces) thus to construct the brain network. The third approach is to combine these two by a neurodynamic model, that is, the neurocomputational approach. This book mainly focuses on spatial analysis with 6 parts and 15 chapters. The first part, Biophysical Basis of EEG, with Chapters 1–3 is about the overall features of EEG, the fundamental electricity behind the brain electric field, and the equivalent source models in current EEG studies; they are the equivalent charge, equivalent dipole, equivalent potential model, and their variants. The second part, Forward Model of EEG, with Chapters 4 and 5 is about the EEG forward problem, including the analytical theory of the EEG forward problem for a homogeneous sphere model, the spherical harmonic series for multilayer sphere, and the numerical solution for a realistic shape head. The third part, Equivalent Distributed Source and High-resolution Imaging, with Chapters 6–9 is about high-resolution EEG imaging technologies, including surface Laplacian approach, equivalent distributed source approach, cortical potential or potential in an infinite homogeneous space, and a spherical harmonic spectrum-based theoretical framework unifying those approaches of high-resolution EEG. The fourth part, Theory and Algorithm of EEG Inverse, with Chapters 10–12 focuses on the EEG inverse problem, such as low-resolution tomography, multi-signal classification, and various linear iteration algorithms. The fifth part, Zero EEG Reference, with Chapters 13 and 14 is about the reference electrode, principally the new zero reference is introduced and proved under the view of a spatial-temporal problem, and its value for various applications is illustrated. The last part, Cloud EEG Platform, with Chapter 15 is the introduction of the EEG cloud platform WeBrain for large-scale EEG and EEG-fMRI datasets, contributed by Li Dong. As a novel EEG/ERP study style, WeBrain is easy to use for users (even with no computer programming skills) to run a study on a modern web browser of any kind.

To our knowledge, although the brain is the most complex system in nature with many miraculous functional manifestations, its operation at all levels is carried out according to known physical laws and their mathematical descriptions, and there are no exceptional reports. Thus, a systematic introduction to the knowledge of mathematics and physics of EEG would be quite valuable for the domain future. However, due to the diversity of readers, those complex mathematics and physics are separately presented in appendices. Appendices A, C, E, G and H focus on the basic mathematics needed to understand the basic logics of the various algorithms in current EEG data processing. Specially, Appendix A focuses on special functions related to the basic Poisson equation of EEG potential. Appendix C focuses on Green function for equivalent distributed source on a closed surface. Appendix E focuses on the basic theory of linear inversion which is the basis of the main body of the EEG inverse studies. Appendix G provides the matrix theory and Bayesian estimate theory of the reference choices. Appendix H focuses on the fundamental knowledge of vector and matrix. Appendices B, D, and F focus on basic electricity-based mathematical derivation to show the physical basis of the formulas in the following EEG spatial analysis. Besides, there is a certain degree of independence between chapters, so it is not necessary to follow the order, they can be read separately for readers with different backgrounds.

It is hoped that the book could bring inspiration for readers whose interest is an in-depth application of EEG or the development of new methods and technologies for mining and analyzing EEG. Accordingly, this book is suitable to be used as a reference book for researchers or a textbook for students in neuroengineering, cognitive neuroscience, cognitive psychology, neurology, psychiatry, neuroelectrophysiology, biomedical engineering, biophysics, applied mathematics, engineering electromagnetics, etc.

The content is funded by STI 2030–Major Projects 2022ZD0208500 and a few grants from the National Natural Science Foundation of China in the past years. The main contents are accumulated in the past 30 years, with contributions from many students in my lab such as Peng Xu, Li Dong, Yiran Zhai, Junpeng Zhang, Xu Lei, Shiang Hu, Yun Qin, Yingchun Zhou, Yongxiu Lai, and Tiejun Liu, etc. and big helps from collaborators such as Pedro Antonio Valdés-Sosa, Paul L. Nunez, Christoph M. Michel, Bin He, and Andrew Chen. Specifically, Ms Yongxiu Lai, Shuang Yu and Ting Zhang made great efforts in editing and smoothing the language translation. I would like to express my gratitude here to all of them and many other collaborators not specifically mentioned.

MATLAB® is a registered trademark of The MathWorks, Inc. For product information, please contact:

The MathWorks, Inc.
3 Apple Hill Drive
Natick, MA 01760-2098 USA
Tel: 508-647-7000
Fax: 508-647-7001
E-mail: info@mathworks.com

Introduction

Exploring the brain, so as to understand the human being itself, is a continuously important topic in multidisciplinary fields. As the information automatically released from mental activity, including normal or abnormal ones, electroencephalogram (EEG) provides a critical window for humans to peep at the brain non-invasively. As a completely non-dangerous and low-cost technology, EEG can be used to probe not only the functional response under external stimuli but also the resting spontaneous functional state. And the research results can be more easily shared by all mankind. Therefore, it is of obvious significance to clarify the mechanism of EEG from the perspective of physics, mathematics, and engineering science, so as to help multidisciplinary people to better understand and use EEG information more effectively. This book attempts to systematically show the general overview of the EEG technology. The book includes 6 parts and 15 chapters. The first part is about the background knowledge of EEG (Chapters 1–3), the second part is for EEG forward (Chapters 4 and 5), the third part is for high-resolution EEG imaging (Chapters 6–9), the fourth part is for EEG inverse problems (Chapters 10 and 12), the fifth part is for EEG reference electrode, and the last part is about the platform for EEG study in the cloud. The 2–4 parts are all for spatial analysis of EEG, and the reference problem in Part 5 looks like a temporal problem, but the root is a spatial problem too. Therefore, this book is mainly for the mathematics and physics in spatial problems of EEG.

This book can be used as reference book or teaching materials for researchers in neuroengineering, cognitive neuroscience, neurology, psychiatry, applied mathematics, and brain-like intelligence.

1 Overview of EEG

Following the discovery of human electroencephalogram (EEG) in 1929, significant progress has been made in its neural mechanism, recording technologies, and signal-processing algorithms. Certainly, various applications are the critical driving forces. Now, EEG has been widely adopted in various fields, such as the diagnosis of neurological and mental disorders, drug-treatment effect evaluation, special talent selection, brain cognitive function study, and recent brain-apparatus communication (BAC) development (including brain–computer interface, brain–living organ interaction, neurofeedback, etc.) (Yao et al., 2020, 2022). There are several excellent monographs on these issues, such as EEG for clinical application (Niedermeyer and Silva, 1999), the event-related potential for brain cognitive function (Luck, 2014), and neurophysics of EEG (Nunez and Srinivasan, 2006; Ilmoniemi and Sarvas, 2019; Knosche and Haueisen, 2022). This book pays more attention to the engineering physics and mathematics of EEG. This chapter will give a brief introduction to the background of EEG, including its history, measurement technology, and the fundamentals of evoked EEG.

1.1 BRIEF HISTORY

The importance of electricity in physiological processes was discovered in 1789 by Luigi Galvani (1737–1798). Richard Caton (1842–1926) has been credited as being the first person to study the electrical activity of the brain. He was a doctor who worked in Liverpool, England, and was commissioned by the British Medical Association to explore the electrical phenomena in the exposed cerebral hemispheres of the rabbit and monkey. Caton first reported his findings to the Association on August 24, 1875, and then published a brief report in British Medical Journal, with 20 lines of words only. In 1877, he published a more detailed report in the same journal. His experiment involved rabbits, cats, and monkeys. Caton used a galvanometer in the experiment and was regarded as the discoverer of the fluctuating potentials that make up the brain waves. He also noted that "when any part of the gray matter is in a functionally active state, the corresponding current usually undergoes a negative change." Therefore, he is also considered a pioneer in evoked EEG. In the same period of Caton, some other physiologists in Europe also independently carried out research on the brain and its electrical activities. One of the discoveries with a significant influence on the neuroscience community was made by Fritsch (1838–1927) and Hitzig (1838–1907) in 1870; it disclosed that the cerebral cortex can be affected by electrical stimulation. The physiologist Ernst Fleischl von Marxow (1846–1891) succeeded in recording the evoked potential on the skin of a frog in 1883. In 1912, Pravdich-Neminsky used photographic paper to record the EEG of

DOI: 10.1201/9781032639260-1

FIGURE 1.1 Human EEG. The first human EEG published by Berger (1929), with his son as the subject.

the dog and created the term "electro-cerebrogram" (Pravdich-Neminsky, 1912). In 1920, Forbes and Thatcher proposed to amplify EEG with a vacuum tube in place of the string galvanometer, and in 1936, this technology became an industry standard (Swartz and Goldensohn, 1998).

Hans Berger (1873–1941), a neuropsychiatrist, is considered the discoverer of the human EEG. He coined the German name "Elektenkephalogram" for EEG. His research was carried out in a small old laboratory. From 1902 to 1910, he studied the electrical activity in the dog's brain with a capillary electrometer. He started using the Einthoven-type string galvanometer in 1910 and later adopted the Edelmann-type string galvanometer. In 1926, he began to use the Siemens dual-coil galvanometer, whose sensitivity was as high as 130 μV/cm. With this instrument and nonpolarizable pad electrodes, he recorded the human EEG and published the result in 1929 (Figure 1.1). Actually, in June 1924, he discovered the oscillations possibly from the lower brain by Edelmann-type linear galvanometer. In 1925, he realized that a skull defect might not necessarily be a favorable condition for effective recording, because of the existence of the thick meninges and postoperative support materials, etc., and on the contrary, an equal or even better record might be got given intact skull and scalp. From 1926 to 1929, he got a perfect α-wave with a dual-coil galvanometer. In the report in 1929, he pointed out the existence of α-rhythm and α-block (Niedermeyer and Silva, 1999).

Following Berger's pioneering work on human EEG, the domain got sustained and rapid development. The following are some important events (Kennard, 1949; Luck, 2014; Niedermeyer and Silva, 1999; Swartz and Goldensohn, 1998; Yao et al., 2019):

1932: Hans Berger and Dietsch used the Fourier transforms to analyze EEG.
1934: Fisher and Lowenbach recorded epileptic discharges.
1930s: Lindsley used EEG to study maturity.
1935: Gibbs, Davis, and Lennox started EEG application in clinical epilepsy.
1935: Foerster and Altenburger collected electrocorticography during surgery.
1937: Hoagland studied the effect of metabolism on EEG.
1936: The first clinical EEG department was established in Massachusetts General Hospital.
1937: Dawson confirmed the evoked potential (EP) by using a photographic superimposition technique.
1938: Schwaub used two cameras to simultaneously record EEG and behavior.
1940: Renshaw proved the possible connection between the slow potential of the neural mass/population and the EEG oscillation.
1949: Lindsley described the impact of midbrain injury on EEG.

1940s: EEG was introduced into psychosomatic medicine such as peptic ulcers and allergic conditions.
1950: Offner and Goldman described the average reference.
1953: Aserinsky and Kleitmean described rapid eye movement sleep EEG (REM).
1958: Jasper led a committee to develop the standard of a 10–20 electrodes system.
1962: Galambos and Sheatz reported the event-related potential (ERP) by using computer average technology.
1964: Grey Walter and his colleagues reported cognitive-related ERP, i.e., contingent negative variation (CNV).
1965: Sutton, Braren, Zubin, and John discovered the P3 component.
1970s: Infant EEG was reported.
1972: Cohen published the first magnetoencephalogram (MEG) based on a superconducting quantum magnetometer (SQUID).
1973: Vidal coined the concept of brain–computer interface (BCI).
1975: Hjorth proposed the Laplace transform of EEG.
1980s: The digital EEG appeared and was accepted, and then appeared the source localization technology.
1993: Ives et al. reported simultaneously EEG-functional magnetic resonance imaging.
1994: Pascual-Marqui et al. published the low-resolution electromagnetic tomography (LORETA).
2001: Yao established zero reference of EEG, namely, the reference electrode standardization technique (REST).
2003: Friston et al introduced dynamic causal modeling (DCM) to EEG/MEG brain imaging.
2000s: Open platforms and softwares for EEG processing appeared, such as EEGLab (Delome and Makeig, 2004), FieldTrip (Oostenveld et al., 2011), WeBrain (Dong et al., 2021), etc.
2010s: Deep learning introduced to EEG source localization (Adler et al., 2017; Bore et al. 2022).
2020s: Yao coined the concept of brain-apparatus conversations/communications (BAC) to fuse brain–machine interaction including BCI, and brain-physiological organ/system interaction including psychosomatic medicine (Yao et al., 2020, 2022), simultaneously EEG-electrocardiogram/Intestinal electrogram/electromyogram, etc. might become a regular option in the future.

1.2 BASIC FEATURES

EEG is the spontaneous bioelectric activity of a neuron population recorded by scalp electrodes and can be expressed in the form of a curve (brain wave), with the potential as the vertical axis and time as the horizontal axis. The current EEG equipment has been fully computerized, with the EEG signals stored in the computer after digitization and displayed on the screen or printed under the control of the computer.

The brain wave is characterized by continuous changes in time and space distribution. The potential (amplitude), time (period), and phase of the brain wave constitute the basic and straightforward characteristics of EEG. The period of the brain wave is slightly different from that of sinusoidal waves in physics. It refers to the duration of time between the bases of one wave and the next wave, expressed in milliseconds (ms). The number of periods that occur per second is called frequency, expressed in hertz (Hz). Superficially, EEG waves are similar to sinusoidal waves; it is actually a composite wave formed by the superposition of multiple periodic and transient components. The amplitude of the brain wave is usually illustrated by a straight line from the wave crest that is perpendicular to the baseline; it meets the line between the bases of the two adjacent waves. The distance from the meeting point to the wave crest is known as the amplitude of the brain wave, expressed in microvolt (μV). The reason for adopting this measurement method is that the baseline of the EEG is often unstable. The amplitude of the brain wave is mainly determined by the intensity of the electrical activity occurring in the brain and the selection of reference electrode. According to the amplitude, the brain wave usually falls into four types: low-amplitude brain wave, below 25 μV; medium-amplitude brain wave, 25–75 μV; high-amplitude brain wave, 75–150 μV; ultrahigh-amplitude brain wave, above 150 μV.

The changes in brain wave amplitude can be roughly divided into three types: (1) a very fast sudden change, such as epileptic waves; (2) a change that takes place in a short period time (tens of ms to several minutes), such as eye-opening during eye closure, or changes caused by external stimuli and thinking activities; (3) a slow change in amplitude that lasts a few days or even several years, taking place during the development process in the child or the degeneration process in aged people. The phase of the brain wave is also known as the polarity of the brain wave. The general rule is that, with the baseline as a line of demarcation, the brain wave with its crest upward is negative, while the brain wave with its crest downward is positive. It should be noted that in brain wave recording, the negative potential is usually recorded above the baseline, while the positive potential is recorded below the baseline. After the computerization of EEG, there have been many reports in which the positive potential is put above the baseline. So, the reader needs to pay attention to the setting of coordinates during reading. According to the phase, the brain wave falls into the following types: monophasic brain wave, biphasic brain wave, and multiphase brain wave. For a simultaneous observation and comparison of the brain waves at two sites, the phase relation between them is an important indicator. When the brain waves at the two sites have the same period and phase at the same time point, they are considered to be in-phase; when the brain waves at the two sites are deflected in the opposite direction of the baseline at the same time, they are considered to be out-of-phase. In the EEG of healthy people, the α-waves at the bilaterally symmetric sites are generally in-phase, especially between the left and right occipital regions, but there may be a phase difference between the left and right parietal regions. Phase inversion can be seen between the occipito-parietal region and the frontal region. The phase relation between the brain waves in different brain areas may be affected by the selection of reference electrode, making it necessary to use zero potential point as the reference as noted in Chapters 13 and 14. For the localization of brain dysfunction, it is of great significance to judge whether the brain waves are in-phase or out-of-phase.

1.3 TYPICAL EEG WAVES

The routine EEG mainly contains brain waves with frequencies of 0.5–30 Hz and is normally classified by frequency to represent various components.

Generally, the δ- and θ-waves, which are slower than the α-wave, are collectively referred to as slow waves, while the β- and γ-waves, which are faster than the α-wave, are collectively referred to as fast waves. In addition, a brain wave different from the above that is prone to appear under certain conditions, such as pathologic conditions, is named according to its waveform characteristics such as the spike-wave, spike and wave complex, vertex sharp transient, etc. At present, the division of EEG frequency bands is subjective, and how to more objectively divide frequency bands, such as using factor decomposition, is still a problem that needs to be developed (Jing et al., 2016).

1.3.1 α-Wave

The average amplitude of the α-wave in healthy adults is 30–50 μV, and the frequency range is roughly 8–13 Hz. It mainly exists in the form of a sinusoidal wave in the parieto-occipital region. The α-wave is the main component of the EEG in most healthy adults. It appears in adults at rest with their eyes closed in waking state. The amplitude of the α-wave decreases and it is replaced by a higher β-wave, which has a higher frequency when adults keep their eyes open in the waking state or concentrate their attention on something. The α-wave changes with the maturation of the brain or as age increases. For children, the frequency of α-waves gradually increase with the development of the brain and stabilize in adulthood. They decrease step by step in elder age. There is an obvious difference in the α-wave from individual to individual. In practice, the frequency range of personalized α-waves can be determined by extending its crest/peak forward and backward for a particular distance, for example, "crest±2" Hz (Yao et al., 2004, 2005). The frequency, amplitude, and spatial distribution of α-waves are important indicators that reflect the functional status of the brain.

So far, there remains a lack of a conclusion on the origin of the α-wave. Some reports implicated a negative relation between head size and the spectral position of the alpha peak (Pα) (Nunez et al., 1978) and proposed evidence of the influence of global boundary conditions on slightly damped neocortical waves. However, some recent researchers attempted to re-examine this finding by computing the correlations of occipital Pα with head size and cortical surface area and did not find such a relation ($p > 0.05$). On the other hand, biophysical models also predicted that white matter architecture, determining time delays and connectivities, could have an important influence on Pα (Valdés-Hernández et al., 2010). In addition, for some people, the alpha waves are splitted. Similarly, the origins of other brainwave rhythm components are also unclear and are worthy of further study.

1.3.2 β-Wave

The frequency range of the β-wave is about 14–30 Hz, and its amplitude is generally 5–30 μV. It spreads throughout the brain, especially in the front area and temporal region. In about 6% of healthy adults, the EEG is dominated by the β-wave activity.

The β-wave is related to gender, personality, age, and mental state. Generally, the β-wave is more common in women than in men and more common in elders than in adults. The amplitude of β-waves may increase in the case of emotional instability or after the use of drugs such as sedatives or hypnotics. β-wave can be further split into β1-wave and β2-wave. The frequency of the β1-wave is about 13–20 Hz. Like the α-wave, it is affected by mental activity. The frequency of the β2-wave is about 20–30 Hz. It appears when the central nervous system (CNS) is highly active.

1.3.3 θ-Wave

The frequency of the θ-wave is about 4–7 Hz, and its amplitude is 10–40 μV. The number of θ-waves gradually decreases from children to adults, while its frequency gradually increases and its amplitude gradually decreases. There are only a few θ-waves scattered in the EEG of healthy adults. The θ-wave primarily appears in the parietal region and temporal region of children. For adults, θ-wave may appear and last for about 20 seconds when they are emotionally depressed, in particular when they feel disappointed or frustrated. The θ-wave increases in number during fatigue or sleep and it exists commonly in the elder age and under pathological conditions.

1.3.4 δ-Wave

The frequency of the δ-wave is about 0.5–3.5 Hz. It appears in sleeping people, infants, and patients with severe organic encephalopathy. The δ-wave can also be recorded in the brain of an experimental animal that has undergone subcortical transection. This type of surgery causes a functional separation between the cerebral cortex and the reticular activating system. Hence, the δ-wave can be generated alone in the cortex.

1.3.5 μ-Rhythm

The μ-rhythm is a comb-shaped rhythm of 8–12 Hz that appears in the central area. It is visible in the central area on one side, and can appear on both sides asynchronously and asymmetrically. The μ-rhythm does not disappear when the eyes are open, but it may be missing during fist clenching (contralateral), mental activity, or tactile stimulation. The μ-rhythm can appear in healthy people and those with neurosis or post-traumatic syndrome. Its function is not yet clear.

1.3.6 Vertex Sharp Transient

The vertex sharp transient, also known as a hump, is a negative sharp wave that appears simultaneously in the parietal region and central area at the beginning of light sleep. It is particularly visible in the parietal region. It can be bi-phasic or tri-phasic and is a negative wave in most cases, but it may be a positive wave in children. The frequency of the peak wave is 3–5 Hz, and its amplitude is 100–300 μV. It is called a "double hump" when appearing in pairs.

1.3.7 Spindle Wave

The sleep spindle wave is also called σ (sigma) rhythm. Its frequency is 12–14 Hz in adults and 10–12 Hz in children. It mainly appears in the parietal region/central area and sometimes can appear widely. The σ-rhythm may be out of sync on the left and right in childhood and is reduced significantly or disappear in people aged above 60. The σ-rhythm is the main brain wave sign of light sleep.

1.3.8 K-complex

The K-complex is a composite wave formed by the vertex sharp transient and σ rhythm. It may appear spontaneously during light sleep or be induced by external sensory stimuli, especially acoustic stimuli. Normally, it appears synchronously on both sides in a symmetric manner.

1.3.9 High-Frequency EEG

As mentioned above, EEG generally refers to signals of 0.5–30 Hz due to the limitation of traditional measurement instrument performance. With the advancement of signal processing technology and electrophysiological signal acquisition technology, the record frequency of EEG signals could reach up to 10,000 Hz, enabling people to analyze and utilize the signals in a wider frequency band, such as high-frequency oscillations (HFOs), which consist of ripples with a frequency of 80–250 Hz and fast ripples with a frequency of 250–500 Hz (Engel et al., 2009). When healthy people are in slow-wave sleep or have their eyes closed for rest, HFOs, especially ripples of 80–250 Hz, appear frequently. For the above reason, HFOs are considered to be involved in neural processes such as brain information encoding, sensory integration, and memory formation (Lachaux et al., 2012).

1.3.10 Ultralow-Frequency EEG

With the DC-EEG or full-band EEG technique (Vanhatalo et al., 2005), it was found that the dominating frequency range of the total spectral power of preterm neonates is much below the conventional EEG frequency band (as low as 0.01–0.1 Hz). Full-band EEG may open new avenues for the monitoring and diagnostics of activity-dependent diseases and malfunctions of the immature brain and potentially benefits the understanding of human intelligence and its development, thus a very promising area to be involved.

1.4 EEG MEASUREMENTS

1.4.1 Electroencephalograph and Recording Electrode

The electroencephalograph is a microvolt-level sophisticated electronic device with a million times magnification. There are strict requirements for its working environment and conditions. Usually, it should be used in a quiet, dark room that is kept out of the sun and electromagnetic interference. The electroencephalograph used in clinical practice should have at least eight channels; the others may be 12, 16, 19, and 32. In cognitive research, 32 and 64 channels are the most popular two choices; the others are 64, 96, 128, or 256 channels. In general, the larger the channel number is, the more abundant the spatiotemporal information of EEG obtained is. However, if more electrodes are adopted, the device will be more expensive, and it will take a longer time to install electrodes. Also, considering the spatial low-pass filtering effect of the skull, an excessively dense electrode array is of no practical significance. Because of this, a rational choice should be made as the case may be.

The electrodes used to record EEG include funnel-shaped electrodes, circular electrodes, and needle electrodes. In addition, there are special electrodes that need to be placed at specific locations, such as sphenoid electrodes, nasopharyngeal electrodes, cortical electrodes, depth electrodes, and ear canal electrodes.

Regarding the position of the scalp electrode, there are many placement methods such as the ones proposed by Montreal, Cohn, and Gibbs, etc. However, the most widely used one is the 10–20 system (Jasper, 1958). See below for how to place 19 electrodes (Figure 1.2):

1. *Anteroposterior position*: Draw a line from the nasion to the occipital tuberosity. Then, make five points on the line from the front to the back, named frontopolar point (F_p), frontal point (F_z), central point (C_z), parietal point (P_z), and occipital point (O), respectively. The length of the distance from F_p to the nasion and the distance from O to the occipital tuberosity account for 10% of the total length of the line, respectively, while the remaining points are all separated by 20% of the total length of the line.
2. *Central transverse position*: Draw a line from the left pre-auricular point to the right pre-auricular point via the central point (C_z). Then, mark the left temporal point (T_3) and right temporal point (T_4), as well as the left central point (C_3) and right central point (C_4), symmetrically on both sides of the line. The length of the distance from points T_3 and T_4 to the corresponding pre-auricular point accounts for 10% of the total length of the

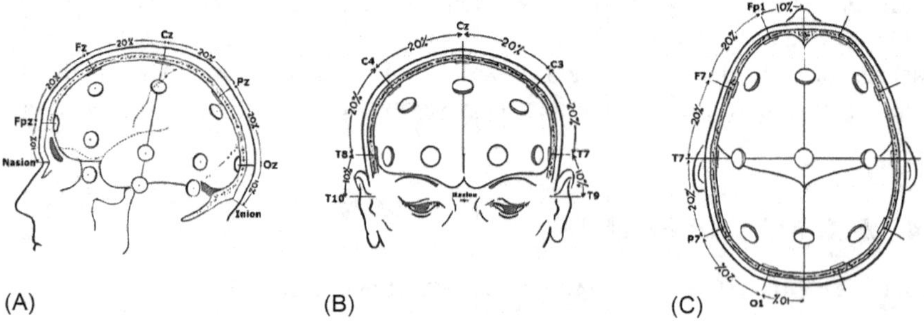

FIGURE 1.2 Schematic diagram of electrode placement for 10–20 system (Seeck et al., 2017). Where A, B and C show the sagittal plane, anterior/coronal plane and transverse plane (top view) with the positioned electrodes, respectively.

line, respectively, while the remaining points are all separated by 20% of the total length of the line.

3. *Lateral position*: Draw a line from the frontopolar point (F_p) backward to the occipital point (O) via T_3 and T_4 on both sides. Then, mark the left frontopolar point (F_{p1}), right frontopolar point (F_{p2}), left anterior-temporal point (F_7), right anterior-temporal point (F_8), left posterior-temporal (T_5), right posterior-temporal (T_6), left occipital point (O_1), and right occipital point (O_2) symmetrically on the line from the front to the back. The length of the distance from F_{p1} and F_{p2} to F_p, as well as the distance from O_1 and O_2 to O, accounts for 10% of the total length of the line, respectively, while the remaining points are all separated by 20% of the total length of the line.
4. The remaining electrode positions, including left frontal point (F_3) and right frontal point (F_4), as well as left parietal point (P_3) and right parietal point (P_4), are located at the center of the line from the frontal point (F_z) to F_7 and F_8, and at the center of the line from parietal point (P_z) to T_5 and T_6, respectively. The earlobe electrodes on the left and right sides are represented by A_1 and A_2, respectively.

In the serial number of the above-mentioned recording electrodes, an odd number is usually used to represent the left side while an even number is used to represent the right side. Totally eight electrodes are placed on the head on both the left and right sides, respectively. Considering that there are already three electrodes at F_z, C_z, and P_z in the anteroposterior position and two earlobe electrodes on the left and right sides, there are a total of 21 electrodes. The characteristic of the 10/20 system is that there are clearly defined electrode positions in the cephalic region and anatomical regions in the cerebral cortex. Moreover, the arrangement of the electrodes is proportional to the size and shape of the head. There are electrodes placed at all the major parts of the cephalic region corresponding to the convex surface of the cerebral cortex. Generally, when 32 electrodes need to be placed, the 10/20 rule still can be adopted. When 64 or more electrodes need to be placed, a corresponding expansion approach is adopted, such as the 10–10 system recommended by the U.S. Electroencephalography Society (Figure 1.3), or an even distribution of electrodes on the scalp surface, such as the Geodesic Sensor Nets adopted by EGI (Electrical Geodesics, Inc.).

Recently, IFCN pointed out that more information about the temporal lobes would be needed to better locate epilepsy. Further, they proposed a system revised from the 10–20 system, with six new electrodes, including F_9, F_{10}, T_9, T_{10}, P_9, and P_{10}, added to the system, expanding the number of the most basic electrodes to 25 (Figure 1.4) (Seeck et al., 2017).

1.4.2 Schemes for EEG Montage

EEG montage means combining electrodes for a certain purpose to keep a record in a certain form, such as the commonly used monopolar montage and bipolar montage.

There is an electrode in the monopolar montage used as the common electrode, such as an earlobe or a vertex electrode, called an "inactive" electrode or reference

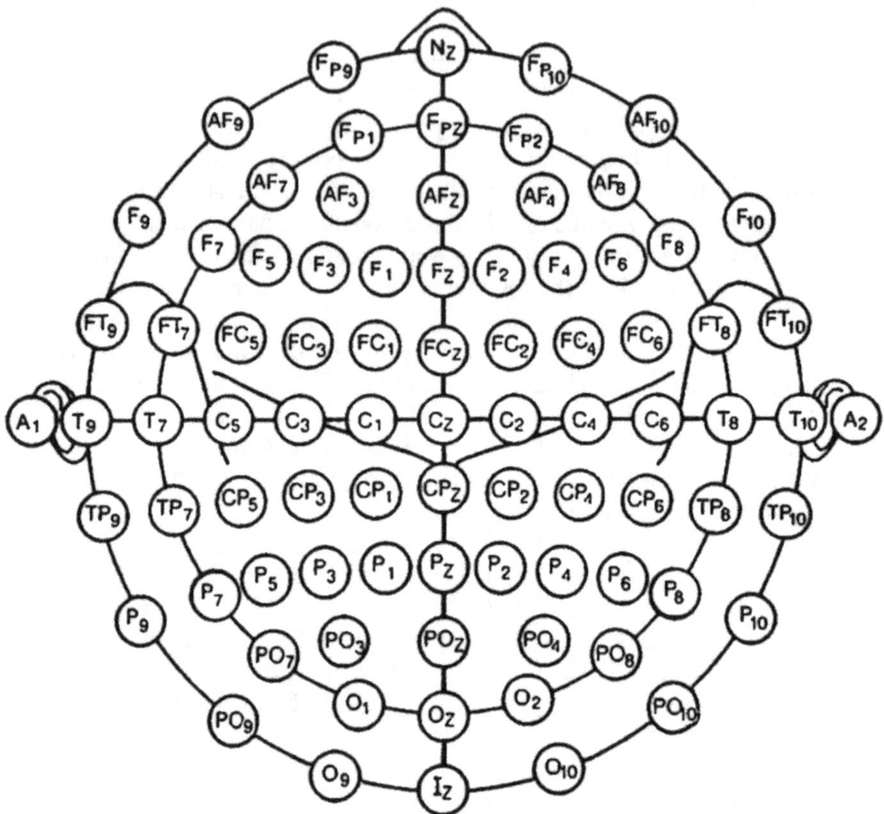

FIGURE 1.3 Schematic diagram of electrode placement for 10–10 system (Seeck et al., 2017).

electrode. Each lead amplifier of the electroencephalograph has two terminals, i.e., G_1 and G_2. In the monopolar montage, the reference electrode is connected to the G_2 (+) terminal, while the active electrode is connected to the G_1 (−) terminal. In this case, the negative potential under the active electrode is recorded as a negative wave with its crest upward, and the earlobe electrode or vertex C_z electrode is widely used as the common reference. The monopolar montage is normally adopted for multichannel recording based on 32 or more electrodes. After the recording is completed, the monopolar montage may be converted into another type of montage, including linked mastoids or linked ears, average reference, or zero reference (REST) (Yao et al., 2019).

There are generally three ways to connect the earlobe reference electrode with the active head electrode: (1) the left and right earlobe electrodes are used as reference electrodes for the active electrode on the ipsilateral hemisphere, respectively; (2) the left and right earlobe electrodes are used as a reference electrode for the electrodes on the contralateral hemisphere, respectively; (3) the left and right earlobe electrodes are

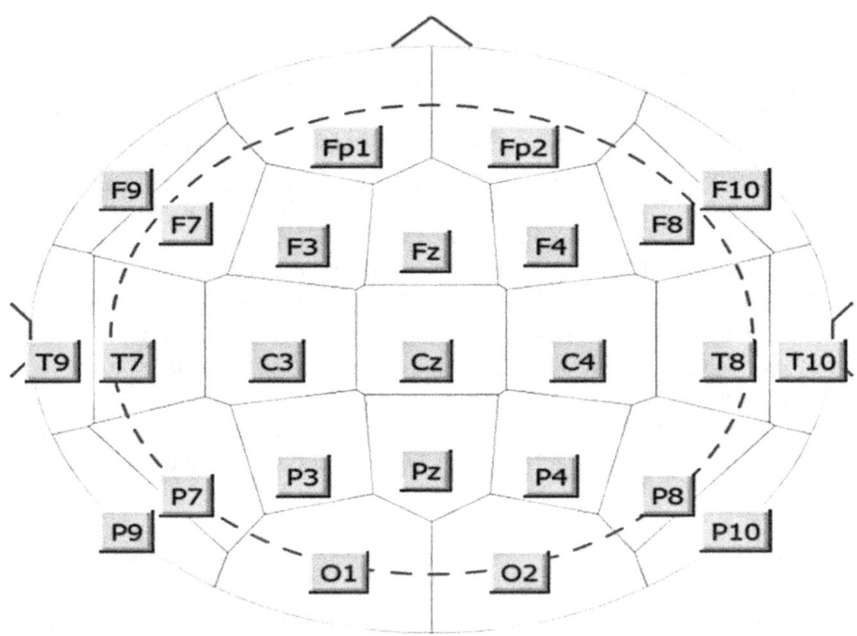

FIGURE 1.4 Modified 10–20 system (Seeck et al., 2017).

linked together first (linked mastoids, LM, or linked ears) as reference electrode for all other electrodes. The online linked-ears scheme is no longer recommended, because the physical connection between the two ears may cause interference to the electric field of the entire brain. Now it is recommended to link the ears together by computer after single-electrode reference recording; apparently, the first and second situations here are not monopolar (unipolar) reference, even so, they can be re-referenced to the unipolar reference by a REST-based algorithm (Dong et al., 2023).

The average common electrode means that the average potential of all electrodes at every moment is used as a common reference electrode. The earlier average reference (AR) was also achieved through a physical connection. It was proposed with inspiration from the central terminal reference of ECG (Offner, 1950; Goldman, 1950). Later, it was realized that this kind of online physical connection might have an impact on the electric field of the brain, thus is stopped for using. In current practice, after the recording with unipolar reference, it is converted to AR-based recordings by computer.

From a physical point of view, for an ideal montage, a real inactive electrode should be adopted as the reference, but such a physical reference point cannot be found during EEG measurements, because there is no point where the potential is zero or a constant on the human body surface or scalp surface all the time. The purpose of both earlobe reference and average reference is to build a reference that is approximately zero, at least in some hypothetical cases. However, recent studies show that this type of reference is unsatisfactory that they are not zero for general senses. At present, the best approximately zero reference is a reference at infinity (Yao, 2001) approximately achieved by the EEG reference electrode standardization technology (REST). The relevant details are described in Chapters 13 and 14.

The bipolar montage means that two active head electrodes are linked to G_1 and G_2 terminals of each EEG amplifier to record the potential difference between the two active electrodes. It essentially forcibly uses an electrode as a reference electrode, and this reference electrode is not fixed but varies according to the amplifier. The bipolar montage still is frequently applied to clinical medicine, such as in epilepsy practice. There are also many different bipolar montages, e.g., nearly two are connected from the front to the back or from left to right (Niedermeyer and Silva, 1999).

Generally speaking, the monopolar montage can be easily converted into another type of montage, including the conversion between different monopolar montages and the conversion from a monopolar montage into a bipolar montage. On the contrary, if the original record is a bipolar montage, it can hardly be converted into a monopolar montage unless the REST-based new algorithm is used as a bridge (Dong et al., 2023).

1.4.3 Basic Requirements for EEG Recording

1. *Amplifier*: The EEG signal is very weak, generally on an order of magnitude of μV. So, the input stage should have the following characteristics: low input noise (≤3 μV PP), high gain (0.5×10^3 to 10×10^4), high common-mode rejection ratio (CMRR) ($K \geq 90$ dB), low drift, and high input impedance (≥10 MΩ), etc. To meet these requirements, a low-noise differential circuit must be adopted. Besides, electric components must also be strictly selected and pay attention to the technological process. The instrument must be well grounded. According to the current level of the instrument, a shielding room is generally unnecessary. But if there is strong electromagnetic interference from the surrounding environment, a shielding room is required.
2. *Electrode*: The recording electrode itself should not generate noise or drift. The experimental results prove that Ag-AgCl or golden circular electrode is excellent. However, owing to the high input impedance of the current amplifier, the electrodes and electrode pastes made of other materials can also achieve good results. The electrode must be kept clean to reduce noise. The interelectrode impedance must be measured routinely before recording, and its value should not exceed 5 kΩ. The electrode impedance needs to be re-measured if artifacts appear during recording. In recent years, great progress has been made in active electrode (Gargiulo et al., 2012), dry electrode, and capacitance-based noncontact measurement technology (Liao et al., 2012), they may become the main electrodes for their convenience in the near future.
3. *Filter*: During routine recording, the low-frequency filtered-out should not be higher than 1 Hz (−3 dB), while the high-frequency filtered-out should not be lower than 70 Hz (−3 dB). Considering that the spike-wave may be distorted or become smaller, a notch cannot be used under normal

circumstances unless the power-line interference of 50 Hz or 60 H (country dependent) fails to be removed by all means.

4. *Recording length*: Clinically, spontaneous EEG recording should comprise a very satisfactory record with a length of at least 20 minutes. If conducting some induction experiments, such as hyperventilation or flash stimulation, etc., a longer record should be kept. The shorter the EEG recording is, the less chance there is for the occurrence of abnormality. For event-related potential studies, the trials interval depends on the task (Luck, 2014).

1.4.4 Artifacts in EEG Recording

Artifacts refer to various nonbrain potential interferences in EEG recording (Table 1.1). The presence of artifacts poses great challenges to the processing and interpretation of EEG. For this reason, artifacts should be reduced as much as possible during recording. Meanwhile, it is necessary to identify artifacts accurately. Of course, it is also currently imperative to develop an advanced artifact identification and removal method, e.g., a method for the removal of ocular artifacts (Liu and Yao, 2006; Wang et al., 2014). In the process of recording, the examiner should observe the state of the subjects meticulously with patience. If any artifact is identified, it should be corrected in time while the cause of the artifact must be recorded.

1.4.5 Clinic Induction Experiment

Owing to the limited recording time, the artifact may be missed and it is often difficult to detect abnormalities by routine recording. If a certain physiological stimulus is given to the patient during the examination, hidden brain wave abnormalities may be revealed or the existing abnormalities may become more obvious. That is what is known as the clinical-inducing test. Now, the clinically induced test has become an important means in the clinical practice of EEG for improving abnormality detection. The following inducing tests are commonly used in clinical practice.

TABLE 1.1
Artifacts in EEG Recording

Artifacts from Human Body	Artifacts from Equipment
Movements of eyelids and eye	Instrument malfunction
Electromyography from muscle contraction (swallowing, chewing, coughing, etc.)	Poor electrode contact or shaking of the electrode wire
ECG	Electrode malfunction
Sweaty skin	AC power interference
Breathing, crying	Poor grounding
Vascular pulsation	Influence of other electronic equipment
…	…

1. *Hyperventilation*: The subject is asked to close his/her eyes and breathe deeply with expiration as much as the subject can for 3 minutes, 20–25 times per minute, then restore calm breathing. The EEG keeps recording the entire course of hyperventilation and the later at least another 3 minutes of calm breath.

   Hyperventilation causes more CO_2 to be removed from the body. The subject gradually enters a state of respiratory alkalosis (complicated by symptoms including dry mouth, numbness of four extremities, trembling, and dizziness), accompanied by reflex cerebral vasoconstriction, cerebral blood flow reduction, O_2 and glucose supply reduction, and changes in biochemical metabolites such as neurotransmitters (e.g., a decrease in the level of γ-aminobutyric acid), leading to changes in the brain wave. The inducing effect of the hyperventilation test can cause the generation of bilateral high-amplitude slow waves in most children and some adults. Paroxysmal abnormal brain waves appear during or after the hyperventilation test. Especially, the waves can be identified to be abnormal when the spike wave, sharp wave, or other seizure waves appear. Hyperventilation has a significant effect on the seizure, particularly absences. It can not only induce a typical 3-Hz spike wave rhythm but also is accompanied by clinical absence seizure. For a patient with organic brain damage, hyperventilation often induces or enhances localized or generalized abnormalities.
2. *Eyes-open and eyes-closed test*: EEG recording is usually made in the waking state with the eyes closed. In the eyes-open and eyes-close test, the subject is asked to keep his/her eyes open for about 5 seconds and then close his/her eyes. This process can be repeated several times, but the time interval should be more than 10 seconds. The mechanism is as follows: The light falls directly on the retina after the eyes are opened, bringing about a stimulus to the optic nerve, through the lateral geniculate nucleus reaching the occipital cortex (primary visual cortex). Then, it reaches the forehead, motor area, and temporo-parietal junction via the striate area in the occipital region, exciting some areas of the cortex, thus exerting an inhibitory effect on the α-wave. It is visible on the EEG that the α-rhythm weakens or disappears. This phenomenon is called visual response. The inhibitory effect on the α-wave in the eyes-open state is related to age and individual differences. Usually, the initial α-wave may have its frequency increase and amplitude enlarged for a short time after the eyes are just from open to close, an obvious amplitude modulation phenomenon, and this is known as recruiting response. The average latency (from eyes open till α-wave disappearing) of healthy adults aged 21–50 is about 356 ms, while the recovery period (from eye close till α-wave reappearing) is about 525 ms. In the event of incomplete inhibition, lengthened latency, or nondisappearing α-wave in one hemisphere or a brain area, it indicates that the conduction of photic stimulation in this hemisphere or area is delayed or abnormal. The delay in the recovery period indicates that the subject's ability decreases to inhibit the cortical excitability caused by photic stimulation. Epileptic waves can be induced in some patients with epilepsy in the eyes-open and eyes-close tests.

3. *Flash stimulation*: The subject is asked to keep awake, but with eyes closed. A flashlight is placed 15–20 cm in front of his/her eyes. The light intensity is usually 100,000 candela and intermittent flash stimulation is presented 1–30 times per second, lasting 0.1–10 ms each flash. A flash stimulation generally is of 5–10 seconds, then with another frequency selected for the next stimulation after an interval of 10 seconds. Generally, the flash frequencies should be applied from low to high.

   The earliest response to flash stimulation is rhythm suppression caused by the attention to the light. A wave of the same frequency as flash stimulation or a harmonic wave of flash stimulation appears in the occipital parietal region. This phenomenon is known as rhythm assimilation or photic driving. An effective frequency that can cause photic driving is usually a frequency close to the dominant rhythm of the background wave. The abnormal wave excited by a flash stimulation mainly is multiple spike and wave complex or spike and wave complex. Generally, it is symmetrical and synchronous on both sides, visible in the frontal region and central area. Sometimes, there is a difference in seizure waves between the left and right sides. If the photic driving effect is asymmetrical on the left and right sides, especially absent on one side, it indicates that this side is abnormal. Flash stimulation is effective for some cases of epilepsy, especially myoclonic seizures. Sometimes, it even induces a clinical epileptic seizure, i.e., photosensitive epilepsy.
4. *Sleep and sleep deprivation*: During sleep, the brainstem ascending reticular activating system has a weaker effect on the cerebral cortex and limbic system, making abnormalities such as seizure waves and asymmetric brain waves to be easily released. The opposite case is sleep prohibition or sleep deprivation. The subject is asked to stay awake for more than 24 hours or sleep for a shorter period time. Then, the EEG is recorded. Sleep deprivation can reduce the threshold of epilepsy, thereby making it easier for seizure waves to appear. Generally, it is suitable for people who have no seizure waves in a routine waking state or those with suspected epilepsy that are prone to have seizure waves during sleep.

   In addition to the above-mentioned inducing tests, there is also acoustic stimulation, carotid artery compression, and bemegride inducing test, etc. To be specific, the intravenous injection of a large-dose spasmogenic drug, such as bemegride or pentylenetetrazole (PTZ), may cause convulsions. So, these drugs are generally not used except under special circumstances.

1.5 EVOKED POTENTIAL

The evoked potential (EP) technology has made great contributions to electrophysiological studies in the sensory system. With the advancement of computer technology and modern signal processing technology in recent years, the EP data mining has been greatly improved. In particular, EP has been used by cognitive scientists as a window for looking into the mysteries of the brain. A specific event is designed to evoke the related potential response, i.e., the event-related potentials (ERP). The latency of the component is praised as the "response time in the twenty-first century" by cognitive psychologists. If the spatial distribution and response intensity on the scalp surface are taken into consideration, it can also be thought of as the "response time and intensity in a 2D space," thus offers additional quite abundant brain information.

1.5.1 Introduction to Evoked Potential

Evoked potential refers to a measurable electrical change generated at any part of the central nervous system (CNS) when any point on a sensory organ, sensory nerve, or sensory pathway is stimulated. The evoked potential is different from the spontaneous one. The nervous system generates spontaneous electrical activity all the time. Therefore, an evoked potential appears over the spontaneous background. In practice, it is a challenge to extract EP from the background EEG. EP itself has the following characteristics:

1. *Latency*: There is a time-locked relationship between the appearance of evoked potentials and the stimulation, i.e., the evoked potential has a definite latency. The length of the latency depends on four factors: (1) The speed of impulse conduction along the nerves; (2) the distance between the stimulation point and recording point; (3) the number of synapses encountered during conduction; (4) synaptic delay. Among the above four factors, the conduction speed, distance, and number of synapses are known, whereas the length of the synaptic delay is related to the functional status of the CNS and the intensity of stimulation. Under the same experimental conditions, the length of the latency should be a constant in the same system.
2. *Response style*: There is a difference in response style from one sensory system to the other sensory system. For example, the visual evoked potential is quite different from the somatosensory evoked potential. However, there should be only one response style in a specific system.
3. *Spatial distribution*: When a peripheral site is stimulated, an evoked potential will just appear at a certain part of the CNS (main response) because sensory information is transmitted along a predefined conduction pathway. To explain the evoked potential, one must be familiar with the detailed structure of the conduction pathway.

   Anesthesia has a significant effect on evoked potentials. For example, barbiturates can often reduce the spatial distribution of evoked potentials in the cerebral cortex. Therefore, when evaluating the spatial distribution of evoked potentials, we should notice whether an anesthetic is already used and what effect it has exerted.
4. *Primary response and secondary response*: The evoked potentials, generated in the cortex when any point in a sensory organ or pathway is stimulated, can be divided into two types: one is primary response, and the other is the post or secondary response. Generally, the primary response corresponds to a fixed latency. It is a phenomenon of strict localization. The secondary response is different from the primary response because it can

appear in most of the cortex. The pathway of the second response may be different from the primary response. Furthermore, some studies suggested that there may be a middle response whose latency is between the primary and second response.

1.5.2 Evoked and Induced Potentials

A widely accepted view is that EP is a signal generated by an activated neuron population that has a time-locked relationship with stimulation. This signal is superimposed on a normal, spontaneous EEG activity. Another view is that EP is part of an ongoing brain electrical activity. The latter model has received special attention following the research work of Sayers et al. (1974). In their research, Fourier analysis was performed on a piece of EEG data recorded immediately after high-/low-intensity acoustic stimulation. As a result, the distribution of the phase spectrum varied significantly with the intensity of stimulation, while there was no such difference in amplitude. So, they drew the following conclusion: It is by re-organizing the phase spectrum of the existing spontaneous EEG activity. At present, some progress has been made in the research on evoked potentials from the perspective of the phase spectrum. In addition, EP has been divided into evoked response and induced response according to whether there is a strict time-locked relationship between the response and stimulation (David et al., 2006). For the evoked response, the signal can be highlighted by superimposed averaging; for the induced response, it can be reflected by energy superposition after the Fourier transform or wavelet transform. Please refer to the literature for more details about the generation mechanism of the evoked response and induced response.

1.5.3 Extraction of Evoked Potentials

When the receptors in the body are stimulated, a change will take place in not only the cerebral cortex but also the brainstem, thalamus, and limbic system. However, under normal living conditions (nonanesthetized), we may only find some EEG rhythm changes because the stimulus acting on the receptors is random and continuous, and the most common case is that the α rhythm is disrupted (α blocking). In addition, when receiving a single artificial stimulus, the EEG changes a little and is buried in the strong spontaneous EEG. Therefore, the earlier researchers always kept the animals in deep anesthesia when recording the evoked potentials of the brain. During deep anesthesia, the spontaneous EEG is suppressed, making it easier to extract evoked potentials. However, deep anesthesia is not a normal condition, after all, and the evoked potentials also may have been altered.

With the application of modern computer and signal processing technology, it has been very easy to extract the potential changes that have a fixed relationship with stimulation, i.e., an evoked potential with time locked. Its basic principle is to have the records obtained from repeated stimulations superimposed together. In this way, all the electrical signals that have a time-locked relationship with stimulation will be strengthened gradually due to in-phase superposition. In contrast, the spontaneous potentials that do not have a time-locked relationship with stimulation will weaken one another in the course of superposition and may approach zero because they appear randomly. By superimposed averaging this way, the time-locked evoked potentials can be highlighted, while the result therefore obtained is called average evoked potential. Random signal analysis shows that after N times of superposition, the signal will be amplified by N times while the noise will be amplified by the square root of N, so the signal-to-noise ratio will be increased by $N/\sqrt{N}$, i.e., $\sqrt{N}$ times. Of course, some loss of details will be inevitable in the averaging. For this reason, the researchers have tried to establish a single-trial or a-few-trials extraction technique (Xu and Yao, 2007) and modify the simple average to an adaptive weighted average, etc. Since the application of the average evoked potential technology to electrophysiology, a great advance has been made, i.e., accurate recording of the evoked potential in the waking state. This method has been quickly applied to clinical and cognitive research practice. Actually, in today's practice of EEG, simple superimposed averaging is the most common one in use, is a robust and acceptable technique, and is a good illustration of the "science being of simplicity."

1.5.4 Event-Related Potentials

ERP (event-related potential) is a type of EP. It refers to potential changes on the scalp caused by a series of EEG activities generated when processing stimulation. It is an evoked potential involving psychological factors and an important way to observe the process of information processing in the brain.

ERPs consist of two components, i.e., exogenous and endogenous components. The endogenous component is closely related to the human cognitive process and also related to human memory, attention, intelligence, etc. It has a long latency, from 100 ms to several seconds, and its waveform changes dramatically. It has nothing to do with the physical parameters of the stimulus, but with the task, the deictic words, and the subject's mental state and effort. The endogenous component can only be extracted under certain conditions, e.g., the subject must be awake and concentrated, and understand the main points of the experiment. There must be more than two types of stimuli, i.e., target stimuli and non-target stimuli; they may be visual, acoustic, digital, linguistic, and graphical stimuli, which need to be programmed by a certain probability.

The exogenous component is the early component of the brain's response to the stimulus. It is affected by the physical properties of the stimulus (intensity, type, frequency, etc.) and is significantly correlated with the stimulating factors. It has a short latency, no longer than 100 ms. So, it is also called the early component. It represents the classical sensory channel, while its latency and amplitude are closely related to the physical parameters of the stimulus.

In the research of ERPs, the most studied component is the large positive slow-wave P300 that appears in 250–450 ms after stimulation. It is related to background information detection, feedback, prediction, memory search, learning, deep processing, and concept formation. The second one is *N*400, which is related to linguistic problems. For more components, please refer to the classical book (Luck, 2014).

The following are the stimulation protocols commonly used in the research of ERPs: (1) odd-ball stimulation, which includes two stimuli, one of which appears occasionally as a target signal; (2) selective attention, which presents a series of stimuli in two or more channels (e.g., binaural, binocular, or audiovisual), and the subject is required to find the occasional target in one of the channels; (3) memory comparison: The subject is required to find the target in short-term memory from stimuli mixed with several nontarget, such as memory search. There are many similar experimental modes. Researchers must choose, improve, or design new stimulation modes reasonably according to the purpose of the research.

At present, the analysis methods of ERPs remain in development (Hu and Zhang, 2019), including the principal component analysis (PCA), independent component analysis (ICA), wavelet analysis, dynamic network analysis, empirical mode decomposition (EMD), and multicomponent decomposition algorithm (Yin et al., 2009). The main parameters adopted include peak value/amplitude, latency, exciting area, spectrum, and topological features of (dynamic) EEG/ERP network, etc.

REFERENCES

Adler J, Öktem O. Solving ill-posed inverse problems using iterative deep neural networks. Inverse Probl. 2017, 33(12): 1–24.

Berger, H. 1929. Ãœber das Elektrenkephalogramm des Menschen. 1st report. *Arch. Psychiat. Nervenkr.* 87:527–570.

Bore, J.C., P. Li, L. Jiang, et al. 2022. A long short-term memory network for sparse spatiotemporal EEG source imaging. *IEEE Trans Med Imaging* 40(12):3787–3800.

David, O., J.M. Kilner, K.J. Friston. 2006. Mechanisms of evoked and induced responses in MEG/EEG. *NeuroImage* 31:1580–1591.

David, O., K.J. Friston. 2003. A neural mass model for MEG/EEG: Coupling and neuronal dynamics. *NeuroImage* 20(3):1743–1755.

Delorme, A., S. Makeig. 2004. EEGLAB: An open source toolbox for analysis of single-trial EEG dynamics including independent component analysis. *J Neurosci Meth* 134(1):9–21.

Dong, L., L. Zhao, Y. Zhang, et al. 2021. Reference Electrode Standardization Interpolation Technique (RESIT): A novel interpolation method for scalp EEG. *Brain Topogr* 34:403–414.

Dong, L., Y. Lai, M. Duan, et al. 2023. Rereferencing of clinical EEGs with nonunipolar mastoid reference to infinity reference by REST. *Clin Neurophysiol* 151:1–9.

Engel Jr, J., A. Bragin, R. Staba, et al. 2009. High-frequency oscillations: What is normal and what is not? *Epilepsia* 50:598–604.

Gargiulo, G.D., G. Cohen, A.L. McEwan, et al. 2012. Active electrode design suitable for simultaneous EIT and EEG. *Electron Lett* 48(25):1583–1584, doi: 10.1049/el.2012.3212.

Goldman, D. 1950. The clinical use of the "average" reference electrode in monopolar recording. *EEG Clin Neurosci* 2:209–212.

Hu, L., Z. Zhang, Eds. 2019. *EEG Signal Processing and Feature Extraction*. Singapore: Springer Press.

Ilmoniemi, R.J., J. Sarvas. 2019. *Brain Signals: Physics and Mathematics of MEG and EEG*. Boston, MA: MIT Press.

Jasper, H. 1958. The 10-20 electrode system of the International Federation. *EEG Clin Neurosci* 10:367–380.

Jing, W., Y. Wang, G. Fang, et al. 2016. EEG bands of wakeful rest, slow-wave and rapid-eye-movement sleep at different brain areas in rats. *Front Comput Neurosci* 10:79.

Kennard, M.A. 1949. Inheritance of electroencephalogram patterns in children with behavior disorders. *Psychosom Med* 11(3):151–157.

Knösche, T.R., J. Haueisen. 2022. *EEG/MEG Source Reconstruction*. Switzerland: Springer Press.

Lachaux, J.P., N. Axmacher, F. Mormann, et al. 2012. High-frequency neural activity and human cognition: Past, present and possible future of intracranial EEG research. *Prog Neurobiol* 98(3):279–301.

Liao, L.D., C.T. Lin, K. McDowellet, et al. 2012. Biosensor technologies for augmented brain-computer interfaces in the next decades. *Proc. IEEE* 100:1553–1566.

Liu, T., D. Yao. 2006. Removal of the ocular artifacts from EEG data using a cascaded spatio-temporal processing. *Comput Meth Prog Bio* 83:95–103.

Luck, S.J. 2014. *An Introduction to the Event-Related Potential Technique*. Boston, MA: MIT Press.

Niedermeyer, E., F. Da Silva. 1999. *Electroencephalography: Basic Principles, Clinical Applications, and Related Fields*. Baltimore, MD: Williams & Wilkins.

Nunez, P.L., L. Reid, R.G. Bickford. 1978. Relationship of head size to alpha frequency with implications to a brain wave model. *EEG Clin Neurosci* 44:344–352.

Nunez, P.L., R. Srinivasan. 2006. *Electric Fields of the Brain: The Neurophysics of EEG*. New York: Oxford University Press.

Offner, F.F. 1950. The EEG as potential mapping: The value of the average monopolar reference. *EEG Clin Neurosci* 2:213.

Oostenveld, R., P. Fries, E. Maris, et al. 2011. FieldTrip: Open source software for advanced analysis of MEG, EEG, and invasive electrophysiological data. *Comput Intel Neurosci* 2011:1–9.

Pravdich-Neminsky, V.V. 1912. E. versuch der registrierung der elektrischen Gehirnerscheinungen. *Zentralbl Physiol* 27:951–960.

Sayers, P., H.A. Beagley, W.R. Henshall. 1974. The mechanisms of auditory evoked EEG responses. *Nature* 247:481–483.

Seeck, M., L. Koessler, T. Bast, et al. 2017. The standardized EEG electrode array of the IFCN. *Clin Neurophysiol* 128:2070–2077.

Swartz, B.E., E.S. Goldensohn. 1998. Timeline of the history of EEG and associated fields. *EEG Clin Neurosci* 106:173–176.

Valdés-Hernández, P.A., A. Ojeda-González, E. Martínez-Montes, et al. 2010. White matter architecture rather than cortical surface area correlates with the EEG alpha rhythm. *NeuroImage* 49(3):2328–2339.

Vanhatalo, S., J. Voipio, K. Kaila. 2005. Full-band EEG (FbEEG): An emerging standard in electroencephalography. *Clin Neurophysiol* 116:1–8.

Wang, Z., P. Xu, T. Liu, et al. 2014. Robust removal of ocular artifacts by combining independent component analysis and system identification. *Biomed Signal Proces* 10:250–259.

Xu, P., D. Yao. 2007. Development and evaluation of the sparse decomposition method with mixed over-complete dictionary for evoked potential estimation. *Comput Biol Med* 37(12):1731–1740.

Yao, D. 2001. A method to standardize a reference of scalp EEG recordings to a point at infinity. *Physiol Meas* 22(4):693–711.

Yao, D., L. Wang, K.D. Nielsen, et al. 2004. Cortical mapping of EEG alpha power using a charge layer model. *Brain Topogr* 17(2):65–71.

Yao, D., L. Wang, R. Oostenveld, et al. 2005. A comparative study of different references for EEG spectral mapping: The issue of the neutral reference and the use of the infnity reference. *Physiol Meas* 26(3):173–184.

Yao, D., Y. Qin, S. Hu, et al. 2019. Which reference should we use for EEG and ERP practice? *Brain Topogr* 32(4):530–549.

Yao, D., Y. Qin, Y. Zhang. 2022. From psychosomatic medicine, brain-computer interface to brain-apparatus communication. *Brain-Appar Commun J Bacomics* 1:66–88.

Yao, D., Y. Zhang, T. Liu, et al. 2020. Bacomics: A comprehensive cross area originating in the studies of various brain-apparatus conversations. *Cognit Neurodyn* 14(4):425–442.

Yin, G., J. Zhang, Y. Tian, et al. 2009. A multi-component decomposition algorithm for event-related potentials. *J Neurosci Meth* 178:219–227.

2 Electromagnetics behind Brain Electric Field

As an important basis for the research of electroencephalogram (EEG) and magnetoencephalogram (MEG), the basic electromagnetism is first introduced in this chapter, including the differential and integral form of Maxwell's equation, the constitutive relations of media, boundary conditions, the electromagnetic waves in conductors, the electrostatic problems and uniqueness theorem, the multipole expansion of the charge distribution and electrostatic potential or electrostatic field, the electrostatic energy and electrostatic interaction, the theorems in electrostatics, the field of a charge in a medium, the duality principle between the field of a charge in media and the steady current field in conductors, etc. Most of the contents are extracted from typical textbooks on electromagnetism, electrodynamics, and field theory (Fu, 1991; Zhang, 2003; Stratton, 1941; Morse and Feshbach, 1953; Panofsky and Phillips, 1962; Landau and Lifshitz, 1975; Jackson, 1999; Salam, 2014), especially from Sections 1.1 and 1.2 in Fu (1991); Sections 2.1, 2.2, 2.4 and 2.6 in Zhang (2003); Wikswo and Swinney (1984, 1985) and Section 1.11 in Jackson (1962). These contents are helpful to the following chapters, especially for those who are interested in further developing EEG methods. However, they are not necessary for most readers who just want to use those EEG methods in various psychological and clinical practices, then, this chapter can be skipped.

2.1 MAXWELL'S EQUATIONS

Before Maxwell, the electrostatic field and magnetic field were treated as independent because the fundamental laws then, i.e., Gauss theorem, Ampere circuit law, the conservativeness of static fields, and continuity principle of magnetic flux, all focussed on the rules of spatial varies. However, Faraday's law of electromagnetic induction reveals that a varying magnetic field can induce an electric field. An important contribution of Maxwell is the addition of displacement current into Ampere's circuital law, thus making the time-varying fields have the property of continuity and revealing that a varying electric field could induce a magnetic field, too. Considering that a varying electric field and a varying magnetic field would induce each other and undergo mutual transformation ceaselessly, Maxwell asseverated the existence of electromagnetic waves and advanced the theory that light is a type of electromagnetic wave. His prediction was confirmed later in Hertz's experiment. As the foundation of electromagnetic theory, Maxwell's equations describe the universal law of macroscopic electromagnetic phenomena and quantitatively describe the relationship between electric and magnetic fields, and their dependency on charges and currents, as well as their spatiotemporal characteristics. Apparently, these equations are also the basis of EEG and MEG. While the main brain waves are of low frequency, they

DOI: 10.1201/9781032639260-2

may be approximated as a static electric or magnetic field (refer to Section 3.2). Here, the analysis and discussion will be started from the general Maxwell's equation so that the logic could get across to the reader.

2.1.1 Maxwell's Equations

The differential form of Maxwell's equations describes the spatiotemporal relationship between the field and field source at any point in space and is only suitable for the points where the physical properties of its medium do not vary suddenly. Seen as follows:

$$\nabla\times\vec{H}=\vec{J}+\frac{\partial\vec{D}}{\partial t} \tag{2.1a}$$

$$\nabla\times\vec{E}=-\frac{\partial\vec{B}}{\partial t} \tag{2.1b}$$

$$\nabla\cdot\vec{B}=0 \tag{2.1c}$$

$$\nabla\cdot\vec{D}=\rho \tag{2.1d}$$

where $\vec{E}$=Electric field strength (V/m), $\vec{H}$=Magnetic field strength (A/m), $\vec{B}$=Magnetic flux density (Wb/m^2), $\vec{D}$=Electric flux density (C/m^2), $\vec{J}$=Current density (A/m^2), ρ=Charge density (C/m^3). Besides, according to the law of conservation of charge, the current continuity equation can be obtained as:

$$\nabla\cdot\vec{J}=-\frac{\partial\rho}{\partial t} \tag{2.2}$$

The equations (2.1a–d) are respectively called as Maxwell-Ampere's law, Faraday's law of electromagnetic induction, Gaussian magnetic field law (the right end is zero since the magnetic monopole has not yet been found in nature), and Gaussian electric field law (conservation of charge). These four equations are not completely independent of each other. For time-varying fields, it can be shown that two of them are not independent. If the two curl equations and the current continuity equation are considered independent, then the two divergence equations can be derived, indicating that these two divergence equations are non-independent. According to Helmholtz's theorem, a vector field that disappears in the far area is uniquely determined by its curl and divergence (a vector field can be expressed as an overlay of a rotation-free divergence field and a divergence-free curl field). It needs to be pointed out that the distinction between independent and dependent equations is relative. We can also take equations (2.1a,b,d) as independent equations, then equations (2.1c) and (2.2) are thus dependent.

2.1.2 Constitutive Relation

As mentioned above, if the two curl equations in Maxwell's equations and the current continuity equation are considered to be independent, there are five vector functions

$\vec{E}$, $\vec{H}$, $\vec{D}$, $\vec{B}$ and $\vec{J}$ and one scalar function r in those three equations. As one vector function can be decomposed into 3 scalar functions, it means that we have 16 scalar functions. However, the three independent equations are of only 7 scalar equations, they cannot determine the whole 16 unknown scalar functions. Therefore, these three independent equations are also called unrestricted forms of Maxwell's equations. The difference between the number of equations and that of unknowns is hidden in the internal connections among the five vectors, they depend on the characteristics of the medium and are known as the constitutive relation or composition relationship. After introducing three such auxiliary vector equations, we get nine additional scalar equations, so that the number of unknown field functions is the same as that of field equations, and then the field equations are solvable, thereby becoming a restricted form of Maxwell's equations.

2.1.2.1 Free Space

For free space (e.g., vacuum or air medium), the composition relationship is the simplest one, i.e., $\vec{D} = \varepsilon_0 \vec{E}$, $\vec{B} = \mu_0 \vec{H}$, $\vec{J} = 0$, where ε_0 and μ_0 represent the permittivity (or medium constant) and magnetic permeability of the free space, respectively. As in the International Units System,

$$\varepsilon_0 = \frac{1}{C^2 \mu_0} = \frac{1}{36\pi} \times 10^{-9} \text{ F/m}, \qquad \mu_0 = 4\pi \times 10^{-7} \text{ H/m},$$

$$C = \frac{1}{(\mu_0 \varepsilon_0)^{1/2}} = 3 \times 10^8 \text{ m/s}.$$

2.1.2.2 Medium Polarization

The medium as a whole should be electrically neutral, i.e., there is an equal number of positive and negative charges. Under natural conditions, electric neutrality is also manifested in every macroscopic area, including a macroscopically infinitesimal region, i.e., a macroscopic point. Under the influence of an external electric field, the local electrical neutrality may be destroyed, resulting in a heterogeneous distribution of positive and negative charges. This is called polarization. A non-zero charge density distribution may exist in a polarized medium, called bound charged density. Bound charge means that the positive and negative charges cannot be truly separated from each other, and neither of them can leave the medium, so it is different from the free charge density. The medium is composed of kinds of molecules in a certain proportion and configuration. For a homogeneous medium, the proportion and configuration are the same everywhere.

In terms of electrical properties, molecules can be divided into two types: polar and nonpolar molecules. A polar molecule has a non-zero electric dipole moment because its positive and negative charges are relatively separated from each other; whereas a nonpolar molecule has a zero electric dipole moment since its positive and negative charges are all uniformly distributed. When exerting an external electric field, the positive electric center of the molecule will move in the direction of the electric field, while the negative electric center will move oppositely. The electric dipole moment of molecules would be changed by the external electric field so that it will no longer be zero for nonpolar molecules. This is known as molecular polarization. Under natural conditions, the inherent electric dipole moment of a polar molecule is distributed isotropically, so that the vector sum of the electric dipole moments of all molecules in any macroscopic area is zero, and the macroscopic medium is nonpolar. The external electric field has a moment-of-force effect on polar molecules. Considering the energy of a dipole in an electric field, the smaller the angle between the direction of the electric dipole moment and the direction of the electric field is, the lower the energy is, the electric dipole moment would be "twisted toward" the direction of the electric field under the action of the external electric field moment of force effect. In a thermally balanced medium, the external electric field drives the electric dipole moments of polar molecules close to the electric field and gets nonpolar molecules polarized in the direction of the electric field. As a result, the vector sum of the electric dipole moments of all molecules at a macroscopic point, a small volume, is no longer zero. The vector sum of electric dipole moments in a volume is the electric dipole moment of the medium. Let the electric dipole moment of the medium in a volume ΔV around a point be $\Delta\vec{P}$. The limit $\vec{P} \equiv \frac{d\vec{P}}{dV} \equiv \lim_{\Delta V \to 0} \frac{\Delta\vec{P}}{\Delta V}$ represents the intensity of polarization at this point. It should be particularly noted that ΔV is approaching a "macro zero", dV is a macroscopically infinitesimal volume, and they are still large enough from a microscopic perspective to have so many molecules.

For a region with all dipoles inside, the total charge quantity is zero. However, for a volume V surrounded by a closed surface S, the total charge inside is $Q = -\oint \vec{P}\cdot\vec{n}\,ds = -\int_V \nabla\cdot\vec{P}\,dV$, while the bound charge density $\rho_m(\vec{r}) = -\nabla\cdot\vec{P}$. As the medium polarization is caused by the external electric field, for a linear approximation case, the intensity of polarization $\vec{P}$ should be proportional to the electric field intensity $\vec{E}$, and then the proportional relationship can be expressed as $P = \chi\varepsilon_0\vec{E}$. An anisotropic complex media χ can be a second-order three-dimensional (3D) tensor. It reduces to a proportionality constant in an isotropic medium called polarizability. For a medium situation, if we take the total charge density as $\rho_t(\vec{r}) = \rho(\vec{r}) + \rho_m(\vec{r})$ as the counterpart of the free charge density $\rho(\vec{r})$ in a vacuum, then a formal vacuum electrodynamic theory can be adopted too. With $\rho_t(\vec{r})$ in Gauss' theorem, we have $\varepsilon_0\nabla\cdot\vec{E} = \rho_t(\vec{r})$, which can also be expressed as follows

$$\nabla\cdot\vec{D} = \rho(\vec{r}) \tag{2.3}$$

Where the electric displacement field is defined as $\vec{D} = \epsilon\vec{E} = \varepsilon_0(1+\chi)\vec{E}$.

2.1.2.3 Homogeneous and Isotropic Linear Medium

For a homogeneous and isotropic linear medium, the movement of internal charges incurs three states: polarized, magnetized, and conductive, noted by the polarization intensity vector $\vec{P}$, magnetization intensity vector $\vec{M}$, and conduction-current density vector $\vec{J}$, respectively. Among them, the polarization intensity vector and magnetization intensity vector are separately defined as $\vec{P} = \lim_{\Delta V \to 0} \frac{\sum \vec{P}_e}{\Delta V}$ C/m^2 and $\vec{M} = \lim_{\Delta V \to 0} \frac{\sum \vec{P}_m}{\Delta V}$ A/m, where $\vec{P}_e$ and $\vec{P}_m$ are called electric dipole moment and magnetic dipole moment, respectively. Obviously, the definition of electric dipole moment is the same as that in the previous section. Therefore, $\vec{D}$ and $\vec{B}$ can be expressed as a compound vector $\vec{D} = \varepsilon_0 \vec{E} + \vec{P}$ and $\vec{B} = \mu_0 \left(\vec{H} + \vec{M} \right)$, respectively. The composition relationship of media is based on an experiment. For a linear medium, $\vec{P}$ is proportional to $\vec{E}$, and so is $\vec{M}$ to $\vec{H}$, i.e., $\vec{P} = \varepsilon_0 \chi_e \vec{E}$ and $\vec{M} = \chi_m \vec{H}$, where χ_e and χ_m represent the relative electric and magnetic polarizability of the medium, respectively. So, the composition relationship can finally be written as follows:

$$\vec{D} = \varepsilon_0 \vec{E} + \varepsilon_0 \chi_e \vec{E} = \left(1 + \chi_e\right) \varepsilon_0 \vec{E} = \varepsilon_r \varepsilon_0 \vec{E} = \varepsilon \vec{E} \tag{2.4a}$$

$$\vec{B} = \left(1 + \chi_m\right) \mu_0 \vec{H} = \mu_r \mu_0 \vec{H} = \mu \vec{H} \tag{2.4b}$$

where ε and μ represent the permittivity and magnetic permeability of the medium, respectively; ε_r and μ_r represent the relative permittivity and relative magnetic permeability of the medium, respectively. Furthermore, a differential form of Ohm's law is introduced as follows:

$$\vec{J} = \sigma \vec{E} \tag{2.4c}$$

where σ represents the conductivity of the medium. Since χ is non-negative, $\varepsilon \geq \varepsilon_0$. In particular, the medium constant will not be zero. The equations (2.4a–c) together are called the composition relationship.

By substituting these constitutive relations into Maxwell's equations, we can obtain the form:

$$\nabla \times \vec{H} = \vec{J} + \frac{\partial}{\partial t}\left(\varepsilon \vec{E}\right) \tag{2.5a}$$

$$\nabla \times \vec{E} = -\frac{\partial}{\partial t}\left(\mu \vec{H}\right) \tag{2.5b}$$

$$\nabla \cdot \vec{J} = -\frac{\partial \rho}{\partial t} \tag{2.5c}$$

and two dependent equations:

$$\nabla \cdot \left(\mu \vec{H}\right) = 0 \tag{2.5d}$$

$$\nabla \cdot \left(\varepsilon \vec{E}\right) = \rho \tag{2.5e}$$

2.1.2.4 Anisotropic Medium

For a complex anisotropic medium, its composition relationship can be expressed in matrix form or component form.

$$[D]=[\varepsilon][E], D_j=\sum_{k=1}^{3}\varepsilon_{jk}E_k;\ \ [B]=[\mu][H], B_j=\sum_{k=1}^{3}\mu_{jk}H_k;$$

$$[J]=[\sigma][E],\ J_j=\sum_{k=1}^{3}\sigma_{jk}E_k$$

$j,k = 1, 2, 3$, or vector/tensor form $\vec{D}=\vec{\vec{\varepsilon}}\cdot\vec{E}$, $\vec{B}=\vec{\vec{\mu}}\cdot\vec{H}$, $\vec{J}=\vec{\vec{\sigma}}\cdot\vec{E}$. In the composition relationship of this kind of medium, the parallel relationship no longer existed anymore between $\vec{D}$ and $\vec{E}$, $\vec{B}$ and $\vec{H}$, or $\vec{J}$ and $\vec{E}$. So the corresponding ratio relations need to be represented by the tensor permittivity $\vec{\vec{\varepsilon}}$, tensor permeability $\vec{\vec{\mu}}$ and tensor conductivity $\vec{\vec{\sigma}}$, respectively. If a medium is described with $\vec{\vec{\varepsilon}}$, it is an anisotropic medium, such as crystal and the plasma in a constant magnetic field; if a medium is described with $\vec{\vec{\mu}}$, it is an anisotropic magnetic medium, such as the ferrite in a constant magnetic field; there are few media described with $\vec{\vec{\sigma}}$, like the rectification boundary in a semiconductor and the plasma in a constant magnetic field, etc. Anyway, these constitutive relations can be used to derive the restricted form of Maxwell's equations for anisotropic media.

2.1.3 Integral Forms and Boundary Conditions

The differential forms of Maxwell's equations apply only to a certain point in space where the physical properties of a medium are continuous everywhere. However, for an obvious multi-zone medium, a sudden change in the medium physical properties is bound to take place at the boundary surface, leading to a sudden change in the field vector at the boundary surface. Therefore, for a point on the surface, the differential forms of Maxwell's equations do not make sense anymore, making it necessary to consider an integral form of Maxwell's equations that is applicable in a limited space.

For volume V surrounded by a closed surface S, equations (2.1a–d) are volumetrically integrated by starting from the unrestricted form of Maxwell's equations and using the vector identities as follows (equations H.32 and H.34 in Appendix H):

$$\int_V \nabla\cdot\vec{F}dV=\oint_S \vec{F}\cdot\vec{n}dS \tag{2.6a}$$

$$\int_V \nabla\times\vec{F}dV=\oint_S \vec{n}\times\vec{F}dS \tag{2.6b}$$

where $\vec{n}$ is the unit vector of the outer normal of the closed surface. The formula (2.6a) is the Gaussian divergence theorem, and the formula (2.6b) is the Stokes theorem. Now applying equation (2.6a) to equations (2.1c, d), and equation (2.6b) to the volume integral of equations (2.1a, b), we obtain:

$$\oint_S \left(\vec{n}\times\vec{H}\right)dS = \int_V \left(\vec{J}+\frac{\partial\vec{D}}{\partial t}\right)dV \tag{2.7a}$$

$$\oint_S \left(\vec{n}\times\vec{E}\right)dS = -\int_V \left(\frac{\partial\vec{B}}{\partial t}\right)dV \tag{2.7b}$$

$$\oint_S \left(\vec{n}\cdot\vec{B}\right)dS = 0 \tag{2.7c}$$

$$\oint_S \left(\vec{n}\cdot\vec{D}\right)dS = \int_V \rho dV \tag{2.7d}$$

Now, we discuss some particular cases below.

2.1.3.1 General Cases

Suppose there is a small closed cylinder with an interface between two different media, and the upper and lower sides of the cylinder are parallel to the interface, as shown in Figure 2.1. Suppose the electromagnetic field enters medium 2 from

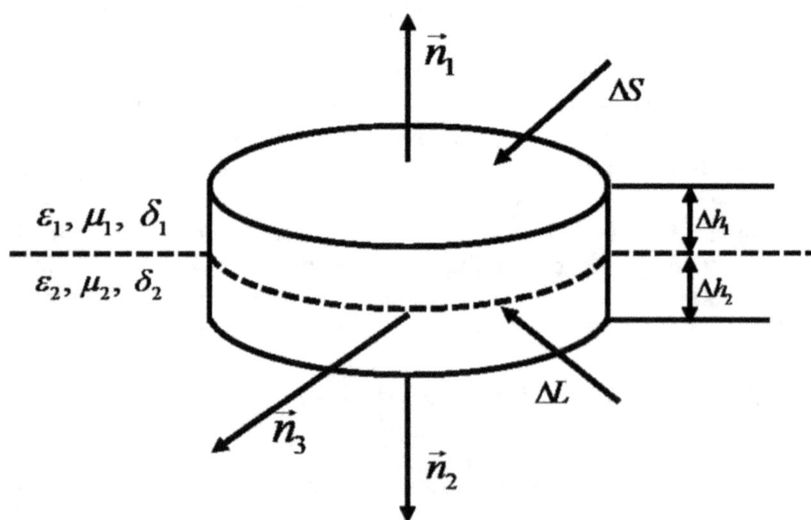

FIGURE 2.1 A closed cylinder cross an interface between two different media.

medium 1, let Δh_1 and $\Delta h_2 \to 0$, the volume of the small closed cylinder shrinks to a point, as long as the electric flux density and magnetic density near the interface are limited, we have

$$\lim_{\Delta h \to 0} \left(\Delta h_1 \vec{D}_1 + \Delta h_2 \vec{D}_2 \right) = 0 \tag{2.8a}$$

$$\lim_{\Delta h \to 0} \left(\Delta h_1 \vec{B}_1 + \Delta h_2 \vec{B}_2 \right) = 0 \tag{2.8b}$$

If there are charges and currents on the interface, the volume charge density ρ and volume current density $\vec{J}$ in Maxwell's equations are singular at the interface. So they need to be replaced with surface charge density ρ_S and surface current density $\vec{J}_S$, they can be defined as follows:

$$\rho_S = \lim_{\Delta h \to 0} \left(\rho_1 \Delta h_1 + \rho_2 \Delta h_2 \right) \ \mathrm{C/m^2} \tag{2.9a}$$

$$\vec{J}_S = \lim_{\Delta h \to 0} \left(\vec{J}_1 \Delta h_1 + \vec{J}_2 \Delta h_2 \right) \ \mathrm{A/m} \tag{2.9b}$$

Equations (2.8) and (2.9) will be applied to equation (2.7a–d). Given a fixed boundary, to obliterate the time derivative terms in equations (2.7a) and (2.7b), the partial derivative can be moved to the outside of integration. The terms directly proportional to Δh can be removed according to equation (2.8). At the same time, among the terms containing $\vec{n}$ cross products and dot products, only those in which $\vec{n}$ is consistent with $\vec{n}_1$ or $\vec{n}_2$ are retained. Therefore, the corresponding forms of equations (2.7a–d) on the boundary surface are transformed into the general boundary conditions for the tangential and normal components of the field as shown by the following equations (2.10).

$$\vec{n}_1 \times \vec{H}_1 + \vec{n}_2 \times \vec{H}_2 = \vec{J}_S \tag{2.10a}$$

$$\vec{n}_1 \times \vec{E}_1 + \vec{n}_2 \times \vec{E}_2 = 0 \tag{2.10b}$$

$$\vec{n}_1 \cdot \vec{B}_1 + \vec{n}_2 \cdot \vec{B}_2 = 0 \tag{2.10c}$$

$$\vec{n}_1 \cdot \vec{D}_1 + \vec{n}_2 \cdot \vec{D}_2 = \rho_S \tag{2.10d}$$

Taking equation (2.7a) as an example, let's first look at the volume integral on the right. When the cylinder thickness approaches zero, the volume integral term of the electric displacement vector is equal to zero due to equation (2.8), and the current density term exists because of equation (2.9). About the surface integral on the left, the lateral surface integral is zero because the cylinder thickness approaches zero, while the integral on the cylinder's upper and lower surfaces may remain there. Finally, both sides are the same surface integral, thus no special consideration is needed for any one of them, and that's the result equation (2.10a). The same is true for other cases.

For the current continuity equation (2.2), calculate the surface integral inside the closed cylindrical surface. Considering the contribution of the current passing through the top, bottom, and side surfaces to the surface integral, we obtain

$$\oint_S \vec{n}_s \cdot \left(\vec{J}_1 \Delta h_1 + \vec{J}_2 \Delta h_2\right) dl + \left(\vec{n}_1 \cdot \vec{J}_1 + \vec{n}_2 \cdot \vec{J}_2\right) \Delta S$$

$$= -\left(\frac{\partial \rho_1}{\partial t} \Delta h_1 + \frac{\partial \rho_2}{\partial t} \Delta h_2\right) \Delta S \tag{2.11}$$

To find the solution to equation (2.11), we need to use 2D divergence (i.e., surface divergence), its definition is as follows:

$$\nabla_S \cdot \vec{A} = \lim_{\Delta S \to 0} \frac{\oint_S \left(\vec{n}_s \cdot \vec{A}\right) dl}{\Delta S}$$

Where $\vec{n}_s$ is the outward unit normal vector on the side of the small closed cylinder. When Δh_1, Δh_2 and $\Delta S \to 0$, based on equation (2.9), equation (2.11) can be transformed into

$$\nabla_S \cdot \vec{J}_S + \left(\vec{n}_1 \cdot \vec{J}_1 + \vec{n}_2 \cdot \vec{J}_2\right) = -\frac{\partial \rho_S}{\partial t} \tag{2.12}$$

It is the boundary condition for $\vec{J}$, including the tangential and normal components, derived from the current continuity equation. In summary, equations (2.10) and (2.12) give all the boundary conditions.

2.1.3.2 Neither of the Two Media Is an Ideal Conductor

If the two media are not ideal conductors, there is no tangential surface current $\left(\vec{J}_S = 0\right)$ on the boundary surface. The boundary conditions for the field and current are as follows:

$$\begin{aligned}
\vec{n}_1 \times \vec{H}_1 + \vec{n}_2 \times \vec{H}_2 &= 0 \\
\vec{n}_1 \times \vec{E}_1 + \vec{n}_2 \times \vec{E}_2 &= 0 \\
\vec{n}_1 \cdot \vec{B}_1 + \vec{n}_2 \cdot \vec{B}_2 &= 0 \\
\vec{n}_1 \cdot \vec{D}_1 + \vec{n}_2 \cdot \vec{D}_2 &= \rho_S \\
\vec{n}_1 \cdot \vec{J}_1 + \vec{n}_2 \cdot \vec{J}_2 &= -\frac{\partial \rho_S}{\partial t}
\end{aligned} \tag{2.13}$$

If there is no charge accumulation on the interface, the surface charge density is zero, and the relevant boundary conditions degrade into continuous normal components of the electric displacement vector; if the number of charges does not increase or decrease with time on the interface, the time derivative of the charge density on the right end of the equation will be zero. This boundary condition reduces to a continuous normal current.

$$\vec{n}_k \cdot \vec{J}_1 = \vec{n}_k \cdot \vec{J}_2, \quad k = 1,2 \tag{2.13a}$$

It is one of the boundary conditions commonly used for the research on EEG in Chapters 3 and 4.

2.1.3.3 Medium 2 Is an Ideal Conductor

If medium 2 is an ideal conductor, there is no field in medium 2, and the boundary condition reduces to:

$$\begin{aligned}
\vec{n}_1 \times \vec{H}_1 &= \vec{J}_S \\
\vec{n}_1 \times \vec{E}_1 &= 0 \\
\vec{n}_1 \cdot \vec{B}_1 &= 0 \\
\vec{n}_1 \cdot \vec{D}_1 &= \rho_S \\
\nabla_S \cdot \vec{J}_S + \vec{n}_1 \cdot \vec{J}_1 &= -\frac{\partial \rho_S}{\partial t}
\end{aligned} \tag{2.14}$$

The physical meaning of the boundary condition is consistent with the integral and differential forms of Maxwell's equations, except that it is necessary to compress the volume to a surface, and the area to a line on the boundary. All these illustrate the physical field-source relationship. So, the boundary condition can be regarded as a special form or limit form of Maxwell's equations on the boundary. Actually, in the boundary condition, only two tangential field conditions are required, while the two normal field conditions can be used to verify the results obtained from the tangential field conditions, just as in the situation where the two curl equations are basic field equations in Maxwell's equations.

2.2 ELECTROMAGNETIC WAVES IN CONDUCTOR

Electromagnetic waves can propagate without attenuation inside a vacuum and ideal insulating medium ($\sigma = 0$), owing to the absence of energy dissipation. Since organisms are good conductors, it's necessary to study the electromagnetic wave in a conductor. As there are free electrons in a conductor, driven by the electromagnetic waves, the free electrons will move to form a conduction current, which generates Joule heat, causing a continuous loss of electromagnetic waves. So, the electromagnetic wave inside the conductor is an attenuating wave, the electromagnetic energy is converted into heat energy during the propagation process.

The propagation of the electromagnetic waves in the conductor is a process in which the alternating electromagnetic field and the moving of free electrons constrain each other. This interaction determines the existence form of the electromagnetic wave in the conductor. In the following, we will start with the characteristics of free charge distribution in the conductor, and then solve Maxwell's equations in the presence of conducting current distribution.

2.2.1 Free Charge and Conductor

Under the electrostatic condition, the conductor is electrically neutral, with free charges distributed on its surface only. In a rapidly varying field, does the free charge distribution of a conductor keep the same as in an electrostatic field?

Suppose there is a distribution of free charges in a certain area inside the conductor, with a density of ρ, then an electric field $\vec{E}$ will be generated, i.e. $\varepsilon\nabla\cdot\vec{E}=\rho$. Under the action of the electric field $\vec{E}$, a conduction current $\vec{J}$ appears in the conductor. With constitutive equation (2.4c), we have $\nabla\cdot\vec{J}=\frac{\sigma}{\varepsilon}\rho$, suggested that when charge density ρ appears somewhere in the conductor, there is a current flowing outward from this point. It is very obvious from a physical point of view, that when there is charge accumulation in a certain area, the charges repel each other there, inevitably generating a current that diverges outward. Due to the outflow of charges, the charge density in the volume decreases. Further, with the current continuity equation (2.2), we have $\frac{\partial\rho}{\partial t}=-\nabla\cdot\vec{J}=-\frac{\sigma}{\varepsilon}\rho$, and the solution as follows:

$$\rho(t)=\rho_0 e^{-\frac{\sigma}{\varepsilon}t}=\rho_0 e^{-\frac{t}{\tau}} \tag{2.15}$$

where ρ_0 represents the charge density at $t=0$. Thus it can be seen that the charge density decays exponentially with time, and the decay time constant τ (the time when the value of ρ decreases to ρ_0/e) is $\tau^{-1}=\sigma/\varepsilon$. So the condition for a good conductor can be defined as follows:

$$\sigma/(\varepsilon\omega)\gg 1 \tag{2.16}$$

Equation (2.16) implies that when the frequency of the electromagnetic wave satisfies $\omega\ll\tau^{-1}=\sigma/\varepsilon$, $\rho(t)=0$ means that no charge exists inside a conductor in one cycle. For a general metal conductor, the order of magnitude of τ is 10^{-17}, and it can be seen as a good conductor as long as the electromagnetic wave frequency is $\ll 10^{17}$. Generally, a metal conductor can be regarded as a good conductor, and its charges can only be distributed on the surface as there is no free charge inside a good conductor. Correspondingly, when $\sigma/(\varepsilon\omega)\ll 1$, it is a good insulator. When $\sigma/(\varepsilon\omega)\sim 100$, it is a semiconductor. For the electrical phenomena in a biological medium, equation (2.16) can be completely satisfied if the biological relevant parameters are substituted into it. Therefore, the biological medium can be regarded as a good conductor, except for the area where the current is generated, i.e., it is unnecessary to consider free charges except for the cellular electrical activity in the excited zone.

2.2.2 Sinusoidal Electromagnetic Waves and Conductor

Because of the Fourier transform, any complex wave can be expressed as a combination of sinusoidal and cosine waves, therefore, electromagnetic wave propagation can be investigated in the form of sinusoidal waves.

Here we will use $e^{i(\omega t-kR)}$ to describe the propagation of a sinusoidal electromagnetic wave. In the formula, ω represents the angular frequency, t is the time, k is the wavenumber, R is the wave propagation distance, i is the imaginary unit. Now, a time-harmonic expression of the electric field can be assumed as follows:

$$\vec{E} = \vec{E}_0 e^{i(\omega t-kR)} \tag{2.17}$$

With Maxwell's equations, we can get the curl on both sides of equation (2.1b) and substitute (2.1a) and the related constitutive equation in the result to derive the following equation: $\nabla \times \nabla \times \vec{E} = -\frac{\partial}{\partial t}\mu\left(\sigma\vec{E} + \varepsilon\frac{\partial}{\partial t}\vec{E}\right)$. By substituting equation (2.17) in it and using the curl formula, we obtain:

$$\nabla \times \nabla \times \vec{E} = \nabla\left(\nabla \cdot \vec{E}\right) - \nabla^2\vec{E} = -\frac{\partial}{\partial t}\mu\left(\sigma\vec{E} + \varepsilon\frac{\partial}{\partial t}\vec{E}\right) = -i\omega\mu(\sigma + i\omega\varepsilon)\vec{E} \tag{2.18}$$

As there is no free charge in a good conductor, the divergence of the electric field is zero, i.e., $\nabla \cdot \vec{E} = 0$. So we get the dispersion relationship:

$$k^2 = -i\omega\mu\sigma\left(1 + \frac{i\omega\varepsilon}{\sigma}\right) = \omega^2\mu\left(\varepsilon - i\frac{\sigma}{\omega}\right) \tag{2.19}$$

and wave equations:

$$\nabla^2\vec{E} + k^2\vec{E} = 0, \quad \left(\nabla \cdot \vec{E} = 0\right) \tag{2.20}$$

This equation is consistent in form with that for non-conductive media. It has a solution in the form of a monochromatic plane wave, i.e., equation (2.17). As can be seen from equation (2.19), the wavenumber k here is a complex number. Considering that the wave propagation has a direction, equation (2.17) can be rewritten as follows:

$$\vec{k} = \vec{\beta} - i\vec{\alpha} \tag{2.21}$$

$$\vec{E} = \vec{E}_0 e^{-\vec{\alpha}\cdot\vec{r}} e^{i(\omega t - \vec{\beta}\cdot\vec{r})} \tag{2.22}$$

The formula (2.22) shows that in the conductive medium ($\sigma \neq 0$), the wave number is complex, so there is attenuation in the propagation process of the wave, among which $e^{-\vec{\alpha}\cdot\vec{r}}$ is the attenuation factor. Usually, the propagation distance r where the amplitude of the wave attenuates to $1/e$ of its initial value is defined as the penetration depth δ, $\delta = 1/\alpha$.

By substituting equation (2.21) in (2.19), we obtain

$$k^2 = \beta^2 - \alpha^2 - i2\vec{\alpha}\cdot\vec{\beta} = \omega^2\mu\left(\varepsilon - i\frac{\sigma}{\omega}\right) \tag{2.23}$$

Let the imaginary and real parts equal respectively, we get

$$\beta^2 - \alpha^2 = \omega^2\mu\varepsilon \tag{2.24a}$$

$$\vec{\alpha}\cdot\vec{\beta} = \frac{1}{2}\omega\mu\sigma \tag{2.24b}$$

If $\vec{\alpha}$ and $\vec{\beta}$ are in different directions, it is a non-uniform plane wave. When they are in the same direction, it is a uniform plane wave (the waves have the same amplitude in the direction perpendicular to wave propagation). For a uniform plane wave, we have

$$\alpha\beta = \frac{1}{2}\omega\mu\sigma \tag{2.25}$$

Solving equations (2.24a) and (2.25) simultaneously, we get:

$$\begin{aligned} \beta &= \omega\sqrt{\mu\varepsilon}\left\{\frac{1}{2}\left[\sqrt{1+\left(\frac{\sigma}{\omega\varepsilon}\right)^2}+1\right]\right\}^{1/2} \\ \alpha &= \omega\sqrt{\mu\varepsilon}\left\{\frac{1}{2}\left[\sqrt{1+\left(\frac{\sigma}{\omega\varepsilon}\right)^2}-1\right]\right\}^{1/2} \end{aligned} \tag{2.26}$$

Now, let's have some discussions on the above analytical results.

For a good conductor, the ratio between the imaginary part and the real part of k^2 in equation (2.19) is $\sigma/(\varepsilon\omega)$. Also, as the good conductor satisfies $\sigma/(\varepsilon\omega) \gg 1$, the real part of k^2 can be ignored. In this case, the following results can be derived from equations (2.23) and (2.24):

$$\alpha \approx \beta \approx \sqrt{\omega\mu\sigma/2}, \quad \delta = 1/\alpha = \sqrt{2/\omega\mu\sigma} \tag{2.27}$$

This equation indicates that the penetration depth is inversely proportional to the square root of conductivity and frequency. Therefore, for a high-frequency electromagnetic wave, the electromagnetic field is only concentrated in a thin layer on the surface of the conductor. Accordingly, the high-frequency current also flows in a thin layer on the surface. This phenomenon is called the skin effect. This effect has important applications in industry, e.g., for electromagnetic shielding, high-frequency quenching, etc. Equation (2.27) also shows that electromagnetic waves of different frequency bands differ from one another in terms of action depth when they interact with an organism. If microwaves and millimeter waves are used to irradiate an organism, the direct action range is limited to the body surface, while the indirect action may cover the humoral system, nervous system, and other transfer systems such as the acupuncture action system, but they still needed to to be confirmed, and on the contrary, if a low-frequency current is applied to the human body, it may spread throughout the body. This is the mechanism adopted by electrical impedance tomography (EIT), low-frequency transcranial alternating current stimulation (tACS), and transcranial direct current stimulation (tDCS). Electromagnetic waves have been widely used in biomedicine, with a very different application mechanism for different frequency bands. This book focuses on the application of low-frequency electromagnetic fields in brain science, where it can be regarded as a quasi-static electric field problem as the propagation effect is negligible on the brain wave bands (Section 3.2).

2.3 ELECTROSTATICS AND UNIQUENESS THEOREM

2.3.1 Electrostatics and Poisson's Equation

Electrostatics focuses on the electrostatic field generated by static charge distribution and its effect on charges, as well as the conditions under which charges stay static. For static charges, the charge density is independent of time, and the current density vector $\vec{J}(\vec{r})=0$. Static charges do not generate a magnetic field, i.e., $\vec{H}=\vec{B}=0$. In the electrostatic field, $\vec{E}(\vec{r})$ and $\vec{D}(\vec{r})$ have nothing to do with time, i.e., $\frac{\partial \vec{D}}{\partial t}=0$. Under these conditions, there are only two Maxwell's equations left, i.e.,

$$\nabla \cdot \vec{D}=\rho, \ \nabla \times \vec{E}=0 \tag{2.28}$$

The second equation just shows that the electrostatic field is a curl-free field and can be defined as the gradient of a potential field $\Phi(\vec{r})$ (because the curl of the gradient is zero, equation (2.28)) as

$$\vec{E}(r)=-\nabla \Phi(\vec{r}) \tag{2.29}$$

In practical electrostatic problem, the medium is often divided into uniform regions, and ε in the constitutive relationship in each region is a constant. By substituting equation (2.29) into (2.28), we obtain

$$\nabla^2 \Phi=-\frac{\rho}{\varepsilon} \tag{2.30}$$

It is Poisson's equation that the electrostatic potential $\Phi(\vec{r})$ satisfies, with the source $\frac{\rho}{\varepsilon}$.

2.3.2 Boundary Conditions

Regarding the boundary conditions that should be met by the electric field on the boundary, the electric field is taken as a static field according to quasi-static

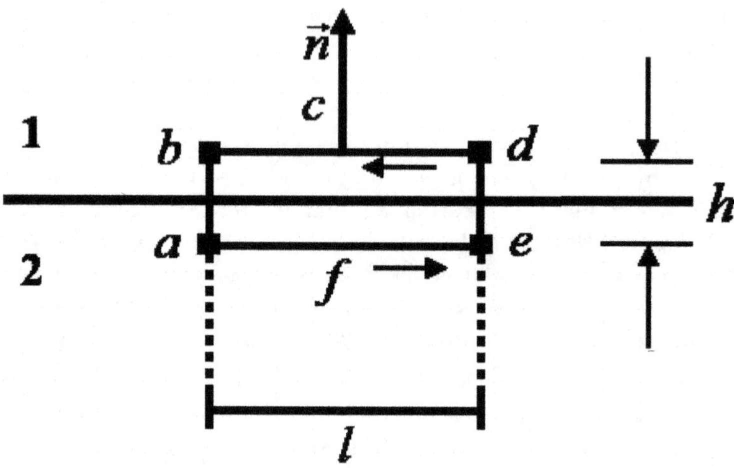

FIGURE 2.2 A rectangular loop cross the interface of different media.

approximation (Section 3.2) and a circuit is constructed on the boundary (Figure 2.2), then according to the circuital theorem of the electrostatic field, we have $\oint \vec{E} \cdot d\vec{l} = 0$.

Considering that the width of the circuit, h, approaches 0, the integral is changed into $\vec{E}_1 \cdot \vec{l}^{\,0} \Delta l - \vec{E}_2 \cdot \vec{l}^{\,0} \Delta l = 0$, i.e., $\vec{E}_{1t} - \vec{E}_{2t} = 0$ or denoted by $\vec{n} \times \left(\vec{E}_1 - \vec{E}_2\right) = 0$, where $\vec{l}^{\,0}$ is the unit vector along the boundary. It shows that the tangential component of electric field intensity on the interface of different media is continuous, i.e., equation (2.13).

For the two points, a and b, selected randomly on both sides close to the interface, their potential difference is:

$$U_{ab} = \Phi_a - \Phi_b = \int_a^b \vec{E} \cdot d\vec{l} \tag{2.31a}$$

Since the electrostatic field is a curl-free field, its line integral has nothing to do with the path. Therefore, the integral of any path extending from a to b can be regarded as an integral along such as the path *afedcb*, and on this integral path, we let h approach 0, and $ab = de$ also approaches 0, so the integral is simplified as follows:

$$\int_a^b \vec{E} \cdot d\vec{l} = \int_{afe} \vec{E} \cdot d\vec{l} + \int_{dcb} \vec{E} \cdot d\vec{l} = \left(E_{2t} - E_{1t}\right) = 0 \tag{2.31b}$$

Substitute equation (2.31b) in (2.31a), and we can get

$$\Phi_a = \Phi_b = \Phi_2 = \Phi_1 \tag{2.32}$$

that is, the potential on the interface is continuous.

This potential continuity and the normal current continuity, (2.32) and (2.13a), are the two most commonly used boundary conditions for EEG research as shown in Chapters 4 and 5.

Another important boundary condition is about the electric displacement vector which satisfies (2.10d). If the surface charge density ρ_S is zero, the normal electric displacement vector is continuity.

$$\vec{n}_{1(2)} \cdot \vec{D}_1 = \vec{n}_{1(2)} \cdot \vec{D}_2 \tag{2.32a}$$

2.3.3 The Boundary between Medium and Vacuum

In practice, the medium always occupies a limited space, while there is a vacuum outside the medium. Thus, the interface between the medium and vacuum must be taken into consideration. It is only necessary to regard vacuum as a medium with a medium constant of ε_0. Poisson's equation (2.30) and the two boundary conditions (2.32) and (2.13a) apply to the vacuum and the interface between the vacuum and medium, respectively. In other words, the boundary condition is the normal current being zero (can not flow out), and this is widely adopted as the scalp surface potential boundary condition.

$$\vec{n}_1 \cdot \vec{D}_1 = 0 \tag{2.32b}$$

2.3.4 Conductor Boundary Conditions and Infinite Boundary Conditions

Because any electric field will induce current in a conductor, the electric field intensity in the conductor must be zero everywhere under the condition of electrostatic equilibrium (zero current). According to equation (2.29), the electrostatic potential Φ in each isolated conductor must therefore be the same everywhere, which is the conductor boundary condition in electrostatics. The electrostatic field can extend to infinity under the condition that the electrostatic system is not completely shielded by a closed conductor surface. Therefore, the boundary condition at infinity is required to find the definite solution of Poisson's equation (2.30). The charge distribution is regarded as a collection of point charges, and Poisson's equation with them as the sources should have a particular solution that can be expressed as the sum of the Coulomb potentials of all these point charges. Accordingly, the electrostatic field strength also can be expressed as the sum of the Coulomb field. This type of electrostatic potential and electrostatic field should approach zero at infinity:

$$\Phi(\vec{r}) \xrightarrow[r\to\infty]{} 0, \quad \vec{E}(\vec{r}) \xrightarrow[r\to\infty]{} 0 \tag{2.33}$$

That's the boundary condition for the electrostatic potential and electrostatic field at infinity. This also means that we should take infinity as the zero reference point for potential (Yao, 2001). The electrostatic field that meets this boundary condition, obtained from the given charge distribution, is taken as the electrostatic field generated by this charge distribution. If the electrostatic system is surrounded by a closed conductor surface, the boundary condition on this conductor surface can be used to replace the boundary condition at infinity (2.33), i.e., the electrostatic potential on this surface should be a constant, and zero is one of the acceptable constants. Because it is always feasible to add or subtract an arbitrary constant to or from the electrostatic potential, the electrostatic potential on the surface of the closed conductor can be set to zero. This is called grounding. In this way, the boundary condition in equation (2.33) on the surface of the closed conductor is taken as:

$$\Phi = 0 \tag{2.34}$$

2.3.5 Maxwell's Equations and EEG

Based on the electrostatic approximation regarding EEG (see Chapter 3), EEG can be considered as steady electric fields (current fields) in a biological conductor. In this situation, EEG is decoupled with MEG, making it only necessary to adopt two of Maxwell's equations and the current continuity equation. Among them, (2.1d) degrades into Poisson's equation (2.30), while (2.1d) and (2.2) degrade into boundary conditions (2.32) and (2.13a). In a word, EEG field theory is a special situation of Maxwell's equations.

2.3.6 Uniqueness Theorem

2.3.6.1 Uniqueness Theorem of Electrostatic Potential

For an electrostatic system, given the charge distribution on each insulating medium, the surface charge distribution on the surface of the medium and the interface between the media, and the potential on each conductor, the problem of electrostatics becomes the solving of Poisson's equation (2.30) under these boundary conditions. This is a well-defined mathematical problem. The uniqueness theorem is that Poisson's equation (2.30) has a unique solution under these boundary conditions.

Solution: Supposing $\Phi(\vec{r})$ and $\Phi'(\vec{r})$ are two solutions to this problem, we can define:

$$F(\vec{r}) = \Phi(\vec{r}) - \Phi'(\vec{r}) \tag{2.35}$$

According to equation (2.32), as both Φ and Φ' are continuous at the interface, so do their difference. Take equations (2.4a), (2.10d) and (2.29) together, by subtracting the boundary conditions (2.10d) that Φ and Φ' satisfies separately, we obtain:

$$\varepsilon_1 \frac{\partial F_1}{\partial n} - \varepsilon_2 \frac{\partial F_2}{\partial n} = 0 \tag{2.36}$$

Where $F_1 = \Phi_1 - \Phi_1'$, $F_2 = \Phi_2 - \Phi_2'$. This formula is the boundary condition of F on the interface between the two media. Since Φ and Φ' are set to the same value on the conductor (given the potential on the surface of each conductor), equation (2.36) means $F = 0$ on the surface, and that's the potential continuity boundary condition widely used in Chapters 4 and 5.

By subtracting the Poisson's equations that Φ and Φ' satisfy, we get

$$\nabla^2 F = 0 \tag{2.37}$$

It indicates that F satisfies Laplace's equation.

Suppose $\Psi_1(r)$ and $\Psi_2(r)$ are two arbitrary second-order differentiable functions, with Gauss' theorem for the surface integral of the vector $\Psi_1\nabla\Psi_2$ on the closed surface S, we have Green's theorem

$$\oint_S \Psi_1(\nabla\Psi_2)\cdot d\vec{s} = \int_V \left[\Psi_1\nabla^2\Psi_2 + (\nabla\Psi_1)\cdot(\nabla\Psi_2)\right]dV \tag{2.38}$$

where the volume integral domain V is the space enclosed by S. This formula is used in the volume V_i occupied by the ith homogeneous medium. Let $\Psi_1 = \varepsilon_1 F, \quad \Psi_2 = F$. According to Laplace's equation (2.37), the following is obtained

$$\oint_{S_i} \varepsilon_i F(\nabla F)\cdot d\vec{s} = \int_{V_i} \varepsilon_i (\nabla F)^2 \cdot dV \tag{2.39}$$

Where S_i is the surface of V_i. Suppose there is a common interface $S_{ii'}$ between the two media i and i', the sum of the two surface integrals, the left surface integral in equation (2.39) for media i and i' on this common interface $S_{ii'}$, is:

$$\int_{S_{ii'}} F\left(\varepsilon_i \frac{\partial F_i}{\partial n} - \varepsilon_{i'}\frac{\partial F_{i'}}{\partial n}\right)ds = 0 \tag{2.40}$$

Here, boundary condition (2.36) and the F continuity on the interface are used, and the normal direction $\boldsymbol{n}$ points to side i' of the interface from side i. Based on equation (2.39), we can sum up all the media, including the vacuum. According to equation (2.40), all the media interface integrals are mutually canceled out, its left end only leaves the surface integral on the conductor surface and that at infinity. Considering the factor $F=0$ (for conductor) in the integrand function, the integral on the surface of a finite conductor is zero. For the surface integral at infinity, let Φ and Φ' approach 0 at infinity in the form of $1/r$ as in the Coulomb potential of a charge q is $\Phi(\vec{r}) = \dfrac{q}{4\pi\varepsilon r}$, and also let F approach 0 in the same manner, then ∇F should approach zero in the form of $1/r^2$. The integrand function of the surface integral on the left of equation (2.39) will approach zero at infinity in the form of $1/r^3$, while the integral area is in the form of r^2. By multiplying the two terms, it is clear that the surface integral approaches zero at infinity in the form of $1/r$, thus the sum on the left of equation (2.39) is zero for all the media, which means the sum on the right is

$$\sum_i \int_{V_i} \varepsilon_i (\nabla F)^2 dV = 0 \tag{2.41}$$

In the medium, $\varepsilon_i \neq 0$ and ε_i is always positive. The sufficient and necessary condition for this equation is that $\nabla F = 0$ everywhere, thus F is a constant. Because $F=0$ at infinity, or $F=0$ on a shielded conductor under the condition that the conductor is shielded by a closed conductor surface, i.e., the constant F is zero. $F=0$ everywhere means everywhere

$$\Phi = \Phi' \tag{2.42}$$

Equation (2.42) means that any two solutions to an electrostatic problem are identical to each other, thus the solution is unique. This theorem shows that all the brain potentials which satisfy Poisson's equation and boundary conditions have a unique solution.

2.3.6.2 Uniqueness Theorem of the Electromagnetic Field

Consider the electromagnetic field in a limited space, the sources are expected to exist inside and outside the space, the electromagnetic field in the space cannot be completely determined on condition that only the sources inside the space are known, as it is necessary to know the impact of the field from sources outside the area, and fortunately, this impact can be replaced by the equivalent field on the boundary surface. Hence, the electromagnetic field in space is uniquely determined by the internal field source and the boundary field value. The uniqueness theorem provides a theoretical basis for the uniqueness of the solution, making no doubt for the correctness of the solution. This shows that the uniqueness theorem is an important theoretical basis for calculating electromagnetic field boundary value problems. So, under what conditions and in what range can a solution be unique? Here the range includes time intervals and space regions, while the conditions involve initial conditions and boundary conditions.

The uniqueness theorem of the electromagnetic field can be summarized as: in a finite region V, given field sources, as well as the initial value of the electric field strength and magnetic field strength at any point when $t = t_0$, and the tangential component of the electric field or the magnetic field on the boundary surface when $t \geq t_0$, the electromagnetic field in region V can be uniquely determined as long as $t > 0$. In other words, there will be a unique field that satisfies the field equation of a given source, the initial conditions, and the boundary conditions. For a time-harmonic field, considering that its initial conditions are replaced with periodicity, the uniqueness of the solution can be ensured in given boundary conditions. For a static field, which can be considered as the case where the frequency is equal to zero, only the boundary conditions are needed.

For unbounded space, it can be assumed that the boundary of the surrounded area approaches infinity. If the boundary is assumed to be spherical, its radius R approaches infinity, so that all the space is enclosed. Therefore, when a time-harmonic source is located in a finite region in unbounded space, there will be only a unique solution as long as the field quantity satisfies Maxwell's equations, as well as the boundary condition in the region and the infinity condition at infinity. The infinity condition (radiation condition) for the field quantity is

$$\lim_{R\to\infty} R\begin{pmatrix} \vec{E} \\ \vec{H} \end{pmatrix} = \text{limited value or } \lim_{R\to\infty} R\begin{pmatrix} \vec{E} \\ \vec{H} \end{pmatrix}\cdot \vec{n} = 0 \tag{2.43}$$

The above electromagnetism uniqueness theorem states that providing boundary conditions for Maxwell's equations uniquely fixes a solution for those equations. This theorem provides the basis for the EEG forward problem. However, it should not be misunderstood as providing boundary conditions (or the field solution itself) uniquely fixes a source distribution (inverse problem). Actually, the EEG inverse problem is nonunique as illustrated by the following theorem in Chapter 10.

2.4 MULTIPOLE EXPANSION OF ELECTROSTATIC POTENTIAL AND FIELD

2.4.1 Point Charge and Dipole

Considering the charge distribution, electrostatic potential, and electrostatic field in a homogeneous medium, the Coulomb potential can be rewritten as follows:

$$\Phi(\vec{r}) = \frac{q}{4\pi\varepsilon|\vec{r}-\vec{r}'|} \tag{2.44}$$

where $\vec{r}'$ is the radial vector of the position where point charge q is located. By moving this point charge to $\vec{r} = \vec{r}' + \vec{l}$, and then placing a point charge $-q$ at $\vec{r}'$, under the condition that displacement $\vec{l}$ is infinitesimal, and retaining the first-order term of Taylor's expansion of $1/|\vec{r}-\vec{r}'|$, the total electrostatic potential of the two-point charges can be expressed as

$$\Phi(\vec{r}) = \frac{q}{4\pi\varepsilon}\vec{l}\cdot\nabla'\frac{1}{|\vec{r}-\vec{r}'|} = -\frac{q}{4\pi\varepsilon}\vec{l}\cdot\nabla\frac{1}{|\vec{r}-\vec{r}'|} \tag{2.44a}$$

$\nabla' = -\nabla$ is the gradient operator for $\vec{r}'$ coordinate. Let $\vec{l} \to 0$ but keep the following vector a finite value

$$\vec{P} \equiv q\vec{l} \tag{2.45}$$

And let $\vec{r}' = 0$, we obtain:

$$\Phi(\vec{r}) = \frac{\vec{P}\cdot\vec{r}_0}{4\pi\varepsilon r^2} \tag{2.46}$$

$\vec{r}_0 \equiv \frac{\vec{r}}{r}$ is the unit vector of $\vec{r}$ direction. The system which consists of two opposite point charges is called an electric dipole, with its dipole moment represented by $\vec{P}$ and its total charge being zero. Equation (2.46) is the electrostatic potential of the electric dipole, and its electrostatic field is:

$$\vec{E}(\vec{r}) = -\nabla\Phi(\vec{r}) = \frac{3(\vec{P}\cdot\vec{r}_0)\vec{r}_0 - \vec{P}}{4\pi\varepsilon r^3} \tag{2.47}$$

The radial vector for the dipole position is still marked as $\vec{r}'$. Then (2.46) can be transformed as follows:

$$\Phi(\vec{r}) = \frac{\vec{P}\cdot(\vec{r}-\vec{r}')}{4\pi\varepsilon|\vec{r}-\vec{r}'|^3} \tag{2.48}$$

Similarly, we can define an electric quadrupole and electric quadrupole moment, an electric octopole and electric octopole moment. Generally, it is practical to define an electric 2^n-pole and electric 2^n-pole moment, where n is a natural number. The electric multipoles defined in this way are all based on point-like charges combination, while the electric multipole moment can be defined for general charge distribution. The electric dipole moment defined in equation (2.45) is just a special case of the general definition. For more information, please refer to Wikswo and Swinney (1984, 1985) or Section 4.1 of this book.

2.4.2 Electric Multipole Moment for Arbitrary Charge Distribution

The function

$$g(\vec{r}-\vec{r}') = \frac{1}{4\pi|\vec{r}-\vec{r}'|} \tag{2.49}$$

satisfies the Poisson's equation of a point source:

$$\nabla^2 g(\vec{r}-\vec{r}') = -\delta(\vec{r}-\vec{r}') \tag{2.50}$$

The function of equation (2.49) is therefore called Green's function of Poisson's equation, from which we can obtain the electrostatic potential that satisfies Poisson's equation (2.30) and tends to zero at infinity.

$$\Phi(\vec{r}) = \int \frac{\rho(\vec{r}')}{4\pi\varepsilon|\vec{r}-\vec{r}'|}\,dV' \tag{2.51}$$

where $\vec{r}'$ is the integral variable for volume integral $\int dV'$. With the generation function of the Legendre polynomial (Zhang, 2003), we have:

$$\begin{aligned}\frac{1}{|\vec{r}-\vec{r}'|} &= \left(r^2+r'^2-2rr'\cos\hat{\theta}\right)^{-1/2} = \frac{1}{r>}\left[1+\left(\frac{r<}{r>}\right)^2-2\frac{r<}{r>}\cos\hat{\theta}\right]^{-1/2} \\ &= \frac{1}{r>}\sum_{l=0}^{\infty}\left(\frac{r<}{r>}\right)^l P_l\left(\cos\hat{\theta}\right)\end{aligned} \tag{2.52}$$

Where $\hat{\theta}$ is the angle between $\vec{r}$ and $\vec{r}'$, $P_l(\xi)$ is the Legendre Polynomial of ξ, $r>$ is the larger one between r and r', $r<$ is the smaller one of them. With the addition theorem of spherical harmonic functions, this formula can be further separated into variables as follows:

$$\frac{1}{|\vec{r}-\vec{r}'|}=\frac{1}{r>}\sum_{l=0}^{\infty}\left(\frac{r<}{r>}\right)^{l}\frac{4\pi}{2l+1}\sum_{m=-l}^{l}Y_{lm}(\theta,\varphi)Y_{lm}^{*}(\theta',\varphi') \tag{2.53}$$

Where (θ,φ) is the azimuth pointed by $\vec{r}$, (θ',φ') is the azimuth pointed by $\vec{r}'$. Substitute these formulas in equation (2.51) and pay attention to the spherical coordinate expression of the volume element $dV' = r'^2dr'd\Omega'$, where $d\Omega' = \sin\theta' d\theta' d\varphi'$ is the solid angle element. By defining $\rho_{lm}(r') = \int Y_{lm}^{*}(\theta',\varphi')\rho(\vec{r}')d\Omega'$, where its integral range is a solid angle of 4π, we obtain:

$$\Phi(\vec{r})=\frac{1}{\varepsilon}\sum_{l=0}^{\infty}\frac{1}{2l+1}\sum_{m=-l}^{l}\left[\begin{array}{l}\frac{1}{r^{l+1}}\int_{0}^{r}\rho_{lm}(r')r'^{l+2}dr' \\ +r^{l}\int_{r}^{\infty}\rho_{lm}(r')r'^{1-l}dr'\end{array}\right]Y_{lm}(\theta,\varphi) \tag{2.54}$$

Assuming that the charge is distributed in a limited space, the origin of the coordinates is set at an appropriate position in this space. There must be a sufficiently large number of R to let $\rho(\vec{r}) = 0$ when $r > R$. As the minimum value of R is called the radius of charge distribution, we use R to represent this radius. For $r > R$, (2.54) becomes as follows:

$$\Phi(\vec{r})=\sum_{l=0}^{\infty}\frac{1}{4\pi\varepsilon r^{l+1}}\sqrt{\frac{4\pi}{2l+1}}\sum_{m=-l}^{l}P_{lm}^{*}Y_{lm}(\theta,\varphi) \tag{2.55}$$

where

$$P_{lm}^{*}\equiv\sqrt{\frac{4\pi}{2l+1}}\int_{0}^{R}\rho_{lm}(r')r'^{l+2}dr'=\sqrt{\frac{4\pi}{2l+1}}\int r'^{l}Y_{lm}^{*}(\theta,\varphi)\rho(\vec{r}')dV' \tag{2.56}$$

The final volume integral covers the space of charge distribution. The origin of the coordinates can be set outside the charge distribution that is at an appropriate position in the passive area. Then, there must be a small enough radius R', to let $\rho(\vec{r}) = 0$ when $r < R'$. The maximum of R' is called the passive zone radius, as can be denoted by R'. For $r < R'$, equation (2.54) becomes as follows:

$$\Phi(\vec{r})=\sum_{l=0}^{\infty}\frac{r^{l}}{4\pi\varepsilon}\sqrt{\frac{4\pi}{2l+1}}\sum_{m=-l}^{l}Q_{lm}^{*}Y_{lm}(\theta,\varphi) \tag{2.57}$$

where

$$Q_{lm}^{*}=\sqrt{\frac{4\pi}{2l+1}}\int_{R'}^{\infty}\rho_{lm}(r')r'^{1-l}dr'=\sqrt{\frac{4\pi}{2l+1}}\int r'^{-l-1}Y_{lm}^{*}(\theta,\varphi)\rho(\vec{r}')dV' \tag{2.58}$$

The final volume integral extends all over the region where charges are distributed. Since the origin is outside this region, and $\vec{r} \neq 0$ in the integral domain, the integrand has no singularity. The expansion based on a spherical harmonic function is called harmonic expansion or multipole expansion. Equation (2.54) is the multipole expansion of electrostatic potential, while (2.55) and (2.57) are its two specific situations, corresponding to the outside ($r > R$) and the inside ($r < R$) of the source region, respectively. Equation (2.55) is the most commonly used one. Generally, the multipole expansion of electrostatic potential refers to this specific situation. For a given l, $m = l, l-1, l-2, \ldots -l$, a total of $2l+1$ different values can be got. They correspond to $2l+1$ quantities, including $P_{ll}, P_{ll-1}, P_{ll-2}, \ldots, P_{l-l}$, which correspond to $2l+1$ components of 2^l-pole moment. The rationality of this appellation is that $r^l Y_{lm}(\theta,\varphi)$ is an l-degree polynomial of x, y, and z on Cartesian coordinates, conforming to the conventional meaning of the 2^l-pole moment.

According to the relationship between Cartesian coordinates (x, y, z) and spherical coordinates (r,θ,φ)

$$\left.\begin{aligned} x &= r\sin\theta\cos\varphi \\ y &= r\sin\theta\sin\varphi \\ z &= r\cos\theta \end{aligned}\right\} \tag{2.59}$$

and the definition of spherical harmonics we have:

$$Y_{00}(\theta,\varphi) = \sqrt{\frac{1}{4\pi}}, \tag{2.60}$$

$$\left.\begin{aligned} rY_{10}(\theta,\varphi) &= \sqrt{\frac{3}{4\pi}}\, z \\ rY_{1\pm1}(\theta,\varphi) &= \mp\sqrt{\frac{3}{8\pi}}\,(x \pm iy) \end{aligned}\right\} \tag{2.61}$$

When $l = 0$, $2^l = 1$, the corresponding electric unipolar moment

$$P_{00} = \int \rho(\vec{r}')\,dV' = Q \tag{2.62}$$

Q is the total charge of this charged system. When $l = 1$, $2^l = 2$, the corresponding electric dipole moments with $2l+1=3$ components are

$$\left.\begin{aligned} P_{10} &= \sqrt{\frac{4\pi}{3}}\int r'Y_{10}(\theta',\varphi')\rho(\vec{r}')\,dV' = \int z'\rho(\vec{r}')\,dV' = P_z \\ P_{1\pm1} &= \sqrt{\frac{4\pi}{3}}\int r'Y_{1\pm1}(\theta',\varphi')\rho(\vec{r}')\,dV' = \pm\frac{1}{\sqrt{2}}\int (x'+iy')\rho(\vec{r}')\,dV' \equiv \mp\frac{1}{\sqrt{2}}(P_x \pm iP_y) \end{aligned}\right\}$$

(2.63)

where P_x, P_y and P_z are the components of the moment vector

$$\vec{P} \equiv \int \vec{r}'\rho(\vec{r}')dV' \tag{2.64}$$

For the electric dipole composed of the point charge $-q$ at the origin and the point charge q at the vector diameter $\vec{d}$, this vector $\vec{P}$ in equation (2.64) is the electric dipole moment defined by equation (2.45), in a word, equation (2.64) can be used as the electric dipole moment vector $\vec{P}$. The general definition of equation (2.63) shows that the P_{11}, P_{10}, and P_{1-1} of the electric dipole moment are the recombinations of the P_x, P_y, P_z components of the electric dipole moment vector. Similarly, the situation of electric quadrupole moments can be discussed.

Equation (2.55) can be expressed as:

$$\left.\begin{aligned} \Phi(\vec{r}) &= \sum_{l=0}^{\infty} \Phi_l(\vec{r}), \\ \Phi_l(\vec{r}) &= \sum_{m=-l}^{l} P_{lm}^{*}\Phi_{lm}(\vec{r}), \\ \Phi_{lm}(\vec{r}) &= \frac{1}{4\pi\varepsilon r^{l+1}}\sqrt{\frac{4\pi}{2l+1}}Y_{lm}(\theta,\varphi). \end{aligned}\right\} \tag{2.65}$$

Utilize (2.60) and (2.62) to get the unipolar electrostatic potential:

$$\Phi_0(\vec{r}) = \frac{Q}{4\pi\varepsilon r} \tag{2.66}$$

It can be regarded as the electrostatic potential of the point charge Q at the origin. Utilize (2.63) and (2.61) to get the dipole electrostatic potential:

$$\Phi_1(\vec{r}) = \frac{\vec{P}\cdot\vec{r}_0}{4\pi\varepsilon r^2} \tag{2.67}$$

Compared with equation (2.46), it can be regarded as the electrostatic potential of the electric dipole $\vec{P}$ at the origin. Here $\vec{r}_0$ notes the unit vector of $\vec{r}$.

The negative gradient of the multi-pole expansion of the electrostatic potential gives the multi-pole expansion of the electrostatic field. For example, from equation (2.65), we obtain:

$$\left.\begin{aligned} \vec{E}(\vec{r}) &= \sum_{l=0}^{\infty} \vec{E}_l(\vec{r}), \\ \vec{E}_l(\vec{r}) &= \sum_{m=-l}^{l} P_{lm}^{*}\vec{E}_{lm}(\vec{r}), \\ \vec{E}_{lm}(\vec{r}) &= -\nabla\Phi_{lm}(\vec{r}) = -\frac{1}{\sqrt{4\pi(2l+1)}\varepsilon}\nabla\left[\frac{1}{r^{l+1}}Y_{lm}(\theta,\varphi)\right] \end{aligned}\right\} \tag{2.68}$$

$\vec{E}_l(\vec{r})$ is called the electrostatic 2^l pole field. There are many ways to analytically calculate the gradient of the final equation, among which the more intuitive one is to express the equation with spherical harmonics in rectangular coordinates. This method has been used to calculate the electrostatic potential (2.66) and (2.67) with $l \le 2$, and their negative gradient is the electrostatic field with $l \le 2$. The calculation shows that the electric monopole field and the electric dipole field generated by arbitrary charge distribution $\rho(\vec{r})$ are as follows respectively.

$$\vec{E}_0(\vec{r}) = \frac{Q}{4\pi\varepsilon r^2}\vec{r}_0 \tag{2.69}$$

$$\vec{E}_1(\vec{r}) = \frac{3(\vec{P}\cdot\vec{r}_0)\vec{r}_0 - \vec{P}}{4\pi\varepsilon r^3} \tag{2.70}$$

It can be regarded as the electrostatic field of a point charge Q and an electric dipole $\vec{P}$ at the origin, respectively.

The charge distribution itself can also be expanded to multiple poles:

$$\rho(\vec{r}) = \sum_{l=0}^{\infty}\sum_{m=-l}^{l}\rho_{lm}(r)Y_{lm}(\theta,\varphi) \tag{2.71}$$

According to the orthogonal normalization of spherical harmonics:

$$\int Y_{lm}^{*}(\theta,\varphi)Y_{l'm'}(\theta,\varphi)d\Omega = \delta_{ll'}\delta_{mm'} = \begin{cases} 1, & \text{if } \ l = l' \ \text{ and } \ m = m' \\ 0, & \text{or else} \end{cases} \tag{2.72}$$

we can obtain:

$$\begin{aligned} \int Y_{lm}^{*}(\theta,\varphi)\rho(\vec{r})d\Omega &= \sum_{l'=0}^{\infty}\sum_{m'=-l'}^{l'}\rho_{l'm'}(r)\int Y_{lm}^{*}(\theta,\varphi)Y_{l'm'}(\theta,\varphi)d\Omega \\ &= \sum_{l'=0}^{\infty}\sum_{m'=-l'}^{l'}\rho_{l'm'}(r)\delta_{ll'}\delta_{mm'} = \rho_{lm}(r) \end{aligned} \tag{2.73}$$

It specifies the coefficient of multipole expansion (2.71), i.e., radial function $\rho_{lm}(r)$.

2.5 ELECTROSTATIC ENERGY AND INTERACTION

An electrostatic field exerts a force on charges, suggesting that there is energy contained in the electrostatic field. Static charge density distribution $\rho(\vec{r})$ is accompanied by electrostatic potential $\Phi(\vec{r})$ and electrostatic field $\vec{E}(\vec{r})$. The electrostatic energy in the volume V is defined as follows (Zhang, 2003)

$$W = \frac{1}{2}\int_V \vec{E}\cdot\vec{D}dV = \frac{1}{2}\int_V \varepsilon(\nabla\Phi)^2 dV \tag{2.74}$$

Assuming that the medium is uniform and ε is a constant, after a partial integral, and with Poisson's equation (2.3), we obtain

$$W = \frac{1}{2}\int_V \rho(\vec{r})\Phi(\vec{r})dV + \frac{1}{2}\int_S \varepsilon\Phi(\nabla\Phi)\cdot d\vec{s} \tag{2.75}$$

where S is the surface of V. If the electrostatic system is shielded by the surface S of a conductor, and the constant potential Φ on the surface can be set to zero (grounded) so that the surface integral on the right end of this equation should be equal to zero. If the system is open, the surface S can be pushed to infinity. Viewed from infinity, finite charge distribution can be regarded as a point charge. As a result, Φ approaches zero at infinity in the form of $1/r$, so $\nabla\Phi$ approaches zero in the form of $1/r^2$, and the area S in the form of r^2. The overall result is that the surface integral in the right end of equation (2.75) approaches zero in the form of $1/r$. Anyway, appropriate boundary conditions make this surface integral equal to zero, the expression (2.75) of electrostatic energy is simplified as follows:

$$W = \frac{1}{2}\int_V \rho(\vec{r})\Phi(\vec{r})dV \tag{2.76}$$

The volume integral covers areas where the charge density $\rho(\vec{r})$ is non-zero.

Suppose there are two charge density distributions, $\rho_1(\vec{r})$ and $\rho_2(\vec{r})$, and their electrostatic potentials, $\Phi_1(\vec{r})$ and $\Phi_2(\vec{r})$, satisfy Poisson's equation:

$$\left.\begin{aligned} \nabla^2\Phi_1(\vec{r}) &= -\frac{1}{\epsilon}\rho_1(\vec{r}) \\ \nabla^2\Phi_2(\vec{r}) &= -\frac{1}{\epsilon}\rho_2(\vec{r}) \end{aligned}\right\} \tag{2.77}$$

When the two charge distributions coexist, the total charge density is

$$\rho(\vec{r}) = \rho_1(\vec{r}) + \rho_2(\vec{r}) \tag{2.78}$$

Adding the two equations in equation (2.77) can obtain the Poisson's equation. The charge density in the right end is the total charge density represented by this equation, and the total electrostatic potential in the left end is $\Phi(\vec{r}) = \Phi_1(\vec{r}) + \Phi_2(\vec{r})$. Substitute this equation and (2.78) into (2.76) to get the Electrostatic energy of the total system:

$$W = W_1 + W_2 + V \tag{2.79a}$$

$$\left.\begin{aligned} W_1 &= \frac{1}{2}\int_V \rho_1(\vec{r})\Phi_1(\vec{r})dV \\ W_2 &= \frac{1}{2}\int_V \rho_2(\vec{r})\Phi_2(\vec{r})dV \end{aligned}\right\} \tag{2.79b}$$

$$V = \frac{1}{2}\int_V \rho_1(\vec{r})\Phi_2(\vec{r})dV + \frac{1}{2}\int_V \rho_2(\vec{r})\Phi_1(\vec{r})dV \tag{2.79c}$$

where W_i, $i = 1, 2$, is the electrostatic energy when charge distribution $\rho_i(\vec{r})$ exists alone, called self-energy. Self-energy can be understood as the magnitude of the work done by the external force to aggregate an infinite number of infinitely dispersed infinitesimal elements in a charged body into this charged body; V is the electrostatic energy that occurs when the two charge distributions exist side by side, known as the electrostatic interaction energy, representing the interaction between two charge distributions, called as mutual energy. It can be understood as the work done by the electrostatic force between the charged bodies in the charged body system when the charged bodies are separated from one another at the current position to infinity. Due to the singularity of the unipolar electrostatic potential at the origin as shown by equation (2.66), the self-energy of a point charge is infinite. Generally, the self-energy of point charges is ignored in electrostatics. As is known from equation (2.77)

$$\begin{aligned}\int_V \rho_1(\vec{r})\Phi_2(\vec{r})dV &= -\varepsilon\int_V \left[\nabla^2\Phi_1(\vec{r})\right]\Phi_2(\vec{r})dV \\ &= -\varepsilon\int_V \Phi_1(\vec{r})\nabla^2\Phi_2(\vec{r})dV + \varepsilon\int_S (\Phi_1\nabla\Phi_2 - \Phi_2\nabla\Phi_1)\cdot d\vec{s} \\ &= \int_V \Phi_1(\vec{r})\rho_2(\vec{r})dV\end{aligned} \tag{2.80}$$

The Green formula (C.14) is used for the second step and (2.77) is used again for the third step. According to the aforementioned method from (2.75) to (2.76), the surface integral can be ascertained to be zero. The two items in equation (2.79c) can be combined to obtain:

$$V = \int_V \rho_1(\vec{r})\Phi_2(\vec{r})dV = \int_V \Phi_1(\vec{r})\rho_2(\vec{r})dV \tag{2.81}$$

By omitting the subscripts 1 and 2, this formula can be written as

$$V = \int_V \rho(\vec{r})\Phi(\vec{r})dV \tag{2.82}$$

Just remember that $\rho(\vec{r})$ is one charge distribution and $\Phi(\vec{r})$ is the electrostatic potential accompanying the other one charge distribution, i.e., the external electric field for $\rho(\vec{r})$. Equation (2.79c) represents the potential energy of charge distribution $\rho(\vec{r})$ under the action of the external electric field. The electrostatic potential $\Phi(\vec{r})$ of the external electric field generated by the charge distribution at a distance can be treated by multipole expansion. When studying the interaction between a charge distribution $\rho(\vec{r})$ and the external electric field, we do not care about how the external electric field is generated by the charge distribution at a distance. If the charge distribution and the electric potential distribution are both treated by orthogonal multipole expansion, the monopole electrostatic effect can be calculated as below:

$$V_0 = Q\Phi(0) \tag{2.83}$$

It is equivalent to the effect of the point charge Q and the electrostatic field at the origin. The effect of the dipole electrostatic is

$$V_1 = -\vec{P}\cdot\vec{E}(0) \tag{2.84}$$

It is the interaction between the electric dipole $\vec{P}$ and the electrostatic field at the origin. By analogy, a quadrupolar electrostatic effect can be obtained. It should be mentioned that reported studies rarely consider EEG problems from the perspective of energy. Perhaps something new may be discovered if problems are analyzed from the perspective of energy, especially under the condition of multi-source interaction dynamics in the future.

2.6 THEOREMS OF ELECTROSTATICS

Theorem 1

The electrostatic potential cannot be assigned a maximum or minimum at any point in any region inside a homogeneous medium where there is no charge distribution.

The potential energy of the test charge q in the electrostatic field is as follows

$$V(\vec{r}) = q\Phi(\vec{r}) \tag{2.85}$$

It is equation (2.83) according to theorem 1, the potential energy is of no minimum in an uncharged region inside a homogeneous medium. This makes it impossible for the test charge to be in a stable and static state at any point in this region. All electric fields in a conductor cause carriers (charged ions) to move. The current should be zero everywhere under electrostatic conditions, so that the electric field intensity in the conductor must be zero everywhere, and so is the electric displacement $\vec{D} = \varepsilon\vec{E}$. As a result, the charge density $\rho(\vec{r}) = \nabla\cdot\vec{D}$ is zero everywhere inside the conductor. Under electrostatic conditions, the charges can only be distributed on the surface of the conductor.

Consider a system, there are N conductors in a homogeneous medium with a medium constant of ε, a free charge density distribution $\rho(\vec{r})$ in the medium, and a charge q_i in the ith conductor. For this system, there are:

Theorem 2: Uniqueness Theorem

The electrostatic field and the charge distribution on the conductors are uniquely determined by the charge on each conductor, q_i, $i = 1, 2, \ldots, N$, and the charge density distribution $\rho(\vec{r})$ in the medium.

Theorem 3: Thomson Theorem

The static distribution of charges on a conductor constantly causes the electrostatic energy to be minimum.

In other words, minimum electrostatic energy is a sufficient and necessary condition for the static distribution of charges on a conductor.

Let's consider placing a neutral conductor in an electrostatic field where there are N conductors. The charge density on this neutral conductor must be zero everywhere before it is placed in the field. If the charge density is maintained at zero everywhere, the neutral conductor will not be affected by any force when it is placed in the field. After it is placed somewhere in the electric field, the charge distribution that is neutral everywhere generally does not minimize the electrostatic energy , so this is not a static distribution. The carriers on it will move to form an electric field to offset the external electric field inside it. That's how it is polarized. The charge distribution on other conductors also changes during polarization. Eventually, the electrostatic energy is minimized, indicating that the neutral conductor will be attracted to the electrostatic field when it approaches the electrostatic field of a conductor system.

2.7 DUALITY PRINCIPLE

The principle of duality is used to describe the formal similarity between two phenomena so that a related formula for a domain can be transformed into one for another domain after simple variable substitution. For example, if the charge Q_0 is replaced with the current source density I_F, the potential formula for the electrostatic field of a point charge in a medium is turned into one for a current source in a conductor. This result is not surprising, because the current control equation can be accurately converted into an electrostatic field control equation after appropriate substitution. This fact means that after simple symbol swapping, the solution to an electrostatic problem in a medium can be equivalently changed into the solution to a steady current field problem in a conductor (vice versa). This is how the principle of duality works. Specifically, between electrostatics and steady current field, if the following substitutions are made $\varepsilon \to \delta$, $\vec{D} \to \vec{J}$, $\rho \to I_F$, then the equations of the steady current field

$$\nabla^2 \Phi = -\frac{I_F}{\delta}, \quad I_F = \nabla \cdot \vec{J}, \quad \vec{J} = \delta \vec{E} \tag{2.86}$$

will be transformed into that in electrostatics, i.e.

$$\nabla^2\Phi = -\frac{\rho}{\varepsilon}, \quad \rho = \nabla\cdot\vec{D}, \quad \vec{D} = \varepsilon\vec{E} \tag{2.87}$$

$\vec{E} = -\nabla\Phi$ are shared by the two cases. Actually, the widely recognized fact is that the mathematical knowledge of a current in a volume conductor corresponds so closely to that of a charge in a dielectric medium. Its advantage is that the knowledge about a subject (e.g., electromagnetism) can be easily transferred to another (e.g., physiology). But it must be remembered that duality is not equivalence, e.g., the medium constant and the conductivity have completely different practical meanings. When it comes to the similarities between them, except $\vec{E} = -\nabla\Phi$, their sources are also the same as each other. The sources are charges in electrostatics, while $I_F = \nabla\cdot\vec{J}_S$ in the current field, i.e., the divergence of current. According to the law of conservation of charge, I_F refers essentially to the increase or decrease in charge quantity, so we see charge being the common source.

It is based on this principle of duality that the theory of medium electrostatics is easily applied to the electric field of the conductive brain in this book. Of course, the electric field of the brain is not exactly a steady field in a conductor. It just changes slowly at low frequencies like a brain wave (see the explanation of quasi-static approximation in Chapter 3), here the brain electric wave like the alpha wave is not an electromagnetic wave propagating at the speed of light, but the physiological wave following the electrophysiological fluctuation, so it is very slow and low frequency. It can be regarded as an approximately steady field at every moment, or rather, the current source density can be considered as a constant at every moment (sampling point). But actually, the current source density cannot be a constant since it changes in neural activity. When treated as a quasi-static field, it treats each sampling point separately so that it can be treated as a static field or a steady field.

Because of this duality principle, in general, EEG books or articles, people often do not mention whether the relevant equations are for medium or conductor. Because the formal same equation will result in the same calculation result, people do not need to care about the details underground, but we should know that it is the duality principle that guarantees the correctness of such carelessness.

In other words, the above discussion also may cause confusion: is the brain a conductor or a dielectric medium? Strictly speaking, it is neither a complete conductor (zero resistance) nor a complete insulating medium. The electromagnetic physical properties of brain tissue include both conductivity and permittivity. In the above introduction we see that if we take the brain as a medium, we have a set of equations (2.87) related to the generation of the electric field by charge. Similarly, if we take it as a conductor, we have another group of equations (2.86) for the generation of electric and magnetic fields by a current. Moreover, these two groups of equations are very similar in form and can be connected through simple dual transformations. Now, what should we say? According to equation (2.16) or (3.4), the brain is a good conductor, so describing the brain based on conductor, current, potential and magnetic field, etc., is more close to the real situation. In fact, the subsequent Quasi-static approximation analysis in Section 3.2 indicates that under certain circumstances, the dielectric (capacitive) effect can be ignored, and in case where it cannot be ignored, we can use complex conductivity to include the effect of the dielectric effect (Nunez and Srinivasan, 2006). Meanwhile, due to the duality principle, we needn't nitpick the acceptance of the equations for charge in the dielectric medium in this book or other EEG literature.

2.8 EQUIVALENCE PRINCIPLE

The field equivalence principle is established on the uniqueness theorem. The basic content of the broadly defined electromagnetic equivalence principle is as follows: The solution of a boundary value problem can be taken as an equivalent problem of another known solution. That is to say, in a space region, two different configurations of sources that can induce the same field are said to be equivalent in this region. When the field or potential in a given space region needs to be found, it is not necessary to know its real source configuration, while an equivalent source configuration can be used as a substitute. The equivalent source configuration can be placed outside the given space region or on its boundary. The equivalent source configuration is not unique. It can be constructed in a variety of ways, and the original boundary conditions must be met. The forward uniqueness theorem offers a one-to-one corresponding relation between the field and the source configuration. As long as the same field in the supposed region can be generated, all the source configurations are equivalent no matter how different they are in nature and distribution. If they are not equivalent source configurations, they excite different fields. The principle of equivalence has been widely used to solve electromagnetic field boundary value problems. Besides, many fundamental theorems in electromagnetics are the inferences and specific applications of the equivalence principle, such as the image theory, Huygens principle, etc. The equivalent principle utilized in Chapter 6, it assumes that the effect of a source on its surrounding region can be generated from an equivalent source distribution on its surrounding curved surface. That's the theoretical basis for the equivalent source method of EEG in Chapters 6 and 7. Another example is the Huygens principle, each point on a wavefront acts as a point source for spherical waves, and the superposition of the new individual waves from each of the point sources on the old wavefront forms a new wave, here the distribution of real sources is replaced by the point-source distribution on the Huygens surface, the given closed surface. The Huygens principle is a special form of the field equivalence principle, and they are both based on the opposite of the uniqueness theorem.

It should be emphasized that the uniqueness theorem mentioned in 2.3 provides a forward theoretical basis for the research on EEG and MEG, rather than the uniqueness of the inverse problem. On the contrary, the equivalent source principle, on the one hand, focuses on the un-uniqueness of the inverse brain electromagnetic problem, i.e., the same observation data may come from different source configurations, including real and equivalent source configurations. This requires us to adopt priori information as much as possible in tackling the EEG inverse problem to compress the multiplicity of the solution. On the other hand, the concept of equivalent principle also provides a theoretical basis for the subsequent research on zero-reference technology and equivalent source-based high-resolution cortical imaging technology.

It should be noted that the equivalence principle is introduced from the perspective of electromagnetic fields here. For the potential equivalence principle commonly applied to EEG research, please refer to the description of the Green function in Chapter 6.

REFERENCES

Fu, G. 1991. *Engineering Electromagnetic Theory and Method (In Chinese)*. Beijing: People's Posts and Telecommunications Press.

Jackson, J.D. 1962. *Classical Electrodynamics*. New York: John Wiley & Sons Press.

Jackson, J.D. 1999. *Classical Electrodynamics*, 3rd ed. New York: John Wiley & Sons Press.

Landau, L.D., E.M. Lifshitz. 1975. *The Classical Theory of Fields*. Oxford: Butterworth-Heinemann, ISBN 978-0-7506-2768-9.

Morse, P.M., H. Feshbach. 1953. *Methods of Theoretical Physics*, 2nd ed., 2 Pts. New York: McGraw Hill Education.

Panofsky, W.K.H., M. Phillips. 1962. *Classical Electricity and Magnetism*. Boston, MA: Adison-Wesley Publishing Company, Inc.

Paul L. Nunez, Ramesh Srinivasan. 2006. Electric Fields of the Brain: The Neurophysics of EEG. Oxford University Press.

Salam, M.A. 2014. *Electromagnetic Field Theories for Engineering*. Berlin: Springer Science & Business Media.

Stratton, J.A. 1941. *Electromagnetic Theory*. New York: McGraw Hill Education.

Wikswo Jr, J.P., K.R. Swinney. 1984. A comparison of scalar multipole expansions. *J Appl Phys* 56(11):3039–3049.

Wikswo Jr, J.P., K.R. Swinney. 1985. Scalar multipole expansions and their dipole equivalents. *J Appl Phys* 57(9):4301–4308.

Yao, D. 2001. A method to standardize a reference of scalp EEG recordings to a point at infinity. *Physiol Meas* 22(4):693–711.

Zhang, Q. 2003. *Classical Field Theory (In Chinese)*. Beijing: Science Press.

3 Biophysics and Source Models of EEG

This chapter first introduces the basic biophysical knowledge of EEG, including the brain structure and the electric phenomena in neurons, and then presents the electrostatic approximation of biological electromagnetic phenomena. Based on these contents, the last two sections discuss the models of electrical sources in the brain for EEG.

3.1 BIOPHYSICAL BASIS

3.1.1 Anatomical Structure of the Brain

The human brain has an average weight of about 1,400 g (~3 pounds) and a surface area of about 2,500 cm^2. This layer of gray matter, with a thickness of 2–4 mm, is in the outermost layer of the brain in the form of a complex fold. The longitudinal sulcus divides the brain into two hemispheres. According to the sulci and fissures on the brain surface, the cerebral hemispheres can be divided into the frontal, temporal, vertex, and occipital lobes. The two hemispheres are connected by nerve fibers. Under the cerebral cortex are the diencephalon and brainstem, which are closely related to the activities of internal organs such as breathing, heartbeat, body temperature, and gastrointestinal peristalsis, as well as the arousal level and emotional state, thus being the main basis of psychosomatic medicine (Yao et al., 2022).

There are about 100 billion neurons in the human brain, about 14 billion of which exist in the cerebral cortex, and each neuron has more than 1,000 connections, forming an extremely complex and huge neural network. The information transmission among neurons is conducted primarily by chemical transmission, so-called chemical synapses, and there are as many as about 50 neurotransmitters involved in chemical transmission. The release and activity level of the neurotransmitters are regulated by neuromodulators, and they jointly lay a biological foundation for the highly complex brain functional activities. In addition, there are also electrical synapses, realized by so-called gap junctions. The channels that cross the membranes of two cells allow for the direct exchange of charge carriers and neurotransmitters. They are thought to play an important role in the fast communication among local groups of neurons and mediate the high-frequency oscillations of membrane potentials (Draguhn et al., 1998). The glial cells are dozens of times as many as neurons. In the past, glial cells have been considered to mainly provide structural support intending to maintain a proper ion concentration and transport nutrients for neurons, now it is known that they very actively participate in neural information processing. A special type of glial cell called oligodendrocyte serves as the insulating sheath of myelin, which surrounds nerve fibers so that

DOI: 10.1201/9781032639260-3

signals can be transmitted faster. Beneath the cerebral cortex are a large number of nerve fibers, which are called white matter or medulla because of their light color (similarly, called as gray matter for the cortical area on the brain surface). White matter plays a role in connecting different cortical areas and other parts to nerve endings. And it is increasingly recognized as equally critical for cognition (Filley and Fields, 2016).

Detailed anatomical knowledge helps understand EEG phenomena: (1) The structure and functional activities at the cellular level help to understand the origin of the current source at the macro level; (2) The distribution law of conductivity on a macro scale is the basis for accurately solving the forward and inverse problems of EEG. For example, if the EEG source is known or can be assumed in the gray matter layer only, and the range of gray matter can be determined by MRI, etc., the original 3D EEG source localization problem can be simplified into a quasi-2D problem to improve the accuracy.

The structure of the head is extremely complex. Outside the brain is Cerebrospinal fluid (CSF), followed by the skull and scalp. The thickness and conductivity of these components differ from position to position. In addition, there are pores and cracks in the skull, and the brain tissues are obviously anisotropic. For example, the electric transmission along the nerve fiber of white matter is about ten times as fast as that along the tangential direction of the fiber (bundle of axons). For brain function study, we are usually dependent on limited structural information. Therefore, the related algorithms are often based on the corresponding simplified models, such as the three-layer or four-layer concentric sphere models, or the CT/MRI-based three-layer or four-layer realistic head models. In these models, the conductivity of each layer is generally supposed to be constant.

3.1.2 Bioelectric Phenomena in Neuron

The following is a brief introduction to the bioelectric phenomena in nerve cells. For more details, please refer to the related literatures, such as Plonsey and Barr (1988) and Knösche and Haueisen (2022).

3.1.2.1 Neuron

A typical section of the visual cortex of a rat's brain is as Cajal drew in 1988. It is usually divided into six layers except for some specific brain areas. Neurons in the fourth layer are mainly composed of pyramidal and stellate cells. Their apical dendrites are parallel to each other while stretching upward, forming the surface layer of gray matter. The superimposed electric activities in this layer are considered to be the main source of brain electromagnetic signals. The size, shape, arrangement density, and interconnection of nerve cells in each of the six layers have certain characteristics. However, the neurons in different parts of the cerebral cortex differ from one another in type and density. The complex synaptic connections among the neurons in the cortex form a variety of short and long circuits. After entering the cortex, most of the afferent fibers first integrate themselves into a complex local circuit formed by interconnected cells in various layers and then act on efferent neurons. For example, the afferent fibers from the thalamus stimulate the astrocytes in the fourth layer at first. The latter stimulates the pyramidal cells and Martinotti cells in the fifth layer simultaneously. Through the recurrent collaterals of the pyramidal cells and the ascending axons of the Martinotti cells, the nervous impulse is finally transmitted to the efferent neurons in the second and third layers (from the cortex surface down). In this way, a local circuit is established between the neurons in each layer and the efferent neurons. Also, the local circuit helps to strengthen or inhibit the activity of the efferent neurons. In addition, recent studies showed that the vertical cortical column is the basic functional unit of the cerebral cortex and the basic structure of the complex neuronal circuit.

For every pyramidal neuron, there is a nucleus and many metabolic structures in its soma. The dendrites stretch upward like tree branches, while the axon is a long fiber that grows out of the soma. On the soma and dendrites, there are normally thousands of synapses. They can be divided into two types: inhibitory and excitatory. Most excitatory synapses are located on the surface of dendrites, while inhibitory synapses are often close to the soma.

3.1.2.2 Resting Potential

Electrophysiological experimental results showed that when neurons are resting, the intramembranous potential is more negative than the extramembranous potential. This type of difference between the intramembranous and extramembranous potentials is called resting potential or membrane potential. Figure 3.1 shows a resting potential of –70 mv. According to a large number of experimental studies, Hodgkin et al. proved in the 1940s that the resting potential is mainly due to K^+ in the cell diffusing outward. Because Na^+, K^+, Cl^-, and organic negative ions are heterogeneously distributed inside and outside the membrane, the intracellular concentration of K^+ is higher than that outside the membrane, while the extracellular concentration of Na^+ is higher than that in the intracellular concentration, organic anions are within cells, the extramembranous concentration of Cl^- is higher than the intramembranous concentration. In the resting state, the cell membrane has selective permeability with respect to the above-mentioned ions, i.e., a high permeability to K^+, a low permeability to Na^+ and Cl^-, thus K^+ diffuses to the outside of the membrane along its concentration gradient, leading to an increase in the number of positive charges outside the membrane, and resulting in a potential difference between the inside and outside of the membrane. Outflow stops at the equilibrium potential of K^+ (equilibrium between the electric field force and diffusive force), with a steady potential difference formed. Thus, the resting potential is approximately equivalent to the equilibrium potential of K^+ (Nernst potential).

3.1.2.3 Action Potential

A conductible potential change that takes place when the neurons are stimulated and therefore get excited is known as an action potential, as seen in Figure 3.1. Microelectrodes can be placed in the cells to record how a change takes place in intramembranous potential when the neurons are excited. This depolarization manifests as the rapid disappearance of the intramembranous resting potential (~–70 mv), followed by the appearance of an ascending branch of positive potential (contra polarization, ~+40 mv). This reversal process, in which the extramembranous potential becomes

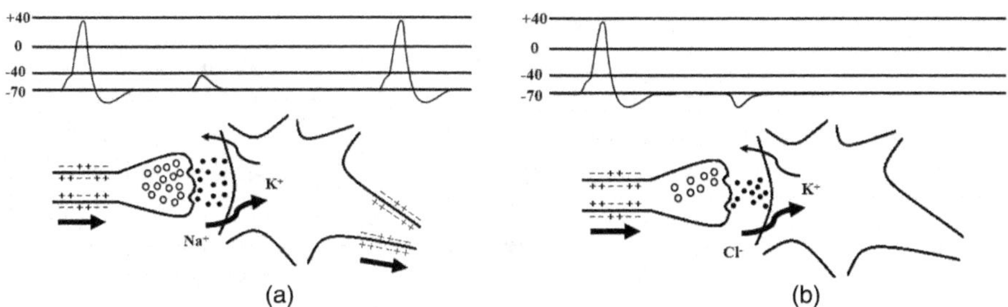

FIGURE 3.1 Schematic diagram of resting potential, action potential, and postsynaptic potential. (a) Excitatory synapse, (b) inhibitory synapse.

negative while the intramembranous potential becomes positive in a flash. Then, the cell membrane will quickly return to its originally polarized state, i.e., being repolarized.

The action potential is generated when the neuron suffers from stimulation with strength above its threshold, activating/opening the Na^+ channel on the cell membrane rapidly, i.e., the cell membrane's permeability to Na^+ increases suddenly, with the permeability of K^+ in the resting state transformed to Na^+. When the permeability of Na^+ exceeds that of K^+, a depolarization process occurs, i.e., the intramembranous potential is positive while the extramembranous potential is negative. The inward flow of Na^+ stops at the equilibrium potential of Na^+. Subsequently, K^+ outflows, getting the membrane potential repolarized, due to the increased permeability of the cell membrane to K^+. Whenever action potential occurs, some Na^+ flows into the membrane, while some K^+ flows out of the membrane, but the total quantity of exchange between K^+ and Na^+ is extremely small. Nevertheless, the cell membrane transports the Na^+ out of the cells with the aid of the Na^+ –K^+ pump during the recovery period and transports the K^+ outside the membrane into the cells. It is due to the concentration difference between Na^+ and K^+ inside and outside the membrane that there exists resting membrane potential under certain conditions, which is the basis for the generation of the next action potential.

Once generated, the action potential can be transmitted to a distant place along the cell membrane. The action potential of an axon does not attenuate during transmission. Also, the action potential cannot be aggregated or generated by the subthreshold stimulus. The action potential is generated only when stimulation reaches the threshold. Moreover, no matter how high the stimulation intensity is, the action potential is unchanged. This is known as a phenomenon of all-or-none, or called 1-or-0 phenomenon. The action potential of the cell body is analogous to that of the axon, but the duration of the former is slightly longer than that of the latter. A change takes place in the excitability of the cell body or axon after the action potential occurs so that they cannot be excited by any stimulus within a period of time. This period is called the absolute refractory period. After that, they will have an excitatory response to a stimulus above the threshold, but not to a subthreshold stimulus. This is called an excitability recovery period, or a supernormal period as the excitability is lower than normal. Finally, the excitability returns to a normal level.

3.1.2.4 Synapse and Postsynaptic Potential

The nerve impulse, as a kind of information, is transmitted along the neuron network. Most of these transmissions are dependent on chemical transmitters, and only a few synapses are electrical (gap junction). The functional contacts among neurons are called synapses, through which the nerve impulse information can be transmitted directionally. As can be seen under an electron microscope, the axon terminal is divided into many sprigs. The sprig ending is ventricose and spherical, called a synaptosome, which adheres to the postsynaptic neuron body or dendrite surface. The membrane of the axon terminal is called the presynaptic membrane, after which a synaptic cleft, with a width of 200–500 Å. The synaptosome contains a large number of aggregated vesicles, called synaptic vesicles, with a diameter of 200–800 Å. The vesicles contain a high concentration of chemical transmitters. Normally, a stimulus can hardly cause dendrites to generate action potential because its lateral spine is thin, while its resistance is large. Instead, its resting potential will decrease or increase, i.e., a postsynaptic potential (PSP) occurs. The postsynaptic potential lasts for a long time, possibly up to 100 ms or longer. As 100 ms corresponds to 10 Hz, PSP is considered to be the main neural basis of scalp surface EEG. According to the contact position, synapses can be divided into a variety of contact forms, such as axon–dendritic synapse, axon–somatic synapse, axon–axonic synapse, etc., these three types of synapses are the most common forms of a functional connection. Among them, axon–dendritic synapse may be effective in changing the excitatory state of neurons; axon–somatic synapse may excite neurons; while axon–axonic synapse may be the structural basis for presynaptic inhibition.

According to its effect on the functional activity of the next neuron, synapses can be divided into two types, i.e., excitatory synapse and inhibitory synapse. Now therefore, the process of synaptic transmission can be summarized as follows: when an electric signal arrives at the synapses, the presynaptic neuron terminal is depolarized, causing Ca^{2+} to flow inward. Special transmitter molecules are released from the synaptic vesicles into the synaptic cleft. These molecules diffuse quickly through the cleft. Some of them attach to the receptor molecules on the surface of the postsynaptic membrane, causing the receptor molecules to change their shape, leading to a change in the permeability of postsynaptic membrane ions. As a result, Na^+, K^+, and Cl^- ions start flowing, getting the postsynaptic membrane depolarized or hyperpolarized, thereby forming excitatory postsynaptic potential (EPSP) or inhibitory postsynaptic potential (IPSP), as shown in Figure 3.1.

EPSP is manifested as partial depolarization of postsynaptic neurons. The principle of its production is generally thought to be that the permeability of the postsynaptic cell membrane to Na^+, K^+, Cl^-, and Ca^{2+} is increased (especially, the Na^+ channel opened) due to a combination of the excitatory chemical transmitters with the postsynaptic membrane receptor, getting the postsynaptic membrane depolarized, generating EPSP. If there are a small number of excitatory synapses, EPSP is only a local potential, which does not excite postsynaptic neurons. If multiple EPSPs appear at the same time, these local potentials will be combined. When the aggregate potential reaches the threshold value, the postsynaptic neuron will generate an action potential. IPSP mainly caused postsynaptic neurons to be hyperpolarized. It means higher difficulty in causing depolarization and excitement. That's what is known as the inhibitory effect.

3.1.3 Mechanism behind EEG

Electroencephalogram refers to potential changes recorded with electrodes on the scalp surface, normally abbreviated as EEG. If the electrodes are placed on the surface of the intracranial cortex, the record is known as an electrocorticogram (ECoG). If an invasive deep electrode is adopted, the record is known as stereoelectroencephalography (SEEG), which is usually adopted along with surgery. In physics, EEG is derived from ECoG through the low-pass filtering of the skull. Despite many studies on the formation principle of EEG/ECoG, it remains not fully elucidated. Listed below are some of the main points of view popular at present.

3.1.3.1 Postsynaptic Potential and EEG

Microelectrodes may be used to record the potential of cerebral cortical neurons with electrodes inside and outside the cell membrane simultaneously in animal experiments. The results showed that EEG is mainly related to the changes in the postsynaptic potential of cerebral cortex neurons, i.e., the changes of the excitatory or inhibitory postsynaptic potential recorded with microelectrodes. This postsynaptic potential has a long duration but low amplitude, thus it can be aggregated. A drug study showed that strychnine can selectively inhibit IPSP, i.e., change takes place in the amplitude of the brain wave on the surface of the animal brain smeared with strychnine. The γ-aminobutyric acid (GABA) can selectively inhibit EPSP that a corresponding change also takes place in the amplitude of brain waves on the brain surface. To form a brain wave change on the cortical surface, a synchronous change must take place in the postsynaptic potentials of many neurons. According to the composition of cerebral cortical neurons, the granulosa cells are arranged discordantly, making it impossible for the cellular electrical activities to jointly form a strong electric field. In contrast, the pyramidal cells are arranged neatly, with their apical dendrites parallel to each other and perpendicular to the cortical surface. When the pyramidal cells excite synchronously, the change in their aggregated potential may be recorded on the surface of the cerebral cortex or the scalp. Therefore, it is generally believed that EEG is mainly formed when the postsynaptic potentials of the apical dendrites of the pyramidal cells in the cerebral cortex are aggregated together. It is worth mentioning that some recent studies have shown that head-surface EEG can be adapted to detect subcortical electrophysiological activity, thus further enlarging the value of EEG and MEG in brain science (Samuelsson et al., 2019; Seeber et al., 2019).

The intensity of the current generated by the postsynaptic potential decreases as it is far away from the synapses. The length coefficient of its exponential attenuation is $\lambda = (g_m r_s)^{-1/2}$, where g_m represents membrane conductance and r_s represents the unit-length resistance of the intracellular fluid (Scott, 1977). For cortical neurons, the typical value λ is 0.1–0.2 mm. At a certain distance, the postsynaptic potential seems to be generated by a current dipole oriented along the dendrites and with the intensity of $P = I\lambda$; where current I can be calculated according to the change in postsynaptic potential: $I = \Delta V/(\lambda r_s)$. Therefore, $P = \Delta V/r_s$. Because $r_s = 4/(\pi d^2\sigma_{\text{in}})$, $P = \pi d^2\sigma_{\text{in}}\Delta V/4$, where d represents the diameter of the dendrite, and σ_{in} represents the conductivity of the intracellular medium. By using the typical values $d = 1\ \mu\text{m}$, $\sigma_{\text{in}} = 1\ \Omega^{-1}\text{m}^{-1}$ and $\Delta V = 25\ \text{mV}$, the equivalent dipole intensity of a single postsynaptic potential (PSP) is about $P = 20\ \text{fA}\cdot\text{m}^{-1}$.

In an MEG study, the current dipole intensity is usually set to 10 nA/m magnitude to explain the magnetic field measured outside the scalp. Thus it can be estimated that during a typical evoked response period, there should be 1 million synapses activated at the same time. Considering that there are about 10^5 pyramidal cells per mm^2 of the cortex, and each neuron contains thousands of synapses, a detectable MEG signal can be generated as long as one-thousandth of the synapses are activated simultaneously on a square millimeter scale (Hamalainen et al., 1993).

In terms of EEG, some literatures hold that about $6\,\text{cm}^2$ of cortical gyrus tissues must be active synchronously so that a few microvolt's potential can be recorded directly on the scalp surface without averaging (Cooper et al., 1965; Ebersole, 1997). It roughly contains a dipole layer formed of 600,000 minicolumns or 60,000,000 neurons. A larger range of activities is required to generate a recordable potential in the fissure and sulci. It is generally believed that the action current mainly contributes to small-scale extracellular recordings, while its contribution to the head-surface potential is much smaller than the synaptic potential. In general, the dendrites of the pyramidal cells are arranged in parallel to form a large dipole layer, which is good for the superposition of all synaptic source fields (Nunez and Srinivasan, 2006).

It should be noted that the above discussions on EEG and MEG, summarized from the literatures, cannot be used as the basis for the spatial resolution of the two technologies, because it is based on specific detection requirements and a specific assumption about the source, and also ignores the effect of noise, etc. Some studies showed that given 128 electrodes and a signal-noise ratio of 5%–10%, the spatial resolution of ECoG can reach up to 1.2–1.4 cm (Wang and He, 1998); another study on EEG source localization showed that the localization error is generally more than 2 cm (Song et al., 2015); some other studies showed that given the same number of sensors, the accuracy of EEG source localization may even be higher than that of MEG (Liu et al., 2002). We believe that EEG and MEG have their sensitive sources of neural activity. It is the complementarity but not the priority which should be emphasized between them. The best is to combine them under favorable conditions. However, under limited resources, EEG should undoubtedly be the first choice for its accessibility for all researchers and subjects in all countries.

3.1.3.2 Rhythm in the Brain

3.1.3.2.1 Thalamus

The brain wave recorded on the surface of the cerebral cortex or the scalp of a person without task is a spontaneous potential change. This spontaneous rhythmic electric phenomenon may be caused by the autonomous electric activity of neurons. Every neuron in the brain is in a network with other neurons. It is now believed that these neuronal circuits, existing in the cerebral cortex, or between the cortex and thalamus, and in the thalamus, play a major role in regulating the rhythmicity of the brain wave (Hughes and Crunelli, 2005). For example, a theory based on feedback inhibition holds that the thalamic neurons, which have fibers projected to the cerebral cortex, can exert about a 100-ms inhibitory effect on themselves by s by its axon's collaterals working with inhibitory cells, thereby making the cerebral cortical impulse rhythmic. Some experimental results showed that the rhythm of brain waves is very likely to originate from the thalamus because the spontaneous rhythmic activity of the cortex disappears once the dorsomedial nucleus of the thalamus is destroyed. Therefore, the dorsomedial nucleus of the thalamus is likely to be the starting point of brain wave synchronization.

3.1.3.2.2 Reticular Formation

Animal experiments showed that there is a reticular formation in the thalamus and even the entire brainstem. The reticular formation of the brainstem is primarily related to the maintenance of wakefulness, while the reticular formation of the thalamus may be a regulatory system involved in high-level differentiated conscious activities. The experimental results indicated that the ascending impulse of the reticular formation of the brainstem can interfere with the rhythmic activity formed by the cortex-thalamus circuit to desynchronize it. If the non-specific projection nucleus of the thalamus suffers from a rhythmic electrical stimulation for 60 times/seconds, the rhythm of an α-like wave in the cerebral cortex will be replaced by a fast rhythmic activity. The reason may be that the synchronization of the thalamus is disturbed by high-frequency electrical stimulation, causing desynchronization to occur, thereby increasing the frequency of the brain wave and decreasing the amplitude, i.e., low-amplitude fast waves appear. The phenomenon of α wave blocking caused by eye-opening in a human EEG may be due to the entry of fiber collaterals into the reticular formation of the brainstem along with afferent sensory stimulation. The above principle causes the desynchronization of α activities. Some animal experiments also showed that the reticular formation of the brainstem is related to rapid eye movement sleep (REMS).

3.1.3.2.3 Limbic System

The cerebral limbic system consists of the hippocampus, hippocampal gyrus, cingulate, and amygdala. Experimental results demonstrated that when a lower animal undergoes awakening stimulation (e.g., a tactile stimulus), the slow waves in its cortical motor area and the visual area will disappear while low-amplitude fast waves will appear, and there will be rhythmic slow waves of 4–6 periods/seconds in the hippocampus. Such rhythmic waves in the hippocampus are most obvious in rabbits, but uncommon in monkeys or humans. The amygdala in cats, dogs, or humans generates rhythmic fast waves of 20–40 periods/seconds in the waking state, and the rhythmic fast waves will be enhanced when given an awakening stimulus. In addition, the limbic system is more susceptible to damage, which forms an epileptogenic focus.

3.1.3.3 Anomalous Brain Wave

EEG abnormalities are probably related to cerebropathy or brain dysfunction accompanied by a physical disease. Generally, when a metabolic change takes place in the cortical neurons, the period and amplitude of the brain wave change, too. The slow waves, generated during sleep, consciousness disorder, or deep anesthesia, indicate that the cerebral cortex is functionally inhibited and biochemical metabolism is slowed down. In addition, neonatal EEG is composed of a great many δ waves, which correspond to functional immaturity of cortical neurons, especially the imperfection and slow metabolism of neuron dendrites. Organic damage to the brain, such as ischemia, hypoxia, edema, and necrosis, are also often accompanied by decelerated neuronal metabolism and abnormal biochemical metabolism, as well as local or diffuse slow waves. The diffuse δ waves are usually related to the non-specific projection system; while local δ slow waves are mainly related to the specific projection system.

The brain wave on the surface of the cerebral cortex is not synchronized with the deep brain wave, but when a paroxysmal brain wave (e.g., spike-wave) appears, there is a high synchronization between the electric activities on the surface of the cortex and in the deep brain. During epileptic seizures or strychnine poisoning, the excitability of cortical neurons becomes very high, and the cortical neurons are likely to affect one another. At this moment, the mutually independent neuron circuits may be excessively synchronized, resulting in a high-amplitude spike wave with an amplitude dozens of times as high as that of α wave. During an epileptic seizure, a single neuron is of a high-frequency (200–900 times/seconds) action potential spike, but the frequency of the seizure wave on the EEG is much lower than the discharge frequency of a single neuron. This is because the seizure in the cortex is a sum of the EPSPs of many neurons. Local cortical lesions can form a local synchronization circuit around them to cause focal brain wave seizures. If the focus is located deep under the cortex or has a strong electric field effect, diffuse wave seizures will be caused.

There are several views on why paroxysmal brain waves stop as follows: (1) After a long period of high-frequency discharge of cortical neurons, the cell membrane is hyperpolarized. At the moment, the Na^+ channel in the cell membrane is closed. Because Na^+ cannot flow inward, the cell membrane is hardly depolarized, thereby putting an end to seizure discharge. Most antiepileptic drugs have the effect of hyperpolarizing the cell membrane, such as γ-aminobutyric acid (GABA); (2) The intracerebral inhibitory mechanisms, such as the caudate nucleus, putamen, cerebellum, and ascending reticular inhibitory system, have an inhibitory effect on the cerebral cortex, causing the cortical neurons to be hyperpolarized; and (3) the ascending activation system of the brainstem reticulum is functionally weakened. For example, phenobarbital can inhibit the ascending activation system of the reticulum, making it possible to suppress the appearance of paroxysmal brain waves. Generally, when the function of brain nerve cells is inhibited, the EEG frequency becomes slow. If the inhibitory effect is extremely significant, brain wave amplitude will decrease, and even all brain electrical activities disappear. Correspondingly, when the nerve cells are excited, the brain wave frequency will increase, and over-excitation is bound to cause paroxysmal activities.

3.2 ELECTROSTATIC APPROXIMATION

As mentioned above, any change of ion concentration in a specific cortical area causes a change in the potential near the cell and leads to the generation of electric current. The current in turn affects the potential of other areas through volume conduction (Eshel, 1993). According to the knowledge from Maxwell's equations to the electric fields of the brain, the key point of the electromagnetic problem in bioelectromagnetics is almost all about the application of Maxwell's equations under quasi-static conditions. This section will make a detailed discussion on the rationality of quasi-static simplification.

3.2.1 Electrical Properties of Biological Tissues

According to the description in Section 2.3, the electromagnetic field in an organism is related to the frequency of the electromagnetic field and the conductivity, permittivity, and magnetic permeability of tissues. There is no need to consider the magnetic permeability of an organism because it is approximately constant. Also, it is generally recognized that the amplitude of any electromagnetic field outside the source region is small enough, making it unnecessary to consider the nonlinear effect of the relevant tissue. In a linear medium, the current density outside the source region is directly proportional to the electric field, i.e., Ohm's law holds. The proportional coefficient is the conductivity, which is a macroscopic constitutive parameter (equation 2.4c). It is a tensor for an anisotropic matter and a scalar for an isotropic matter.

Since the beginning of the last century, researchers have been making measurements of the conductivity and permittivity of humans and vertebrates. However due to the difference in measurement methods and conditions, a wide range of values has been obtained, suggesting that the measurement conditions have a great influence on the results. In addition, most of the existing measurements are made in vitro, but the electrical properties of organisms change after the organisms die. For example, the conductivity increases within 3 minutes to 2 hours after death because the insulative of the cell membrane gradually decreases after death. Also, tissue hemorrhage changes the electrical properties and that is why bones are usually soaked in a salt solution for measurement. Under this circumstance, the percentage of salt in the solution also has a great effect on the conductivity and permittivity. Apparently, in vivo measurement should be adopted to provide more reliable data, however, in vivo measurement is limited to animals or parts of the human body, such as fingers or limbs. It has also been discovered that the conductivities of some tissues, especially body fluids, are related to temperature.

It is not easy to compare the results in the literatures, because the frequency range adopted for measurement is often different. Most of the existing values are for the radio frequency range (RF), with frequencies up to 100 MHz. Such frequencies are much higher than the main frequencies involved in biological electromagnetic phenomena. For neural electrical activity, the interested range is below 3 kHz. Since human tissues are composed of cells, conductivity and permittivity are highly related to frequency. For example, a three-fold change takes place in skin conductivity when the frequency rises from 10 to 1,000 Hz. This is because the cell membrane is thin and poorly conductive, and its conductivity is dependent on capacitive conduction.

As reviewed by Lai et al. (2005), the reported brain conductivity ranges from 0.12 to 0.48/Ωm, and the human skull from 0.006 to 0.015/Ωm. Such a large variation made it difficult for practical selection. Fortunately, it is sufficient to specify the conductivity ratios of the brain, the scalp, and the skull, if one is only concerned with solving an EEG inverse problem. Rush and Driscoll (1968) used the electrolytic tank to measure the impedance of the human skull, and they argued that it is about 80 times as high as that of the brain and scalp. Cohen and Cuffin (1983) measured the potential and magnetic field under the same stimulus and used them for dipole localization in a standard three-layer head model. The results showed that consistent localization results can be achieved when the conductivity of the brain, skull, and scalp is set to 1:1/80:1. Homma et al. (1995) also conducted a study on dipole localization in an actual three-layer head model, concluding that the above ratio is the best one. This ratio has been widely used in the forward and inverse research of EEG since then. However, different opinions have advanced over the past few decades. Oostendorp et al. (2000) conducted an experiment in vivo and discovered that the ratio should be 15. By recording intracranial and extracranial potentials in epilepsy patients, Lai et al. (2005) analyzed the ratio using the concentric sphere model and the cortical imaging technique, finding that it should be 24.8 ± 6.6. Zhang et al. (2006) suggested that the ratio should be 18.7 ± 2.1 for the finite element head model.

As mentioned above, the reported conductivities of human brain tissues are quite different, and they are different from person to person, researchers are inevitably concerned about the effect of conductivity on brain electromagnetic inversion. Stok (1987) argues that the conductivities of different tissues may have a great effect on localization. Haueisen (1996) used the finite element method to simulate the effect of conductivity on the scalp potential and magnetic field. They considered a total of 13 different tissues and found that the scalp potential is mainly affected by the conductivity of the skull and scalp, and less affected by white matter, gray matter, and CSF. On the contrary, the magnetic field is affected mainly by white matter, gray matter, and CSF, but barely by other tissues. This also indicates that the intracranial volume current is the main contributor to MEG. That's also why MEG models often consider the intracranial area only. This result also shows that a simple sphere model is not appropriate, because the sphere model implies that the magnetic field has nothing to do with the conductivity.

Recently, Akhtari et al. (2016) measured the conductivities of human brain tissues, concluding that the difference in conductivities among the samples is primarily caused by the concentration of sodium cations. Therefore, if the sodium ion concentration can be measured in vivo locally in the brain tissues, the result can be used to estimate local conductivity in vivo. These studies also showed that the individual difference in conductivity is also dynamic. So, it is of particular importance to make a real-time measurement or estimation. Considering that the error of conductivity may affect EEG source localization, it is very important to estimate the location of the EEG source and the conductivity at the same time. Acar et al. (2016) used EEG and MRI data to estimate the ratio of conductivity in two persons, concluding that the ratio is 34 and 54, respectively. Gutierrez et al. (2004) assumed a spherical head model and known dipole position, where the position may be due to knowledge of specific evoked response and event-related experiments, or be estimated by other imaging modalities such as MEG or fMRI. They take 1.79/Ωm as the conductivity of the cerebrospinal fluid (CSF) and then estimate the conductivities of the brain, skull, and scalp, respectively. The results are 0.3046, 0.0112, 0.6755/Ωm for one subject, and 0.3213, 0.0128, and 0.8239/Ωm for another. Baysal and Haueisen (2004) did the same work on nine persons, obtaining a mean value of 23. Vallaghe et al. (2007) assumed that the somatosensory evoked potential (SEP) source was located on the cortex, and then found that the ratios were 89 and 81 for SEP of the left hand and the right hand, respectively. Besides, electrical impedance tomography (EIT) or magnetic resonance electrical impedance tomography (MREIT) are another two important methods to estimate conductivities. These results show that there is not only a great difference in conductivity from person to person, but also some difference from part to part in one person. So, if a real-time conductivity estimation method is developed, it will help the EEG source localization. Moreover, conductivity itself is an important parameter with physiological, pathological, and psychological significance.

A recent meta-analysis shows that the conductivity in the head model reported in the literatures does vary greatly with measurement methods, ages, and health conditions, making it necessary to build a head model with great care. Ideally, a personalized head model should be built. It is also pointed out that if a personalized head model fails to be built, the following parameters are recommended: scalp 0.41 S/m (siemens per meter), skull 0.02 S/m, (if the skull is represented by three layers, spongiform 0.048 S/m, intracranial compact layer 0.007 S/m and extracranial compact layer 0.005 S/m), CSF 1.71 S/m, gray matter 0.47 S/m, white matter 0.22 S/m, and the ratio of the brain to the skull should be set to 50.4 if gray matter and white matter are considered to be one (McCann et al., 2019).

3.2.2 Quasi-static Approximation of Bioelectromagnetic Phenomena

Maxwell's equations describe the temporal-spatial processes, and two of them involve the time derivative. If it can be proved that the relevant time dependence can be ignored, the related equations and boundary conditions can be simplified. The following analysis shows that the feasibility of simplification is dependent on signal frequency and medium properties. Here, the organism is considered an isotropic linear medium, so the field can be seen as a superposition of sinusoidal waves of different frequencies. Relevant studies show that if the following four conditions are met, the time dependence and wave propagation effect can be ignored for the equations.

3.2.2.1 Propagation Effect

If the amplitude of a wave changes insignificantly on the biological medium scale r, the propagation effect of the wave can be ignored. As seen from equation (2.22), its condition is $|\vec{\alpha} \cdot \vec{r}| \ll 1$, or simply $|kr| \ll 1$. We may have the relevant parameters of a biological material substituted into the equation for verification, for example, if r is set to 1m and the EEG frequency is set to <1,000 Hz, the deviation caused by $e^{-\vec{\alpha} \cdot \vec{r}}$ is below 4%, suggesting that this condition is met (Plonsey, 1969).

3.2.2.2 Capacitance Effect

If the contribution of the displacement current in equation (2.1a) is much smaller than that of the ohmic current, the capacitance effect of the medium can be ignored. Equations (2.4a) and (2.4c) hold if the medium is of linear dielectric and conductivity.

If the electromagnetic signal is a sinusoidal wave with frequency *f*, the derivative of dielectric flux density is

$$\partial\vec{D}/\partial t = i2\pi f\vec{D} = i\omega\vec{D} \quad (\omega = 2\pi f) \tag{3.1}$$

Correspondingly, formula (2.1a) becomes

$$\nabla\times\vec{H} = \sigma\vec{E} + i\omega\varepsilon\vec{E} \tag{3.2}$$

Take the divergence at both sides of the equation (3.2). Since the divergence of the curl is zero, the left of the equation is zero, and the right becomes

$$\nabla\cdot\left(\sigma\vec{E} + i2\pi\varepsilon f\vec{E}\right) = 0 \tag{3.3}$$

The first term on the left of the formula (3.3) is derived from the conductive effect, and the second term is derived from the dielectric (capacitance) effect. Obviously, the condition for ignoring the dielectric effect is

$$2\pi f\varepsilon/\sigma \ll 1 \tag{3.4}$$

Take the reported typical actual values into the above equation to get $2\pi 30\times 8.9\times 10^{-6}/1.538 = 0.001 \ll 1$, here the number 30 is the assumed maximum physiological frequency of a typical EEG, and the conductivity of CSF is set to 1.538. $\varepsilon = \varepsilon_r\varepsilon_0 = \varepsilon_r 8.9\times 10^{-12}$. Because $\varepsilon_r < 10^6$ (van den Broek, 1997), the maximum value of ε can be set to 8.9×10^{-6}. This equation shows that ignoring the dielectric effect is a good approximation. Actually, equation (3.4) is exactly the good conductor condition shown in equation (2.16). It shows that in a good conductor, the capacitance effect of the medium can be ignored. However, if the upper limit parameters of the skull and scalp are substituted into equation (3.4), the capacitance effect cannot be ignored, so the impedance of these tissues may become frequency-dependent to some extent (van den Broek, 1997). For a monochromatic signal, retaining capacitance effect, the volume conductor can be represented by a complex conductivity, and it acts like a resistor in parallel with a capacitor, and the resistance is replaced by impedance (Nunez and Srinivasan, 2006).

3.2.2.3 Magnetic Induction Effect

According to Maxwell's equations, an electric field and a magnetic field may excite each other. This is the basic principle of electromagnetic wave propagation. Equation (3.2) shows that changing the electric field induces a magnetic field. Therefore, ignoring the capacitance effect or displacement current means neglecting the effect of a changing electric field on the magnetic field. Here, based on Faraday's law, i.e., equation (2.1b), the contribution of a changing magnetic field to an electric field is considered. Now, let's consider the condition for ignoring the effect of a changing magnetic field on the electric field. Obviously, after these two conditions are met, the interaction between the electric field and the magnetic field disappears, i.e., they are only associated with their sources, separately, thus each one looks like a steady field. Then they can be described by a method completely similar to the one for a static field.

Based on an experimentally confirmed constitutive equation like (2.4a), etc., we can derive the conditions under which Faraday's law of induction can be ignored reasonably. Considering that $\nabla \cdot \vec{B} = 0$ and the divergence of the curl is zero, we can define a magnetic vector potential $\vec{B} = \nabla \times \vec{A}$, $\nabla \cdot \vec{A} = 0$. Accordingly, equation (2.1b) is turned into

$$\nabla \times \vec{E} = -i\omega \nabla \times \vec{A},\ \nabla \times \left(\vec{E} + i\omega \vec{A}\right) = 0 \tag{3.5}$$

Since the curl of the gradient is zero, a scalar potential Φ can be defined as $\vec{E} + i\omega \vec{A} = -\nabla \Phi$. Then

$$\vec{E} = -i\omega \vec{A} - \nabla \Phi \tag{3.6}$$

This equation shows that the electric field can be divided into two parts: (1) According to $\nabla \cdot \vec{D} = \rho$, one of Maxwell's equations, the charge can excite a curl-free electric field, which exactly corresponds to the second term of equation (3.6) (its curl is zero, but its divergence is not zero); (2) According to Faraday's law of electromagnetic induction (2.1b), a changing magnetic field can only excite curl electric field (the curl of the magnetic vector potential in equation (3.6) is not zero, but its divergence is zero). Therefore, the contribution of magnetic induction to the total electric field cannot be ignored unless $\left|i\omega \vec{A}\right| \ll \left|\nabla \Phi\right|$ (Plonsey and Heppner, 1967).

Take the curl on both sides of Maxwell's equation (2.1b) to get

$$\nabla \times \nabla \times \vec{E} = -\partial\left(\nabla \times \vec{B}\right)/\partial t \tag{3.7}$$

Then combine (2.1a) and (2.4b) to get

$$\nabla \times \nabla \times \vec{E} = -\mu \partial\left(\vec{J} + \partial \vec{D} / \partial t\right)/\partial t \tag{3.8}$$

Suppose the field is a sinusoidal signal with a frequency of *f*, then

$$\partial \vec{D}/\partial t = i2\pi f \vec{D} \tag{3.9}$$

By using (2.4a) and Ohm's law (2.4c), equation (3.8) becomes

$$\nabla \times \nabla \times \vec{E} = -i2\pi f \mu\left(\sigma + i2\pi f \varepsilon\right)\vec{E} \tag{3.10}$$

or

$$\left(\nabla \times \nabla \times + i2\pi f \mu\left(\sigma + i2\pi f \varepsilon\right)\right)\vec{E} = 0$$

If the following conditions are met

$$\left|i2\pi f \mu\left(\sigma + i2\pi f \varepsilon\right)\right| / \left|\nabla \times \nabla\right| \ll 1 \tag{3.11}$$

We may approximately assume that the curl component caused by magnetic induction in the electric field of the brain can be ignored, i.e., the effect of the first term at the right end of equation (3.6) can be ignored, that is, the contribution of the changing magnetic field to the electric field can be ignored.

Now, let's evaluate this condition by typical actual parameters. First of all, the denominator of this equation has $1/L^2$ orders of magnitude. For the head, its maximum characteristic length is equal to the distance between the root of the nose and the inion, with an average value of about 0.3 m. The magnetic permeability μ of biological material is the same as that of free space $\mu = 4\pi \cdot 10^{-7}\ H/\text{m}$. The maximum conductivity of the brain, i.e., the conductivity of the most conductive brain area (CSF), is $\sigma = 1/65\ \Omega^{-1}\text{cm}^{-1} = 1.538\ \Omega^{-1} \cdot \text{m}^{-1}$. The maximum value of the permittivity ε is set to $10^6 \varepsilon_0$, while the maximum physiological frequency of a typical EEG is set to 30 Hz. Considering the condition under which the capacitance effect can be ignored, i.e., $\sigma \gg 2\pi f\varepsilon$, then equation (3.11) changes to

$$2\pi \cdot 30 \cdot 1.2566 \cdot 10^{-7} \cdot 1.538 / (1/(0.3 \cdot 0.3)) \ll 1 \tag{3.12}$$

It can be seen that ignoring the magnetic induction effect is a good approximation when calculating the electric field of the brain (van den Broek, 1997). In this case, Maxwell's second equation becomes

$$\nabla \times \vec{E} = 0 \tag{3.13}$$

Since the curl of the gradient of any function F is zero.

$$\nabla \times (\nabla F) = 0 \tag{3.14}$$

the electric field can be expressed as a scalar function

$$\vec{E} = -\nabla \Phi \tag{3.15}$$

Equation (3.15) is a static description of the potential Φ generated by a static source. Therefore, equation (3.13) can also be called a quasi-static approximation.

3.2.2.4 Quasi-static Approximation of Boundary Conditions

With equations (3.8)-(3.10), we have

$$\vec{J}_t = \vec{J} + \frac{\partial}{\partial t}\vec{D} = (\sigma + i\omega\varepsilon)\vec{E} = \left(\nabla \times \nabla \times \vec{E}\right)/(-i\omega\mu) \tag{3.16}$$

This equation is an expression for the total current, including the ohmic current and the displacement current. According to the law of conservation of charge, the normal component of the total current at the interface is required to be continuous (i.e., assuming that there is no free charge at the interface), i.e.,

$$\vec{J}_{t1} \cdot \vec{n} = \vec{J}_{t2} \cdot \vec{n} \tag{3.17}$$

Combine (3.16) to get

$$(\sigma_1 + i\omega\varepsilon_1)\vec{E}_1 \cdot \vec{n} = (\sigma_2 + i\omega\varepsilon_2)\vec{E}_2 \cdot \vec{n} \tag{3.18}$$

where $\vec{n}$ is the unit vector of the interface normal. According to equation (3.18), we can have

When the static field $\omega = 0$ or the quasi-static condition $\sigma/(\omega\varepsilon) \gg 1$ is met, we can have $\sigma_1\vec{E}_1 \cdot \vec{n} = \sigma_2\vec{E}_2 \cdot \vec{n}$. In this case, there is no displacement current, but only ohmic current. Then from equation (3.15), we can have

$$\sigma_1(-\nabla\Phi_1)\cdot\vec{n} = \sigma_2(-\nabla\Phi_2)\cdot\vec{n} \tag{3.19}$$

Equations (3.18) and (3.19) show that the current is continuous on the boundary. At this scene, it is similar to the five of the group equation (2.13) in the case when the time derivative of the surface charge density is zero.

In a non-conductive (insulating) medium, $\sigma_1 = \sigma_2 = 0$. Therefore,

$$\varepsilon_1\vec{E}_1 \cdot \vec{n} = \varepsilon_2\vec{E}_2 \cdot \vec{n}, \quad \vec{D}_1 \cdot \vec{n} = \vec{D}_2 \cdot \vec{n} \tag{3.20}$$

In this situation, there is no ohmic current and the displacement current is continuous, similar to the forth of the group equation (2.13) in the case when the surface charge density is zero.

At the boundary between biological medium and insulating medium or vacuum, $\sigma_2 = 0$, and the boundary condition is

$$(\sigma_1 + i\omega\varepsilon_1)\vec{E}_1 \cdot \vec{n} = i\omega\varepsilon_2\vec{E}_2 \cdot \vec{n} \tag{3.21}$$

If the displacement current in the biological medium is negligible (satisfying the condition of a good conductor), then

$$\sigma_1\vec{E}_1 \cdot \vec{n} = i\omega\varepsilon_2\vec{E}_2 \cdot \vec{n} \tag{3.22}$$

Equation (3.22) means that the ohmic current in the biological medium is turned into the displacement current in region 2 after passing through the boundary. If $|\varepsilon_2\omega/\sigma_1| \ll 1$, the displacement current in region 2 can be further ignored. The measured results also showed that the electric field outside the head is very weak and decays rapidly, so it can be ignored. In other words, there is no normal current flowing out of region 1, and this is the common fact used for outer boundary conditions for biological medium such as the brain.

$$\sigma_1\nabla\Phi_1 \cdot \vec{n} = \sigma_1\frac{\partial\Phi_1}{\partial n} = 0 \tag{3.23}$$

To sum up, all criteria but the capacitance effect are valid for all tissues. The capacitance effect can be considered in a quasi-static equation by using a complex conductivity. However, it is generally not considered in current EEG practice.

In addition, as equation (2.32) proved in Section 2.4, the potential at the interface is continuous under the condition of quasi-static approximation.

The boundary conditions commonly used in EEG research are as follows: the normal current at the interface is continuous (3.19), the potential at the interface is continuous (2.32), and the normal current on the outer boundary is zero (3.23). They will be repeatedly used in the subsequent analysis of the electric fields of the brain. For further discussion on the quasi-static approximation of EEG, please refer to Plonsey and Heppner (1967), Gulrajani (1998), Nunez and Srinivasan (2006), etc.

3.3 BIOELECTRICAL SOURCE MODELS

3.3.1 Overview of Bioelectrical Sources

According to the analysis in Section 3.2, the problem of bioelectricity can be considered in a quasi-static situation. Neuroelectrophysiological studies show that the scalp potential mainly comes from the postsynaptic potential of the pyramidal cells. A pyramidal cell can be summarized by its two characteristic structures, i.e., the cell body and the apical dendrites. In the resting state, the entire cell membrane is uniformly polarized, and the intracellular domain is about 70 mV lower than the extracellular domain. When an excitatory input causes the apical dendrites to be completely depolarized, i.e., the membrane potential becomes zero, the pyramidal cell will behave like a dipole: a long and thin structure has different potentials at its two ends. Because the cellular protoplasm and extracellular fluid contain diffusible ions, there is a current generated between the normally polarized cell body and the end of the apical dendrites. The dipole is not an electrostatic dipole, but a current dipole.

In the following, Maxwell's equations will be utilized to analyze related issues. In the EEG source region, the related neuron excitement behavior needs to be described with the non-linear Hodgkin-Huxley (H-H) equation or another neuron model. As the scalp surface EEG is a behavior outside the source region, under the hypotheses of quasi-static approximation, the time terms in Maxwell's equations may be reasonably ignored for the region far away from the source.

Now, let's explain the electric current in the brain. First, the bioelectric current is generated from the source current density $\vec{J}_s(x,y,z,t)$ (also called primary current or impressed current), which is driven by the electrochemical process of excitable cells. In other words, it is a non-conservative current, which is a bioelectric activity of nerve cells where the electrical energy is converted from chemical energy. Such a bioelectrical source behaves like a current dipole. Therefore, the source current density is behaviorally equal to the dipole moment density of the source. It should be noted that $\vec{J}_s(x,y,z,t)$ is zero everywhere outside excitatory cells (Plonsey, 1969).

For a homogeneous and infinite large-volume conductor with the conductance of σ, the source current $\vec{J}_s(x,y,z,t)$ excites an electric field $\vec{E}$ and a conducting current $\sigma\vec{E}$. Therefore, the total current density $\vec{J}$ can be expressed as (Geselowitz, 1967).

$$\vec{J} = \vec{J}_s + \sigma\vec{E} \tag{3.24}$$

Current $\sigma\vec{E}$ is often referred to as return current, ohmic current, or secondary current. The return current is a necessity for preventing charge accumulation. Because it is quasi-static, the electric field $\vec{E}$ can be expressed as the negative gradient of the potential Φ at each moment, $\vec{E} = -\nabla\Phi$, so equation (3.24) can be written in the following form:

$$\vec{J} = \vec{J}_s - \sigma\nabla\Phi \tag{3.25}$$

According to the law of conservation of charge, the spatial change of $\vec{J}$ determines the temporal change rate of the local charge, $\nabla \cdot \vec{J} = -\partial\rho/\partial t$. Furthermore, as the tissue capacitance is ignorable (quasi-static condition), or in other words, if the source changes, the charge will be redistributed within a negligible time, then the change of charge density is guaranteed to be zero within a negligible time too, so that the spatial divergence of the total current is zero, i.e., the total current $\vec{J}$ is spiral and winded into a closed circle along the direction of the current. Now, therefore, equation (3.25) can be simplified as Poisson's equation:

$$\nabla \cdot \vec{J}_s = \nabla \cdot \sigma\nabla\Phi + \nabla \cdot \vec{J} = \sigma\nabla^2\Phi \tag{3.26}$$

Equation (3.26) is a partial differential equation of Φ, where $\nabla \cdot \vec{J}_s$ is a source function (or forcing function).

In summary, if we consider the intracellular current and the extracellular current separately during bioelectricity research at the cell scale, the extracellular current seems to appear at one place or disappear at another place when it passes through cell membranes (current source/sink). From a macroscopic perspective, no longer distinguish between intracellular and extracellular domain, the current can still be divided into source current and ohmic current as in equation (3.24), they together form a continuous current as a whole, while the invisible cell membrane plays the role of a gas station for this continuous current though the active transmembrane current driven by voltage-gated ion channels. The behavior of the complex cell membrane is briefly expressed as a source current density. The divergence of the total current density is zero, and this is true for the extracellular domain (static approximation), too. For the intracellular domain, the current flows in and out of it in equal quantity, so it is also true that the divergence of the total current density is zero. Actually, for the region in a cell, it is feasible to assume that the divergence of the current is zero under the condition of static approximation. For a point on the cell membrane, we may phenomenally ignore the microscopic nonlinear behavior, and just take it as a simple medium wall, so the divergence of the incoming and outgoing current is zero, too, as the incoming and outgoing are equal to keep a steady state. The current is called source current in the cell membrane and ohmic current outside which may arrive at the scalp surface. According to equation (3.26), at the macro scale, EEG is phenomenally generated by source current which is caused by the underlying cell membrane biochemical process.

For a homogeneous infinite region, the solution of equation (3.26) is

$$4\pi\sigma\Phi(\vec{r}) = \int_v \frac{1}{|\vec{r} - \vec{r}'|} \nabla \cdot \vec{J}_S(\vec{r}')\,dv(\vec{r}') \tag{3.27}$$

Since the source element $\nabla \cdot \vec{J}_s dv$ in equation (3.27) behaves like a point source, i.e., its field varies with the reciprocal of the distance. So we can take $-\nabla \cdot \vec{J}_s$ as the primary current source density or equivalently regard it as an equivalent point charge (duality principle in Section 2.7). Correspondingly, (3.27) may be transformed into

$$4\pi\sigma\,\Phi(\vec{r}) = -\int_v \frac{I_F}{|\vec{r}-\vec{r}'|} dv(\vec{r}') \tag{3.28}$$

Using vector identities equation (H.15) in Appendix H, $\nabla \cdot \left(\vec{J}_s / r\right) = \nabla(1/r) \cdot \vec{J}_s + (1/r)\nabla \cdot \vec{J}_s$, equation (3.27) can also be transformed into

$$4\pi\sigma\,\Phi(\vec{r}) = \int_v \nabla\left(\frac{1}{|\vec{r}-\vec{r}'|}\right) \cdot \vec{J}_S(\vec{r}')dv(\vec{r}') - \int_v \nabla \cdot \left(\frac{\vec{J}_S(\vec{r}')}{|\vec{r}-\vec{r}'|}\right) dv(\vec{r}') \tag{3.29}$$

Apply the divergence (Gauss) theorem to the second term at the right end of the equation (3.29), and notice that what we want is a solution for field points outside the range of the volume source, which means $\vec{J}_s = 0$ on a closed surface outside the volume source. So the second term at the right end of the equation is zero, and equation (3.27) is transformed into

$$4\pi\sigma\,\Phi(\vec{r}) = \int_v \nabla\left(\frac{1}{|\vec{r}-\vec{r}'|}\right) \cdot \vec{J}_S(\vec{r}')dv(\vec{r}') \tag{3.30}$$

This equation indicates that potential Φ can also be directly explained to be caused by the primary current in an infinite, homogeneous volume conductor with conductance of σ. Here, $\vec{J}_s dv$ comes off as a dipole (the field generated from it changes according to the square of the reciprocal distance), so the source current density $\vec{J}_s$ can also be equivalently interpreted as a dipole moment density according to the duality principle in Section 2.7.

Similarly, have a curl on the Maxwell equation (2.1a) under static approximation, we have

$$\nabla \times \nabla \times \vec{H} = \nabla \times \vec{J} \tag{3.31}$$

With the vector equation (H.23)$\nabla \times \nabla \times \vec{H} = \nabla\left(\nabla \cdot \vec{H}\right) - \nabla \cdot \nabla\vec{H}$, (2.4b), and (2.1c), we get

$$\nabla^2 \vec{B} = -\mu \nabla \times \vec{J} \tag{3.32}$$

Equation (3.32) is a Poisson's equation and its solution is

$$\vec{B}(\vec{r}) = \frac{\mu_0}{4\pi} \iiint_V \vec{J}\left(\vec{r}'\right) \times \nabla'\left(\frac{1}{\vec{r}-\vec{r}'}\right) dv(\vec{r}') \tag{3.33}$$

It is assumed here that the medium has a homogeneous magnetic permeability. This equation is called the Biot-Savart law. It is the basis for the Magnetoencephalogram (MEG). The equation indicates that a magnetic field is related not only to the source current but also to the volume current. However, if the region is a homogeneous, infinitely large conductor, the contribution of the volume current is zero (Hamalainen et al., 1993). For a spherically symmetrical conductor head model (including a multi-layered sphere model), if the primary current is radial, too, both radial and tangential components of the magnetic field on the scalp surface are bound to disappear. The current source at the center of the sphere will never generate a magnetic field outside the sphere; a current perpendicular to the plane in a planar layered medium does not generate a magnetic field, because this is the limiting case of a multilayered sphere (Hamalainen et al., 1993). This conclusion holds for any axisymmetric current in an axisymmetric medium (Grvnszpan and Geselowitz, 1973). Therefore, for a spherically symmetric conductor model, MEG is only sensitive to the tangential component of the primary current. This phenomenon also determines the inherent non-uniqueness of the MEG inverse problem, i.e., the radial component of the primary current cannot be determined.

3.3.2 Source in an Inhomogeneous Volume Conductor

In this section, an inhomogeneous region is approximated as a combination of some homogeneous sub-regions, each of which is homogeneous, resistive, and isotropic. In such a piecewise homogeneous region model, the theoretical model of each sub-region is the same as that in the previous section, while the only difference is that the field is required to meet the boundary conditions. Suppose the boundary of region j is S_j then the mathematical form of the continuous normal current component and the potential on the boundary S_j is

$$\Phi'(S_j) = \Phi''(S_j) \tag{3.34}$$

$$\sigma_j' \nabla \Phi'(S_j) \cdot \vec{n}_j = \sigma_j'' \nabla \Phi''(S_j) \cdot \vec{n}_j \tag{3.35}$$

where one prime and two primes respectively represent the two sides of the boundary. The unit normal vector $\vec{n}_j$ is from one prime region to the two primes region. Take dv as a volume element, and Ψ and Φ are two scalar functions that have good mathematical properties in each homogeneous region, then according to Green's second theorem (C.14), we have

$$\begin{aligned}
&\sum_j \int_{Sj} \left[\sigma_j'(\Psi' \nabla \Phi' - \Phi' \nabla \Psi') - \sigma_j''(\Psi'' \nabla \Phi'' - \Phi'' \nabla \Psi'')\right] \cdot d\vec{S_j} = \\
&\sum_j \int_{Vj} \left(\Psi \nabla \cdot (\sigma_j \nabla \Phi) - \Phi \nabla \cdot (\sigma_j \nabla \Psi)\right) dV_j
\end{aligned} \tag{3.36}$$

Let $\Psi = \dfrac{1}{|\vec{r} - \vec{r}'|}$, where $|\vec{r} - \vec{r}'|$ is the distance from any field point $\vec{r}$ to the voxel or surface element $\vec{r}'$ in the integral, and Φ is the electric potential. Substituting equations (3.34) and (3.35) into equation (3.36), the following results can be obtained (Geselowitz, 1967).

$$4\pi\sigma\Phi(\vec{r}) = \int_V \vec{J}_S(\vec{r}')\cdot\nabla\frac{1}{|\vec{r}-\vec{r}'|}dV(\vec{r}') + \sum_j \int_{S_j} (\sigma_j''-\sigma_j')\Phi(\vec{r}')\nabla\left(\frac{1}{|\vec{r}-\vec{r}'|}\right)\cdot d\vec{S}_j(\vec{r}') \tag{3.37}$$

This equation represents the potential at any point $\vec{r}$ in an inhomogeneous volume conductor containing an internal volume source. The first term at the right end of equation (3.37), containing $\vec{J}_S$ corresponds to equation (3.30), representing the contribution of the source current. The second integral represents the effect of inhomogeneity. This equation $(\sigma_j''-\sigma_j')\Phi\vec{n}_j$ represents the size of an equivalent dipole source due to the interface caused by polarized double-layer, where $\vec{n}_j$ is the unit normal vector of $d\vec{S}_j$; the direction of $\vec{n}_j$ is also the direction of the equivalent layer dipole source and $d\vec{S}_j$. This can be demonstrated by rewriting equation (3.37).

$$4\pi\sigma\Phi(\vec{r}) = \int_V \vec{J}_S(\vec{r}')\cdot\nabla\frac{1}{|\vec{r}-\vec{r}'|}dV(\vec{r}') + \sum_j \int_{S_j} (\sigma_j''-\sigma_j')\Phi(\vec{r}')\vec{n}_j\cdot\nabla\left(\frac{1}{|\vec{r}-\vec{r}'|}\right)dS_j(\vec{r}') \tag{3.38}$$

Equation (3.38) shows that the potential expressions related to vectors $\vec{J}_S$ and $(\sigma_j''-\sigma_j')\Phi\vec{n}_j$ are totally the same as each other, except that the former is a volume current source density (volume integral), while the latter is a surface source density (surface integral). In equations (3.37) and (3.38) and those mentioned earlier, the gradient operator is for the source coordinates, $\nabla'\left(\frac{1}{|\vec{r}-\vec{r}'|}\right) = \frac{\vec{a}_r}{|\vec{r}-\vec{r}'|^2}$, where $\vec{a}_r$ is the unit vector from the source point to the field point. The volume source $\vec{J}_S$ is a primary source, while the surface source is caused by the field generated by the primary source, so it can be called a secondary source. These formulae are the bases of the indirect boundary element method (IBEM) in Section 5.4.

3.3.3 Equivalent Source Models

Here, we are making a discussion from an engineering perspective as to what kind of source model can or should be used for EEG research. As can be seen from equations (3.28) to (3.30) above, the same potential can be understood as the result of either a point source (primary current source density) or a dipole source (primary current). Here, we will consider this problem in the sense of mathematics. The EEG source model is a very fundamental issue as it is the starting point for such as source localization (EEG forward and inverse problem), high-density EEG imaging, and EEG zero reference technology.

According to equation (3.26), the EEG problem can be expressed as Poisson's equation. By referring to its solution in an unbounded homogeneous space, its solution in an inhomogeneous space can be formally expressed as

$$\Phi(\vec{r}) = -\int_V \nabla \cdot \vec{J}_s(\vec{r}')G(\vec{r},\vec{r}')dV \tag{3.39}$$

where the unknown function G is called the Green function and is also known as a scalar lead field in electrocardiology. V represents the source region and $\nabla \cdot \vec{J}_s(\vec{r}') = I_F(\vec{r}')$ represents the primary current source density or primary volume source density. The Green function contains information about several aspects including the geometric properties, boundary conditions, and electrical properties of the model. Since the primary current is limited to the brain area, equation (3.39) can be, by reference to the evolution from equation (3.29) to (3.30), rewritten as

$$\Phi(\vec{r}) = \int_V \vec{J}_s(\vec{r}') \cdot \vec{L}(\vec{r},\vec{r}')dV \tag{3.40}$$

Where

$$\vec{L}(\vec{r},\vec{r}') = \nabla_{r'} G(\vec{r},\vec{r}') \tag{3.41}$$

It is called (vector) lead field. Equation (3.40) is the basis for all EEG problems based on the dipole model. As can be seen from the comparison of equation (3.39) with equation (3.40), equation (3.39) reveals that an eddy current does not contribute to the head-surface potential because the divergence of the curl is zero, but this fact is not clearly stated in equation (3.40). According to the field theory, any vector field such as $\vec{J}_s$ can be decomposed into a rotational part $\vec{J}_r$ and an irrotational part $\vec{J}_i$, as well as the gradient field $\vec{J}_h$ of a harmonic function (Grave de Peralta Menendez et al., 2000), i.e.,

$$\vec{J}_s = \vec{J}_r + \vec{J}_i + \vec{J}_h \tag{3.42}$$

The relevant part satisfies

$$0 = \nabla \cdot \vec{J}_r = \nabla \times \vec{J}_i, \vec{J}_h = \nabla H \tag{3.43}$$

H is a Harmonic function in V, therefore, $\nabla \cdot \vec{J}_h = \nabla^2 H = 0$. Substitute (3.43) into (3.39) to get

$$\begin{aligned} \Phi(\vec{r}) &= -\int_V \left\{ \nabla \cdot \vec{J}_r(\vec{r}')G(\vec{r},\vec{r}') + \nabla \cdot \vec{J}_i(\vec{r}')G(\vec{r},\vec{r}') + \nabla \cdot \vec{J}_h(\vec{r}')G(\vec{r},\vec{r}') \right\} dV \\ &= -\int_V \nabla \cdot \vec{J}_i(\vec{r}')G(\vec{r},\vec{r}')dV \end{aligned} \tag{3.44}$$

This equation indicates that the introduction of the rotational part $\vec{J}_r$ and harmonic function H does not change the essence of equation (3.39), or in other words, only the irrotational part $\vec{J}_i$ of $\vec{J}_s$ contributes to head-surface potential recording. It means that the source corresponding to head-surface potential recording is just the irrotational part $\vec{J}_i$ of the actual source $\vec{J}_s$. If $\vec{J}_i$ is simply understood as the "authorized representative" of $\vec{J}_s$, then they satisfy

$$0 = \nabla \times \vec{J}_s \Leftrightarrow \vec{J}_s = \nabla \Phi_p \tag{3.45}$$

The assumed irrotational $\vec{J}_s$ is defined as the gradient of a scalar function Φ_p at the right end of the equation (3.45) according to the field theory (the curl of the gradient is zero). From the perspective of a steady current field, Φ_p is a quantity proportional to the potential field corresponding to the irrotational part of the current field $\vec{J}_s$. Now, equations (3.40) and (3.45) can be combined to derive equation (3.46) (Grave de Peralta Menendez et al., 2000).

$$\Phi(\vec{r}) = \int_V \nabla\Phi_p(\vec{r}') \cdot \nabla G(\vec{r},\vec{r}') dV = \int_V \{\nabla G(\vec{r},\vec{r}') \cdot \nabla\} \Phi_p(\vec{r}') dV \tag{3.46}$$

To sum up, three equivalent expressions are derived as shown by equations (3.39), (3.40), and (3.46) for head-surface potential recording. However, they are slightly different from one another in physics: (3.39) shows that the rotational part of $\vec{J}_s$ does not contribute to the potential, and neither does the current source density or volume source density obtained by inversion contain the rotational current; equation (3.45) is directly used to define $\vec{J}_s$ in equation (3.46), thus turning off the rotational part, which does not contribute to the potential. In contrast, (3.40) does not offer a clear indication of whether the rotational part is contributive or existent.

Equations (3.39), (3.40), and (3.46) are equivalent to one another in terms of head-surface potential recording, indicating that any one of them can be selected as the starting point of the EEG inverse problem. Most of the early work started from equation (3.40), i.e., the primary current source (equivalent dipole). In form, the equation does not exclude the contribution of the rotational current, so there may be a rotational current component irrelevant to head-surface potential in the solution, increasing the uncertainty of the solution. Uzunoglu et al. (1991), carried out an inversion study on equivalent surface charge distribution. Almost simultaneously, we carried out a series of studies, including high-resolution EEG cortical imaging based on the point charge source/current source density (Yao, 1995, 1996; Yao et al., 2004a), head-surface equivalent charge distribution imaging (Yao et al., 2004b), and 3D tomography (Yao and He, 1998; He et al., 2002; Xu et al., 2008). Grave de Peralta Menendez et al. (2000) developed a 3D tomography method according to equation (3.46), where the source is a quasi-potential quantity Φ_p.

The above three equations provide three basic starting points for the study of the spatial problem of EEG. The three forms of sources are all equivalent sources for generating electric fields in the brain and on the scalp surface, rather than exploring the true physiological neural source. The true neurophysiological source is the synchronized electrical activity of a huge number of neurons, and each neuron needs to be described with a complex model such as the H-H equation. It is impossible or meaningless to establish a direct connection between them and the electric fields of the brain on the scalp surface. In EEG practice, it depends on an algorithm design which equivalent source should be used. The three are all equivalent sources, and what is told to us is nothing but the "location" and "intensity" of the equivalent sources, it is unnecessary to be entangled with the selection of a specific equivalent source but to see the accuracy of the "location" and the relative "intensity" for different cases. From the perspective of algorithm implementation, the number of unknown quantities in the equivalent charge source approach is the same as that in the potential source approach and just one-third of the dipole source approach. Meanwhile, like the dipole source, an equivalent charge source can be used as a discrete or distributed source, while the potential source cannot be taken as a "discrete" point source. So, we believe that the equivalent charge source is more advantageous in practice though it is crushed by the dipole source in current practice.

The above analysis may lead to the following conclusion (Grave de Peralta Menendez et al., 2000): EEG cannot provide the whole story underneath the surface because it can only provide information about the irrotational part, whereas, in principle, there may exist both rotational and irrotational currents in excitable tissues. If so, the following strategies may be helpful: (1) With prior knowledge to provide the information of the rotational part that cannot be detected by EEG; (2) With MEG to make up the information of the rotational part.

3.3.4 Typical Sources and Their Electrical Fields in Infinite Medium

The source and field equations are already introduced from the perspective of field theory in Chapter 2. Here, an introduction will be given in a more straightforward form, thus making it convenient for EEG users to read. As mentioned above, the EEG source can be a monopole point source (3.39), a bipolar dipole (3.40), or a quasi-potential source (3.46). Next, the fields generated from a monopole point source and dipole source will be introduced to enhance the direct impression.

3.3.4.1 Monopole Point Source

The simplest source is a point source or monopole source in the world. Suppose there is a point current source density I_F in an infinitely large homogeneous conductor with conductance of σ. All current lines from it are homogeneously distributed and divergent radially. Thus, for a concentric sphere with a radius of r, according to the law of conservation of charge, the current density J on the surface must be homogeneous and equal to I_F divided by the entire surface. Because the current is radial, the current density can also be expressed as a vector, i.e., equation (3.47) below

$$J = \frac{I_F}{4\pi r^2}, \quad \vec{J} = \frac{I_F}{4\pi r^2}\vec{a}_r \tag{3.47}$$

where $\vec{a}_r$ is the unit radial vector and its origin is at the point source. Associated with the current field defined by equation (3.47) is a scalar electric potential field Φ.

Considering that the current field is radial at every position, there should be no change in the potential on a surface where r is a constant. In other words, the equipotential surface comprises a series of concentric layers surrounding the point source, and the potential of each layer decreases with the increase of r. According to the field theory, there is an equation (3.15) linking the electric field $\vec{E}$ and scalar potential Φ, while there is Ohm's law (2.4c) between the electric field and the current. Equations (2.4c) and (3.15) are substituted into equation (3.47), so

$$\vec{J} = \frac{I_F}{4\pi r^2}\vec{a}_r = -\sigma\nabla\Phi \tag{3.48}$$

To satisfy the equation (3.48), $\nabla\Phi$ can only have a radial component. Thus

$$\frac{I_F}{4\pi r^2} = -\sigma\frac{d\Phi}{dr} \tag{3.49}$$

Integrate (3.49) from ∞ to r to get

$$\Phi = \frac{I_F}{4\sigma\pi r} \tag{3.50}$$

As inferred above, (3.50) shows that Φ is a constant on the spherical surface with radius r, i.e., the concentric sphere with r as a constant is an equipotential surface. Equation (3.50) shows that when $r \to \infty$, the potential approaches zero, consistent with the fact that the integral constant in equation (3.50) is assumed to be zero. So, here, a point at infinity is selected as the reference point, and its potential is zero (this problem will also be discussed in Chapter 13, where the historical reference electrode problem is tackled). As can be seen from equation (3.50), the magnitude of the potential around the point current source density I_F is inversely proportional to the radius (the starting point is at the monopole source). Equation (3.50) is also the definition of point charge potential in physics. As can be seen, the selection of a point at infinity as the zero reference point is an axiomatic assumption in physics, and therefore it is also the ideal reference point for electrophysiological studies such as EEG research. Many encouraging results have been achieved in the past studies on the equivalent monopole source, including cortical potential imaging (Yao, 1995, 1996), cortical equivalent monopole source imaging (Yao et al., 2004a), scalp surface imaging (Yao et al., 2004b), and 3D tomography of equivalent monopole source (Yao and He, 1998, He et al., 2002, Xu et al., 2008), and the methods have been integrated in the EEG specific platform WeBrain (Dong et al., 2021).

3.3.4.2 Dipole

In bioelectricity, it is argued that an isolated monopole current source will never be found because of charge conservation, and this is also the reason why people intuitively accepted dipole rather than charge as the source in the main body of current EEG practice. In fact, both the equivalent dipole and charge at the macro EEG scale is a phenomenal description, the underneath physical sources all are various charges or ions and their flow. Here, we think it is a subjective choice problem, so we don't worry about this problem anymore but look at the characteristics of the dipole field.

For dipole, we may suppose that there are many positive and negative single (charge) sources with a total charge of zero. The simplest positive-negative single source combination is the dipole. A dipole consists of two monopoles ("source"/"sink") opposite in sign and equal in value. The two monopoles are separated by a very small distance d. The strict definition is $d \to 0$, $I_F \to \infty$, with $p = dI_F$ finite. The quantity p is called the dipole moment. A dipole is a vector, whose direction is defined to be from the sink (negative) to the source (positive). In this sense, a dipole is not a true bioelectrical source, but a representation of a special charge source combination. Actually, the monopole is the most basic source model in electricity, and it is the basis for all composite source models such as dipole, quadrupole, tripole, and distributed sources. As a matter of fact, if $\vec{d}$ is the displacement from the sink to the source, while $\vec{a}_d$ is a unit vector in this direction, the dipole can be abbreviated as

$$\vec{p} = I_F\vec{d} = I_F d\vec{a}_d \tag{3.51}$$

Suppose there is a dipole of arbitrary orientation, and the origin of the coordinate system is at the negative monopole. If the positive monopole is also at the source point, they will cancel each other out, causing the field to be zero. Therefore, the field that appears as the positive monopole moves from the origin to a certain location $\vec{d}$ is the field of the dipole. Accordingly, it can be obtained by checking the formula of the positive monopole potential and estimating the potential change that may take place when it is moved from the origin to its current location. This can be easily done with the Taylor series.

$$\Phi_d = \frac{\partial\left(\frac{I_F}{4\pi\sigma r}\right)}{\partial d}d,\ \Phi_d = \nabla\left(\frac{I_F}{4\pi\sigma r}\right)\cdot\vec{d},\ \Phi_d = \frac{p}{4\pi\sigma}\nabla\left(\frac{1}{r}\right)\cdot\vec{a}_d \tag{3.52}$$

Here the dipole moment is $p = dI_F$ defined in a strict situation of $d \to 0,\ I_F \to \infty$. For the Taylor expansion, when $d \to 0$, all the higher-order terms tend to zero, and the zero-order term cancels out the field of the negative monopole originally located at the origin, then what remains is the first-order term represented by equation (3.52), therefore, it is the exact mathematical expression of a dipole potential.

Because the gradient operator acts on the source coordinates, $\nabla\left(\frac{1}{r}\right) = \frac{1}{r^2}\vec{a}_r$, where $\vec{a}_r$ is the unit vector from the source point to the field point, the field of the dipole can be written in more detail as

$$\Phi_d = \frac{p}{4\pi\sigma r^2}\vec{a}_r\cdot\vec{a}_d,\ \Phi_d = \frac{p\cos\theta}{4\pi\sigma r^2} \tag{3.53}$$

In equation (3.53), θ is a polar angle between $\vec{a}_r$ and $\vec{a}_d$. As can be seen from a comparison between the dipole field and the above monopole field, the dipole field changes in the form of $1/r^2$ while the monopole field changes in the form of $1/r$.

In addition, the equipotential surface of the dipole is not a concentric sphere, because there is a factor $\cos\theta$ in its expression. Given an r value, the maximum potential of the dipole appears on the polar axis of the dipole. In terms of practical application, the dipole model is the most common one. Referring to the information given in the previous sections, a dipole can be an equivalent representative of a source current or the boundary effect of an inhomogeneous medium.

A reason why people prefer to use the dipole model is that the brain is recognized to be electrically neutral on the whole. So, it is believed that there is no isolated monopole source. However, the above analysis indicates that overall electric neutrality does not necessarily mean electric neutrality everywhere at any scale. For example, If we analyze a practical approximate dipole on a smaller scale, we will find that there are two monopole sources. In the coming Chapter 5 of some numeric algorithm implementations, the dipole model has been assumed to be a combination of a point source and sink. Therefore, although the current spatial analysis of EEG forward and inversion are mainly based on the dipole model, we still believe that the key to the selection of a proper equivalent source lies in the practical equivalence of the source as well as its sensitivity and robustness in inversion imaging, and it's a problem valuable for further studies.

3.3.4.3 Other Source Models

In addition to the monopole and dipole models, higher-level equivalent models, such as the quadrupole model, have been studied and tried during the research of bioelectricity (Geselowitz, 1960; Yao, 1995). However such research has not yet gotten enough acceptance, for the physiological significance of the quadrupole and higher-level poles remains unclear. Also, the field and potential of the quadrupole source decay faster than the dipole with distance, harder to effectively detect on the scalp surface. So, it is difficult for a high-level source to play a key role alone. Instead, it is usually used as a correction term for the dipole model.

From a physiological perspective, the equivalent dipole model is also very rough. The activity of distributed excitatory cells is not carried out in isolation, adjacent cells are activated synchronously to some extent. So, a more realistic model should be a distributed source model. There are two kinds of distributed source models, i.e., surface distribution and volume distribution. There are three types of specific sources, including the monopole, dipole, and quasi-potential distribution as introduced above. The distribution may be uniform or Gaussian (Lei et al., 2009). The main difficulty in applying a distributed source model is as follows: the number of source parameters increases sharply, posing a huge challenge to subsequent problem-solving. Hence, the specific form of distributed sources is usually considered with a relevant algorithm. For example, a volumetrically distributed source is adopted for 3D tomography, while a spherical surface distributed source is adopted for cortical imaging (Yao, 1996). Besides, in EEG source localization, the adopted model included an instantaneous dipole model for a specific time point or time window, a quasi-stationary dipole model in which the number and location of sources do not change but with its intensity changeable, and a model in which two sources appeared simultaneously in the symmetrical anatomical structure of the brain, etc. (Scherg, 1993). Of course, these considerations for a dipole in practice can also be utilized for other source models such as the monopole source.

It should be noted that because a biological medium is considered to be linear, the principle of superposition holds, the forward or linear inversion of any source model composed of dipoles and/or monopoles can be performed by the superposition rule. Under this circumstance, it is clear that the most fundamental issue is the forward and inverse of the simplest monopole and dipole.

3.3.4.4 Source Model of Spontaneous EEG

Conceptually, the above-mentioned source models can be used for both spontaneous EEG and evoked EEG. But in practice, it is thought that there are a finite number of sources for evoked EEG, so it can be expressed as a combination of finite sources. For spontaneous EEG, it seems to be active throughout the brain, so De Munck and Van Dijk (1999) and De Munck et al. (1992) proposed a random dipole distribution model. Actually, spontaneous EEG is quite irregular, making it impossible to predict its future trend according to its past trend. Of course, this does not mean that spontaneous EEG is completely random, for the reason that under certain circumstances, there are often some specific rhythm signals and spatial patterns appearing, such as the alpha wave which appears when the eyes are closed. However the precise shape of these signals is difficult to predict, and detailed knowledge about them is generally not of great use in current practice. On the contrary, their statistical properties have certain application values. e.g., it can be used to improve dipole fitting and localization in ERP. For more information, please refer to De Munck et al. (1992).

3.3.4.5 Neural Computational Source Model

Based on the duality principle (Section 2.7), the above unipole, dipole, and quadrupole sources for the medium are adopted to equivalently represent the electric phenomena of actual neural current in the conductive brain. Now there are also more and more efforts to use the computational dynamic model in EEG, they model the brain dynamics using biophysical characteristics of neural excitation and inhibition. For example, the neural mass model (NMM) has been used to model human alpha rhythm, evoked potentials, and epileptic activities (Breakspear, 2017; Sun et al., 2022). The Wilson-Cowan model has been adopted in modeling the stable steady-state visual evoked potentials (SSVEPs) (Zhang et al., 2021). The dynamic causal model (DCM) was used for auditory-motor task-evoked potential (Friston, 2003; Liu et al., 2017). In general, for the neural mechanism study of EEG, these models themselves can be used directly, for example, the neural activities at some ROIs can be reconstructed from the scalp recordings, and each ROI is modeled by tuning the parameters of an NMM (Zavaglia et al., 2006). Another way, the neural computational model output may be taken as a spatio-temporal source model in EEG forward (Sun et al., 2022).

REFERENCES

Acar, Z.A., C.E. Acar, S. Makeig. 2016. Simultaneous head tissue conductivity and EEG source location estimation. *NeuroImage* 124:168–180.

Akhtari, M., D. Emin, B.M. Ellingson, et al. 2016. Measuring the local electrical conductivity of human brain tissue. *J Appl Phys* 119(6):064701.

Baysal, U., J. Haueisen. 2004. Use of a priori information in estimating tissue resistivities: Application to human data in vivo. *Physiol Meas* 25(3):737–748.

Breakspear, M. 2017. Dynamic models of large-scale brain activity. *Nat Neurosci* 20:340–352.

Cajal, S.R. 1888. Estructura de los centros nerviosos de las aves. *Rev Trim Histol Norm Patol* 1:1–10.

Cohen, D., B.N. Cuffin. 1983. Demonstration of useful differences between magnetoencephalogram and electroencephalogram. *EEG Clin Neurosci* 56(1):38–51.

Cooper, R., A.L. Winter, H.J. Crow, et al. 1965. Comparison of subcortical, cortical, and scalp activity using chronically indwelling electrodes in man. *EEG Clin Neurosci* 18:217–228.

David, O., K.J. Friston. 2003. A neural mass model for MEG/EEG: Coupling and neuronal dynamics. *NeuroImage* 20(3):1743–1755.

DeMunck, J.C., B.W. Van Dijk. 1999. *The Spatial Distribution of Spontaneous EEG and MEG*. Berlin Heidelberg: Springer.

DeMunck, J.C., P.C.M. Vijn, F.H. Lopes da Silva. 1992. A random dipole model for spontaneous activity. *IEEE Trans Biomed Eng* 39:791–804.

Dong, L., L. Zhao, Y. Zhang, et al. 2021. Reference Electrode Standardization Interpolation Technique (RESIT): A novel interpolation method for scalp EEG. *Brain Topogr* 34:403–414.

Draguhn, A., R.D. Traub, D. Schmitz, et al. 1998. Electrical coupling underlies high-frequency oscillations in the hippocampus in vitro. *Nature* 394:189–192.

Ebersole, J.S. 1997. Defining epileptogenic foci: Past, present, future. *J Clin Neurophysiol* 14:470–483.

Eshel, Y. 1993. Correlations between anatomical asymmetries in the head and scalp potential amplitude asymmetry: A mathematical model. PhD Dissertation, Tel-Aviv University.

Filley, C.M., R.D. Fields. 2016. White matter and cognition: Making the connection. *J Neurophysiol* 116(5):2093–2104.

Geselowitz, D.B. 1960. Multiple representation for an equivalent cardiac generator. *Proc IRE* 48(1):75–79.

Geselowitz, D.B. 1967. Bioelectric potentials in an in homogeneous volume conductor. *Biophys J* 7(1):1–11.

Gravede Peralta Menendez, R., S.L. Gonzalez Andino, S. Morand, et al. 2000. Imaging the electrical activity of the brain: ELECTRA. *Hum Brain Mapp* 9:1–12.

Grynszpan, F., D.B. Geselowitz. 1973. Model studies of the magnetocardiogram. *Biophys J* 13(9):911–925.

Gulrajani, R. 1998. *Bioelectricity and Biomagnetism*. New York: John Wiley & Sons Press.

Gutierrez, D., A. Nehorai, C. Muravchik. 2004. Estimating brain conductivities and dipole source signals with EEG arrays. *IEEE Trans Biomed Eng* 51:2113–2122.

Hamalainen, M., R. Haria, R.J. Ilmoniemi, et al. 1993. Magnetoencephalgraphy theory, instrumentation and applications to noninvasive studies of the working human brain. *Rev Mode Phys* 65(2):414–487.

Haueisen, J. 1996. Methods of numerical field calculation for neuromagnetic source location. PhD Dissertation, Techniche Universitat Ilmenau.

He, B., D. Yao, J. Lian, et al. 2002. An equivalent current source model and Laplacian weighted minimum norm current estimates of brain electrical activity. *IEEE Trans Biomed Eng* 49(4):277–288.

Homma, S., T. Musha, Y. Nakajima, et al. 1995. Conductivity raions of the Scalp-skull-brain model in estimating equivalent dipole sources in human brain. *Neurosci Res* 22:51–55.

Hughes, S.W., V. Crunelli. 2005. Thalamic mechanisms of EEG alpha rhythms and their pathological implications. *Neuroscientist* 11(4):357–372.

Knösche, T.R., J. Haueisen. 2022. *EEG/MEG Source Reconstruction*. Switzerland: Springer Press.

Lai, Y., W. van Drongelen, L. Ding, et al. 2005. Estimation of in vivo human brain-to-skull conductivity ratio from simultaneous extra- and intra-cranial electrical potential recordinings. *Clin Neurophysiol* 116(2):456–465.

Lei, X., P. Xu, A. Chen, et al. 2009. Gaussian source model based iterative algorithm for EEG source imaging. *Comput Biol Med* 39:978–988.

Liu, A.K., A.M. Dale, J.W. Belliveau. 2002. Monte Carlo simulation studies of EEG and MEG localization accuracy. *Hum Brain Mapp* 16:47–62.

Liu, T., F. Li, Y. Jiang, et al. 2017. Cortical dynamic causality network for auditory-motor tasks. *IEEE Trans Neural Syst Rehabil Eng* 25(8):1092–1098.

McCann, H., G. Pisano, L. Beltrachini. 2019. Variation in reported human head tissue electrical conductivity values. *Brain Topogr* 32:825–858. DOI: 10.1007/s10548-019-00710-2.

Nunez, P.L., R. Srinivasan. 2006. *Electric Fields of the Brain: The Neurophysics of EEG*. New York: Oxford University Press.

Oostendorp, T.F., J. Delbeke, D.F. Stegeman. 2000. The conductivity of the human skull: Results of in vivo and in vitro measurements. *IEEE Trans Biomed Eng* 47(11):1487–1492.

Plonsey, R. 1969. *Bioelectric Phenomena*. New York: Mcgraw Hill Education.

Plonsey, R., D.B. Heppner. 1967. Considerations of quasi-stationarity in electrophysiological systems. *Bull Math Biophys* 29:657–664.

Plonsey, R., R.C. Barr. 1988. *Bioelectricity: A Quantitative Approach*. Berlin: Springer Science & Business Media.

Rush, S., D.A. Driscoll. 1968. Current distribution in the brain from surface electrodes. *Anesth Analg* 47(6):717–723.

Samuelsson, J.G., S. Khan, P. Sundaram, et al. 2019. Cortical Signal Suppression (CSS) for detection of subcortical activity using MEG and EEG. *Brain Topogr* 32:215–228.

Scherg, M., J.S. Ebersole. 1993. Models of brain sources. *Brain Topogr* 5(4):419–423.

Scott, A.C. 1977. *Neurophysics*. New York: John Wiley & Sons Press.

Seeber, M., L.M. Cantonas, M. Hoevels, et al. 2019. Subcortical electrophysiological activity is detectable with high-density EEG source imaging. *Nat Commun* 10:753.

Song, J., C. Davey, C. Poulsen, et al. 2015. EEG source localization: Sensor density and scalp surface coverage. *J Neurosci Meth* 256:9–21.

Stok, C.J. 1987. The Influence of model parameters on EEG/MEG single dipole source estimation. *IEEE Trans Biomed Eng* 34:289–296.

Sun, R., A. Sohrabpour, G.A. Worrell, et al. 2022. Deep neural networks constrained by neural mass models improve electrophysiological source imaging of spatiotemporal brain dynamics. *PNAS* 119(31):e2201128119.

Uzunoglu, N.K., E. Ventouras, C. Papageorgiou, et al. 1991. Inversion of simulated evoked potentials to charge distribution inside the human brain using an algebraic reconstruction technique. *IEEE Trans Med Imaging* 10(3):479–484.

Vallaghe, S., M. Clerc, J.M. Badier. 2007. In vivo conductivity estimation using somatosensory evoked potentials and cortical constraint on the source. *The 2007 4th IEEE International Symposium on Biomedical Imaging: From Nano to Macro*, Arlington, VA, pp. 1036–1039.

vanden Broek, B. 1997. Volume conducting effects in EEG and MEG. PhD Dissertaion, University of Twente, The Netherland.

Wang, Y., B. He. 1998. A computer simulation study of cortical imaging from scalp potentials. *IEEE Trans Biomed Eng* 45(6):724–735.

Xu, P., Y. Tian, X. Lei, et al. 2008. Equivalent charge source model based iterative maximum neighbor weight for sparse EEG source localization. *Ann Biomed Eng*. DOI: 10.1007/s10439-008-9570-4.

Yao, D. 1995. Three dimensional mapping of EEG by equivalent source technique. *J Biomed Eng* 12(4):332–341.

Yao, D. 1996. The equivalent source technique and cortical imaging. *EEG Clin Neurosci* 98:478–483.

Yao, D., B. He. 1998. The Laplacian weighted minimum estimate of three dimensional equivalent charge distribution in the brain. *Proceedings of the 20th Annual International Conference of the IEEE EMBS*, Hong Kong, China, vol. 4, pp. 2108–2111.

Yao, D., L. Wang, K.D. Nielsen, et al. 2004a. Cortical mapping of EEG alpha power using a charge layer model. *Brain Topogr* 17(2):65–71.

Yao, D., Y. Qin, Y. Zhang. 2022. From psychosomatic medicine, brain-computer interface to brain-apparatus communication. *Brain-Appar Commun J Bacomics* 1(1):66–88.

Yao, D., Z. Yin, X. Tang, et al. 2004b. High-resolution electroencephalogram (EEG) mapping: Scalp charge layer. *Phys Med Biol* 49(22):5073–5086.

Zavaglia, M., L. Astolfi, F. Babiloni, et al. 2006. A neural mass model for the simulation of cortical activity estimated from high resolution EEG during cognitive or motor tasks. *J Neurosci Meth* 157:317–329.

Zhang, G., Y. Cui, Y. Zhang, et al. 2021. Computational exploration of dynamic mechanisms of steady state visual evoked potentials at the whole brain level. *NeuroImage* 237:118166.

Zhang, Y., W. van Drongelen, B. He. 2006. Estimation of in vivo brain-to-skull conductivity ratio in humans. *Appl Phys Lett* 89(22):223903.

4 Brain Electrical Field in Regular Head Model

The early studies on the relation between the electric source in the brain and the surface EEG were made by Brazier (1949) for the homogeneous spherical model, Geisler and Gerstein (1961) for two concentric spheres, Paicer, Sances, and Larson (1967) and Rush and Driscoll (1969) for three concentric spheres. Definitely, the ideal forward model should be a realistically shaped head model which can only be solved by a numerical method, as illustrated in Chapter 5, and owing to the implicit approximation of the numerical method itself, there exist unavoidable errors in the calculation results. Furthermore, the realistic head model is barely easy to popularize for wide use due to its high computational complexity and complex modeling procedure. What is worse is that the numerical calculation is not helpful for the analysis of causality. Therefore, various approximate head models, particularly spherical head model, and multilayer spherical head model have been extensively studied and widely used. Actually, during the algorithm development for forward modeling of the electric field of the realistic brain, the analytical theory of regular shape is widely used to analyze the correctness and effectiveness of the various numerical methods. Among the existing analytical theories of various regular shapes, only the homogeneous spherical head model has a closed analytic solution, while all the multilayer spherical head models and ellipsoidal models have series solutions only. This chapter will focus on the homogeneously layered spherical head models.

4.1 MULTIPOLE EXPANSION OF ELECTRICAL POTENTIAL IN INFINITE MEDIUM

As pointed out in Chapter 3 equation (3.26), the EEG follows the Poisson's equation:

$$\nabla \cdot \vec{J}_s/\sigma = \nabla^2\Phi(\vec{r}) = -s(\vec{r}) \tag{4.1}$$

It has a solution as

$$\Phi(\vec{r}) = \frac{1}{4\pi}\int_v \frac{s(\vec{r}')}{|\vec{r}-\vec{r}'|}dv(\vec{r}') \tag{4.2}$$

The integrals in equation (4.2) spread all over the domain where the source s is nonzero. If the source distribution range v can be surrounded by a closed surface S, the potential of the areas outside of S, $\Phi(\vec{r})$, would satisfy the Laplace equation

$$\nabla^2\Phi(\vec{r}) = 0,\ \vec{r} \notin v \tag{4.3}$$

DOI: 10.1201/9781032639260-4

Equation (4.3) can be solved by multipole expansion. If the surface S is spherical, it will be easier to solve the equation by multipole expansion based on spherical coordinates. If the boundary S is a long spheroid, then the expansion of the long spheroid is more convenient (Law et al., 1993). If the boundary S is cubic, the rectangular harmonic analysis can be adopted (Alldredge, 1981). If the boundary is a spherical cap, spherical cap harmonic analysis can be applied (De Santis et al., 1990). Obviously, this classification takes the field matching on the boundary S into consideration. There are in-depth studies on these methods in geophysics such as geomagnetic field and gravitational field. Readers might as well refer to the literature in geophysics. Here, the boundary S is only treated as a spherical surface with a radius of a, as follows:

$$\nabla^2\Phi(\vec{r}) = 0, \quad r > a \tag{4.4}$$

There are several equivalent series expansions for this situation:

1. Directional cosine expansion
2. Taylor series expansion and its tensor form (Wikswo and Swinney, 1984; Yao, 1998)
3. Spherical harmonic multipole expansion

Now a detailed discussion will be conducted on spherical harmonic multipole expansion, i.e.,

$$4\pi\Phi(\vec{r}) = \sum_{n=0}^{\infty}\sum_{m=0}^{n}\frac{1}{r^{n+1}}\left(a_{nm}\cos m\phi + b_{nm}\sin m\phi\right)P_n^m(\cos\theta) \tag{4.5}$$

In such a spherical harmonic series expansion, there are $(2n+1)$ coefficients at each level n, independent of one another, and no redundancy is existed (Wikswo and Swinney, 1984). There are $3n$ coefficients in cosine expansion and 1/2 $(n+1)$ $(n+1)$ in Taylor series expansion, suggesting that there is a redundancy problem with these two types of expansion. Although the number of coefficients can be reduced to $(2n+1)$ by using the corresponding redundancy conditions, there will be an additional equation that has to be solved. Therefore, spherical harmonic expansion is widely taken for general problem. The theoretical relationships between the coefficients related to Taylor series expansion and its tensor form and the spherical harmonic coefficients are summarized in the literature (Wikswo and Swinney, 1984). The first part of this section is primarily referenced from Wikswo and Swinney (1984), while the second part is primarily referenced from Wikswo and Swinney (1985). More details can be found in these two articles. In addition, you can also refer to Section 2.5 and Appendix A of this book.

4.1.1 Spherical Harmonic Spectra Expansion of Potential

For the convenience of application, we will now give a detailed introduction to spherical harmonic series expansion. The equation (4.5) can be rewritten as

$$\Phi(\vec{r}) = \sum_{n=0}^{\infty}\sum_{m=0}^{n}\left(a_n^m\Phi_{nm}^e(\vec{r}) + b_n^m\Phi_{nm}^o(\vec{r})\right) \tag{4.6}$$

where

$$\Phi^e_{nm}(\vec{r}) = \frac{1}{4\pi r^{n+1}} Y^e_{nm}(\theta,\phi), \quad \Phi^o_{nm}(\vec{r}) = \frac{1}{4\pi r^{n+1}} Y^o_{nm}(\theta,\phi) \tag{4.7}$$

is called a unit even or odd nm-multipole potential, and

$$Y^e_{nm}(\theta,\phi) = \cos(m\phi) P_n^m(\cos\theta), \quad Y^o_{nm}(\theta,\phi) = \sin(m\phi) P_n^m(\cos\theta) \tag{4.8}$$

is called even or odd spherical harmonic functions, while $P_n^m(\cos\theta)$ is the first type of associated Legendre function, $n=0$ corresponds to a monopole term, $n=1$ corresponds to a dipole term, which has three components, $n=2$ corresponds to a quadrupole term, which has five components, and the n-th multipole term has $(2n+1)$ components. It should be noted that the multipole term is singular at the origin of coordinates ($r=0$). So, the spherical harmonic series expansion must be applied under the condition that $r' < a < r$, where r' represents the maximum radial coordinates of the source region. According to Euler's formula, multipole strength can be expressed as a complex number.

$$a_{nm} + jb_{nm} = \varepsilon_m \frac{(n-m)!}{(n+m)!} \int_0^{2\pi} e^{jm\phi'} d\phi' \int_0^{\pi} P_n^m(\cos\theta') \sin\theta' d\theta' \int_0^a s(\vec{r}') r'^{n+2} dr' \tag{4.9}$$

where ε is the Neumann factor: when $m=0$, $\varepsilon=1$; when $m \neq 0$, $\varepsilon=2$. Both $Y^o_{n0}(\theta,\phi)$ and b_{n0} are always zero.

The unit potential, multipole potential (4.8), and multipole strength (4.9) of the above spherical harmonic expansion can also be expressed in rectangular coordinates, as shown in Table 4.1. The potential generated by each multipole component is obtained from the product of the multipole intensity and the corresponding unit potential.

The corresponding relationship with quasi-static charge distribution (duality principle in Section 2.7) is $s(\vec{r}) = 1/\varepsilon_0 \rho(\vec{r}) = \nabla \cdot \vec{E}(\vec{r})$. If charge distribution is further assumed to be composed of some isolated point charges q, source-sink images can be drawn for all components of the spherical harmonic multipole expansion (Figure 4.1), where the source-sink location and intensity are shown in Table 4.2. The source-sink images in Figure 4.1 are mutually orthogonal, and each one has a unique nonzero moment, respectively. The theoretical definition of multipole is based on the limiting conditions that the distance a approaches zero while the monopole intensity q approaches infinity, so that aq is a constant for the dipole, and while a^2q is a constant for the quadrupole.

Based on the source-sink images in Figure 4.1, the dipole and quadrupole can be represented by a combination of monopoles. So, if we find the solution to monopole in a specific conductor model, the solutions to the other situations can be developed by superposition without the need to specifically derive the solution. Actually, the direct derivation of a high-level multipole is rather complicated although the idea is similar to that for monopole derivation. Moreover, in terms of application, the

TABLE 4.1
The Cartesian Coordinate Representation of the Multipole Expansion in Spherical Harmonic Analysis (Wikswo and Swinney, 1984)

Monopole

$$\Phi_{00}^{e} = \frac{1}{4\pi}\frac{1}{r} \qquad a_{00} = \int_{v} s(\vec{r}')dv(\vec{r}')$$

Dipole

$$\Phi_{10}^{e} = \frac{1}{4\pi}\frac{z}{r^3} \qquad a_{10} = \int_{v} s(\vec{r}')z'\,dv(\vec{r}')$$

$$\Phi_{11}^{e} = \frac{1}{4\pi}\frac{x}{r^3} \qquad a_{11} = \int_{v} s(\vec{r}')x'\,dv(\vec{r}')$$

$$\Phi_{11}^{o} = \frac{1}{4\pi}\frac{y}{r^3} \qquad b_{11} = \int_{v} s(\vec{r}')y'\,dv(\vec{r}')$$

Quadrupole

$$\Phi_{20}^{e} = \frac{1}{4\pi}\frac{3z^2 - r^2}{2r^5} \qquad a_{20} = 1/2\int_{v} s(\vec{r}')(3z'^2 - r'^2)dv(\vec{r}')$$

$$\Phi_{21}^{e} = \frac{1}{4\pi}\frac{3xz}{2r^5} \qquad a_{21} = \int_{v} s(\vec{r}')x'z'\,dv(\vec{r}')$$

$$\Phi_{21}^{o} = \frac{1}{4\pi}\frac{3yz}{2r^5} \qquad b_{21} = \int_{v} s(\vec{r}')y'z'\,dv(\vec{r}')$$

$$\Phi_{22}^{e} = \frac{1}{4\pi}\frac{3(x^2 - y^2)}{r^5} \qquad a_{22} = 1/4\int_{v} s(\vec{r}')(x'^2 - y'^2)dv(\vec{r}')$$

$$\Phi_{22}^{o} = \frac{1}{4\pi}\frac{6xy}{r^5} \qquad b_{22} = 1/2\int_{v} s(\vec{r}')y'x'\,dv(\vec{r}')$$

The solution of the Poisson's equation (4.1) is (4.2). Suppose the maximum range of the source is a, then outside this source region, the Poisson's equation is simplified to Laplace equation (4.3), and its solution can be expressed as a spherical harmonic series (4.6) for $r > a$, the detail of each variable is shown in equations (4.7)–(4.9). This table lists the representations of monopole, dipole, and quadrupole in rectangular coordinates.

monopole, dipole, and quadrupole can be simply treated as a vector with one component, three components, and five components, respectively, with no need to deal with the source-sink image of each component.

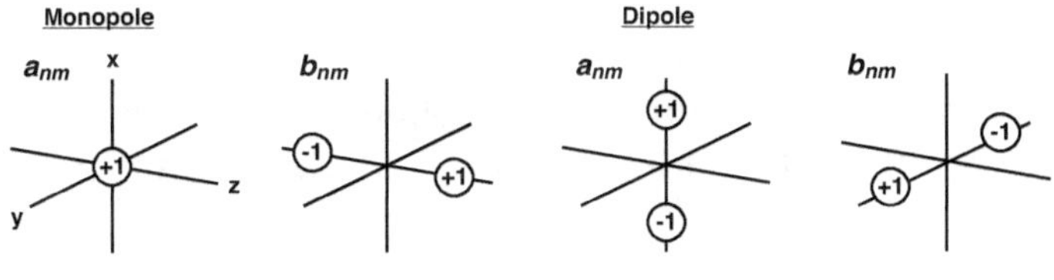

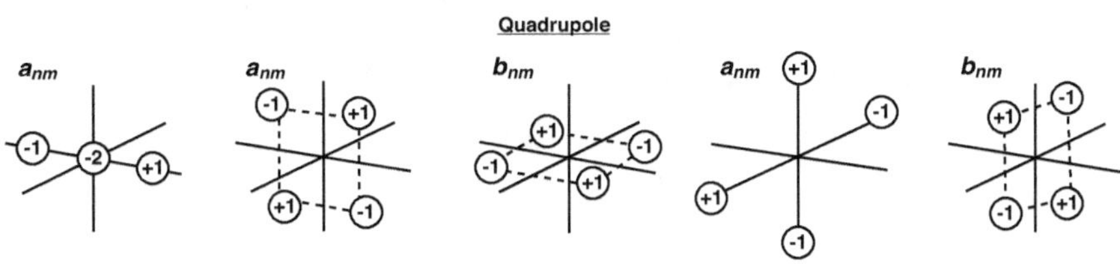

FIGURE 4.1 The source-sink image of multipole expansion in spherical harmonic analysis (monopole, dipole, quadrupole) (Wikswo and Swinney, 1984).

TABLE 4.2
Source-Sink Coordinates of Spherical Harmonic Multipole Expansion (Wikswo and Swinney, 1984)

$b = a/\sqrt{2};\ c = b/\left(2\sqrt[3]{6}\right);\ d = a/\sqrt[3]{2};\ e = a/\sqrt[3]{4}$

Spherical Harmonic Coefficient	Monopole Strength (Unit *q*)	Cartesian Coordinates		
		X	*Y*	*Z*
	Dipole			
a_{10}	+1	0	0	+a/2
	−1	0	0	−a/2
a_{11}	+1	+a/2	0	0
	−1	−a/2	0	0
b_{11}	+1	0	+a/2	0
	−1	0	−a/2	0
	Quadrupole			
a_{20}	−2	0	0	0
	+1	0	0	+b
	+1	0	0	−b
a_{21}	+1	+a/2	0	+a/2
	+1	−a/2	0	−a/2
	−1	+a/2	0	−a/2
	−1	−a/2	0	+a/2
b_{21}	+1	0	+a/2	+a/2
	+1	0	−a/2	−a/2
	−1	0	+a/2	−a/2
	−1	0	−a/2	+a/2

(*Continued*)

TABLE 4.2 (*Continued*)
Source-Sink Coordinates of Spherical Harmonic Multipole Expansion (Wikswo and Swinney, 1984)

$b = a/\sqrt{2};\ c = b/\left(2\sqrt[3]{6}\right);\ d = a/\sqrt[3]{2};\ e = a/\sqrt[3]{4}$

Spherical Harmonic Coefficient	Monopole Strength (Unit q)	Cartesian Coordinates		
		X	*Y*	*Z*
	Quadrupole			
a_{22}	+1	+a	0	0
	+1	−a	0	0
	−1	0	+a	0
	−1	0	−a	0
b_{22}	+1	+b	+b	0
	+1	−b	−b	0
	−1	+b	−b	0
	−1	−b	+b	0

4.1.2 Spherical Harmonic Spectra Expansion for Dipole Distribution

Among them, those variables with "priming" are the multipole expansion relative to the origin and those without are the multipole expansion relative to (x_0, y_0, z_0) (Wikswo and Swinney, 1985).

Figure 4.1 shows the source-sink image of the spherical harmonic expansion of potential in the above; here monopole is the fundamental source. However, the monopole is a more underlying concept than the dipole which is more commonly used in practice. This section will focus on the distributed dipole source. According to equations (4.1) and (4.2), there is

$$\Phi(\vec{r}) = \frac{1}{4\pi\sigma}\int_v \frac{-\nabla'\cdot\vec{J}_s(\vec{r}')}{|\vec{r}-\vec{r}'|}dv(\vec{r}') \tag{4.10}$$

Equations (4.1) and (4.10) show that the source s is the divergence of a certain vector field. Therefore, it is possible to use the field theory formula to obtain another representation of the spherical harmonic coefficients in Table 4.1. For the sake of generality, we take the following general form:

$$A_{nm}^{p} = \int_{v} s(\vec{r}')f(\vec{r}')dv(\vec{r}') \tag{4.11}$$

where p is either e or o. $A_{nm}^{e} = a_{nm}$, $A_{nm}^{o} = b_{nm}$, $f(\vec{r}')$ is a weighting function, and its specific form depends on different moment components. Now let $s(\vec{r}) = -\nabla\cdot\vec{S}$, then equation (4.11) becomes

$$A_{nm}^{p} = -\int_{v} \nabla \cdot \vec{S}(\vec{r}') f(\vec{r}') dv(\vec{r}') \tag{4.12}$$

Use vector equation $f\nabla \cdot \vec{S} = \nabla \cdot \left(f\vec{S}\right) - \vec{S} \cdot \nabla f$ to transform equation (4.12) into

$$A_{nm}^{p} = -\int_{v} \nabla \cdot \left(f\vec{S}(\vec{r}')\right) dv(\vec{r}') + \int_{v} \vec{S}(\vec{r}') \cdot \nabla f dv(\vec{r}') = \int_{v} \vec{S}(\vec{r}') \cdot \nabla f dv(\vec{r}') \tag{4.13}$$

where the first term disappears due to $\vec{J}_s = 0$ on the integral surface when the integral surface is located outside the source region after the volume integral is changed to the surface integral by utilizing the Gauss theorem. The $f(\vec{r}')$ is obtained by reference to Table 4.1, which is now transformed into Table 4.3, where $\vec{S}$ represents the corresponding dipole moment density.

As an application of Table 4.3, Table 4.4 listed the quadrupole coefficients for a dipole located at (x_0, y_0, z_0) with the dipole moment of (m_x, m_y, m_z). It shows that if

TABLE 4.3
The Rectangular Coordinate Representation of the Multipole Expansion in Spherical Harmonic Analysis Based on Dipole Density (Wikswo and Swinney, 1985)

	Monopole
$\Phi_{00}^{e} = \frac{1}{4\pi}\frac{1}{r}$	$a_{00} = \int_{v} \nabla \cdot \vec{S}(\vec{r}') dv(\vec{r}') = 0$
	Dipole
$\Phi_{10}^{e} = \frac{1}{4\pi}\frac{z}{r^3}$	$a_{10} = \int_{v} S_z(\vec{r}') dv(\vec{r}')$
$\Phi_{11}^{e} = \frac{1}{4\pi}\frac{x}{r^3}$	$a_{11} = \int_{v} S_x(\vec{r}') dv(\vec{r}')$
$\Phi_{11}^{o} = \frac{1}{4\pi}\frac{y}{r^3}$	$b_{11} = \int_{v} S_y(\vec{r}') dv(\vec{r}')$
	Quadrupole
$\Phi_{20}^{e} = \frac{1}{4\pi}\frac{3z^2 - r^2}{2r^5}$	$a_{20} = \int_{v} \left[3z'S_z(\vec{r}') - \vec{r}' \cdot \vec{S}(\vec{r}')\right] dv(\vec{r}')$
$\Phi_{21}^{e} = \frac{1}{4\pi}\frac{3xz}{2r^5}$	$a_{21} = \int_{v} \left[x'S_z(\vec{r}') + z'S_x(\vec{r}')\right] dv(\vec{r}')$
$\Phi_{21}^{o} = \frac{1}{4\pi}\frac{3yz}{2r^5}$	$b_{21} = \int_{v} \left[y'S_z(\vec{r}') + z'S_y(\vec{r}')\right] dv(\vec{r}')$

(Continued)

TABLE 4.3 (*Continued*)
The Rectangular Coordinate Representation of the Multipole Expansion in Spherical Harmonic Analysis Based on Dipole Density (Wikswo and Swinney, 1985)

Quadrupole

$\Phi_{22}^{e} = \frac{1}{4\pi}\frac{3(x^2 - y^2)}{r^5}$	$a_{22} = 1/2\int_v [x'S_x(\vec{r}') - y'S_y(\vec{r}')]dv(\vec{r}')$
$\Phi_{22}^{o} = \frac{1}{4\pi}\frac{6xy}{r^5}$	$b_{22} = 1/2\int_v [x'S_y(\vec{r}') + y'S_x(\vec{r}')]dv(\vec{r}')$

The Poisson's equation $\nabla^2\Phi(\vec{r}) = -s(\vec{r})$ has the solution (4.10). For biological conductive media, $s(\vec{r}) = -1/\sigma\nabla\cdot\vec{j}_s$, $\vec{s}(\vec{r}) = 1/\sigma\vec{j}_s$. Suppose the range of the source is a, then outside the source region, the Poisson's equation is simplified to the Laplace equation (4.4), and its solution can be expressed by a spherical harmonic series $\Phi(\vec{r}) = \sum_{n=0}^{\infty}\sum_{m=0}^{n}\left(a_{nm}\Phi_{nm}^{e}(\vec{r}) + b_{nm}\Phi_{nm}^{o}(\vec{r})\right) = \sum_{n=0}^{\infty}\sum_{m=0}^{n}\sum_{p}A_{nm}^{p}\Phi_{nm}^{p}(\vec{r})$, $p = e\ \&\ o$, r>a. The details of each variable are shown in Table 4.1. This table gives the rectangular coordinate representations of the monopole, dipole, and quadrupole, with $A_{nm}^{p} = \int_v \vec{s}(\vec{r}')\cdot\nabla f\, dv(\vec{r}')$, and $\sigma\vec{s}(\vec{r}) = \vec{j}_s(\vec{r})$ as the density of the dipole moment.

TABLE 4.4
Offset Equations

Dipole	Quadrupole
$a'_{10} = a_{10} = m_z$	$a'_{20} = a_{20} = 2z_0m_z + x_0m_x + y_0m_y$
$a'_{11} = a_{11} = m_x$	$a'_{21} = a_{21} - z_0m_x - x_0m_z$
$b'_{11} = b_{11} = m_y$	$b'_{21} = b_{21} - z_0m_y - y_0m_z$
	$a'_{22} = a_{22} - 1/2(x_0m_x - y_0m_y)$
	$a'_{22} = a_{22} - 1/2(x_0m_x - y_0m_y)$

there is only one dipole, the diploe moment can be obtained from the dipole term in the spherical harmonic multipole expansion, and its eccentric position can be revealed according to the quadrupole term. If there are two dipoles, the octopole term must be involved, which contains the second power of the coordinates, suggesting that a nonlinear problem needs to be solved. This shows that although it is theoretically feasible to get information on dipoles by the spherical harmonic coefficients, it is difficult in practice (Wikswo and Swinney, 1985; Geselowitz, 1965).

4.2 DIPOLE IN HOMOGENEOUS SPHERE

The research on the analytical solution for the forward problem of EEG was started by Wilson and Bayley (1950). They first derived a formula for the dipole potential in a homogeneous conducting sphere. Then, Frank developed a slightly different form in 1952. However, there are some problems with the formulas proposed by Wilson and Frank, such as division by zero and other meaningless situations at some special locations. Later, Brody et al. derived a strict formula suitable for the point on the spherical surface based on the concept of lead field, but this formula cannot be used to calculate the potential at points inside the sphere. Based on these efforts, we derived an analytical solution for the potential of a dipole in a homogeneous conducting sphere model, and this solution is non-singular at all, thus getting this problem solved perfectly (Yao, 2000a).

Suppose the rectangular coordinates of the field point $\vec{r}$ and the position $\vec{r}_0$ of the dipole $\vec{P}$ are (x, y, z) and (x_0, y_0, z_0), respectively, and the length of the displacement $\vec{r}_p = \vec{r} - \vec{r}_0$ between these two points is

$$r_p = \left(r^2 + r_0^2 - 2rr_0\cos\phi\right)^{1/2} \tag{4.14}$$

where $\cos\varphi$ is the cosine of the angle between the vectors $\vec{r}$ and $\vec{r}_0$. Suppose $\vec{P} = P_x\vec{e}_x + P_y\vec{e}_y + P_z\vec{e}_z$, where (P_x, P_y, P_z) are the three coordinate components of the dipole moment, and $(\vec{e}_x, \vec{e}_y, \vec{e}_z)$ are the unit vectors of the three coordinate axis directions of the rectangular coordinate system. R is the radius of the sphere. We can have (Yao, 2000a)

$$\Phi = \frac{\vec{P}}{4\pi\sigma}\cdot\left\{\frac{\vec{r}-\vec{r}_0}{r_p^3} + \frac{\left(\vec{r} - \frac{r^2}{R^2}\vec{r}_0\right)}{R^3 r_{pi}^3} + \frac{1}{R^3 r_{pi}}\left[\vec{r} + \frac{\vec{r}\frac{r_0 r}{R^2}\cos\phi - \frac{r^2}{R^2}\vec{r}_0}{r_{pi} + 1 - \frac{r_0 r}{R^2}\cos\phi}\right]\right\} \tag{4.15}$$

where $r_{pi} = \left(1 + \left(\frac{r_0 r}{R^2}\right)^2 - 2\frac{r_0 r}{R^2}\cos\varphi\right)^{1/2}$. Figure 4.2 shows a potential distribution inside the brain calculated by equation (4.15).

For the surface of a spherical brain, $r = R$, $Rr_{pi} = r_p$, we have

$$\Phi = \frac{\vec{P}}{4\pi\sigma}\cdot\left\{2\frac{\vec{R}-\vec{r}_0}{r_p^3} + \frac{1}{R^2 r_p}\left[\vec{R} + \frac{\vec{R}r_0\cos\varphi - R\vec{r}_0}{R + r_p - r_0\cos\varphi}\right]\right\} \tag{4.16}$$

Equation (4.16) is exactly the formula given by Brody et al. (1973). Obviously, for equation (4.15), the dipole can be located anywhere in the sphere, and the electrode can also be located anywhere inside the sphere or on the surface of the sphere. Therefore, it is the final answer to this classic problem, and it has been utilized as the theoretical standard for evaluating various numeric algorithms for EEG forward.

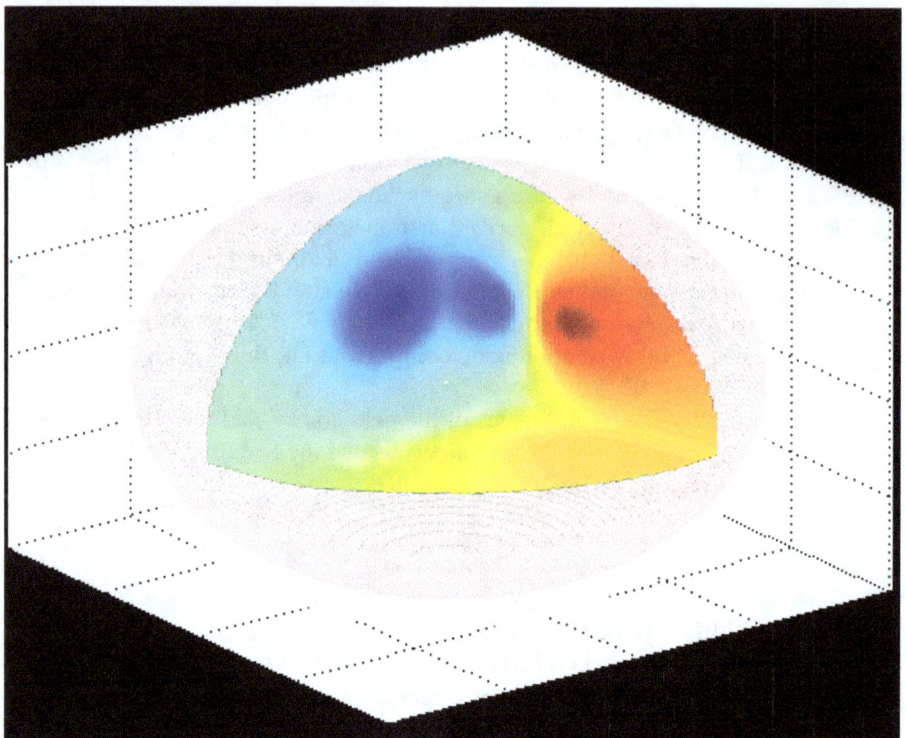

FIGURE 4.2 The potential in a sphere generated by a dipole inside the sphere according to equation (4.15).

4.3 2^n POLE IN HOMOGENEOUS SPHERE

Section 4.2 derived the potential formula for a dipole in a homogeneous sphere (Yao, 2000a). For an actual problem with EEG/ECG, it is sometimes necessary to know the potential formulas for a quadrupole or an octopole, etc. (Yao, 1998).

As we know, any 2^n-pole is composed of point sources. Of course, a 2^n-pole can be said to be composed of 2^{n-1}-poles (but not vice versa). Suppose the potential of a point source with unit strength (monopole source) in a conductor model is $u(\vec{r},\vec{r}_0)$, where $\vec{r}$ and $\vec{r}_0$ represent the position vectors of the field and source points, respectively. Let $S = \vec{r} \cdot \vec{r}_0$, and the definitions of other quantities are the same as those in Section 4.2.

Suppose the origin of the rectangular coordinate system is at the center of the sphere. Now suppose there is a point source in a small region V near $\vec{r}_0$, and the range size of V is l, while the density of the point source is i (scalar). According to the principle of superposition, the potential generated from the distributed source in the homogeneous conductive sphere is

$$\Phi = \iiint_V \frac{i(\vec{r}_0')}{4\pi\sigma} u(\vec{r},\vec{r}_0') dv(\vec{r}_0') \tag{4.17}$$

According to the knowledge of multivariate function differentiation, if $u(\vec{r},\vec{r}_0)$ has continuous partial derivatives of any order relative to $\vec{r}_0$, and $|\vec{r}_0' - \vec{r}| \geq l$, then it can be expanded at $\vec{r}_0$ into the Taylor series in the following tensor form:

$$u(\vec{r},\vec{r}_0') = u(\vec{r},\vec{r}_0) + (\vec{r}_0' - \vec{r}_0)\cdot\nabla^0 u(\vec{r},\vec{r}_0')\,|_{\vec{r}_0'=\vec{r}_0} + \frac{1}{2!}(\vec{r}_0' - \vec{r}_0)(\vec{r}_0' - \vec{r}_0) : \nabla^0\nabla^0 u(\vec{r},\vec{r}_0')\,|_{\vec{r}_0'=\vec{r}_0} + \cdots \tag{4.18}$$

where $\vec{r}_0$ is a point selected in V, and ∇^0 means the derivative of $\vec{r}_0'$. Substitute (4.18) into (4.17) to get

$$\begin{aligned}\Phi(\vec{r}) &= \iiint_V \frac{i(\vec{r}_0')}{4\pi\sigma} dv(\vec{r}_0') u(\vec{r},\vec{r}_0) + \iiint_V \frac{i(\vec{r}_0')}{4\pi\sigma}(\vec{r}_0' - \vec{r}_0) dv(\vec{r}_0')\cdot\nabla^0 u(\vec{r},\vec{r}_0')\,|_{\vec{r}_0'=\vec{r}_0} \\ &+ \frac{1}{2!}\iiint_V \frac{i(\vec{r}_0')}{4\pi\sigma}(\vec{r}_0' - \vec{r}_0)(\vec{r}_0' - \vec{r}_0) dv(\vec{r}_0') : \nabla^0\nabla^0 u(\vec{r},\vec{r}_0')\,|_{\vec{r}_0'=\vec{r}_0} + \cdots \\ &= \frac{I}{4\pi\sigma} u(\vec{r},\vec{r}_0) + \frac{1}{4\pi\sigma}\vec{P}\cdot\nabla^0 u(\vec{r},\vec{r}_0')\,|_{\vec{r}_0'=\vec{r}_0} + \frac{1}{24\pi\sigma}\bar{\bar{Q}} : \nabla^0\nabla^0 u(\vec{r},\vec{r}_0')\,|_{\vec{r}_0'=\vec{r}_0} + \cdots \\ &= \frac{I}{4\pi\sigma} u(\vec{r},\vec{r}_0) + \frac{1}{4\pi\sigma}\vec{P}\cdot\vec{\Theta}_{\vec{P}}(\vec{r},\vec{r}_0) + \frac{1}{24\pi\sigma}\bar{\bar{Q}} : \nabla^0\vec{\Theta}_{\vec{P}}(\vec{r},\vec{r}_0')\,|_{\vec{r}_0'=\vec{r}_0} + \cdots\end{aligned} \tag{4.19}$$

where

$$I = \iiint_V i(\vec{r}_0') dv(\vec{r}_0'), \quad \vec{P} = \iiint_V i(\vec{r}_0')(\vec{r}_0' - \vec{r}_0) dv(\vec{r}_0'),$$

$$\bar{\bar{Q}} = \iiint_V 3i(\vec{r}_0')(\vec{r}_0' - \vec{r}_0)(\vec{r}_0' - \vec{r}_0) dv(\vec{r}_0')$$

According to the theory of multipole expansion of potential, I, $\vec{P}$, and $\bar{\bar{Q}}$ represent the intensity of the monopole at $\vec{r}_0$ (point source), the moment of the dipole, and the moment tensor of the quadrupole, respectively. Thus, it can be seen that the potential generated from any point source near $\vec{r}_0'$ can always be synthesized from the contributions of the monopole, dipole, quadrupole … 2^n-pole at $\vec{r}_0$. Generally, a biological medium is always electrically neutral except in some local regions at the cellular level, i.e., the sum of all monopole sources is zero ($I=0$). Therefore, the contribution of the monopole source, i.e., the first term in equation (4.19), can be ignored on the macroscale of EEG. For the terms that follow, a potential formula can be derived for any 2^n-pole by simple differential operations as long as the potential formula of the dipole is known.

From equation (4.15), we have $\vec{\Theta}_{\vec{P}}$, which satisfies

$$\Phi = \frac{\vec{P}}{4\pi\sigma}\cdot\vec{\Theta}_{\vec{P}} \tag{4.20}$$

Therefore, equations (4.19) and (4.20) give the calculation formula for the potential of any 2^n-pole. Some special cases will be discussed below.

4.3.1 The Potential of a Dipole at the Center of a Sphere

As can be seen from equation (4.20), when the dipole is located at the center of the sphere, the corresponding potential is

$$\Phi = \frac{1}{4\pi\sigma}\vec{P}\cdot\vec{\Theta}_{\vec{P}} = \frac{3P\cos\theta}{4\pi\sigma R^2} \tag{4.21}$$

where θ is the angle between the dipole moment $\vec{P}$ and the field point vector $\vec{R}$. This formula shows that the potential on the spherical surface generated by the dipole at the sphere center is three times as the value as that generated on the same spherical surface in the unbounded space. This phenomenon fully demonstrates that the finite conductor model has a great effect on the potential of the dipole.

4.3.2 The Potential Formula for a Quadrupole

A quadrupole can be composed of dipoles according to the rules in Figure 4.1. For example, Q_{33} is composed of a pair of dipoles, $+\vec{P}$, and $-\vec{P}$, which are oriented along the z-axis and infinitely close to each other. According to equation (4.19), the potential of the quadrupole can be expressed as

$$\Phi(\vec{r}) = \frac{1}{24\pi\sigma}\bar{\bar{Q}}:\nabla^0\vec{\Theta}_{\vec{P}}(\vec{r},\vec{r_0'})\mid_{\vec{r_0'}=\vec{r_0}} = \frac{1}{24\pi\sigma}\sum_{i,j}Q_{ij}\Phi_{ij}(\vec{r},\vec{r_0}) \tag{4.22}$$

where

$$\begin{aligned}\Phi_{xy} = {} & \frac{3(x-x_0)(y-y_0)}{r_p^5} + \frac{3\left(xR^2-r^2x_0\right)\left(yR^2-r^2y_0\right)}{R^9r_{pi}^5} \\ & + \frac{\left(R^2\left(r_{pi}+1\right)x-r^2x_0\right)\left(R^2\left(r_{pi}+1\right)y-r^2y_0\right)}{R^5r_{pi}^2\left(R^2\left(r_{pi}+1\right)-S\right)^2} \\ & + \frac{\left(r^2x_0-xR^2\right)\left(r^2y_0-yR^2\right)}{R^7r_{pi}^3\left(R^2\left(r_{pi}+1\right)-S\right)}\end{aligned} \tag{4.23}$$

and

$$\Phi_{zz} = \frac{3(z-z_0)^2}{r_p^5} + \frac{3(zR^2 - r^2 z_0)^2}{R^9 r_{pi}^5} + \frac{\left(R^2(r_{pi}+1)z - r^2 z_0\right)^2}{R^5 r_{pi}^2\left(R^2(r_{pi}+1)-S\right)^2}$$
$$+ \frac{(r^2 z_0 - zR^2)^2}{R^7 r_{pi}^3\left(R^2(r_{pi}+1)-S\right)} - \frac{1}{r_p^3} - \frac{r^2}{R^5 r_{pi}^3} - \frac{r^2}{R^3 r_{pi}\left(R^2(r_{pi}+1)-S\right)} \tag{4.24}$$

The other components of Φ_{ij} can be derived similarly from equation (4.19). For points on the sphere, $r = R$, $Rr_{pi} = r_p$, the equations (4.23) and (4.24) can be simplified to

$$\Phi_{xy} = \frac{6(x-x_0)(y-y_0)}{r_p^5} + \frac{\left((R^{-1}r_p+1)x - x_0\right)\left((R^{-1}r_p+1)y - y_0\right)}{Rr_p^2\left(r_p + R - R^{-1}S\right)^2}$$
$$+ \frac{(x_0 - x)(y_0 - y)}{Rr_p^3\left(r_p + R - SR^{-1}\right)} \tag{4.25}$$

$$\Phi_{zz} = \frac{6(z-z_0)^2}{r_p^5} + \frac{\left((R^{-1}r_p+1)z - z_0\right)^2}{Rr_p^2\left(r_p + R - SR^{-1}\right)^2} + \frac{(z_0 - z)^2}{Rr_p^3\left(r_p + R - SR^{-1}\right)}$$
$$- \frac{2}{r_p^3} - \frac{1}{Rr_p\left(r_p + R - SR^{-1}\right)} \tag{4.26}$$

If it is further assumed that the quadrupole is located at the center of the sphere, then

$$\Phi_{xy} = \frac{15xy}{2R^5},\ \Phi_{zz} = \frac{15z^2}{2R^5} - \frac{5}{2R^3} \tag{4.27}$$

Specifically, the potential generated from Q_{33} on the sphere is

$$\Phi = \frac{1}{24\pi\sigma} Q_{33}\left(\frac{15z^2}{2R^5} - \frac{5}{2R^3}\right) \tag{4.28}$$

And the potential generated from Q_{33} in unbounded space is

$$\Phi_\infty = \frac{1}{24\pi\sigma} Q_{33}\left(\frac{3z^2}{R^5} - \frac{1}{R^3}\right) \tag{4.29}$$

As can be seen from the two equations (4.28) and (4.29), there is a linear relationship between them, and the proportional coefficient is 5/2. This result also indicates the great effect of the finite conductor model on multipole potential. It is worth mentioning that the proportional coefficient of a quadrupole is 2.5 rather than 3 for a dipole as mentioned in equation (4.21). This shows that the finite conductor model has different effects on different source models.

4.4 DIPOLE IN THREE-LAYER CONCENTRIC SPHERE

The concentric three-layer sphere model is the most commonly used head model, in which the human head is treated as three concentric spheres including the brain, skull, and scalp (Rush and Driscoll, 1969; Arthur and Geselowitz, 1970; Kavanagk et al., 1978). In this model, the inner and outer radiuses of the skull are generally set to 8 and 8.5 cm, respectively, while the radius of the scalp surface is set to 9.2 cm. The electrical resistance is usually set to 2.22 Ωm for the brain and scalp, and $80\times 2.22\ \Omega\text{m} = 177\ \Omega\text{m}$ for the skull. In recent years, the resistivity has been re-estimated with different methods, as shown in Section 3.3.

Because of the characteristics of symmetry and simplicity, it is very easy to build an electrolyte model or a mathematical and computer model. In terms of actual usefulness, although such a simple model does not consider the anisotropy or heterogeneity of brain tissues and the skull, its results are well consistent with the real situation. In simulation study, the radius and conductance constant are usually normalized into their relative values. For example, the radius is normalized by the radius of the head, so the inner and outer surface radius of the skull and the surface radius of the head are changed to 0.87, 0.92, and 1.0, respectively; the resistance is normalized by the resistance of the brain; eventually, the resistance is transferred to 80.0 for the skull and 1.0 for the brain and scalp. For a common multilayer sphere model, a closed solution cannot be obtained, so what is given here is a spherical harmonic series solution.

For simplicity, we consider a radial dipole p_r and a tangential dipole p_t on the z-axis (p_r and p_t can be obtained from a decomposition of a dipole on the z-axis, by rotating the coordinate system, any dipole can be placed on the z-axis).

For p_r in homogeneous space, similar to the practices in Sections 4.2 and 4.3, let $K_r = \dfrac{p_r}{4\pi\sigma}$, then

$$
\begin{aligned}
\Phi_{r\infty} &= K_r \frac{r\cos\theta - r_0}{r^3} \sum_{l=1}^{\infty} \left(\frac{r_0}{r}\right)^{l-1} P_l(\cos\theta) \\
&= K_r \sum_{l=1}^{\infty} \frac{r_0^{l-1}}{r^{l+1}} \cos\theta P_l(\cos\theta) - \frac{r_0^l}{r^{l+2}} P_l(\cos\theta), \quad (r > r_0)
\end{aligned}
\tag{4.30}
$$

where θ is the angle between r- and z-axis, r_0 is the position of the dipole on z-axis, and σ is the conductivity.

If the head is a homogeneous sphere, by utilizing the sphere surface boundary condition, we have

$$\Phi_r = K_r \sum_{l=1}^{\infty} \left(\frac{l+1}{l} \frac{r^l}{R^{2l+1}} + \frac{1}{r^{l+1}} \right) r_0^{l-1} \cos\theta P_l' - \left(\frac{l+2}{l+1} \frac{r^{l+1}}{R^{2l+3}} + \frac{1}{r^{l+2}} \right) r_0^l P_l'$$

$$= K_r \sum_{l=1}^{\infty} \left(\frac{l+1}{l} \frac{r^l}{R^{2l+1}} r_0^{l-1} \cos\theta P_l' - \frac{l+2}{l+1} \frac{r^{l+1}}{R^{2l+3}} r_0^l P_l' + \frac{r_0^{l-1}}{r^{l+1}} \cos\theta P_l' - \frac{r_0^l}{r^{l+2}} P_l' \right) \tag{4.31}$$

and with $P_{l-1}' = xP_l' - lP_l$, $xP_l' = P_{l-1}' + lP_l$. Considering $P_0' = 0$, we can get

$$\Phi_r = K_r \sum_{l=1}^{\infty} \left(\frac{l+1}{l} \frac{r^l}{R^{2l+1}} \right) r_0^{l-1} lP_l + \frac{r_0^{l-1}}{r^{l+1}} lP_l$$

$$= K_r \sum_{l=1}^{\infty} \left(\frac{l+1}{l} \frac{r^l}{R^{2l+1}} lr_0^{l-1} + l \frac{r_0^{l-1}}{r^{l+1}} \right) P_l \tag{4.32}$$

For tangential dipole p_t in homogeneous space, let $K_t = \frac{p_t}{4\pi\sigma}$. Since $\vec{p}_t \cdot \vec{r}_0 = 0$, and $\vec{p}_t \cdot \vec{r} = p_t r \sin\theta \cos\phi$, where ϕ is the azimuth of $\vec{r}$ on the XOY plane, then

$$\Phi_{t\infty} = K_t \frac{\sin\theta\cos\phi}{r^2} \sum_{l=1}^{\infty} \left(\frac{r_0}{r} \right)^{l-1} P_l' = K_t \cos\phi \sum_{l=1}^{\infty} \frac{r_0^{l-1}}{r^{l+1}} P_l^{\prime}, \ (r > r_0) \tag{4.33}$$

$P_l' = \sin\theta P_l\ (\cos\theta)$ is applied in equation (4.33). According to the boundary conditions of the scalp surface for a single sphere, we can get

$$\Phi_t = K_t \cos\phi \sum_{l=1}^{\infty} \left(\frac{l+1}{l} \frac{r^l}{R^{2l+1}} + \frac{1}{r^{l+1}} \right) r_0^{l-1} P_l^1 \tag{4.34}$$

Now, we specifically consider a concentric three-layer sphere model. According to the multipole expansion forms of potential in equations (4.32) and (4.34), the solution of the concentric three-layer sphere model can be set as

$$\Phi_1 = \frac{1}{4\pi\sigma} \sum_{n=1}^{\infty} \left(\frac{A_n}{r^{n+1}} + B_n r^n \right) \Bigg\rangle \begin{matrix} P_n^0(\cos\theta) \\ P_n^1(\cos\theta)\cos\phi \end{matrix}, \ f < \frac{r}{R} \le b \tag{4.35}$$

$$\Phi_2 = \frac{1}{4\pi\sigma k} \sum_{n=1}^{\infty} \left(\frac{C_n}{r^{n+1}} + D_n r^n \right) \Bigg\rangle \begin{matrix} P_n^0(\cos\theta) \\ P_n^1(\cos\theta)\cos\phi \end{matrix}, \ b \le \frac{r}{R} \le c \tag{4.36}$$

$$\Phi_3 = \frac{1}{4\pi\sigma} \sum_{n=1}^{\infty} \left(\frac{E_n}{r^{n+1}} + F_n r^n \right) \Bigg\rangle \begin{matrix} P_n^0(\cos\theta) \\ P_n^1(\cos\theta)\cos\phi \end{matrix}, \ c \le \frac{r}{R} \le 1 \tag{4.37}$$

where σ represents the conductivity of the first and third layers; $k\sigma$ represents the conductivity of the second layer; R represents the radius of the scalp surface. The dipole is located at fR on the z-axis, the radius of the inner sphere is bR, and the radius of the middle layer is cR. The spherical coordinates are (r,θ,ϕ). The upper term P_n^0 corresponds to the radial dipole parallel to the z-axis, while the lower term P_n^1 corresponds to the tangential dipole in the x-direction ($\cos\phi$ needs to be turned into $\sin\phi$ for the tangential dipole in the y-direction).

According to the boundary conditions on each interface, i.e., on $r = bR$ and cR, both the potential and radial current density are continuous, but the radial current density will disappear on $r = R$, thereby we can get five equations:

$$\frac{A_n}{(bR)^{n+1}} + B_n (bR)^n = \frac{1}{k}\left[\frac{C_n}{(bR)^{n+1}} + D_n (bR)^n\right] \tag{4.38}$$

$$\frac{1}{k}\left[\frac{C_n}{(cR)^{n+1}} + D_n (cR)^n\right] = \frac{E_n}{(cR)^{n+1}} + F_n (cR)^n \tag{4.39}$$

$$\frac{-(n+1)A_n}{(bR)^{n+2}} + B_n (bR)^{n-1} n = -\frac{(n+1)}{(bR)^{n+2}} + n(bR)^{n-1} D_n \tag{4.40}$$

$$\frac{-(n+1)C_n}{(cR)^{n+2}} + D_n (cR)^{n-1} n = -\frac{(n+1)}{(cR)^{n+2}} E_n + n(cR)^{n-1} F_n \tag{4.41}$$

$$\frac{-(n+1)E_n}{(R)^{n+2}} + F_n (R)^{n-1} n = 0 \tag{4.42}$$

Finally, it should be noted that the sources are either radial or tangential dipoles, and their potential in an unbounded conductive medium is (the second term of equations (4.32) and (4.34))

$$\begin{aligned}\Phi_1^r &= \frac{p_r}{4\pi\sigma}\sum_{n=1}^{\infty} n(fR)^{n-1}\frac{1}{r^{n+1}}P_n(\cos\theta),\\ \Phi_1^t &= \frac{p_t}{4\pi\sigma}\sum_{n=1}^{\infty}(fR)^{n-1}\frac{1}{r^{n+1}}P_n(\cos\theta)\cos\phi, \quad (r > fR)\end{aligned} \tag{4.43}$$

To compare (4.43) with (4.35), it can be seen

$$A_n = \begin{cases} n(fR)^{n-1} p_r \\ (fR)^{n-1} p_t \end{cases} \tag{4.44}$$

Combining (4.38)–(4.42) and (4.44), we can solve all those coefficients: B_n, C_n, D_n, E_n, F_n.

For the scalp surface, substitute $r = R$ into Φ_3 to get

$$\Phi_3(R,\theta) = \frac{1}{4\pi\sigma R^2}\sum_{n=1}^{\infty}\frac{(2n+1)^3 kf^{n-1}}{(n+1)nd}\left\langle \begin{array}{c} p_r n P_n^0(\cos\theta) \\ p_t P_n^1(\cos\theta)\cos\phi \end{array}\right. \tag{4.45}$$

where

$$\begin{aligned} d &= (k-1)(kn+k+n)\left(c^{2n+1}-b^{2n+1}\right) \\ &+\frac{(kn+k+n)(kn+n+1)}{n+1} - (1-k)^2 n\left(\frac{b}{c}\right)^{2n+1}, \quad k = \sigma_2/\sigma, \end{aligned} \tag{4.46}$$

For the uniform sphere model, it is equivalent to $k = 1$, then $d = \frac{(2n+1)^2}{n+1}$, and so we can get

$$\Phi_3(R,\theta) = \frac{1}{4\pi\sigma R^2}\sum_{n=1}^{\infty}\frac{(2n+1)f^{n-1}}{n}\left\langle \begin{array}{c} p_r n P_n^0 \\ p_t P_n^1 \cos\phi \end{array}\right. \tag{4.47}$$

By comparing (4.45) with (4.47), the influence of the skull layer on the electric field distribution can be discussed.

4.5 DIPOLE AND UNIPOLE IN ONE-TO-THREE LAYER CONCENTRIC SPHERE

4.5.1 Potential in an Infinite Volume Conductor

The following equation gives the potential of a dipole (current density) at any position and any orientation in an infinite medium (Yao, 2000b).

$$\Phi_\infty = \frac{1}{4\pi\sigma}\vec{P}\cdot\nabla_{r_0}\left(\frac{1}{r_p}\right) = \frac{1}{4\pi\sigma}\left\{P_r\frac{\partial}{\partial r_0} + P_\phi\frac{1}{r_0\sin\theta_0}\frac{\partial}{\partial\phi_0} + P_\theta\frac{1}{r_0}\frac{\partial}{\partial\theta_0}\right\}\frac{1}{r_p} \tag{4.48}$$

where $\vec{P}$ is decomposed into three components (P_r, P_θ, P_ϕ) according to a spherical coordinate system; the definitions of r_p, r_0 are the same as in Section 4.2.

For a point current source density or point charge in an infinite medium, there is

$$\Phi_\infty = \frac{1}{4\pi\sigma}\frac{I}{r_p} \tag{4.49}$$

where I is the unipolar source (equivalent charge) intensity (current source density). Using the associated Legendre function, we can get

$$\frac{1}{r_p} = \sum_{l=0}^{\infty}\sum_{m=0}^{l}\frac{r_0^l}{r^{l+1}}\left(2-\delta_m^0\right)\frac{(l-m)!}{(l+m)!}P_l^m(\cos\theta_0)P_l^m(\cos\theta)\cos m(\phi-\phi_0) \tag{4.50}$$

$$\frac{\partial P_l^m}{\partial \theta_0} = -\frac{1}{2}\left[(l-m+1)(l+m)P_l^{m-1}(\cos\theta_0) - P_l^{m+1}(\cos\theta_0)\right],$$
$$\frac{\partial P_l}{\partial \theta_0} = P_l^1(\cos\theta_0) \tag{4.51}$$

Among them, (r_0, θ_0, ϕ_0), (r, θ, ϕ) are the standard spherical coordinates of the source point $\vec{r}_0$ and the field point $\vec{r}$, r_p is the distance from the source point to the field point, and P_n^m is the associated Legendre function. Using equations (4.50 and 4.51), equation (4.48) can be written as

$$\Phi_\infty = \sum_{l=1}^{\infty}\sum_{m=0}^{l} \frac{K_l^m(\vec{P})}{r^{l+1}}\left(Z_l^m \cos m\phi + Y_l^m \sin m\phi\right)P_l^m(\cos\theta)$$
$$= \sum_{l=1}^{\infty}\sum_{m=0}^{l} \frac{K_l^m(\vec{P})}{r^{l+1}} S_l^m(\vec{P}, \theta, \phi) \tag{4.52}$$

where

$$S_l^m(\vec{P}, \theta, \phi) = \left(Z_l^m \cos m\phi + Y_l^m \sin m\phi\right)P_l^m(\cos\theta) \tag{4.53}$$

$$K_l^m(\vec{P}) = \frac{r_0^{l-1}}{4\pi\sigma}\left(2 - \delta_m^0\right)\frac{(l-m)!}{(l+m)!} \tag{4.54}$$

$$Z_l^m = lP_r P_l^m(\cos\theta_0)\cos m\phi_0 + \frac{mP_\phi}{\sin\theta_0} P_l^m(\cos\theta_0)\sin m\phi_0$$
$$-\frac{P_\theta}{2}\left((l-m+1)(l+m)P_l^{m-1}(\cos\theta_0) - P_l^{m+1}(\cos\theta_0)\right)\cos m\phi_0 \tag{4.55}$$

$$Y_l^m = lp_r P_l^m(\cos\theta_0)\sin m\phi_0 + \frac{mP_\phi}{\sin\theta_0} P_l^m(\cos\theta_0)\cos m\phi_0$$
$$-\frac{P_\theta}{2}\left((l-m+1)(l+m)P_l^{m-1}(\cos\theta_0) - P_l^{m+1}(\cos\theta_0)\right)\sin m\phi_0 \tag{4.56}$$

When both l and m are large, $\frac{(l-m)!}{(l+m)!}$ decreases quickly and P_l^m increases quickly. To prevent overflow in the calculation, we can consider combining $\frac{(l-m)!}{(l+m)!}$ with P_l^m, that is, using the normalized associate Legendre function $\tilde{P}_l^m$ instead of the associate Legendre function P_l^m. That is to define $\tilde{P}_l^m = \sqrt{\frac{(l-m)!}{(l+m)!}}P_l^m(x)$, then

$$(l-m)\tilde{P}_l^m = x(2l-1)\sqrt{\frac{l-m}{l+m}}\tilde{P}_{l-1}^m - (l+m-1)\sqrt{\frac{(l-m-1)(l-m)}{(l+m-1)(l+m)}}\tilde{P}_{l-2}^m$$

$$\tilde{P}_m^m = (-1)^m\sqrt{\frac{1}{(2m)!}}(2m-1)!!\left(1-x^2\right)^{m/2}, \quad \tilde{P}_{m+1}^m = x\sqrt{2m+1}\tilde{P}_m^m$$

Rewrite the previous formula to get

$$\tilde{S}_l^m\left(\vec{P},\theta,\phi\right) = \left(\tilde{Z}_l^m \cos m\phi + \tilde{Y}_l^m \sin m\phi\right)\tilde{P}_l^m(\cos\theta) \tag{4.57}$$

$$\tilde{K}_l^m\left(\vec{P}\right) = \frac{r_0^{l-1}}{4\pi\sigma}\left(2-\delta_m^0\right) \tag{4.58}$$

$$\begin{aligned}\tilde{Z}_l^m &= lP_r\tilde{P}_l^m(\cos\theta_0)\cos m\phi_0 - \frac{mP_\phi}{\sin\theta_0}\tilde{P}_l^m(\cos\theta_0)\sin m\phi_0 \\ &-\frac{P_\theta}{2}\left(\sqrt{(l-m+1)(l+m)}\tilde{P}_l^{m-1}(\cos\theta_0) - \sqrt{(l-m)(l+m+1)}\tilde{P}_l^{m+1}(\cos\theta_0)\right)\cos m\phi_0\end{aligned} \tag{4.59}$$

$$\begin{aligned}\tilde{Y}_l^m &= lP_r\tilde{P}_l^m(\cos\theta_0)\sin m\phi_0 - \frac{mP_\phi}{\sin\theta_0}\tilde{P}_l^m(\cos\theta_0)\cos m\phi_0 \\ &-\frac{P_\theta}{2}\left(\sqrt{(l-m+1)(l+m)}\tilde{P}_l^{m-1}(\cos\theta_0) - \sqrt{(l-m)(l+m+1)}\tilde{P}_l^{m+1}(\cos\theta_0)\right)\sin m\phi_0\end{aligned} \tag{4.60}$$

$\tilde{Z}_l^m$ and $\tilde{Y}_l^m$ are called the normalized spherical harmonic coefficients. When $m \le 0$, use the formula $P_l^{-m}(x) = (-1)^m\frac{(l-m)!}{(l+m)!}P_l^m(x)$ to transform the factor $\sqrt{(l-m+1)(l+m)}\tilde{P}_l^{m-1}(\cos\theta_0)$ of equations (4.58) and (4.59) into $(-1)^{1-m}\sqrt{(l-m+1)(l+m)}\tilde{P}_l^{1-m}(\cos\theta_0)$. Then use equation (4.51) to transform equation (4.49) into

$$\begin{aligned}\Phi_\infty &= \sum_{l=1}^{\infty}\sum_{m=0}^{l}\frac{K_l^m(I)}{r^{l+1}}\left(\cos m\phi_0\cos m\phi + \sin m\phi_0\sin m\phi\right)P_l^m(\cos\theta_0)P_l^m(\cos\theta) \\ &= \sum_{l=1}^{\infty}\sum_{m=0}^{l}\frac{K_l^m(I)}{r^{l+1}}S_l^m(I,\theta,\phi)\end{aligned} \tag{4.61}$$

where

$$S_l^m(I,\theta,\phi) = (\cos m\phi_0 \cos m\phi + \sin m\phi_0 \sin m\phi) P_l^m(\cos\theta_0) P_l^m(\cos\theta) \quad (4.62)$$

$$K_l^m(I) = \frac{Ir_0^l}{4\pi\sigma}\left(2-\delta_m^0\right)\frac{(l-m)!}{(l+m)!} \quad (4.63)$$

Do a similar normalization process to get

$$\Phi_\infty = \sum_{l=1}^{\infty}\sum_{m=0}^{l} \frac{\tilde{K}_l^m(I)}{r^{l+1}} \tilde{S}_l^m(I,\theta,\phi) \quad (4.64)$$

where

$$\tilde{S}_l^m(I,\theta,\phi) = (\cos m\phi_0 \cos m\phi + \sin m\phi_0 \sin m\phi) \tilde{P}_l^m(\cos\theta_0) \tilde{P}_l^m(\cos\theta) \quad (4.65)$$

$$\tilde{K}_l^m(I) = \frac{Ir_0^l}{4\pi\sigma}\left(2-\delta_m^0\right) \quad (4.66)$$

Apparently, the formulas are the same in the form before and after normalization. So for the sake of simplicity, we will no longer make a distinction between notations in the following derivation.

$S_l^m(\vec{P},\theta,\phi)$ in equation (4.57) is determined by the position of the dipole and the dipole moment, and has nothing to do with the radius r. $K_l^m(\vec{P})$ in equation (4.58) is a constant, and its magnitude is completely determined by the dipole. The same is true for a monopolar (point) current source density. Based on the similarities between (4.61) and (4.52), the common form below is adopted to simplify the following derivation:

$$\Phi_\infty = \sum_{l,m} \frac{K_l^m}{r^{l+1}} S_l^m(\theta,\phi) \quad (4.67)$$

The above two formulas are conceived as two special cases of this equation. In other words, for a dipole, $K_l^m = K_l^m(\vec{P})$, $S_l^m(\theta,\phi) = S_l^m(\vec{P},\theta,\phi)$, while for a point current source, $K_l^m = K_l^m(I)$, $S_l^m(\theta,\phi) = S_l^m(I,\theta,\phi)$. We will no longer list them in detail during the following derivation. Here, to further simplify the notations, we can also let $S_l^m = K_l^m S_l^m$, and the common formula is turned into

$$\Phi_\infty = \sum_{l,m} \frac{1}{r^{l+1}} S_l^m(\theta,\phi) \quad (4.68)$$

According to this equation, $K_l^m = 1$ in the subsequent equation (4.69). Of course, the substantive results are the same as those achieved based on equation (4.67) with K_l^m assigned a value of equations (4.58) and (4.66).

The boundary conditions require that the potential should be continuous across the boundary, as shown in equation (4.73) below. A concentric multilayer sphere $S_l^m(\theta,\phi)$ can be selected as a common function for all the regions.

4.5.2 Potential in a Homogeneous Single-Layer Sphere

Suppose the potential of a single sphere is represented by Φ_1, then

$$\Phi_1 = \sum_{l,m}\left[A_l^m \cdot r^l + K_l^m \cdot \frac{1}{r^{l+1}}\right]\cdot S_l^m \tag{4.69}$$

According to the boundary condition at $r=R$ for the scalp surface potential, the normal current is zero, and we can obtain

$$lA_l^m \cdot R^{l-1} - (l+1)K_l^m \cdot \frac{1}{R^{l+2}} = 0$$

$$A_l^m = \frac{l+1}{l}\cdot\frac{K_l^m}{R^{2l+1}} \tag{4.69a}$$

Substituting (4.69a) into the potential formula (4.69), and taking $r=R$, we can get

$$\Phi_1 = \sum_{l,m}\frac{2l+1}{lR^{l+1}}K_l^m \cdot S_l^m \tag{4.70}$$

Of course, the single-layer spherical model already has the closed solution as introduced in Section 4.2. The solution of this spherical harmonic series does not have any practical significance, but it is helpful to consider the relations among the one-, two-, and multilayer models as follows.

4.5.3 Potential in a Concentric Two-Layer Sphere

Assuming that the potentials of each layer from the inside to the outside are expressed with Φ_1 and Φ_2, respectively, then they can be written as follows:

$$\Phi_1 = \sum_{l,m}\left[A_l^m \cdot r^l + K_l^m \cdot \frac{1}{r^{l+1}}\right]\cdot S_l^m \tag{4.71}$$

$$\Phi_2 = \sum_{l,m}\left[C_l^m \cdot r^l + D_l^m \cdot \frac{1}{r^{l+1}}\right]\cdot S_l^m \tag{4.72}$$

The corresponding boundary conditions are

$$\Phi_1 = \Phi_2 \,|_{r=r_1} \tag{4.73}$$

$$\sigma_1 \frac{\partial \Phi_1}{\partial r} = \sigma_2 \frac{\partial \Phi_2}{\partial r}\bigg|_{r=r_1} \tag{4.74}$$

$$\frac{\partial \Phi_2}{\partial r}\bigg|_{r=r_2} = 0 \tag{4.75}$$

Substitute (4.71)–(4.72) into (4.73)–(4.75) to get

$$\begin{aligned} lC_l^m \cdot r_2^{l-1} - (l+1)D_l^m \cdot \frac{1}{r_2^{l+2}} = 0 \\ C_l^m = \frac{l+1}{l} \frac{D_l^m}{r_2^{2l+1}} \end{aligned} \tag{4.76}$$

Based on equation (4.74), we can have

$$\begin{aligned} l\sigma_2 C_l^m \cdot r_1^{l-1} - \sigma_2(l+1) D_l^m \cdot \frac{1}{r_1^{l+2}} &= l\sigma_1 A_l^m \cdot r_1^{l-1} - \sigma_1 (l+1) K_l^m \cdot \frac{1}{r_1^{l+2}} \\ &= D_l^m \left((l+1)\sigma_2 \cdot \frac{r_1^{l-1}}{r_2^{2l+1}} - \sigma_2(l+1) \cdot \frac{1}{r_1^{l+2}} \right) \\ &= D_l^m \sigma_2 (l+1) \cdot r_1^{l-1} \left(\frac{1}{r_2^{2l+1}} - \frac{1}{r_1^{2l+1}} \right) \end{aligned} \tag{4.77}$$

$$D_l^m \sigma_2 \left(\frac{1}{r_2^{2l+1}} - \frac{1}{r_1^{2l+1}} \right) = A_l^m \sigma_1 \frac{l}{l+1} - K_l^m \sigma_1 \cdot \frac{1}{r_1^{2l+1}} \tag{4.78}$$

Let $\sigma_{21} = \dfrac{\sigma_2}{\sigma_1}$

$$D_l^m \sigma_{21} \left(\frac{1}{r_2^{2l+1}} - \frac{1}{r_1^{2l+1}} \right) = A_l^m \frac{l}{l+1} - K_l^m \frac{1}{r_1^{2l+1}} \tag{4.79}$$

According to equation (4.73)

$$A_l^m \cdot r_1^l + K_l^m \cdot \frac{1}{r_1^{l+1}} = C_l^m \cdot r_1^l + D_l^m \cdot \frac{1}{r_1^{l+1}} \tag{4.80}$$

Substitute (4.76) in (4.80) to get

$$A_l^m \cdot r_1^{\,l} + K_l^m \cdot \frac{1}{r_1^{\,l+1}} = D_l^m \left(\frac{l+1}{l} \cdot \frac{r_1^{\,l}}{r_2^{\,2l+1}} + \frac{1}{r_1^{\,l+1}} \right) = D_l^m \frac{1}{r_1^{\,l+1}} \left(1 + \frac{l+1}{l} \left(r_1 / r_2 \right)^{2l+1} \right)$$

$$A_l^m \cdot r_1^{\,l} + K_l^m \cdot \frac{1}{r_1^{\,l+1}} = D_l^m \frac{1}{r_1^{\,l+1}} \left(1 + \frac{l+1}{l} \left(r_1 / r_2 \right)^{2l+1} \right)$$

$$A_l^m \cdot \frac{l}{l+1} + K_l^m \cdot \frac{1}{r_1^{\,2l+1}} \frac{l}{l+1} = D_l^m \left(\frac{l}{l+1} \frac{1}{r_1^{\,2l+1}} + \frac{1}{r_2^{\,2l+1}} \right) \tag{4.81}$$

Combine (4.79) with (4.81) to get

$$K_l^m \cdot \frac{1}{r_1^{\,2l+1}} \frac{2l+1}{l+1} = D_l^m \left(\frac{l}{l+1} \frac{1}{r_1^{\,2l+1}} + \frac{1}{r_2^{\,2l+1}} \right) - D_l^m \sigma_{21} \left(\frac{1}{r_2^{\,2l+1}} - \frac{1}{r_1^{\,2l+1}} \right)$$

$$D_l^m \left(\frac{l}{l+1} \frac{1}{r_1^{\,2l+1}} + \frac{1}{r_2^{\,2l+1}} \right) - D_l^m \sigma_{21} \left(\frac{1}{r_2^{\,2l+1}} - \frac{1}{r_1^{\,2l+1}} \right) = K_l^m \cdot \frac{1}{r_1^{\,2l+1}} \frac{2l+1}{l+1}$$

$$D_l^m = K_l^m \cdot \frac{1}{r_1^{\,2l+1}} \frac{2l+1}{l+1} \Big/ \left\{ \left(\frac{l}{l+1} \frac{1}{r_1^{\,2l+1}} + \frac{1}{r_2^{\,2l+1}} \right) - \sigma_{21} \left(\frac{1}{r_2^{\,2l+1}} - \frac{1}{r_1^{\,2l+1}} \right) \right\} \tag{4.82}$$

$$= K_l^m (2l+1) \Big/ \left\{ \left(l + (l+1) \frac{r_1^{\,2l+1}}{r_2^{\,2l+1}} \right) - \sigma_{21} \left(\frac{1}{r_2^{\,2l+1}} - \frac{1}{r_1^{\,2l+1}} \right) \right\}$$

$$D_l^m = K_l^m (2l+1) \Big/ \left\{ \left(l + (l+1) \frac{r_1^{\,2l+1}}{r_2^{\,2l+1}} \right) - \sigma_{21} \left(\frac{1}{r_2^{\,2l+1}} - \frac{1}{r_1^{\,2l+1}} \right) \right\}$$

Substitute (4.81) in (4.78) to get

$$K_l^m (2l+1) \Big/ \left\{ \left(l + (l+1) \frac{r_1^{\,2l+1}}{r_2^{\,2l+1}} \right) - \sigma_{21} \left(\frac{1}{r_2^{\,2l+1}} - \frac{1}{r_1^{\,2l+1}} \right) \right\} \sigma_{21} \left(\frac{1}{r_2^{\,2l+1}} - \frac{1}{r_1^{\,2l+1}} \right) + K_l^m \frac{1}{r_1^{\,2l+1}}$$

$$= \mathrm{A}_l^m \frac{1}{l+1}$$

Further sort out to get

$$A_l^m = K_l^m \frac{l+1}{l} \left[(2l+1) \Big/ B\sigma_{21} \left(\frac{1}{r_2^{\,2l+1}} - \frac{1}{r_1^{\,2l+1}} \right) + \frac{1}{r_1^{\,2l+1}} \right] \tag{4.83}$$

Where $B = \left\{ \left(l + (l+1) \frac{r_1^{\,2l+1}}{r_2^{\,2l+1}} \right) - \sigma_{21} \left(\frac{1}{r_2^{\,2l+1}} - \frac{1}{r_1^{\,2l+1}} \right) \right\}$

$$C_l^m = \frac{l+1}{l r_2^{2l+1}} K_l^m (2l+1) \Bigg/ \left\{ \left(l + (l+1) \frac{r_1^{2l+1}}{r_2^{2l+1}} \right) - \sigma_{21} \left(\frac{1}{r_2^{2l+1}} - \frac{1}{r_1^{2l+1}} \right) \right\} \tag{4.84}$$

$$D_l^m = K_l^m (2l+1) \Bigg/ \left\{ \left(l + (l+1) \frac{r_1^{2l+1}}{r_2^{2l+1}} \right) - \sigma_{21} \left(\frac{1}{r_2^{2l+1}} - \frac{1}{r_1^{2l+1}} \right) \right\} \tag{4.85}$$

The potential of any position in the two-layer model can be calculated by directly substituting the corresponding coefficients into equations (4.71) and (4.72). Finally, for the scalp surface, $r_2 = R$, the potential formula can be obtained:

$$\Phi_2 = \sum_{l,m} \left[C_l^m \cdot R^l + D_l^m \cdot \frac{1}{R^{l+1}} \right] \cdot S_l^m \tag{4.86}$$

with

$$C_l^m \cdot R^l + D_l^m \cdot \frac{1}{R^{l+1}} = \left[\frac{1+1}{l R^{l+1}} K_l^m (2l+1) + \frac{1}{R^{l+1}} K_l^m (2l+1) \right] \Bigg/ \left\{ \left(l + (l+1) \right) \frac{r_1^{2l+1}}{R^{2l+1}} - \sigma_{21} \left(\frac{1}{R^{2l+1}} - \frac{1}{r_1^{2l+1}} \right) \right\}$$

$$= \frac{21+1}{l R^{l+1}} K_l^m (2l+1) \Bigg/ \left\{ \left(l + (l+1) \right) \frac{r_1^{2l+1}}{R^{2l+1}} - \sigma_{21} \frac{1}{R^{2l+1}} \left(1 - \frac{R^{2l+1}}{r_1^{2l+1}} \right) \right\} \tag{4.87}$$

Equations (4.86) and (4.87) are the final solutions of the two-layer model.

4.5.4 Potential in a Concentric Three-Layer Sphere

Similarly, a solution to the scalp surface potential can be developed for any dipole in the three-layer sphere model (Yao, 2000b) and for any current source density (equivalent charge source) in the three-layer sphere model (Yao and He, 1998; He et al., 2002). After a similar derivation, we have the following equation (4.88) for the three-layer sphere model:

$$\Phi(i) = \sum_{l,m} \frac{W_l(i)}{r^{l+1}} S_l^m(\theta, \phi) \tag{4.88}$$

$$a \ge r > r_0,\ i = 1; \quad b \ge r \ge a,\ i = 2; \quad c \ge r \ge b,\ i = 3$$

where $i = 1, 2, 3$, respectively, represents the inner-sphere region (brain region with radius a), the skull (outer radius b), and the scalp region (radius c), r_0 is the maximum radius of the source region, and

$$\begin{aligned} W_l(1) &= A_l r^{2l+1} + 1 \\ W_l(2) &= B_l\left(r^{2l+1} + \gamma\right) \\ W_l(3) &= E_l\left(r^{2l+1} + \chi\right) \end{aligned} \tag{4.89}$$

$$A_l = \frac{(2l+1)\left(1+\gamma a^{-(2l+1)}\right)}{l(1-s)a^{2l+1} + \gamma\left(l + s(l+1)\right)} - a^{-(2l+1)} \tag{4.90}$$

$$B_l = \frac{2l+1}{l(1-s)a^{2l+1} + \gamma\left(l + s(l+1)\right)} \tag{4.91}$$

$$E_l = B_l\left(b^{2l+1} + \gamma\right)\frac{l+1}{(l+1)b^{2l+1} + lc^{2l+1}} \tag{4.92}$$

$$s = \frac{\sigma_2}{\sigma_1} = \frac{\sigma_2}{\sigma_3},\quad f = b/c,\quad \chi = \frac{l}{l+1}c^{2l+1} \tag{4.93}$$

$$\alpha = f^{2l+1}(1-s) - \left(1 + \frac{ls}{1+l}\right),\quad \beta = f^{-(2l+1)}(1-s) - \left(1 + \frac{l+1}{l}s\right),\quad \gamma = \alpha / \beta \tag{4.94}$$

Especially, when $r = a$, $\Phi(1)$ = cortical surface potential, and when $r = c$, $\Phi(3)$ = scalp surface potential.

According to a similar derivation process, a potential calculation formula can be derived for any multilayer model. Appendix B presents a detailed derivation for the four-layer model and proves that starting from the four-layer model, the formula of the three-layer model can be derived by assuming that the third and fourth layers are the same as each other in radius or conductivity. Thus, it can be seen that the formula and software for the four-layer sphere model can be simultaneously applied to the calculation of the 1~3-layer model only by a simple setting of the parameters, which can be achieved by amalgamating the radii or eliminating the difference of conductivity.

4.6 DIPOLE IN MULTILAYER NONCONCENTRIC SPHERE

This section mainly presents the research of Cuffin (1991) on the nonconcentric sphere model, with a basic idea analogous to the above, i.e., a spherical harmonic series expansion and each component fitting the boundary conditions. However, different spherical centers are involved here, so it is necessary to reveal the relationship between spherical harmonic series derived with different sphere centers, thus leading to a more complicated operation than the above. In the original paper, Cuffin (1991) investigated three cases: the shift of the innermost sphere, the shift of the two inner spheres, and the nonconcentric bulb. Here, only the first case is introduced as shown in Figure 4.4.

According to equation (B.2) in Yao (2000a), we have

$$\Phi_\infty = \begin{cases} K\dfrac{r\cos\phi - r_0\cos\phi_0}{r^3}\displaystyle\sum_{l=1}^{\infty}\left(\frac{r_0}{r}\right)^{l-1} P_l'(\cos\varphi), & r > r_0 \\ K\left(r\cos\phi - r_0\cos\phi_0\right)\displaystyle\sum_{l=1}^{\infty}\frac{r^{l-1}}{r_0^{l+2}} P_l'(\cos\varphi), & r < r_0 \end{cases} \tag{4.95}$$

where φ is the angle between vector $\vec{r}$ and $\vec{r}_0$, ϕ/ϕ_0 is the angle between vector $\vec{P}$ and $\vec{r}/\vec{r}_0$.

As shown in Figure 4.3, consider a tangential dipole located on the z-axis, that is, $P_x\vec{e}_x$, then $\phi_0 = \dfrac{\pi}{2}$, $\vec{p}\cdot\vec{r} = pr\cos\phi = \cos(\pi/2-\theta) = pr\sin\theta$, then the formula (4.95) becomes

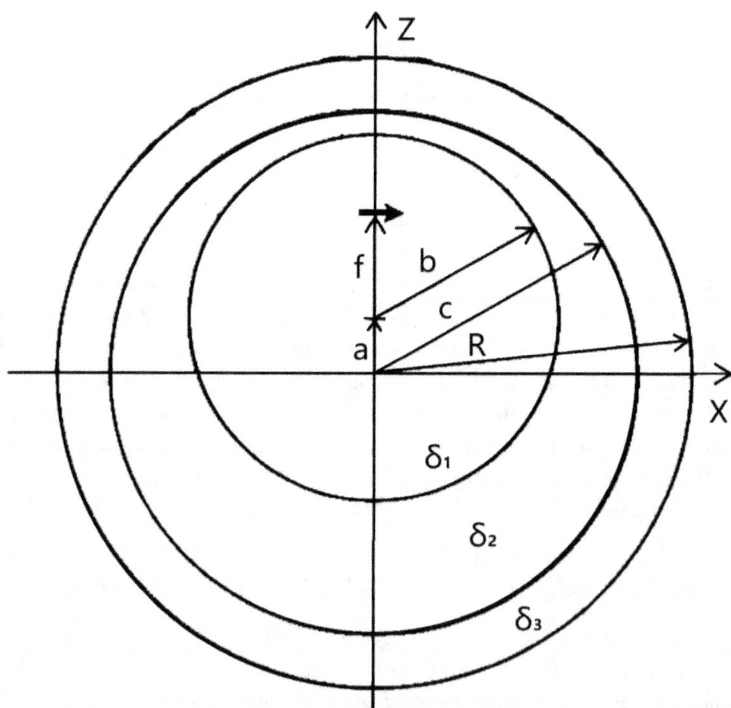

FIGURE 4.3 Multilayer nonconcentric spherical model. The case for the shift of the innermost sphere, where *a* is the eccentricity of the sources located in the inner layer of the three figures.

$$\Phi = \begin{cases} K \sin\theta \sum_{l=1}^{\infty} \frac{r_0^{l-1}}{r^{l+1}} P_l'(\cos\theta), & r > r_0 \\ K \sin\theta \sum_{l=1}^{\infty} \frac{r^l}{r_0^{l+2}} P_l'(\cos\theta), & r < r_0 \end{cases} = \begin{cases} K \sum_{l=1}^{\infty} \frac{r_0^{l-1}}{r^{l+1}} P_l^1(\cos\theta), & r > r_0 \\ K \sum_{l=1}^{\infty} \frac{r^l}{r_0^{l+2}} P_l^1(\cos\theta), & r < r_0 \end{cases} \tag{4.96}$$

Based on this equation, the general solution of the three regions in the nonconcentric sphere can be written as follows (Cuffin, 1991):

$$\Phi_1 = \frac{P_x \cos\phi}{4\pi\sigma_1} \sum_{n=1}^{\infty} P_n^1(\cos\theta') \left(\frac{f^{n-1}}{r'^{n+1}} + B_n r'^n \right) \tag{4.97}$$

$$\Phi_2 = \frac{P_x \cos\phi}{4\pi\sigma_2} \sum_{n=1}^{\infty} P_n^1(\cos\theta') \left(\frac{C_n}{r'^{n+1}} + D_n r'^n \right) \tag{4.98}$$

$$\Phi_3 = \frac{P_x \cos\phi}{4\pi\sigma_3} \sum_{s=1}^{\infty} P_s^1(\cos\theta) \left(\frac{E_s}{r^{s+1}} + F_s r^s \right) \tag{4.99}$$

where P_n^1, P_s^1 is the associated Legendre polynomial. The coordinates with "priming" are used for areas 1 and 2, and the coordinates without "priming" are used for area 3. The origin of the coordinates with "priming" is located at $z = a$. Due to the geometric characteristics of the azimuth angle, $\phi = \phi'$. To meet the boundary conditions at $r = c$, it is necessary to express Φ_2 in unprimed coordinates. According to the transformation formula by Morse and Feshbach (1953), Φ_2 becomes

$$\Phi_2 = \frac{P_x \cos\phi}{4\pi\sigma_2} \sum_{n=1}^{\infty} \left[\frac{C_n}{a^{n+1}} \sum_{s=n}^{\infty} \left(\frac{a}{r}\right)^{s+1} \frac{P_s^1(\cos\theta)(s-1)!}{(n-1)!(s-n)!} + D_n a^n \sum_{s=1}^{n} \left(\frac{r}{a}\right)^s \frac{(-1)^{n-s}(n+1)!P_s^1(\cos\theta)}{(n-s)!(1+s)!} \right] \tag{4.100}$$

Exchange the order of summation to get

$$\Phi_2 = \frac{P_x \cos\phi}{4\pi\sigma_2} \sum_{s=1}^{\infty} P_s^1(\cos\theta) \left[\left(\frac{a}{r}\right)^{s+1} (s-1)! \sum_{n=1}^{s} \frac{C_n}{a^{n+1}(n-1)!(s-n)!} + \left(\frac{r}{a}\right)^s \frac{1}{(s+1)!} \sum_{n=s}^{\infty} \frac{(-1)^{n-s}(n+1)!a^n D_n}{(n-s)!} \right] \tag{4.101}$$

Coefficients C_n, B_n, etc., can be obtained by using the following boundary:

$$\begin{aligned}
&\Phi_1 = \Phi_2, r' = b \\
&\sigma_1 \frac{\partial \Phi_1}{\partial r'} = \sigma_2 \frac{\partial \Phi_2}{\partial r'}, r' = b \\
&\Phi_2 = \Phi_3, r = c \\
&\sigma_2 \frac{\partial \Phi_2}{\partial r} = \sigma_3 \frac{\partial \Phi_3}{\partial r}, r = c \\
&\frac{\partial \Phi_3}{\partial r} = 0, r = R
\end{aligned} \tag{4.102}$$

Solve (4.102) to get the equation of D_n:

$$\begin{aligned}
&\sum_{n=1}^{s} \frac{n(1-k_1)b^{2n+1}D_n}{(nk_1+n+1)a^{n+1}(n-1)!(s-n)!} - \frac{s^2}{[(s+1)!]^2}\left(\frac{c}{a}\right)^{2s+1}\frac{[C_{1s}+(s+1)C_{2s}]}{(C_{1s}-sC_{2s})} \\
&\sum_{n=s}^{\infty} \frac{a^n(-1)^{n-s}(n+1)!D_n}{(n-s)!} = -\sum_{n=1}^{s} \frac{(2n+1)f^{n-1}}{(nk_1+n+1)a^{n+1}(n-1)!(s-n)!}
\end{aligned} \tag{4.103}$$

where

$$C_{1s} = k_2\left[sR^{2s+1} + (s+1)c^{2s+1}\right], \quad C_{2s} = R^{2s+1} - c^{2s+1}, \quad k_1 = \frac{\sigma_1}{\sigma_2}, \quad k_2 = \frac{\sigma_2}{\sigma_3}$$

Use D_n to get the express of Φ_3 (Cuffin, 1991)

$$\Phi_3 = \frac{P_x \cos\phi}{4\pi\sigma_3} \sum_{s=1}^{\infty} \frac{(2s+1)^2 R^s a^{s+1} s! P_s^1(\cos\theta)}{s^2\left[C_{1s} + (s+1)C_{2s}\right]} \sum_{n=1}^{s} \frac{(2n+1)f^{n-1} + n(1-k_1)b^{2n+1}D_n}{(nk_1+n+1)a^{n+1}(n-1)!(s-n)!} \tag{4.104}$$

In practice, the infinite series in formula (4.103) needs to be truncated at S_{max}. D_n is calculated by the truncation S_{max}. Substitute S_{max} and D_n into (4.104) to calculate Φ_3 by increasing S_{max} until there is no significant change in Φ_3.

The potential generated from P_z can be deduced in the same way, and its general solutions in the three regions are

$$\Phi_1 = \frac{P_z}{4\pi\sigma_1} \sum_{n=1}^{\infty} P_n^0(\cos\theta')\left(\frac{f^{n-1}}{r'^{n+1}} + B_n r'^n\right) \tag{4.105}$$

$$\Phi_2 = \frac{P_z}{4\pi\sigma_2} \sum_{n=1}^{\infty} P_n^0(\cos\theta')\left(\frac{C_n}{r'^{n+1}} + D_n r'^n\right) \tag{4.106}$$

$$\Phi_3 = \frac{P_z}{4\pi\sigma_3}\sum_{s=1}^{\infty} P_s^0(\cos\theta)\left(\frac{E_s}{r^{s+1}} + F_s r^s\right) \tag{4.107}$$

The equation of D_n is (Cuffin, 1991)

$$\sum_{n=1}^{s}\frac{(1-k_1)b^{2n+1}D_n}{(nk_1+n+1)a^{n+1}(n-1)!(s-n)!} - \frac{s(s+1)}{[(s+1)!]^2}\left(\frac{c}{a}\right)^{2s+1}\frac{[C_{1s}+(s+1)C_{2s}]}{(C_{1s}-sC_{2s})}$$
$$\cdot\sum_{n=s}^{\infty}\frac{a^n(-1)^{n-s}n!D_n}{(n-s)!} = -\sum_{n=1}^{s}\frac{(2n+1)f^{n-1}}{(nk_1+n+1)a^{n+1}(n-1)!(s-n)!} \tag{4.108}$$

The potential on the surface of the model is

$$\Phi_3 = \frac{P_z}{4\pi\sigma_3}$$
$$\sum_{s=1}^{\infty}\frac{(2s+1)^2 R^s a^{s+1}(s-1)!P_s^0(\cos\theta)}{[C_{1s}+(s+1)C_{2s}]}\sum_{n=1}^{s}\frac{(2n+1)f^{n-1}+(1-k_1)b^{2n+1}D_n}{(nk_1+n+1)a^{n+1}(n-1)!(s-n)!} \tag{4.109}$$

Cuffin (1991) also conducted a numerical analysis. The results showed that for this model, if the innermost sphere is shifted upward by 0.25, the peak potential of P_z will be 22% larger than that in the case of concentricity, and this amounts to a reduction of the skull thickness in the peak area; if it is shifted downward by 0.25, the peak value will drop by 12%, and this amounts to an increase of the skull thickness in the peak area; these two kinds of shifting have no obvious effect on the spatial distribution of potential; if the innermost sphere is shifted in the same way for P_x, the peak potential will increase by 20% and decrease by 12%, respectively, and there will be a slight effect on the spatial distribution of potential.

Finally, what is studied above is just about the dipole located on the shift axis. For a more general case, i.e., the dipole is located at any position with arbitrary orientation, a new coordinate system can be established with the new z-axis passing through the position of the dipole. Also, the new coordinate system can be rotated so that the dipole should only have two components, i.e., z and x, so that the above study can be adopted in the new coordinate system. A similar idea is shown in Section 4.4.

4.7 SPHERICAL CAP HARMONIC ANALYSIS

In the previous spherical harmonic analysis, the basic assumption is that the whole brain is observable, but actually, we cannot observe the whole brain, only part of the brain surface. In geophysics, a spherical cap harmonic analysis method has been developed in response to this situation (Haines, 1985; De Santis et al., 1990). For this, it is necessary to establish an orthogonal spherical harmonic function in the spherical cap area (Figure 4.4).

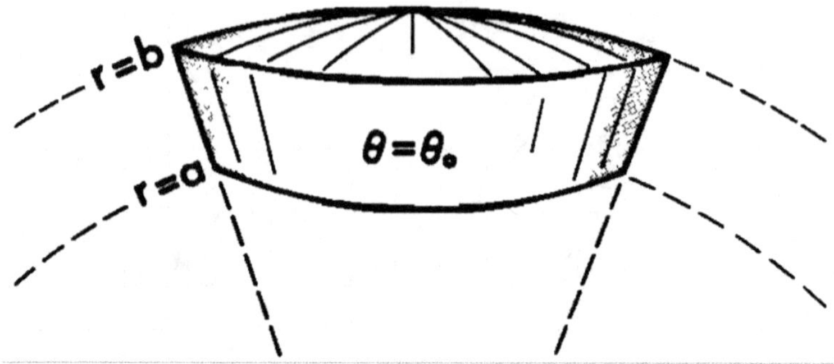

FIGURE 4.4 Schematic diagram of the spherical cap. θ is latitude, and r is the radial radius. The volume of the spherical cap analysis is composed of $r=a$, $r=b$ and $\theta=\theta_0$, =where θ_0 is the maximum half angle of the spherical cap. (From Haines (1985).)

The spherical coordinate solution of the Laplace equation in the spherical cap is

$$\Phi=\sum_{m=0}^{\infty}\sum_{k=m}^{\infty}a\left(\frac{a}{r}\right)^{n_k(m)+1}\left(g_k^m\cos m\phi+h_k^m\sin m\phi\right)P_{n_k(m)}^m(\cos\theta) \tag{4.110}$$

At the same time, it needs to meet the boundary conditions at the center of the spherical cap

$$\frac{\partial\Phi_n^m(r,0,\phi)}{\partial\theta}=0,\quad m=0 \tag{4.111}$$

$$\Phi_n^m(r,0,\phi)=0,\quad m\neq 0 \tag{4.112}$$

and the boundary conditions at the edge of the spherical cap,

$$\frac{\partial\Phi_n^m(r,\theta_0,\phi)}{\partial\theta}=0 \tag{4.113}$$

or

$$\Phi_n^m(r,\theta_0,\phi)=0 \tag{4.114}$$

When only (4.113) needs to be satisfied, the value of potential on the boundary can be an arbitrary function, such as the observed value on the boundary; when (4.114) needs to be satisfied, the slope of potential on the boundary can be an arbitrary function. These two functions cannot be required to be zero at the same time.

For all m, equation (4.110) satisfies (4.111/4.112) and (4.113) or (4.114), so the degree n of Legendre functions $P_{n_k(m)}^m(\cos\theta)$ becomes a real number rather than an integer, denoted as $n_k(m)$, where k is the serial number, k and m are integers.

For the calculation of $n_k(m)$, as well as the computational details of $P^m_{n_k(m)}(\cos\theta)$, please refer to the original literature (Haines, 1985). Obviously, equation (4.110) makes it possible to fit the EEG observation area on the scalp surface. On that basis, by utilizing the filtering operation of the harmonic function, it is feasible to obtain a high-density EEG for various scalp surfaces, such as scalp surface Laplacian over the cap. Compared with the traditional spherical harmonic analysis method, the present method is more rigorous conceptually.

Similarly, if the observation area is a rectangle, such as intracranial observation or body surface ECG observation, a similar rectangular harmonic analysis method developed in geophysics can be adopted (Alldredge, 1981).

4.8 DIPOLE IN ANISOTROPIC MULTILAYER SPHERE AND ELLIPSOID

De Munck (1988) deduced the theoretical formula for the series expansion of the potential distribution of anisotropic multilayered concentric sphere and confocal ellipsoid. Zhou and Oosterom (1992) further proposed a robust numerical calculation method for the formulas, and Zhang (1995) proposed a fast algorithm. De Munck (1988) used a specific method to derive the Green function for current source density first, then found the gradient of the Green function with respect to the source point coordinate, and finally derived a formula for the dipole based on the dot product with the dipole moment vector. This idea is much simpler than directly starting from the dipole formula. A similar practice has been illustrated in Section 4.3. Besides, Nieminen and Stenroos (2016) proposed an expression for the magnetic field in the multilayered concentric sphere and confocal ellipsoidal model.

Existing experimental results show that anisotropy exists in the skull, cerebral cortex, and white matter. The anisotropy rate of the skull and white matter is up to 10%, while the anisotropy rate of the cortical tissues is more than 2%. Therefore, the isotropic approximation may cause systematic errors in source localization. To study the impact of anisotropy on EEG and MEG, Haueisen et al. (2002) directly used the finite element method to analyze the impact of brain tissue anisotropy on EEG and MEG. The results show that anisotropy has little effect on the topographic map of the scalp surface, making it proper to speculate that it has little effect on source localization, but has an obvious effect on the magnitude of EEG and MEG. As a consequence, it may affect the estimated source intensity. Later, Güllmar et al. (2006) investigated the rabbit brain and found that if anisotropy is ignored, the source position may drift by <1.3 mm, with an average of 0.3 mm; the average dipole orientation error is up to 10%, while the average amplitude error is 29%. In general, anisotropy does not have a great effect on the localization error but has a significant impact on the orientation and amplitude. Thus, it is valuable to consider anisotropy for source estimation. Generally, the theoretical formulas are quite complex, so direct numerical analysis may be a more reasonable choice (Güllmar et al., 2010).

REFERENCES

Alldredge, L.R. 1981. Rectangular harmonic analysis applied to the geomagnetic field. *J Geophys Res-Sol Ea* 86(B4):3021–3026.

Arthur, R.M., D.B. Geselowitz. 1970. Effect of inhomogeneities on the apparent location and magnitude of a cardiac current dipole source. *IEEE Trans Biomed Eng* 17(2):141–146.

Brazier, M.A.B. 1949. The electrical fields at the surface of the head during sleep. *EEG Clin Neurosci* 1:195–204.

Brody, D.A., F.H. Terry, R.E. Ideker. 1973. Eccentric dipole in a spherical medium: Generalized expression for surface potentials. *IEEE Trans Biomed Eng* 20:141–143.

Cuffin, B.N. 1991. Eccentric spheres models of the head. *IEEE Trans Biomed Eng* 38(9):871–878.

DeMunck, J.C. 1988. The potential distribution in a layered anisotropic spheroidal volume. *J Appl Phys* 64(2):464–470.

DeSantis, A., O. Battelli, D.J. Kerridge. 1990. Spherical cap harmonic analysis applied to regional field modelling for Italy. *J Geomagn Geoelec* 42(9):1019–1036.

Frank, E. 1952. Electric potential produced by two point current sources in a homogeneous conducting sphere. *J Appl Phys* 23(11):1225–1228.

Geisler, C.D., G.L. Gerstein. 1961. The surface EEG in relation to its sources. *Clin Neurophysiol* 13:927–934.

Geselowitz, D.B. 1965. Two theorems concerning the quadrupole applicable to electrocardiography. *IEEE Trans Biomed Eng* 12:164–168.

Güllmar, D., J. Haueisen, J.R. Reichenbach. 2010. Influence of anisotropic electrical conductivity in white matter tissue on the EEG/MEG forward and inverse solution. *NeuroImage* 51(1):145–163.

Güllmar, D., J. Haueisen, M. Eiselt, et al. 2006. Influence of anisotropic conductivity on EEG source reconstruction: Investigations in a rabbit model. *IEEE Trans Biomed Eng* 53(9):1841–1850.

Haines, G.V. 1985. Spherical cap harmonic analysis. *J Geophys Res* 90(B3):2583–2591.

Haueisen, J., D.S. Tuch, C. Ramon, et al. 2002. The influence of brain tissue anisotropy on human EEG and MEG. *NeuroImage* 15:159–166.

He, B., D. Yao, J. Lian, et al. 2002. An equivalent current source model and Laplacian weighted minimum norm current estimates of brain electrical activity. *IEEE Trans Biomed Eng* 49(4):277–288.

Kavanagk, R.N., T.M. Darcey, D. Lehmann, et al. 1978. Evaluation of methods for three-dimensional localization of electrical sources in the human brain. *IEEE Trans Biomed Eng* 25(5):421–429.

Law, S.K., P.L. Nunez, R.S. Wijesinghe. 1993. High resolution EEG using spline generated surface laplacians on spherical and ellipsoidal surfaces. *IEEE Trans Biomed Eng* 40:145–153.

Morse, P.M., H. Feshbach. 1953. *Methods of Theoretical Physics*, 2nd ed., 2 Pts. New York: Mcgraw Hill Education.

Nieminen, J.O., M. Stenroos. 2016. The magnetic field inside a layered anisotropic spherical conductor due to internal sources. *J Appl Phys* 119(2):023901.

Paicer, P.L., A. Sances, S.J. Larson. 1967. Theoretical evaluation of cerebral evoked potentials. *Proceedings of the 20th ACEMB,* Boston, 9:14.

Rush, S., D.A. Driscoll. 1969. EEG electrode sensitivity: An application of reciprocity. *IEEE Trans Biomed Eng* 16:15–22.

Wikswo Jr, J.P., K.R. Swinney. 1984. A comparison of scalar multipole expansions. *J Appl Phys* 56(11):3039–3049.

Wikswo Jr, J.P., K.R. Swinney. 1985. Scalar multipole expansions and their dipole equivalents. *J Appl Phys* 57(9):4301–4308.

Wilson, F.N., R.H. Bayley. 1950. The electric field of an eccentric dipole in a homogeneous spherical conducting medium. *Circulation* 1:84–92.

Yao, D. 1998. Close solution of electric potential produced by 2n th-order multipole in a homogeneous conducting sphere. *Chin J Biomed Eng* 17(2):97–102 (in Chinese).

Yao, D. 2000a. Electric potential produced by a dipole in a homogeneous conducting sphere. *IEEE Trans Biomed Eng* 47(7):964–966.

Yao, D. 2000b. High-resolution EEG mappings: A spherical harmonic spectra theory and simulation results. *Clin Neurophysiol* 111:81–92.

Yao, D., B. He. 1998. The Laplacian weighted minimum estimate of three dimensional equivalent charge distribution in the brain. *Proceedings of 20th Annual International Conference on IEEE EMBS,* Hong Kong, China, vol. 4, pp. 2108–2111.

Zhang, Z. 1995. A fast method to compute surface potentials generated by dipoles within multilayer anisotropic spheres. *Phys Med Biol* 40(3):335–349.

Zhou, H., A.V. Oosterom. 1992. Computation of the potential distribution in a four-layer anisotropic concentric spherical volume conductor. *IEEE Trans Biomed Eng* 39(2):154–158.

5 Brain Electrical Field in Realistic Head Model

With the Green second theorem, a relation between sources and surface potential for the inhomogeneous volume conductor has developed 50 years ago (Geselowitz, 1967). It provided a way to tackle the electric problem of a realistic head model. In general, the forward problem of a realistic head volume conductor model needs to be solved by a numerical method so that inhomogeneous and anisotropic models can be processed. The following involves a brief review of the establishment of a multi-region head model in Section 5.1, a discussion of the finite element method (FEM) in Section 5.2, an introduction to the meshless method in Section 5.3, the boundary element method (BEM) in Section 5.4, and the finite difference method (FDM) and finite volume method (FVM) in Section 5.5. This chapter mainly focuses on the basic mathematical and physical principles of the relevant methods, whereas for the specific implementation and software please refer to the related literatures or software platforms.

In current practice, the potential of a realistic head model is mostly calculated by BEM and FEM, while FDM, FVM, and meshless methods are only used occasionally. These methods differ from one another in the expression of potential distribution. They are the same as one another in the requirement that the potential should satisfy Poisson's equation (active region), Laplace's equation (passive region), or its weak formulation, and meet the boundary conditions so that equations can be established and solved to obtain a potential solution.

5.1 HEAD MODEL

The volume conductor head models vary in complexity, such as the single-layer spherical model (Yao, 2000a), hemispherical model (Yao, 2017), concentric spheres model, non-concentric spheres model, confocal ellipsoid, realistic head model, etc. A numerical method must be used to obtain the solution of a realistic volume conductor. To apply these methods, the volume conductor or interface (for BEM) must be divided into discrete units. Generally speaking, the smaller the grid size is, the better the approximation is. However, a small grid size means a large number of grids, high computation complexity, and more detailed biological data are needed.

A common realistic head model is the multi-region model, in which the head is divided into many homogeneous parts with different conductivities. The multi-region head model is usually based on CT or MR images, and the related geometric errors may lead to errors in the source estimation. In this section, we will briefly discuss how to establish a multi-region head model and estimate regional conductivity. For conductivity measurement, please also refer to the discussions in Section 3.2.1.

 DOI: 10.1201/9781032639260-5

5.1.1 Multi-Region Head Model

The human head can be well approximated by a four- or three-region model. These regions correspond to the scalp, skull, CSF, and brain tissues, respectively, and each part can be approximated as a homogeneous conductor. The CSF under the skull is a thin conductive layer, and it is the difference between the four-region and three-region models. From the perspective of equivalent dipole localization, the three-region model (brain-skull-scalp, BSS) may provide a reasonable accuracy if the geometric shape of each region is close enough to the real shape. The geometry of the scalp can be obtained by a 3-D digitizer or CT, and the shape of the inner and outer surfaces of the skull can be detected through CT or MRI slice reconstruction. Owing to the sufficiently large contrast of bone CT images, it is technically easy to extract the surface of the skull from CT slices. Now some software can be used for boundary extraction and numerical model establishment. Therefore, the following content just provides a conceptual description.

5.1.1.1 Extraction of the Boundary between Cortex and Skull

As the skull looks dark on MRI slices, simple image threshold processing does not help to extract the skull boundary. So some more complex algorithms are required (Nielsen et al., 2018), such as a combination of mathematical morphology and threshold (Dogdas et al., 2005). Sometimes, manual operation is also needed. The base structure of the skull is complicated, but it can be replaced with a smooth boundary. The cortex boundary can be extracted in the same way. There are sulci in the cortex, and its boundary is irregularly shaped, but it can also be replaced with a smooth surface as an approximate boundary of the cortical region. In addition, if there is no suitable image data, the geometry of the skull can be roughly estimated according to the shape of the scalp surface (Haque et al., 1998). During this process, it is necessary to specify the coordinates system. In an EEG study, it is usually stipulated that the y-axis passes from right to left through the two points on the upper part of the left and right tragi, while the x-axis stretches forward from the back of the head through the nasion and is perpendicular to the y-axis. Markers for MRI/CT imaging can be placed at the nasion and the two tragis, to integrate the result of MRI/CT with that of the surface head digitizer.

5.1.1.2 3D Interpolation and Coordinate Transformation

In forward calculation, the number of nodes in each region is normally inconsistent with the pixels of CT/MR images, making it necessary to build a connection by interpolation. Considering that the conductivity of the skull is much lower than other parts, the accuracy of its geometry would affect the localization of the dipoles. A CT is more reliable than an MRI in skull model construction. Regarding the inverse problem of EEG, the number of nodes should be minimized without reducing the quality of the solution to optimize its solving efficiency. If the active area of the brain is approximately predicted in advance, the nodes can be appropriately concentrated on the possible source region to improve the accuracy of source estimation.

5.1.2 Regional Conductivity Estimation

Section 3.2 presents a discussion on conductivity measurement and offers some recent literatures. In EEG practice, the conductivity values listed in literatures are often used for a multi-regional head model, but the actual conductivity varies from person to person, even for a different state of one person. Inaccurate regional conductivity causes an estimate deviation of the dipole position. Homma et al. (1995) implanted a pair of electrodes in the brain of a patient with epilepsy to generate an artificial current dipole and then studied the effect of the conductivities of the scalp and skull on current dipole estimation. The adopted BSS model was based on CT images. Suppose that the scalp and the brain had the same conductivity, they studied the scalp–skull conductivity ratio in a range of 1:1 to 1:1/120. The result showed that the best dipole estimation is at 1:1/80, which is consistent with the recommendation of many literatures (for the debate on conductivity, please refer to Section 3.2).

Actually, the dipole estimation may not be accurate even if correct conductivity is adopted. This is because the geometry of the BSS head model is somewhat different from the realistic geometry of the subject's head, and this difference is bound to cause a deviation from the estimated dipole position. Musha and Okamoto (1999) introduced a method to optimize the conductivity in a well-constructed BSS head model.

Suppose that a current is injected into the head through the scalp electrode pair k, the potential distribution generated by this current on the scalp is denoted by Φ_k^{meas}. Besides, based on a BSS model constructed with the conductivity σ_{scalp}, σ_{skull} and σ_{brain}, the potential distribution Φ_k^{cal} generated by this current is calculated. For K different electrode pairs, we can thereby obtain the estimation values and measured values for K different potential distributions. The quadratic sum of the difference between the two kinds of values is denoted by

$$O\left(\sigma_{\text{scalp}},\sigma_{\text{skull}},\sigma_{\text{brain}}\right) \equiv \sum_{k=1}^{K} \left\| \Phi_k^{\text{meas}} - \Phi_k^{\text{cal}}\left(\sigma_{\text{scalp}},\sigma_{\text{skull}},\sigma_{\text{brain}}\right) \right\|^2 \tag{5.1}$$

Minimize O to get the optimal conductance value.

Considering the extremely low conductivity of the skull, most of the current injected from the electrode pair is reflected from the skull, while a very small portion of the current reaches the cerebral cortex. Therefore, the current distribution on the scalp is not sensitive to the value of σ_{brain}. Some studies showed that if the values of σ_{scalp} and σ_{skull} are exactly known, the reliability of dipole position estimation will not be reduced by the error of σ_{brain} (Musha and Okamoto, 1999). Therefore, they proposed a strategy by giving a value of σ_{brain}, such as $\bar{\sigma}_{\text{brain}}$, the optimal σ_{scalp} and σ_{skull} can be estimated by minimizing O.

$$O\left(\sigma_{\text{scalp}},\sigma_{\text{skull}}\right) \equiv S\left(\sigma_{\text{scalp}},\sigma_{\text{skull}},\bar{\sigma}_{\text{brain}}\right) \tag{5.2}$$

Musha and Okamoto (1999) reported the results of a computer simulation study on the dependence of these conductivity values on σ_{brain}. Nineteen electrodes were placed on the surface of the three-region model according to the international 10–20 standard. Four different pairs of electrodes were used for the current injection. Given a value of $\bar{\sigma}_{\text{brain}}$ 0.25–4 times as high as the classical value 0.33 S/m, the O value in equation (5.2) was minimized by iterating σ_{scalp} and σ_{skull} for 16 different S/N ratios. The results showed that when the value of $\bar{\sigma}_{\text{brain}}$ was >0.33 S/m, the estimated value of σ_{scalp} and σ_{skull} was close to the actual value.

Actually, an approximate sensitivity matrix can be derived for the relationship between the change in conductivity and the change in scalp surface potential (Gençer and Acar, 2004). Accordingly, a method can be established to iteratively solve the source distribution and conductivity (Acar et al., 2016). Another recent study showed that on the premise of not knowing accurate skull conductivity, the uncertainty of skull conductivity can be treated as an error term to improve the accuracy of EEG source localization by optimizing the error term in the Bayes model (Rimpiläinen et al., 2019). In a word, the main purpose of confirming the conductivity is to better locate the source. In fact, source localization is affected by many factors, and the effect of a single factor such as conductivity can be reduced to a certain extent if the effects of various factors are comprehensively taken into consideration.

5.2 FINITE-ELEMENT METHOD

The finite element method (FEM) is a numerical method derived from elasticity mechanics. It has been widely used to solve electrical, thermal, geological, biological, and fluid problems. So far, it has become one of the main methods of EEG forward simulation, and there is some available software (Vorwerk et al., 2018). FEM can be used for both isotropic and anisotropic inhomogeneous media. The only difference is that the conductivity is a scalar in isotropic media and a tensor in anisotropic media. The information about the tensor can be obtained from magnetic resonance diffusion tensor imaging (DTI). Some studies showed that the anisotropy of the brain has an insignificant effect on the scalp surface EEG and thereby on EEG source localization, but may have an effect on the amplitude of EEG and MEG signals (Haueisen et al., 2002). Below is an introduction to isotropic media, which can be directly extended to anisotropic media (Beltrachini, 2019).

Suppose the solution Φ of Poisson's equation can be approximated by $\tilde{\Phi}$, which is a function of nodal values. Based on nodes, a volume element is generated first, in which $\tilde{\Phi}$ is the interpolation of the nodal values. In the scheme of linear interpolation, only the angular points of the volume element are needed to be taken as nodes. For higher-order interpolation, some nodes on the surface of the element and within the element are required. The interpolation can be denoted by

$$\tilde{\Phi}(\vec{r}) = \sum_i \Phi_i Q_i(\vec{r}) \tag{5.3}$$

where Φ_i is the solution of node i; Q_i is a shape function, which describes the contribution of the value of this node to the potential of point $\vec{r}$. The shape function depends on the shape of the element and the interpolation method.

There are many ways to derive a finite element formula. The most common one is the Galerkin's weighted residual method. The following is its mathematical description. Substitute the interpolation function $\tilde{\Phi}$ into the Poisson's equation. The equation cannot be fully satisfied in any sub-volume v_k, thus there is a residual R_1.

$$\nabla\cdot\left(\sigma\nabla\tilde{\Phi}\right)-\nabla\cdot\vec{J}_s=R_1 \tag{5.4}$$

Also on each surface between the sub-volumes, there are

$$\vec{n}\cdot\left(\sigma_2\nabla\tilde{\Phi}_2\right)\Big|_{S_k}-\vec{n}\cdot\left(\sigma_1\nabla\tilde{\Phi}_1\right)\Big|_{S_k}=R_2 \tag{5.5}$$

where subscripts 1 and 2 represent the two sides of the interface S_k; $\vec{n}$ is the unit normal vector from medium 1 to 2 on the interface. The relevant equation is multiplied by the non-trivial weight, and integrated along all the sub-volumes and interfaces. Then, all the parts are added up to get

$$\begin{aligned}&\sum_{k}^{N_v}\int_{V_k} w\left(\nabla\cdot\left(\sigma\nabla\tilde{\Phi}\right)\right)dV+\sum_{k}^{N_v}\int_{S_k} w\left(\vec{n}\cdot\left(\sigma_2\nabla\tilde{\Phi}_2-\sigma_1\nabla\tilde{\Phi}_1\right)\right)dS\\&-\sum_{k}^{N_v}\int_{V_k} w\nabla\cdot\vec{J}_s dV=\sum_{k}^{N_v}\int_{V_k} wR_1dV+\sum_{k}^{N_v}\int_{V_k} wR_2dV\end{aligned} \tag{5.6}$$

where N_v represents the number of sub-volumes. For the true solution of Poisson's equation, R_1 and R_2 should be zero. Therefore, the solution $\tilde{\Phi}$ to be sought is the one that causes the right end of equation (5.6) to disappear. If we directly apply the function w to both sides of Poisson's equation and assume the surface of each sub-volume v_k is closed, then integrate the equation, we get

$$-\int_V(\nabla w)\cdot\left(\sigma\nabla\tilde{\Phi}\right)dV=\int_V w\nabla\cdot\vec{J}_s dV \tag{5.7}$$

where V is the sum of all the sub-volumes v_k. Equation (5.7) is called the weak expression of the problem because only the first derivative appears in the equation. So, compared with the original differential equation, it has a weaker requirement for $\tilde{\Phi}$, i.e., only requiring the existence of the first derivative of the approximate solution in the element, and then a linear interpolation can be performed. Equation (5.7) should hold for all functions w selected in an appropriate function space. In the Galerkin method, the shape function Q_i is used as a weight function w_i. Equation (5.3) can be substituted into (5.7) to obtain the equation corresponding to each function w_i.

$$\int_V(\nabla Q_i)\cdot\left(\sigma\nabla\left(\sum_{j}^{N_n}Q_j\tilde{\Phi}_j\right)\right)dV=\int_V Q_i\nabla\cdot\vec{J}_s\,dV \tag{5.8}$$

where N_n is the number of nodes. The volume integral is completed by summing the integral over each element. Equation (5.8) can be written in the following matrix form

$$A\vec{\phi} = \vec{f} \tag{5.9}$$

where A is a $N_n \times N_n$ matrix, $\vec{\phi}$ is an unknown vector composed of $\tilde{\Phi}_i$, and $\vec{f}$ is the source function vector. The element A_{ij} in matrix A is only related to the shape function and conductivity. When node j contributes to the potential of a certain element connected to node i, A_{ij} is not zero. If linear interpolation is used, the potential value in the volume of the element is determined by the nodes of the element. Therefore, matrix A is very sparse and can be stored and solved efficiently by the iterative method.

FEM is one of the most important methods for engineering calculation. It has been improved and developed in such as the selection of weight function (or called as a test function) w, derived such as continuous Galerkin (CG), conforming Galerkin (CG) method, discontinuous Galerkin (DG) where the weight function is piecewise continuous, etc. In current practice, almost all of the existing EEG FEMs are CG, however, DG newly got attention (Nußing et al., 2016; Engwer et al., 2017). Apparently, shape function is the crucial point, it has an impact on the complexity and simulation accuracy.

One of the problems with the EEG FEMs is how to deal with the strong singularity introduced by the dipole current source model. Many methods have been proposed. (1) Point charge sources at the grid points around the dipole source position are combined as the approximation of a dipole (Vorwerk et al., 2019). (2) Partial integration method (Yan et al., 1991), where every dipole source is approximated by two adjacent point sources. (3) Subtraction-based FEM, i.e., Poisson's equation is divided into two parts (refer to Section 5.5 below), with one being an equation for homogeneous media containing dipole source term. Under such a circumstance, the dipole source has an analytical solution, so that the problem of the singularity of the source can be avoided (Bertrand et al., 1991; Beltrachini, 2019).

In recent years, some studies tried a mixed FEM, which started from the previous step of Poisson's equation derivation, i.e., equation (3.25) in Section 3.3, which is a first-order equation set.

$$\begin{aligned} \vec{J} + \sigma\nabla\Phi &= \vec{J}^p \\ \nabla\cdot\vec{J} &= 0 \end{aligned} \tag{5.10}$$

where $\vec{J}^p$ is the primary current, and the total current $\vec{J}$ comprises the primary current and the conducting current (ohmic current, second current). Here, both current and potential are unknown (Vorwerk et al., 2017).

In FEM, flexible grid division is both an advantage and one of the difficulties. In general, all of its implementation methods are affected by grid division. Extensive research shows that the grid should be denser for the region near the source and where conductivity changes drastically (Lee and Im, 2019). An ideal algorithm should be an adaptive one that can adjust the grid adaptively according to the problem and result. The existing adaptive grid adjustment methods include moving nodes, refining elements, changing the order of basic functions, and combinations of the above.

5.3 MESHLESS METHOD

In general, it is easy to determine nodes if the geometric information of the brain is known, then we have to establish the connection among the nodes to determine the elements. This process is also a difficult and time-consuming operation, especially for FEM. The characteristic of the meshless method is that only node information is needed, while it is unnecessary to build elements. Because of this characteristic, the meshless method has become a new choice of engineering and scientific calculation. For the meshless method of the EEG forward problem, there still exists the problem of singularity introduced by the dipole source. In this regard, the widely used subtraction approximation method in FEM can be adopted, i.e., the effect of the dipole source is treated alone with an analytic solution in a homogeneous conductor model. See Section 5.5 below for details.

5.3.1 Finite Points Mixed Method

Below is a description of the finite points mixed method (FPMM) proposed by Von Ellenrieder et al. (2005). The reason why it is defined as a mixed method is that the approximate solution does not strictly satisfy Laplace's equation and does not strictly meet the boundary conditions, but is a result of a comprehensive tradeoff. Assume that at any layer $i, i = 1, \ldots, M$, the boundary of the region $\Omega_i, \partial\Omega_i = S_i \cup S_{i-1}, S_0 = \text{null}$, suppose there are a group of nodes $\vec{x}_j$ on the boundary, then the potential of any point $\vec{x}$ in Ω_i can be expressed as

$$\Phi(\vec{x}) = \Phi^a(\vec{x}) = \sum_{k=1}^{K} p_k(\vec{x}) a_k = \vec{p}^T(\vec{x})\vec{a} \tag{5.11}$$

Among them $\Phi^a(\vec{x})$ is the approximate solution of the potential $\Phi(\vec{x})$, $\vec{p}(\vec{x})$ is a vector composed of K approximate basis functions, and $\vec{a}$ is an undetermined coefficient vector.

For example, we can take

$$\vec{p}(\vec{x}) = \vec{p}(x, y, z) = \left[1, x, y, z, xy, xz, zy, x^2, y^2, z^2\right]^T \tag{5.12}$$

That is, it is composed of $K = 10$ complete monomial bases. Furthermore, suppose there are N nodes $\vec{x}_j$ around $\vec{x}$, $\tilde{\Phi}_j = \Phi(\vec{x}_j)$ is the potential value of node $j = 1, \ldots, N$. As will be seen in the following (5.17), $\tilde{\Phi}_j$ is the coefficient of the global approximate equation of the potential. Now, let's first link the vector $\vec{a}$ in the local approximation function (5.11) to $\tilde{\Phi}_j$. The $\vec{a}$ related to point $\vec{x}$ can be obtained by minimizing the following target function.

$$O(\vec{x}) = \sum_{j=1}^{N} w_j(\vec{x})\left(\tilde{\Phi}_j - \vec{p}^T(\vec{x}_j)\vec{a}\right)^2 \tag{5.13}$$

The target function is the weighted sum of the errors between local approximation function (5.11) and the potential $\tilde{\Phi}_j$ on $j = 1, \dots, N$ nodes. Equation (5.13) is known as a "moving least-squares" problem regarding $\vec{a}$. The reason why it is called "moving" is that it is related to $\vec{x}$.

The weighting function can be defined as

$$w_j(\vec{x}) = \begin{cases} e^{-\lambda_j / \left(1 - \left(|\vec{x} - \vec{x}_j| / r_j\right)^2\right)}, & |\vec{x} - \vec{x}_j| \le r_j \\ 0, & |\vec{x} - \vec{x}_j| \ge r_j \end{cases} \tag{5.14}$$

Here, the support range of the weighting function is defined as a spherical area and r_j represents its influencing scope. It is the influencing scope r_j that determines the number N of nodes $\vec{x}_j$ around $\vec{x}$. Different weighting functions can be used, and accordingly, there will be a difference in properties such as continuity. The λ_j is a quantity that controls the attenuation rate of the weighting function.

Define $P = \left[\vec{p}(\vec{x}_1), \dots, \vec{p}(\vec{x}_N)\right]^T$, $W(\vec{x}) = \mathrm{diag}\left\{w_1(\vec{x}), \dots, w_N(\vec{x})\right\}$, where diag{…} represents the diagonal matrix, and then the solution of equation (5.13) can be obtained as

$$\vec{a}(\vec{x}) = \left(P^T W(\vec{x}) P\right)^{-1} P^T W(\vec{x}) \tilde{\Phi} \tag{5.15}$$

where $\tilde{\Phi}_j$ represents the potential value of N nodes. N should be at least equal to K in equation (5.11) to get the unique least squares solution. Also, the node location should not be too special to ensure that P is of full column rank. For the process of the derivation from equations (5.13) to (5.15), please also refer to spatial weighting in Section 12.1.

Equation (5.15) just comprises an expression, where $\tilde{\Phi}_j$, $j = 1, \dots, N$ still is unknown. In fact, what our real concern is $\tilde{\Phi}_j$. To obtain $\tilde{\Phi}_j$, a global approximation of the potential will be constructed below according to equation (5.11). Let us firstly define

$$\vec{u}^T(\vec{x}) = \vec{p}^T(\vec{x}) \left(P^T W(\vec{x}) P\right)^{-1} P^T W(\vec{x}) \tag{5.16}$$

According to equations (5.11) and (5.15), we can get

$$\Phi^a(\vec{x}) = \vec{u}^T(\vec{x}) \tilde{\Phi} = \sum_{j=1}^{N} u_j(\vec{x}) \tilde{\Phi}_j \tag{5.17}$$

where $u_j(\vec{x})$ is the shape function of node j. It shows that the potential approximation at point $\vec{x}$ is a linear combination of the shape function related to N adjacent nodes, and the influence scopes of these nodes cover $\vec{x}$, while the coefficient of the linear combination is the potential $\tilde{\Phi}_j$ of these nodes. In other words, the potential at point $\vec{x}$ seems to be the interpolation combination of the potentials at N adjacent nodes. As can be seen from equation (5.16), the support range of the shape function is the same as that of the weighting function. In addition, as $u_j(\vec{x}_i) \ne 0$, $j \ne i$, the left of equation

(5.17) is just an approximation of the node potential. Next, we will further derive a method to solve $\tilde{\Phi}_j$, which is to require these functions to satisfy the electrical equations and boundary conditions.

Compare the meshless method here with FEM, both of them utilize the interpolation approximation, the meshless method is totally based on the nodes without mesh thus no element is involved, while FEM relies on both nodes and the element.

5.3.2 Boundary Conditions

Starting from equation (5.17), we can get

$$\nabla\Phi^a(\vec{x})\cdot\vec{n} = \sum_{j=1}^{N}\nabla u_j(\vec{x})\cdot\vec{n}\tilde{\Phi}_j \tag{5.18}$$

Now, the boundary conditions can be evaluated with (5.18) and (5.17) at the boundary nodes. For any node on the boundary, it involves the coefficient $\tilde{\Phi}$ of the nodes nearby, so the following linear system, which corresponds to equation (5.17/5.18), is formed for the potential and normal current continuity conditions with respect to every layer.

$$\Phi = U_S\tilde{\Phi} \tag{5.19}$$

$$\nabla\Phi\cdot\vec{n} = F\tilde{\Phi} \tag{5.20}$$

where

$\Phi = \left[\Phi^a(\vec{x}_1),\ldots,\Phi^a(\vec{x}_{Ns})\right]^T$, $\tilde{\Phi} = \left[\tilde{\Phi}_1,\ldots,\tilde{\Phi}_{N_T}\right]^T$, N_T are the total number of nodes. The boundary condition is only for the points N_S on the boundary, so it is a subset. Us and F are rectangular matrices because the boundary conditions only apply to nodes on the surfaces, they are determined by the boundary condition equations (Von Ellenrieder et al., 2005). Integrating the boundary condition equations of all the boundaries, we can finally get the boundary condition equations as follows.

$$D\tilde{\Phi} = b \tag{5.21}$$

5.3.3 Laplace Equation

With the Laplacian operator to equation (5.17), we have

$$\nabla^2\Phi^a(\vec{x}) = \sum_{j=1}^{N}\nabla^2 u_j(\vec{x})\tilde{\Phi}_j \tag{5.22}$$

Considering that the shape function does not fit the Laplacian equation, the Laplacian of the approximation function $\Phi^a(\vec{x})$ is not zero, either. Therefore, $\Phi^a(\vec{x})$ does not strictly satisfy Laplace's equation throughout the area, but we still can force it to satisfy the weighted residual integral formula (Von Ellenrieder et al., 2005).

5.3.4 Linear Equation System

The above operation should be applied to each layer. Finally, all the equations can be put together to derive a complete linear equation system, $\nabla^2\tilde{\Phi} = 0$. As mentioned above, $\tilde{\Phi}$ does not strictly meet the boundary conditions or strictly satisfy Laplace's equation, so it is a mixed method. One of the ways is to solve the following constrained optimization problem

$$\tilde{\Phi} = \min_{D\tilde{\Phi}=b} \left\| \nabla^2\tilde{\Phi} \right\| \tag{5.23}$$

i.e., Laplace equation under boundary constraints. The specific solving process can refer to literature (Von Ellenrieder et al., 2005; Mauricio Pohl-Alfaro et al., 2008; Ala et al., 2015). In a word, based on nodes, the meshless method works by establishing a numerical equation using a unique shape function. The equation cannot be solved in the same way as other numerical methods. Many specific implementation methods have been derived for the meshless method in other disciplines, and some of them would be also effective for EEG.

5.4 BOUNDARY ELEMENT METHOD

5.4.1 Overview

BEM is based on the integral equation. Different BEMs can be established on different integral equations, and so BEM is also called the boundary integral equation method. Here, we will begin with an introduction to BEM with respect to the simplest homogeneously piecewise and isotropic conductor. Assume a current source generating a potential Φ in a volume conductor Ω with the conductivity of σ, the source is represented by the current density (dipole moment) $\vec{J}_s$ (A/m^2), and there is the following theoretical relationship in region Ω (equation (3.26) in Section 3.3).

$$\nabla \cdot (\sigma \nabla \Phi) = \nabla \cdot \vec{J}_s \tag{5.24}$$

The human head is usually approximated by a bounded conductor. In this volume conductor, the potential satisfies equation (5.24), and on the surface Γ, the potential and current meet the corresponding boundary conditions. Utilizing Green's theorem (Appendix C), the differential equation (5.24) is converted into an integral equation on the boundary. The basic idea of BEM is to disperse the integral equation and then solve the equation group. BEM is mainly related to a homogeneous conductor as an inhomogeneous conductor can be broken down into several approximate homogeneous regions. In practice, the head can be divided into the scalp, skull, cerebrospinal fluid (CSF), and brain, each one can be approximately treated as a homogeneous conducting region.

BEM can take many forms. The two representative forms are the direct BEM (DBEM) (Barr et al., 1977; Hamalainen and Sarvas, 1989) and the indirect BEM (IBEM) (Rush et al., 1966). DBEM is based on a potential integral equation and its normal derivative on the boundary. By matching the boundary conditions on the common boundary of adjacent regions, an equation is established and solved. In IBEM, the current source is considered to be located in an infinite homogeneous conductor, and the boundary conditions are replaced by the action of the secondary sources located on the boundary. The derived integral equation includes the action of the secondary sources on the boundary, and finally, the unknown potential is indirectly obtained by these sources. In IBEM, the unknown variable is an auxiliable related to the secondary source on the boundary, while in DBEM, the unknown parameters are potentials and their normal derivative on the boundary.

BEM turns a continuous integral equation into a discrete equation system by two approaches. The first one is based on point collocation, which is at present the mainstream method and a method highlighted in the present section. It requires that the corresponding integral equation and boundary conditions should be met at the discrete grid points. The other one is the Galerkin method, which requires the weighted difference between the two sides of the equation to be minimized. The weight function is also called the shape function (Mosher et al., 1999; Stenroos and Haueisen, 2008; Liu et al., 2011). As a relatively new method, the Galerkin method is expected and worthwhile for further study (Stenroos and Nenonen, 2012). This section is mainly based on Musha and Okamoto (1999). Readers who are interested in the details can also refer to other literatures (Meijs et al., 1989; Fuchs et al., 1998; Frijns et al., 2000; Lynn and Timlake, 1968; Fischer et al., 2002; Cai and Yao, 2003).

5.4.2 Indirect Boundary Element Method

Suppose the head model is multilayered (Figure 5.1), and a non-zero conductivity σ_0 is assumed for the volume conductor region Ω_0 outside the head (it will eventually be zero). For a piecewise homogeneous conductor, equation (5.24) can be rewritten as

$$\nabla \cdot \sigma \nabla \Phi = \nabla^2 \sigma_i \Phi = \nabla \cdot \vec{J}_s \tag{5.25}$$

For each region Ω_i, $i = 1, 2, 3, 4, 0$.

Here, on the all outsides of Ω_1, $\vec{J}_s = 0$. The areas that connect with each other share common boundary conditions

$$\begin{cases} \Phi|_{\Gamma_i^-} = \Phi|_{\Gamma_i^+} \\ \sigma_i^- \partial_n \Phi|_{\Gamma_i^-} = \sigma_i^+ \partial_n \Phi|_{\Gamma_i^+} \end{cases} \tag{5.26}$$

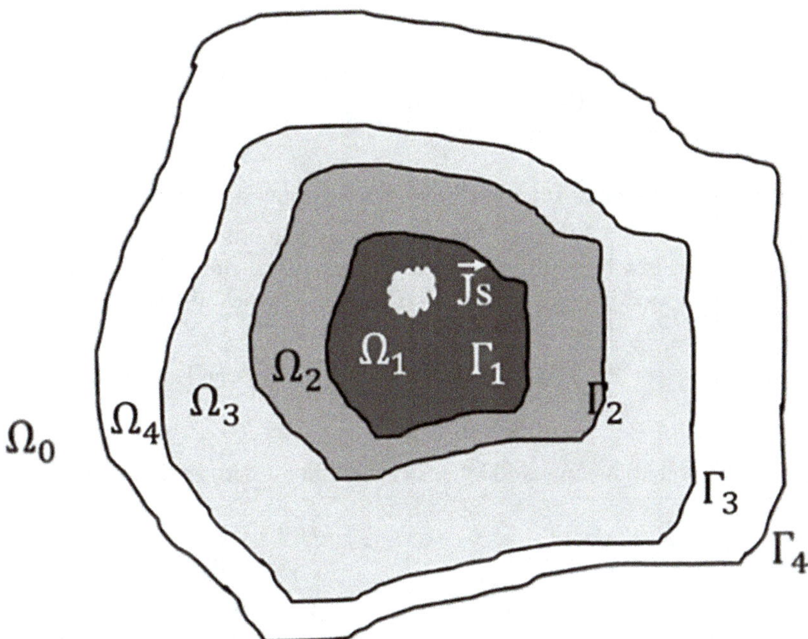

FIGURE 5.1 Schematic diagram of the multilayer head model.

Here, the subscript $i = 1$, 2, 3, and 4 indicate an inside-out boundary. Suppose the potential infinitely far away from the head is zero. So, equations (5.25) and (5.26) uniquely determine the distribution of the potential. Here, Γ_i^+ and Γ_i^- represent the outside and inside of the boundary Γ_i; σ_i^- and σ_i^+ represent the inner and outer conductivity of Γ_i, respectively, such as $\sigma_1^+ = \sigma_2$, $\sigma_4^- = \sigma_4$.

According to the electromagnetic field theory, the potential on the boundary is continuous, but its normal derivative is not continuous. So, an auxiliary variable, Ψ, is introduced here.

$$\Psi = \sigma_i \Phi \tag{5.27}$$

From equation (5.25), we can get

$$\nabla^2 \Psi = \nabla \cdot \vec{J}_s \tag{5.28}$$

For each region Ω_i, $i = 1, 2, 3, 4, 0$.

From equation (5.26), we can get

$$\begin{cases} \Psi|_{\Gamma_i^+} - \Psi|_{\Gamma_i^-} = \lambda \\ \partial_n \Psi|_{\Gamma_i^+} - \partial_n \Psi|_{\Gamma_i^-} = 0 \end{cases} \quad (i = 1, 2, 3, 4) \tag{5.29}$$

where

$$\lambda = \left(\frac{\sigma_i^+}{\sigma_i^-} - 1\right)\Psi\bigg|_{\Gamma_i^-} = \left(\sigma_i^+ - \sigma_i^-\right)\frac{\Psi}{\sigma_i^-}\bigg|_{\Gamma_i^-} = \left(\sigma_i^+ - \sigma_i^-\right)\Phi\Big|_{\Gamma_i} \tag{5.30}$$

The continuity of potential is applied in the above equation, i.e., $\Phi|_{\Gamma_i^+} = \Phi|_{\Gamma_i^-} = \Phi|_{\Gamma_i}$. Equations (5.28) and (5.29) can be understood as potentials generated by $\vec{J}_s$ and a current dipole layer λ in an infinite homogeneous conductor with unit conductivity of $\sigma = 1$, i.e., the equation (3.37) in Section 3.3 (Geselowitz, 1967).

$$\Psi(\vec{r}) = \Psi_\infty(\vec{r}) + \sum_{i'=1,2,3,4} \frac{-1}{4\pi}\int_{\Gamma_{i'}} d\Omega_{r'}\lambda(\vec{r}') \tag{5.31}$$

here Ψ_∞ is the potential generated by $\vec{J}_s$ in an infinite homogeneous conductor area.

$$\Psi_\infty(\vec{r}) = \frac{-1}{4\pi}\int dv' \frac{\nabla\cdot\vec{J}_s(\vec{r}')}{\left\|\vec{r} - \vec{r}'\right\|} \tag{5.32}$$

In these equations, $d\Omega_{r'}$ is the solid angle formed by ds' at $\vec{r}'$ with respect to $\vec{r}$. According to equation (5.27) and we substitute equation (5.30) into (5.31), we get (equation 3.38)

$$\Phi(\vec{r}) = \frac{\Psi_\infty(\vec{r})}{\sigma_r} + \sum_{i'=1,2,3,4} \frac{\left(\sigma_{i'}^- - \sigma_{i'}^+\right)}{4\pi\sigma_r}\int_{\Gamma_{i'}} d\Omega_{i'}\Phi(\vec{r}'),\quad (\vec{r}\in\Omega_i) \tag{5.33}$$

In this equation, when the field point $\vec{r}$ approaches point $\vec{r}_i$ on the boundary Γ_i in the region Ω_i, a singular point will appear in the integration of equation (5.33). Under this circumstance, the Cauchy principal value integral can be adopted (Guiggiani, 1991). As shown in Figure 5.2a, let's assume that $\vec{r}'$ is located on the boundary Γ. Now, change the boundary at point $\vec{r}'$, i.e., an arc ε with a radius of ε is drawn with $\vec{r}'$ as the center of the circle. As shown in Figure 5.2a, the original boundary Γ is divided into two parts: Γ′ and ε. Thus, for the new boundary $\Gamma' + \varepsilon$, point $\vec{r}'$ remains inside the region, so equation (5.33) is still applicable. When $\varepsilon \to 0$, $\Gamma' + \varepsilon \to \Gamma$. Taking a single-layer medium as an example, (5.33) can be written as

$$\begin{aligned}
\Phi(\vec{r}_i) &= \frac{\Psi_\infty(\vec{r}_i)}{\sigma_i^-} + \frac{\left(\sigma_i^- - \sigma_i^+\right)}{4\pi\sigma_i^-}\lim_{\varepsilon\to 0}\int_{\Gamma_i'+\varepsilon} d\Omega_{r'}\Phi(\vec{r}') \\
&= \frac{\Psi_\infty(\vec{r}_i)}{\sigma_i^-} + \frac{\left(\sigma_i^- - \sigma_i^+\right)}{4\pi\sigma_i^-}\left(\int_{\Gamma_i'} d\Omega_{r'}\Phi(\vec{r}') + \lim_{\varepsilon\to 0}\int_{\varepsilon} d\Omega_{r'}\Phi(\vec{r}_i)\right) \\
&= \frac{\Psi_\infty(\vec{r}_i)}{\sigma_i^-} + \frac{\left(\sigma_i^- - \sigma_i^+\right)}{4\pi\sigma_i^-}\left(\int_{\Gamma_i'} d\Omega_{r'}\Phi(\vec{r}') + \Omega_\varepsilon\Phi(\vec{r}_i)\right)
\end{aligned} \tag{5.33a}$$

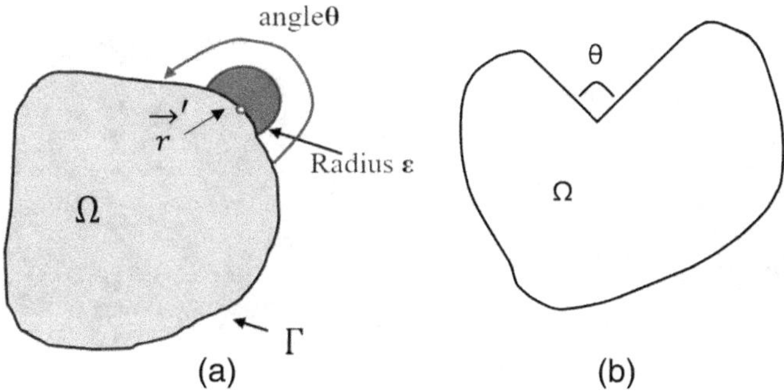

FIGURE 5.2 Illustration of the Cauchy principal value integral. (a) The singular point on the boundary is surrounded by part of a small sphere and lets the radius tend to zero later. For a whole sphere, the solid angle is 4π. For a smooth boundary, the angle θ (Ω_ε) is $4\pi / 2$. (b) For the other boundary, the solid angle needs to be estimated independently.

The above equation (5.33) becomes

$$C_{r_i}\Phi(\vec{r}_i) = \frac{\Psi_\infty(\vec{r}_i)}{\sigma_i^-} + \sum_{i'=1,2,3,4} \frac{\left(\sigma_{i'}^- - \sigma_{i'}^+\right)}{4\pi\sigma_i^-} P\int_{\Gamma_{i'}} d\Omega_{r_{i'}} \Phi(\vec{r}'), \quad \vec{r}_i \in \Gamma_i \tag{5.34}$$

Here, $P\int$ means taking the principal part, i.e., specifically, the boundary at $\vec{r}_i$ is integrated according to equation (5.33a), that is, this boundary point is excluded. The coefficient C_{r_i} in the left end of the equation is related to the solid angle $(4\pi - \Omega_\varepsilon)$ formed by Γ_i at point $\vec{r}_i$. When the boundary at the boundary point is smooth, Ω_ε is 2π. If the boundary is not smooth, its value should be determined independently as in the case shown in Figure 5.2b.

$$C_{r_i} = 1 - \frac{\Omega_\varepsilon}{4\pi} \frac{\sigma_i^- - \sigma_i^+}{\sigma_i^-} \tag{5.35}$$

For equation (5.34), when $\sigma_0 \to 0$, then $\sigma_0 = \sigma_4^+ = 0$. So equation (5.31) can be changed to

$$C_{r_i}\Phi(\vec{r}_i) = \frac{\Psi_\infty(\vec{r}_i)}{\sigma_i^-} + \frac{1}{4\pi} P\int_{\Gamma_4} d\Omega_{r'}\Phi(\vec{r}') + \sum_{i'=1,2,3} \frac{(\sigma_{i'}^- - \sigma_{i'}^+)}{4\pi\sigma_i^-} P\int_{\Gamma_{i'}} d\Omega_{r'}\Phi(\vec{r}'),$$

$$(\vec{r}_i \in \Gamma_i, \quad i = 1,2,3,4) \tag{5.36}$$

This equation can be discretized to establish a standard expression for IBEM (Barnard et al., 1967a,b).

Figure 5.3 shows the simulation results of a three-layer realistic head model based on IBEM (Cai and Yao, 2003). The conductivities of the three layers are set to relative values, i.e., 1.0 (brain), 1/80 (skull), and 1.0 (scalp). The three-layer head model is constructed from MRI images. The number of nodes in each layer is 1281, 963,

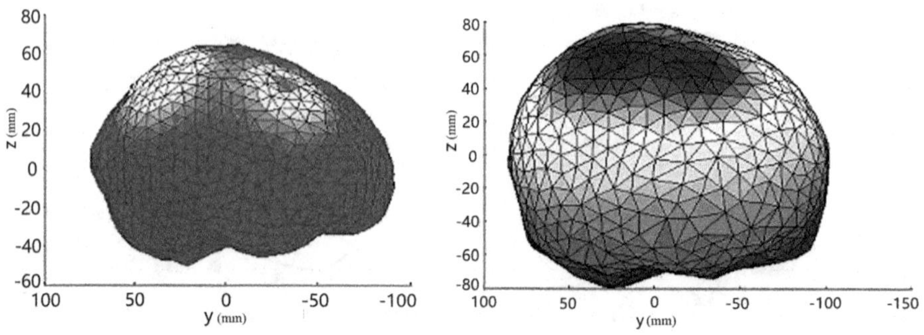

FIGURE 5.3 The IBEM calculation for potential in a three-layer realistic head model.

and 1,086, respectively, while the number of triangles in each layer is 2,558, 1,922, and 2,168, respectively. The dipole sources are located at (−30, −25, −35) mm and (−30, 25, 35) mm, while the dipole moments are (−0.7, −0.5, 0.5) nAm and (−0.6, 0.7, 0.9) nAm. The left figure is the potential of the cortex surface, while the right figure is the potential of the scalp surface. The negative directions of y and z in the figures correspond to the front and underside of the head, respectively. According to the comparison between the figures on the left and right, the cortical potential has a higher spatial resolution, while the difference in spatial resolution between the cortical potential and the scalp potential reflects the blurring effect of the skull layer.

5.4.3 Direct Boundary Element Method

5.4.3.1 Solution of the Laplace Equation in the Passive Region Ω

For region Ω with a boundary of Γ, Φ is potential in the region which satisfies Laplace's equation. Let $\vec{r}_0$ be any a point in region Ω, and ψ is the Green's function, which represents the potential generated by unit point source at point $\vec{r}_0$ in an infinite homogeneous conductor with unit conductivity.

$$\Psi(\vec{r};\vec{r}_0) \equiv {}^{1}\!\!\diagup_{4\pi\|\vec{r}-\vec{r}_0\|} \tag{5.37}$$

Apply Green's theorem (equation C.14) to Φ_∞ and Ψ to get

$$\int_\Gamma ds\left(\Psi\frac{\partial}{\partial_n}\Phi-\Phi\frac{\partial}{\partial_n}\Psi\right)=\int_\Omega dv\left(\Psi\nabla^2\Phi-\Phi\nabla^2\Psi\right)=\int_\Omega dv\Phi\delta^3(\vec{r}-\vec{r}_0)=\Phi(\vec{r}_0)$$

or

$$\begin{aligned}\Phi(\vec{r}_0)&=\frac{1}{4\pi}\int_\Gamma ds\frac{1}{\|\vec{r}-\vec{r}_0\|}\frac{\partial}{\partial_n}\Phi(\vec{r})+\frac{1}{4\pi}\int_\Gamma d\Omega_{\vec{r}_0}\Phi(\vec{r}),\\ d\Omega_{\vec{r}_0}&=ds\frac{\vec{n}\cdot(\vec{r}-\vec{r}_0)}{\|\vec{r}-\vec{r}_0\|^3}\end{aligned} \tag{5.38}$$

where $d\Omega_{\vec{r}_0}$ is the solid angle formed by the surface element $\vec{n}ds$ at $\vec{r}$ with respect to $\vec{r}_0$; $\vec{n}$ is the outward unit normal vector on the boundary of the volume conductor region Ω. Equation (5.38) shows that the potential at a point in a closed passive region is co-determined by the monopolar layer and bipolar layer on the surface. In other words, if the interface potential and normal potential gradient are known, the potential in the region is determined. When $\vec{r}_0$ approaches $\vec{r}_s$ on Γ, we get the following equation, just like deriving equation (5.34) from equation (5.33).

$$\left(1-\frac{\Omega_{r_s}}{4\pi}\right)\Phi(\vec{r}_s)=\frac{1}{4\pi}\int_\Gamma ds\frac{1}{\|r-r_s\|}\frac{\partial}{\partial_n}\Phi(r)+\frac{1}{4\pi}P\int_\Gamma ds\frac{\vec{n}\cdot(\vec{r}-\vec{r}_s)}{\|\vec{r}-\vec{r}_s\|^3}\Phi(\vec{r}) \tag{5.39}$$

Or taking the following expression

$$H(\vec{r}_s,\vec{r})\equiv\left(1-\frac{\Omega_{r_s}}{4\pi}\right)\delta(\vec{r}_s-\vec{r})-\frac{1}{4\pi}P\vec{n}\cdot\frac{(\vec{r}-\vec{r}_s)}{\|\vec{r}-\vec{r}_s\|^3},\quad G(\vec{r}_s,\vec{r})\equiv\frac{1}{4\pi}\frac{1}{\|\vec{r}-\vec{r}_s\|} \tag{5.40}$$

expression (5.39) changes to

$$\int_\Gamma dsH(\vec{r}_s,\vec{r})\Phi(\vec{r})=\int_\Gamma dsG(\vec{r}_s,\vec{r})\frac{\partial}{\partial_n}\Phi(\vec{r}),\qquad(\vec{r}_s\in\Gamma) \tag{5.41}$$

Discretize it to get

$$Hu=Gq \tag{5.42}$$

Here u is an N-dimensional vector, and its element is the potential at N nodes on Γ; q is also an N-dimensional vector, and its element is $\frac{\partial}{\partial_n}\Phi$; H and G are an Nth-order square matrix related to the geometry of Ω.

5.4.3.2 Solution of Poisson's Equation in the Active Region

Let the potential Φ_∞ be a potential generated in an infinite homogeneous conductor with conductivity of σ. It satisfies Poisson's equation $\nabla^2\Phi_\infty=\nabla\cdot\vec{J}_s/\sigma$. The following equation is derived from Green's theorem (equation C.14) of Φ_∞ and Ψ in form (5.37).

$$\frac{1}{4\pi\sigma}\int_\Omega dv\frac{\nabla\cdot\vec{J}_s(\vec{r})}{\|\vec{r}-\vec{r}_0\|}+\Phi_\infty(\vec{r}_0)=\frac{1}{4\pi}\int_\Gamma ds\frac{1}{\|\vec{r}-\vec{r}_0\|}\frac{\partial}{\partial_n}\Phi_\infty(\vec{r})+\frac{1}{4\pi}\int_\Gamma d\Omega_{\vec{r}_0}\Phi_\infty(\vec{r}) \tag{5.43}$$

When all sources are contained in Ω, the first term in the left end of this equation will be equal to $-\Phi_\infty(r)$ (refer to equation 2.50/2.51). Accordingly, we have a form corresponding to equation (5.41).

$$-\Phi_\infty(\vec{r}_s)+\int_\Gamma dsH(\vec{r}_s,\vec{r})\Phi_\infty(r)=\int_\Gamma dsG(\vec{r}_s,\vec{r})\frac{\partial}{\partial_n}\Phi_\infty(\vec{r}),\quad(\vec{r}_s\in\Gamma) \tag{5.44}$$

Discretize it to get

$$-u_\infty+Hu_\infty=Gq_\infty \tag{5.45}$$

These results are applied to the head model in Figure 5.1. The potential Φ in the region Ω_1 surrounded by the boundary Γ_1 satisfies

$$\nabla^2\Phi = \nabla \cdot \vec{J}_s / \sigma_1 \tag{5.46}$$

The electric potential Φ_∞ generated from the primary current source $\vec{J}_s$ in an infinite homogeneous conductor with conductivity of σ_1 also satisfies this equation. The potential difference $\Phi - \Phi_\infty$ is the solution of Laplace's equation in Ω_1. Now, therefore, we have the difference solution satisfying equation (5.42) as follows.

$$H^1\left(u_1^1 - u_{\infty,1}\right) = G^1\left(q_1^1 - q_{\infty,1}\right) \tag{5.47}$$

where the N_1-dimensional vectors u_1^1 and q_1^1 are the Φ and $\frac{\partial}{\partial_n}\Phi$ of the N_1 nodes on Γ_1. H^1 and G^1 are N_1th-order square matrices, which depend on the geometry of Ω_1. The superscript 1 indicates that these quantities correspond to the region Ω_1, while quantities related to the boundary Γ_1 in the infinite conductor marked by ∞, such as $u_{\infty,1}$, $q_{\infty,1}$, here the subscript 1 indicates the variable corresponds to the boundary 1. Since the current source is limited to Ω_1, equation (5.45) also holds for $\Gamma = \Gamma_1$. So, it can be substituted into equation (5.47) to get

$$H^1u_1^1 - G^1q_1^1 = u_{\infty,1} \tag{5.48}$$

There is no current source in the region Ω_2 limited by Γ_1 and Γ_2, so the potential in it is the solution of Laplace's equation. $H^2u^2 = G^2q^2$ can be derived in the same way as equation (5.42). Here, u^2 and q^2 are decomposed into u_1^2 and q_1^2, as well as u_2^2 and q_2^2, according to Γ_1 and Γ_2. Also, H^2 and G^2 are broken down similarly, and then we can get

$$\begin{pmatrix} H_{11}^2 & H_{12}^2 \\ H_{21}^2 & H_{22}^2 \end{pmatrix}\begin{pmatrix} u_1^2 \\ u_2^2 \end{pmatrix} = \begin{pmatrix} G_{11}^2 & G_{12}^2 \\ G_{21}^2 & G_{22}^2 \end{pmatrix}\begin{pmatrix} q_1^2 \\ q_2^2 \end{pmatrix} \tag{5.49}$$

Suppose N_2 are nodes on Γ_2. So, u_2^2 and q_2^2 are N_2-dimensional vectors. Similarly, u_1^2 and q_1^2 are N_1-dimensional vectors. Correspondingly, relevant local matrices H_{11}^2, H_{12}^2, …, G_{21}^2, can be determined, e.g., H_{21}^2 is a $N_2 \times N_1$ matrix. Similarly, for regions Ω_3 and Ω_4, we have

$$\begin{pmatrix} H_{22}^3 & H_{23}^3 \\ H_{32}^3 & H_{33}^2 \end{pmatrix}\begin{pmatrix} u_2^3 \\ u_3^3 \end{pmatrix} = \begin{pmatrix} G_{22}^3 & G_{23}^3 \\ G_{32}^3 & G_{33}^3 \end{pmatrix}\begin{pmatrix} q_2^3 \\ q_3^3 \end{pmatrix} \tag{5.50}$$

and

$$\begin{pmatrix} H_{33}^4 & H_{34}^4 \\ H_{43}^4 & H_{44}^4 \end{pmatrix}\begin{pmatrix} u_3^4 \\ u_4^4 \end{pmatrix} = \begin{pmatrix} G_{33}^4 & G_{34}^4 \\ G_{43}^4 & G_{44}^4 \end{pmatrix}\begin{pmatrix} q_3^4 \\ q_4^4 \end{pmatrix} \tag{5.51}$$

Equations (5.48)–(5.51) are independent of each other, and they are interconnected because they share the boundary conditions, i.e., the potential continuity on the boundary and the continuity of the normal component of the current passing through the boundary

$$u_1^1 = u_1^2, \quad u_2^2 = u_2^3, \quad u_3^3 = u_3^4 \tag{5.52}$$

and

$$\sigma_1 q_1^1 = -\sigma_2 q_1^2, \quad \sigma_2 q_2^2 = -\sigma_3 q_2^3, \quad \sigma_3 q_3^3 = -\sigma_4 q_3^4, \quad q_4^4 = 0 \tag{5.53}$$

Note that there is a minus sign in equation (5.53) because it is due to the derivative of the same boundary in the opposite direction in the two regions. The potential of each interface can be calculated by solving equations (5.48)–(5.51) of the four regions simultaneously. It should be noted that the derivation here is carried out based on the nested multi-region model, but this idea can also be applied to the non-nested multi-region model (Musha and Okamoto, 1999).

The known quantities are H and G, determined by the head model boundary. In the specific solving process, what coexists with the singularity of the equation is the arbitrariness of the reference potential. So a potential reference needs to be defined to avoid this issue. For example, a study made the assumption that the sum of scalp surface potentials is zero (Fischer et al., 2002), definitely, we can also take a specific point as reference such as C_z as in practical EEG recordings. However, these choices are only for the calculation by computer or measurement by differential amplifier, they do not have strong physical support (Yao, 2017), and we need to re-reference it to the true zero after the calculation or measurement according to Chapter 13.

Compared with IBEM, in which only the boundary potential is an unknown vector, the potential gradient on the boundary is also an unknown quantity in DBEM. However, the coefficient matrix is sparse, and many parameters may not appear in the final equation as in the case of the multilayered shell model below. Besides, in DBEM, we only need to know the relationship between the potential on the boundary of each region and its normal derivative, and we can even use FEM to establish this relationship in case of inhomogeneous subregions. Obviously, such a combination of FEM and BEM is a very promising approach to solving the inverse problem of a complex inhomogeneous medium.

5.4.4 Multilayer Shell Model and Direct Boundary Element Method

When DBEM is applied to a multilayered shell model as shown in Figure 5.1, the unknown parameters related to all other boundaries than the innermost and outermost boundaries can be eliminated. Without loss of generality, suppose the current source $\vec{J}_s$ is located in the innermost region Ω_1, and the conductivity outside the head Ω_{N+1} is zero. Since there is no current source in $\Omega_2, \Omega_3, \ldots$ or Ω_N, as shown in equations (5.49) and (5.50) above, we have

$$\begin{pmatrix} H_{n-1,n-1}^{n} & H_{n-1,n}^{n} \\ H_{n,n-1}^{n} & H_{n,n}^{n} \end{pmatrix}\begin{pmatrix} u_{n-1} \\ u_{n} \end{pmatrix}=\begin{pmatrix} G_{n-1,n-1}^{n} & G_{n-1,n}^{n} \\ G_{n,n-1}^{n} & G_{n,n}^{n} \end{pmatrix}\begin{pmatrix} -q_{n-1}^{n} \\ q_{n}^{n} \end{pmatrix}, \quad n=2,3,\ldots,N \tag{5.54}$$

The superscript n in q_{n-1}^{n} and q_{n}^{n} indicate which region it is in; the subscripts n and $n-1$ indicate which boundary it belongs to. For example, region n has two boundaries, Γ_{n-1} and Γ_{n}. Owing to the continuity of the potential u_n, it is only related to the edge n, so the superscript in the mark is removed. The normal vector on Γ_n is from Ω_n to Ω_{n+1}. When Γ_n is taken as a boundary of Ω_{n+1}, the normal vector points inward, which is reflected in the minus sign before q_{n-1}^{n}. The current density j_n on Γ_n satisfies the equation below

$$q_{n-1}^{n}=-j_{n-1}/\sigma_{n}, \quad q_{n}^{n}=j_{n}/\sigma_{n} \tag{5.55}$$

According to the uniqueness theorem of electrostatic theory, when the u_{n-1} on Γ_{n-1} and the q_{n}^{n} on Γ_n are known, the potential in Ω_n is uniquely determined, so the q_{n-1}^{n} on Γ_{n-1} and the u_n on Γ_n are also determined. The value of q_{n-1}^{n} and u_n can be obtained from equation (5.54), and then combined with equation (5.55) we can get

$$\begin{pmatrix} j_{n-1} \\ u_{n} \end{pmatrix}=\begin{pmatrix} T_{n-1,n-1}^{n} & T_{n-1,n}^{n} \\ T_{n,n-1}^{n} & T_{n,n}^{n} \end{pmatrix}\begin{pmatrix} u_{n-1} \\ j_{n} \end{pmatrix} \tag{5.56}$$

where

$$\begin{pmatrix} T_{n-1,n-1}^{n} & T_{n-1,n}^{n} \\ T_{n,n-1}^{n} & T_{n,n}^{n} \end{pmatrix}\equiv\begin{pmatrix} \sigma_n^{-1}G_{n-1,n-1}^{n} & -H_{n-1,n}^{n} \\ \sigma_n^{-1}G_{n,n-1}^{n} & -H_{n,n}^{n} \end{pmatrix}^{-1}\begin{pmatrix} H_{n-1,n-1}^{n} & \sigma_n^{-1}G_{n-1,n}^{n} \\ H_{n,n-1}^{n} & \sigma_n^{-1}G_{n,n}^{n} \end{pmatrix} \tag{5.57}$$

Replace n in equation (5.57) with $n+1$ to get the expression of area Ω_{n+1}. combine the expressions for Γ_{n+1} and Γ_n, and eliminate u_n and j_n, we have

$$\begin{pmatrix} j_{n-1} \\ u_{n+1} \end{pmatrix}=\begin{pmatrix} T_{n-1,n-1}^{n\sim n+1} & T_{n-1,n+1}^{n\sim n+1} \\ T_{n+1,n-1}^{n\sim n+1} & T_{n+1,n+1}^{n\sim n+1} \end{pmatrix}\begin{pmatrix} u_{n-1} \\ j_{n+1} \end{pmatrix} \tag{5.58}$$

where

$$\begin{cases} T_{n-1,n-1}^{n\sim n+1}\equiv T_{n-1,n-1}^{n}+T_{n-1,n}^{n}Q_nT_{n,n}^{n+1}T_{n,n-1}^{n} \\ T_{n-1,n+1}^{n\sim n+1}\equiv T_{n-1,n}^{n}Q_nT_{n,n+1}^{n+1} \\ T_{n+1,n-1}^{n\sim n+1}\equiv T_{n+1,n}^{n+1}P_nT_{n,n-1}^{n} \\ T_{n+1,n+1}^{n\sim n+1}\equiv T_{n+1,n}^{n+1}P_nT_{n,n}^{n}T_{n,n+1}^{n+1}+T_{n+1,n+1}^{n+1} \end{cases}, \quad \begin{cases} P_n\equiv\left(I_n-T_{n,n}^{n}T_{n,n}^{n+1}\right)^{-1} \\ Q_n\equiv\left(I_n-T_{n,n}^{n+1}T_{n,n}^{n}\right)^{-1} \end{cases} \tag{5.59}$$

Repeat this process to combine Ω_2 to Ω_n together, and all the parameters related to the boundaries from Γ_2 to Γ_{N-1} are eliminated, and finally, we can get

$$\begin{pmatrix} j_1 \\ u_N \end{pmatrix} = \begin{pmatrix} T_{1,1}^{2\sim N} & T_{1,N}^{2\sim N} \\ T_{N,1}^{2\sim N} & T_{N,N}^{2\sim N} \end{pmatrix} \begin{pmatrix} u_1 \\ j_N \end{pmatrix} \tag{5.60}$$

In the area Ω_1, it includes all current sources. So we have

$$H_1^1 u_1 = G_1^1 q_1^1 + u_\infty = -\sigma_1^{-1} G_1^1 j_1 + u_\infty \tag{5.61}$$

Eliminate j_1 from equations (5.60) and (5.61) to get

$$Hu_1 = -\sigma_1^{-1} G_1^1 T_{1,N}^{2\sim N} j_N + u_\infty, \qquad H \equiv H_1^1 + \sigma_1^{-1} G_1^1 T_{1,1}^{2\sim N} \tag{5.62}$$

When there is no current flowing into the head, $j_N = 0$; when there is a current flowing into the head, the value of j_N is known. The u_∞ is easy to calculate from $\vec{J}_s$. If a reference is pre-selected such as C_z, equation (5.62) has a corresponding unique solution u_1, and u_N can be found from equation (5.60). As for the above layer elimination method, when the volume conductor region Ω is surrounded by Γ_1 and Γ_2, if there is no current source inside, the potentials on Γ_1 and the normal derivative of potentials on Γ_2 are known, the potential distribution in Ω is well-determined. Accordingly, the normal derivative of the potential on Γ_1 and the potential on Γ_2 is determined, too.

5.4.5 Combination of Boundary Element Method and Other Methods

Let's assume that Figure 5.1 is about the case of two layers only, and the head is divided into the homogeneous brain region Ω_1 and inhomogeneous part Ω_2. Given a potential on Γ_1, the potential distribution in Ω_1 and the normal derivative of the potential on Γ_1 are determined (Musha and Okamoto, 1999). Because, according to the BEM equation (5.61), there is

$$q_1^1 = \left(G_1^1\right)^{-1}\left(H_1^1 u_1 - u_\infty\right) \tag{5.63}$$

Also, for the inhomogeneous region Ω_2, given Dirichlet boundary conditions on Γ_1 and Neumann boundary conditions on Γ_2, the potential distribution in Ω_2 is determined, making it practicable to adopt FEM or other methods to handle inhomogeneous regions separately. Accordingly, q_1^2 on Γ_1 and u_2 on Γ_2 are determined, too. As a result, there exists the following linear relationship between q_1^2 and u_2, as well as between u_1 and q_2^2 as equation (5.56).

$$\begin{pmatrix} -q_1^2 \\ u_2 \end{pmatrix} = \begin{pmatrix} F_{11} & F_{12} \\ F_{21} & F_{22} \end{pmatrix} \begin{pmatrix} u_1 \\ q_2^2 \end{pmatrix} \tag{5.64}$$

The minus sign before q_1^2 is because the normal vector on Γ_1 is pointed from Ω_1 to Ω_2. According to the continuity of the current density on Γ_1, we have

$$q_1^2 = Wq_1^1 \tag{5.65}$$

where W is a matrix determined by the ratio of the conductivity on both sides of Γ_1. If Ω_2 is a homogeneous area, W will be a matrix proportional to the identity matrix. From equations (5.63)–(5.65), we can get

$$\left\{F_{11} + W\left(G_1^1\right)^{-1} H_1^1\right\} u_1 = W\left(G_1^1\right)^{-1} u_\infty - F_{12} q_2^2 \tag{5.66}$$

When there is no current flowing into the head, $q_2^2 = 0$. When there is a current flowing into the head, the value of q_2^2 is known. u_∞ can be calculated from $\vec{J}_s$. Given a potential reference such as C_z, u_1 can be uniquely determined by equation (5.66), while the surface potential u_2 can be determined according to u_1 and equation (5.63).

5.4.6 Discussion

As mentioned above, for DBEM, both single and double layers are involved in the final equation. For IBEM, with an auxiliary variable, just the double layer is involved, which is similar to the case in Section 3.3. Based on the *Representation Theorem* of integral equations, methods also can be established which use a single layer alone, or use both single and double layers simultaneously called the symmetric method (Kybic et al., 2005), a comparative study argued that the symmetric method may offer better results (Gramfort et al., 2010), and its robustness has been improved recently (John et al., 2018), and it has been applied to the case of anisotropy (Rahmouni et al., 2017; Pillain et al., 2018, 2019). Besides, under a quasi-static approximation, it is also practicable to solve the electric field problem of the brain by equating it with the local charge generated from nerve cells and the equivalent charge on the medium interface, thereby constructing a "charge" based algorithm (Solis and Papandreou-Suppappola, 2019).

5.5 OTHER METHODS

5.5.1 Finite Difference Method

In the finite difference method (FDM), all the derivatives are expressed by the change of variables over the distance between two or more grids. A matrix equation can be established by substituting these approximations into Poisson's equation. For example, in the one-dimensional case, the Taylor series expansion of the potential at adjacent points x_{i-1} and x_{i+1} is as follows

$$\Phi(x_{i\pm1}) = \Phi(x_i) \pm h\Phi'(x_i) + 1/2h^2\Phi''(x_i) \pm 1/6h^3\Phi'''(x_i) + O(h^4) \tag{5.67}$$

where h is the distance between adjacent points. From equation (5.67) we can get

$$\Phi''(x_i) = \frac{\Phi(x_{i+1}) - 2\Phi(x_i) + \Phi(x_{i-1})}{h^2} + O(h^2) \tag{5.68}$$

In this formula, the order of error is $O(h^2)$. Applying (5.68) to each grid point in Poisson's equation, we can establish a system of equations corresponding to the grid points. Also based on

$$\Phi'(x_i) = \frac{\Phi(x_{i+1}) - \Phi(x_{i-1})}{2h} + O(h^2) \tag{5.69}$$

we can get the boundary conditions of the body surface (equation 3.23)

$$\Phi(x_{n+1}) - \Phi(x_{n-1}) = 0 \tag{5.70}$$

where x_n is a point on the boundary. This equation implies that a virtual additional point outside the volume is required for the outer surface (Pruis et al., 1993). Another method is to introduce the first-order approximation of $\Phi'(x_i)$, i.e., the difference between n and $n-1$. The overall accuracy will therefore be reduced. The points on the boundary surface of two subregions with different conductivities are also specially treated. As for the dipole sources of EEG, the method is similar to the FVM below. The dipole in Poisson's equation is represented by source and sink, but they cannot be placed in the same discrete unit, instead, they can be placed in adjacent units (Mohr and Vanrumste, 2003). FDM has been well used in anisotropic heterogeneous media (Cuartas Morales et al., 2019). In general, it is a bit difficult for FDM to deal with complex boundary shapes and to design a discrete network to be conformal to the boundary (Pruis et al., 1993). More information about finite difference can referred to in the research by Vanrumste et al. (2001) and Hallez et al. (2005).

5.5.2 Finite Volume Method

In the finite volume method (FVM), a volume conductor is divided into small volume elements that can be of any shape. Then, the Gauss divergence theorem is applied to each of the volume elements, i.e.,

$$\oint_S \sigma(\nabla\Phi) \cdot d\vec{S} = \int_V \nabla \cdot \vec{J}_s \, dV = \int_V I_V \, dV \tag{5.71}$$

This integral equation indicates that the current flowing out of a volume element must be equal to the current source in the volume element. The current density at the points inside and outside the volume element and on its surface can be estimated from the potential. The accuracy depends on the number of points utilized in the numerical integration. By stipulating the current density to be continuous on the surface of the volume element, the equations of all the volume elements can be combined into a unified matrix equation, in which the potential value is unknown. On the boundary with different conductivities on both sides, an equivalent conductivity can be derived by assuming that the currents in the two neighbor elements are the same as each other (Eshel et al., 1995), so that the current flowing across the boundary can be calculated in the same form as that in other homogeneous region. For the outer boundary, it is assumed that the current in the case of first-order approximation (between the boundary point and the inner point) is zero. Equation (5.71), I_V represents the volume current source density, which can be understood as an equivalent point source (charge). Based on the dipole hypothesis, the integral for the right end of the equation (5.71) will be zero. To solve this singular problem, any dipole source can be approximated by a point source and a point sink placed in two adjacent volume elements (Abboud et al., 1994).

FVM was developed from fluid mechanics (Rosenfeld et al., 1991). Its advantage is the use of an integral equation rather than a difference equation, and Gauss' law holds in the entire solution space. This property is of great importance for biomedical problems with drastic changes in conductivity, which actually limits the computational accuracy of BEM. In addition, a well-known advantage of FEM is that they can be flexibly used to handle complex geometries, and FVM can do too. Another advantage of FVM is that its matrix is sparser than that of FEM in many practical situations, therefore, fewer resources are required, thus making it possible to obtain better spatial resolution by further subdivision.

5.5.3 Method Selection

FVM, FEM, and FDM are collectively referred to as the volume element method (VEM) and can be used to deal with anisotropic conductivity, whereas BEM can only be used to deal with anisotropy in just a few cases (Tonon et al., 2001; Potse et al., 2009), but with low computational complexity because of a small number of elements. For VEM, far more nodes and much higher computational complexity are needed to achieve the same accuracy. However, the matrix of VEM to be solved is usually very sparse because the potential at a certain node is usually just related to a few nodes. According to this feature, the time used for solving these equations can be greatly reduced, especially for a region where the potential does not change quickly, only a small number of nodes are required to achieve the preset accuracy.

BEM and FEM are based on planar triangular elements (i.e., a surface is approximated by a set of plane triangles) and a set of hexahedrons or tetrahedrons, respectively. Hexahedral elements can also be used in FVM. For any given point on a surface, a mosaic structure based on triangular elements can be obtained; and for any given set of points in a 3D region, a tetrahedron-based mosaic structure can be obtained.

The boundary conditions of the outer boundary and the interface between different conductance regions involve a directional derivative normal to the surface. Difficulties may appear in FDM and FVM because the derivative on an element surface is usually calculated according to the values on both sides of the surface. If the two sides of the surface differ from each other in medium properties, the derivative can be approximated at a lower order, or an equivalent conductivity to be used, but both these two approaches will result in a less accurate forward solution. Another way is to use an extra point on the same side of the interface, the accuracy can be the same as the ipsilateral area, but the grid complexity is increased. There is no such problem with FEM because the boundary conditions of the element surface are automatically met. For a region with conductivity significantly different from that of the surrounding areas, the accuracy of BEM is also affected.

The fundamental purpose of EEG forward is to solve the inverse problem, i.e., determining the source that can be used to explain the observed potential, it is necessary to solve the forward problem of different source combinations until the forward solution of the source combination is consistent enough with the observed scalp recordings. Therefore, a fast-forward algorithm is required because many source combinations need to be iteratively checked. That's one of the fundamental reasons why BEM is the most widely used method.

In recent years, many specific meshless methods have been developed. Various weighted residual methods are mainly different from one another in weighted residual function and approximation function. The commonly used approximation functions include the moving least square approximation, kernel and reproducing kernel approximation, a partition of unity approximation, radial basis approximation, etc. The weighted residual methods include the Galerkin form, collocation form, local weak form, weighted least squares form, boundary integral form, etc. The biggest advantage of the meshless method is that attention is paid to the node rather than the grid, thus leaving out the calculation process of generating grid cells from the nodes. The meshless method is very popular in other disciplines but is rarely used for EEG and MEG, it is worthy of further tests.

5.5.4 Methods for Removing Source Singularity

As we know, the electromagnetic fields of the brain originate from the activity of the neuron population and are therefore distributed. However, as shown in Section 3.1, the potential generated from any distributed source can be equivalently represented by a multipole series expansion, such as a spherical harmonic series. The expansion is usually conducted with a point in the source region as the origin of coordinates, each term of the series can be interpreted as a multipole located at this origin position. According to the law of conservation of charge, the current monopole source term is generally not considered. A quadrupole can be interpreted as two dipoles close together, while an octopole can be interpreted as two quadrupoles close together, and so on. In other words, any source can be assumed to be a combination of dipoles. Further, when the distance between the observer/recording and the source region is far enough, a source can be approximated by a dipole, i.e., the contribution of higher-order poles is left out. Therefore, it is generally only necessary to calculate the electric potential of the dipole source, namely, the primary current density. Of course, since a dipole can be seen as two monopole sources very close to each other but opposite in sign, all the problems of bioelectricity can be represented by dipoles based on the monopole source theory and the physiological constraint that the algebraic sum of monopole sources is zero. Actually, in FEM, FVM, and FDM, such as equation (5.71), Poisson's equation is generally used as the starting point, and the assumed dipole source is represented by two monopole sources located in adjacent elements in algorithm practice. This is called the St. Venant method in FEM.

The dipole field is singular at the position of the dipole. There are several ways to deal with singularity. First of all, if a dipole is simulated to be of two monopole sources close to each other, equal in size and opposite in sign, i.e., the source and sink, the singularity can be eliminated. The second way is to set an equivalent source on the boundary around the source region. The third way is to divide the potential into two parts, one of which is the potential Φ_∞ generated from the source in an infinite homogeneous medium, while the other part is the solution of Laplace's equation in the volume conductor. This is also called the subtraction approach. The basic idea is as follows.

The moment of a dipole is defined as

$$\vec{p}\left(p_x, p_y, p_z\right) = \lim_{l \to 0; I \to \infty} I\vec{l} \tag{5.72}$$

$\vec{l}$ is the distance vector between two unipolar sources. It acts like a unipolar source with an intensity of I and a nearby unipolar sink with an intensity of $-I$. Suppose

$$\Psi = \Phi - \Phi_\infty \tag{5.73}$$

where Φ_∞ is the special solution of Poisson's equation when the volume conductor is an infinite homogeneous medium. Φ_∞ that satisfied the following Poisson's equation of an infinite medium with a conductivity of σ_s.

$$\nabla \cdot \left(\sigma_s \nabla \Phi_\infty\right) = \nabla \cdot \vec{J}_s \tag{5.74}$$

If the Cartesian coordinate system is chosen so that the source is exactly at the origin, then the solution of this equation is

$$\Phi_\infty\left(\vec{r}\right) = \frac{\vec{r} \cdot \vec{p}}{4\pi\sigma_s r^3} \tag{5.75}$$

where the rectangular coordinate of $\vec{r}$ is (x, y, z), and the dipole is represented by its three coordinate components. Here the potential is assumed to be zero at infinity, that is the infinity zero reference hypothesis in physics. If the origin of the coordinates is at $\vec{r}_0$, then the above solution becomes

$$\Phi_\infty\left(\vec{r}\right) = \frac{1}{4\pi\sigma_s} \frac{\vec{p} \cdot \left(\vec{r} - \vec{r}_0\right)}{\left|\vec{r} - \vec{r}_0\right|^3} \tag{5.76}$$

It represents the potential generated at point $\vec{r}$ by the dipole located at point $\vec{r}_0$.

It is quite a challenge to directly solve equation (5.74) for a realistic head model because the current source term in the equation causes the potential to change rapidly near the source and even approximate ∞, as shown in equation (5.75/5.76). The accuracy of any numerical approximation algorithm is insufficient for such a singular point. After separation by equation (5.73), the potential Φ_∞ is singular near the source, but it has an analytical solution (5.76). The second term Ψ is the solution of Laplace's equation in the realistic head model, and the singularity in the source region is eliminated. Substituted equation (5.73) into Poisson's equation (5.74) $\nabla \cdot \sigma \nabla \Phi = \nabla \cdot \vec{J}_S$, we have Ψ that satisfies the following differential equation.

$$\nabla \cdot (\sigma \nabla \Psi) = \begin{cases} 0, & \vec{r} \in V_s \\ -\nabla \cdot (\sigma \nabla \Phi_\infty), & \vec{r} \notin V_s \end{cases} \quad (5.77)$$

V_s is a region containing the source. In this region, the term at the right end of the equation (5.77) is zero. In the region that does not contain the source, (5.74) shows that the Laplacian of Φ_∞ is zero, so the right end will also disappear for a homogeneous region. In other words, the second term of equation (5.77) is only non-zero where the conductivity is inhomogeneous, i.e., where there is a boundary. The boundary condition of Ψ in the regions different in conductivity meets the following equation

$$\vec{n} \cdot (\sigma_1 \nabla \Psi_1) = \vec{n} \cdot (\sigma_2 \nabla \Psi_2) + \vec{n} \cdot (\sigma_2 - \sigma_1) \nabla \Phi_\infty \quad (5.78)$$

This equation is established by substituting equation (5.73) into the relevant boundary conditions. The subscripts 1 and 2 indicate both sides of the interface.

The solution Ψ is also the solution of Laplace's equation under the Newman boundary conditions (Jackson, 1975), where the source is fully reflected in the boundary condition.

Now we can calculate Φ by solving (5.77) and (5.78) and then adding Φ_∞ in equation (5.76) to it, and Ψ can be calculated by the meshless method or FEM.

REFERENCES

Abboud, S., Y. Eshel, S. Ievy, et al. 1994. Numerical calculation of the potential distrubution due to dipole source in a spherical model of the head. *Comput Biomed Res* 27:441–445.

Acar, Z.A., C.E. Acar, S. Makeig. 2016. Simultaneous head tissue conductivity and EEG source location estimation. *NeuroImage* 124:168–180.

Ala, G., E. Francomano, G.E. Fasshauer, et al. 2015. A meshfree solver for the MEG forward problem. *IEEE Trans Magn* 51(3):1–4.

Barnard, A.C.L., I.M. Duck, M.S. Lynn, et al. 1967a. The application of electromagnetic theory to electro cardiology I. *Biophys J* 7(5):443–462.

Barnard, A.C.L., I.M. Duck, M.S. Lynn, et al. 1967b. The application of electromagnetic theory to electro cardiology II. *Biophys J* 7(5):463–493.

Barr, R.C., M. Ramsey, M.S. Spach. 1977. Relating epicardial to body surface potential distributons by means of neuromagnetic data. *IEEE Trans Biomed Eng* 24(1):1–11.

Beltrachini, L. 2019. The analytical subtraction approach for solving the forward problem in EEG. *J Neural Eng* 16(5):16056029.

Bertrand, O., M. Thevenet, F. Perrin. 1991. "3-D finite element method in brain electrical activity studies," in J. Nenonen, H.M. Rajala, T. Katila (Eds.), *Biomagnetic Localization and 3-D Modeling*, Report Department of Technical Physics, Helsinki University, pp. 154–171.

Cai, Y., D. Yao. 2003. A method to calculate source potential in realistic head model and its effectiveness. *J UEST China* 32(2):149–154.

Cuartas Morales, E., C.D. Acosta Medina, G. Castellanos Dominguez, et al. 2019. A Finite-difference solution for the EEG forward problem in inhomogeneous anisotropic media. *Brain Topogr.* doi: 10.1007/s10548-018-0683-2.

Dogdas, B., D.W. Shattuck, R.M. Leahy. 2005. Segmentation of skull and scalp in 3-D human MRI using mathematical morphology. *Hum Brain Mapp* 26(4):273–285.

Engwer, C., J. Vorwerk, J. Ludewig, et al. 2017. A discontinuous galerkin method to solve the EEG forward problem using the subtraction approach. *SIAM J Sci Comput* 39(1):B138–B164.

Eshel, Y., S. Levy, M. Rosenfeld, et al. 1995. Correlation between skull thickness asymmetry and scalp potential estimated by a numerical model of the head. *IEEE Trans Biomed Eng* 42:242–249.

Fischer, G., B. Tilg, R. Modre, et al. 2002. On modeling the Wilson Terminal in the boundary and finite element method. *IEEE Trans Biomed Eng* 49:217–225.

Frijns, J.H.M., S.L. De Snoo, R. Schoonhoven. 2000. Improving the accuracy of the boundary element method by the use of second-order interpolation fumctions. *IEEE Trans Biomed Eng* 47:1336–1346.

Fuchs, M., R. Drenckhahn, H. Wischmann, et al. 1998. An improved boundary element method for realistic volume-conductor medeling. *IEEE Trans Biomed Eng* 45:980–997.

Gençer, N.G., C.E. Acar. 2004. Sensitivity of EEG and MEG measurements to tissue conductivity. *Phys Med Biol* 49(5):701–717.

Geselowitz, D.B. 1967. Bioelectric potentials in an in homogeneous volume conductor. *Biophys J* 7(1):1–11.

Gramfort, A., T. Papadopoulo, E. Olivi, et al. 2010. OpenMEEG: Opensource software for quasistatic bioelectromagnetics. *Biomed Eng Online* 9(1):45.

Guiggiani, M. 1991. The evaluation of cauchy principal value integrals in the boundary element method: A review. *Math Comput Model* 15(3–5):175–184.

Hallez, H., B. Vanrumste, P. Van Hese, et al. 2005. A finite difference method with reciprocity used to incorporate anisotropy in electroencephalogram dipole source localization. *Phys Med Biol* 50(16):3787–3806.

Hamalainen, M.S., J. Sarvas. 1989. Realistic conductivity geometry model of the human head for interpretation of neuromagnetic data. *IEEE Trans Biomed Eng* 36:165–171.

Haque, H.A., T. Musha, M. Nakajima. 1998. Numerical estimation of skull and cortical boundaries from scalp geometry. *Proceedings of the 20th Annual International Conference on IEEE EMBS*, Hong Kong, China, 20: 2171–2174.

Haueisen, J., D.S. Tuch, C. Ramon, et al. 2002. The influence of brain tissue anisotropy on human EEG and MEG. *NeuroImage* 15:159–166.

Homma, S., T. Musha, Y. Nakajima, et al. 1995. Conductivity raions of the scalp-skull-brain model in estimating equivalent dipole sources in human brain. *Neurosci Res* 22:51–55.

Jackson, J.D. 1975. *Classical Electrodynamics*, 2nd ed. New York: John Wiley & Sons Press.

John E. Ortiz G., Axelle Pillain, Lyes Rahmouni, Francesco P. Andriulli. A. 2018 Calderon Regularized Symmetric Formulation for the Electroencephalography Forward Problem. *Journal of Computational Physics*. 2018: 1–24.

Kybic, J., M. Clerc, T. Abboud, et al. 2005. A common formalism for the integral formulations of the forward EEG problem. *IEEE Trans Med Imaging* 24(1):12–28.

Lee, C., C.H. Im. 2019. New strategy for finite element mesh generation for accurate solutions of electroencephalography forward problems. *Brain Topogr* 32:354–362.

Liu, Y.J., S. Mukherjee, N. Nishimura, et al. 2011. Recent advances and emerging applications of the boundary element method. *Appl Mech Rev* 64(3):400549. doi: 10.1115/1.400549.

Lynn, M.S., W.P. Timlake. 1968. Theues of multiple deflations in the numerical solution of singular systems of equations, with applications to potential theory. *SIAM J Numer Anal* 5(2):303–322.

Meijs, W.H., O.W. Weier, M.J. Peters, et al. 1989. On the numerical accuracy of boundary element method. *IEEE Trans Biomed Eng* 36:1038–1049.

Mohr, M., B. Vanrumste. 2003. Comparing iterative solvers for linear systems associated with the finite difference discretisation of the forward problem in electro-encephalographic source analysis. *Med Biol Eng Comput* 41:75–84.

Mosher, J., R.M. Leahy, P.S. Lewis. 1999. EEG and MEG: Forward solutions for inverse methods. *IEEE Trans Biomed Eng* 46:245–259.

Musha, T., Y. Okamoto. 1999. Forward and inverse problems of EEG dipole localization. *Crit Rev Biomed Eng* 27(3–5):189–239.

Nielsen, J.D., K.H. Madsen, O. Puonti, et al. 2018. Automatic skull segmentation from MR images for realistic volume conductor models of the head: Assessment of the state-of-the-art. *NeuroImage* 174:587–598.

Nußing, A., C.H. Wolters, H. Brinck, et al. 2016. The unfitted discontinuous galerkin method for solving the EEG forward problem. *IEEE Trans Biomed Eng* 63(12):2564–2575.

Pillain, A., L. Rahmouni, F. Andriulli. 2019. Handling anisotropic conductivities in the EEG forward problem with a symmetric formulation. *Phys Med Biol* 64(3):035022.

Pillain, A., L. Rahmouni, F.P. Andriulli. 2018. A Calderon regularized symmetric formulation for the electroencephalography forward problem. *J Comput Phys* 375:291–306.

Pohl-Alfaro, M., O. Yáñez-Suárez, J.R. Jimennez-Alaniz, et al. 2008. Realistic meshless conductor model for EEG inverse problems. *Int J Bioelectromagn* 10(3):176–189.

Potse, M., B. Dubé, A. Vinet. 2009. Cardiac anisotropy in boundary-element models for the electrocardiogram. *Med Biol Eng Comput* 47(7):719–729.

Pruis, G.W., B.H. Gilding, M.J. Peters. 1993. A comparison of different numerical methods for solving the forward problem in EEG and MEG. *Physiol Meas* 14(4A):1–9.

Rahmouni, L., A. Pillain, A. Merlini, et al. 2017. Integral equation modelling of brain fibers for handling white matter anisotropies in the EEG forward problem. *The 2017 International Conference on Electromagnetics in Advanced Applications (ICEAA),* Verona, Italy, IEEE, pp. 1809–1812.

Rimpiläinen, V., A. Koulouri, F. Lucka, et al. 2019. Improved EEG source localization with Bayesian uncertainty modelling of unknown skull conductivity. *NeuroImage* 188:252–260.

Rosenfeld, M., D. Kwak, M. Vinokur. 1991. A fractional step solution mehtod for the unsteady incompressible Navier stokes equations in generalized coordinate system. *J Comput Phys* 94(1):102–137.

Rush, S., A.H. Turner, A.H. Cherin. 1966. Computer solution for time-invariant electric fields. *J Appl Phys* 37(6):2211–2217.

Solis, F.J., A. Papandreou-Suppappola. 2019. Power dissipation and surface charge in EEG: Application to eigenvalue structure of integral operators. *IEEE Trans Biomed Eng* 67(5):1232–1242.

Stenroos, M., J. Haueisen. 2008. Boundary element computations in the forward and inverse problem of electrocardiography: Comparison of collocation and Galerkin weightings. *IEEE Trans Biomed Eng* 55:2124–2133.

Stenroos, M., J. Nenonen. 2012. On the accuracy of collocation and Galerkin BEM in the EEG/MEG forward problem. *Int J Bioelectromagn* 14(1):29–33.

Tonon, F., E. Pan, B. Amadei. 2001. Green's functions and boundary element method formulation for 3D anisotropic media. *Comput Struct* 79(5):469–482.

Vanrumste, B., G. Van Hoey, R. Van de Walle, et al. 2001. The validation of the finite difference method and reciprocity for solving the inverse problem in EEG dipole source analysis. *Brain Topogr* 14(2):83–92.

Von Ellenrieder, N., C.H. Muravchik, A. Nehorai. 2005. A meshless method for solving the EEG forward problem. *IEEE Trans Biomed Eng* 52(2):249–257.

Vorwerk, J., A. Hanrath, C.H. Wolters, et al. 2019. The multipole approach for EEG forward modeling using the finite element method. *NeuroImage* 201:116039.

Vorwerk, J., C. Engwer, S. Pursiainen, et al. 2017. A mixed finite element method to solve the EEG forward problem. *IEEE Trans Med Imaging* 36(4):930–941.

Vorwerk, J., R. Oostenveld, M.C. Piastra, et al. 2018. The field trip SimBio pipeline for EEG forward solutions. *Biomed Eng OnLine* 17:37.

Yan, Y., P.L. Nunez, R.T. Hart. 1991. Finite-element model of the human head: Scalp potentials due to dipole sources. *Med Biol Eng Comput* 29:475–481.

Yao, D. 2000a. Electric potential produced by a dipole in a homogeneous conducting sphere. *IEEE Trans Biomed Eng* 47(7):964–966.

Yao, D. 2000b. High-resolution EEG mappings: A spherical harmonic spectra theory and simulation results. *Clin Neurophysiol* 111:81–92.

Yao, D. 2017. Is the surface potential integral of a dipole in a volume conductor always zero? A cloud over the average reference of EEG and ERP. *Brain Topogr* 30:161–171.

6 Theory of Equivalent Distributed Sources

In current EEG and ERP spatial analysis, equivalent disrinured sources based techniques played important roles in such as high-resolutoion imaging and zero-reference. In the following section 6.1, the general physical models of equivlent dipoloe layer and charge layer are derived respectivley, and the analytic solutions of them are also presented when the layer is a spherical surface. And then in section 6.2 the spherical series solution of a spherical equivalent charge layer is shown, and in section 6.3 the spherical series solution of a spherical equivalent dipole layer in three concentric spheres head model is presented. These different derivations also confirmed that the equivalent sources are determined by the actual sources and layer shape, but has nothing to do with the medium outside the distribution layer.

6.1 PHYSICAL MODEL OF EQUIVALENT DISTRIBUTED SOURCES

As one of the main high-resolution EEG imaging techniques, the cortical imaging technique (CIT) is based on a dipole layer (Sidman et al., 1992) or a charge layer (Yao, 1996), which is supposed to generate the same scalp potential as the true sources underneath the layer; thus, they are the equivalent dipole layer (EDL) (Sidman et al., 1992) or the equivalent charger layer (ECL) of the actual sources (Yao, 1995, 1996). Actually, the equivalent source technique was utilized in earth exploration many years ago (Dampney, 1969), and the equivalent source principle was introduced in Chapter 2. Here, EDL and ECL may be collectively referred to as equivalent source layers (ESL). Certainly, both EDL and ECL can be adopted simultaneously in EEG imaging. ESL on one side can be adopted to generate (reconstruct) the cortical potential, and on the other side, it can also be used directly as an imaging measure. Intuitively, the rationale for ESL imaging is based on the fact that scalp EEGs are mainly generated by cortical sources. Therefore, by approximating cortical sources using a dipole layer normally oriented with respect to the local cortical surface (EDL), or a charge layer (ECL), one may obtain a fairly good approximate representation of brain electrical activities under the non-invasive scalp EEG. The basis is that the pyramidal neurons, assumed to be the main source of the scalp potential, are parallel to each other and normal to the cortical surface, and for each pyramidal cell, its macroperformance is similar to that of a dipole (Section 3.1). However, if considering their apical dendrites, they look like sinks that are similar to negative charges. If the scalp recordings are generated not only by cortical sources but also by neurons located in subcortical regions, both EDL and ECL provide an equivalent source representation of the actual sources underneath, as shown below because either of them may generate the same potential outside the layer as the actual sources do. In the following sections, the focus is on the equivalent physical model of EDL and ECL to confirm the mathematical rationale.

DOI: 10.1201/9781032639260-6

6.1.1 ESL for Cortical Imaging Technique

As shown in Figure 6.1, both EDL and ECL consist of the construction of an equivalent source (dipole or charge) layer (ESL) in a volume conductor that simulates the head so that the potentials generated by ESL would take the values measured on the scalp surface, i.e.,

$$\sum_{i=0}^{N-1} u_i G\left(\vec{e}_i, \vec{e}_j\right) = \Phi\left(\vec{e}_j\right), \quad j = 1, \ldots, M. \tag{6.1}$$

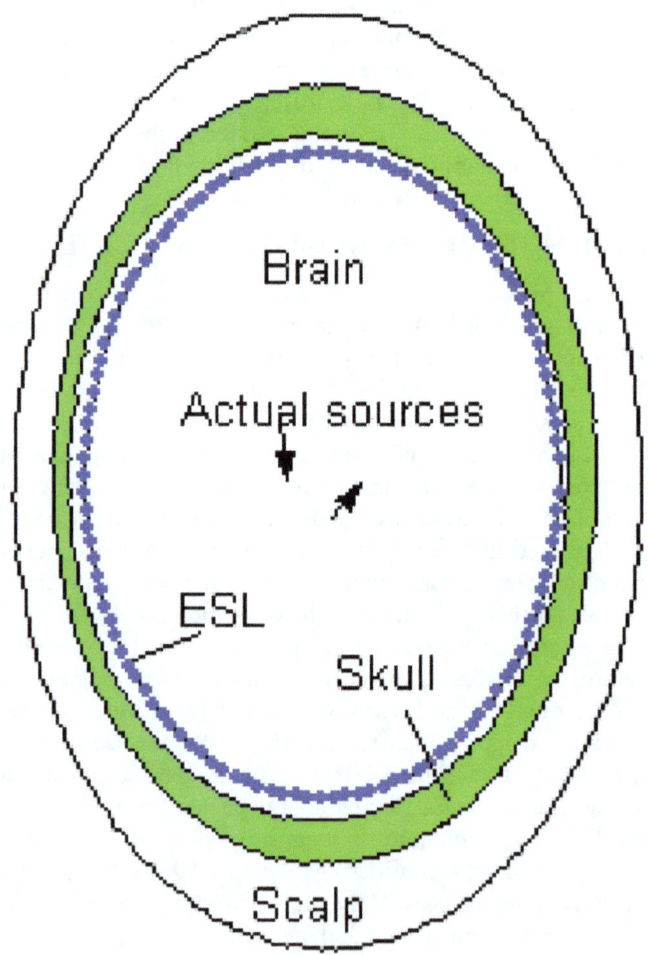

FIGURE 6.1 Illustration of the equivalent source layer model. A three-layer head model is assumed. The dotted line is the assumed ESL, which generates the same potential as the actual sources outside the ESL, including on the scalp surface. (From Yao and He (2003).)

where $\Phi(\vec{e}_j)(j=1,\ldots,M)$ is the scalp surface potentials, which are produced by actual neural electric sources, M is the number of electrodes, N is the number of the discrete sources on the equivalent layer, and u_i is the strength of a discrete equivalent source, $G(\vec{e}_i,\vec{e}_j)$ $(i=0, \ldots, N-1; j=1,\ldots, M)$ is an element of the leadfield matrix for a source located at $\vec{e}_i$ to an electrode at $\vec{e}_j$, which is determined by the source type such as a dipole or a charge and the volume conductor model. The u_i may be obtained by singular value decomposition (SVD) (Sidman et al., 1992) or other regularized inverses. Then the potential outside the equivalent layer produced by ESL is assumed to be approximately the same as the actual potential produced by the actual neural sources, which are located inside the layer. In this way, ESL (EDL or ECL) can be utilized as an intermediate step to get the desired epicortical potential, which is of higher spatial resolution than that of the scalp surface potential, and a theoretical solution of ESL based on a proper physical model will benefit from evaluating the efficiency of the inverse of ESL (Yao, 2000a). Meanwhile, ESL can also be utilized directly as imaging modalities (Yao, 2000a; He et al., 2002b), and an equivalent physical model would be useful in guiding such applications.

The following sections will show the equivalent physical model of ESL and give its closed solution for a dipole enclosed by a spherical layer.

6.1.2 Physical Model of ESL

6.1.2.1 Equivalent Source Layer and Integral Solution of an Electrostatic Problem

According to Section 3.2, the propagation, capacitance, and magnetic induction effects are all neglectable in the governing Maxwell equation over the frequencies of EEG and ECG. Accordingly, the head can be considered a volume conductor only having the factor of conductivity, and the governing equation formally becomes the same as that obtained under the static condition (Gulrajani, 1998).

With the Green function (Appendix C), the formal integral solution of the electrostatic boundary-value problem is (Plonsey, 1969; Gulrajani, 1998; Yao and He, 2003)

$$\Phi(\vec{r}) = \int_V \rho(\vec{r}')G(\vec{r},\vec{r}')d^3\vec{r}' + \oint_S \left[G(\vec{r},\vec{r}')\sigma\frac{\partial\Phi(\vec{r}')}{\partial n'} - \sigma\Phi(\vec{r}')\frac{\partial G(\vec{r},\vec{r}')}{\partial n'} \right] ds' \tag{6.2}$$

The potential $\Phi(\vec{r})$ at $\vec{r}$ in the conductor region V, i.e., the whole region except the source region V_0 bounded by the integral surface S, is composed of two terms: the first one is produced by the source with charge density $\rho(\vec{r})$ in region V, and the second integral term is the contribution of sources in the source region V_0. The conductivity is σ for region V and σ_0 for region V_0. $G(\vec{r},\vec{r}')$ is the Green function, i.e., the potential solution at a field point $\vec{r}$ of a point source (charge) at $\vec{r}'$ in a volume conductor model. For example, for a unit charge in an infinite conductor, $G(\vec{r},\vec{r}') = \dfrac{1}{4\pi\sigma|\vec{r}-\vec{r}'|}$ and $\left(-\dfrac{\partial G}{\partial n'}\right)$ are the Green function of a dipole, S is the boundary surface between V and

V_0, and $\frac{\partial}{\partial n'}$ is the normal derivative at the surface S (directed outside from V to V_0). For a neural electric problem, suppose all the actual sources are enclosed by the supposed ESL, for example, if a layer located on the surface of the cortex encloses all the neural electric sources inside, the first term vanishes. Then we have

$$\Phi(\vec{r})=\oint_s\left[G(\vec{r},\vec{r}')\sigma\frac{\partial\Phi(\vec{r}')}{\partial n'}-\sigma\Phi(\vec{r}')\frac{\partial G(\vec{r},\vec{r}')}{\partial n'}\right]ds', \quad \text{for } \vec{r}' \text{ on } S \tag{6.3}$$

Equation (6.3) shows that the contribution of sources in V_0 to the potential in V may be reproduced by a charge layer and a dipole layer, and for the charge layer, its strength density is $\sigma\frac{\partial\Phi(\vec{r}')}{\partial n'}$ which is actually the normal current density on the surface, and for the dipole layer, its strength density is $\sigma\Phi(\vec{r}')$ which is a quantity proportional to the surface potential $\Phi(\vec{r}')$. In convenience for the following discussions, $\Phi(\vec{r}')$ in the integral of equation (6.3) is re-noted as $\Phi'(\vec{r}')$, then we have

$$\Phi(\vec{r})=\oint_s\left[G(\vec{r},\vec{r}')\sigma\frac{\partial\Phi'(\vec{r}')}{\partial n'}-\sigma\Phi'(\vec{r}')\frac{\partial G(\vec{r},\vec{r}')}{\partial n'}\right]ds', \text{ for } \vec{r}' \text{ on } S \tag{6.4}$$

6.1.2.2 Pseudo-Boundary in a Homogeneous Region

Equation (6.4) is valid for any electrostatic boundary, such as a physical boundary with a similar or very different conductivity or even a pseudo-boundary in a homogeneous region. Apparently, a virtual boundary is assumed in this special case for ESL imaging. Therefore, $\frac{\partial\Phi'(\vec{r}')}{\partial n'}=\frac{\partial\Phi(\vec{r}')}{\partial n'}$, $\Phi'(\vec{r}')=\Phi(\vec{r}')$, and both EDL and ECL are adopted to represent the actual contribution of the inner sources to $\Phi(\vec{r})$. Based on equation (6.4), we may get an equivalent source layer technique or imaging method similar to that of EDL or ECL but including both a charge layer and a dipole layer at the same spatial position (Figure 6.1). To implement such a technique, the discrete unknown variables will be double those of either the EDL or ECL cases. So, more computation is needed, and such a cumbersome technique has never been suggested in the literature.

6.1.2.3 Boundary with a Zero Outside Conductivity

According to equation (6.4), if we have

$$\frac{\partial\Phi'(\vec{r}')}{\partial n'}=0 \tag{6.5}$$

Then

$$\Phi(\vec{r})=\oint_s\left(-\sigma\Phi'(\vec{r}')\right)\frac{\partial G(\vec{r},\vec{r}')}{\partial n'}ds', \quad \text{for } \vec{r}' \text{ on } S \tag{6.6}$$

Equations (6.5) and (6.6) show that if the normal current on the boundary is zero, then the general integral solution reduces to the continuous EDL form, whose discrete form was given by equation (6.1). In other words, the strength density of EDL is proportional to the surface potential produced by the actual sources inside the surface with zero outside conductivity (Figure 6.2), and the boundary condition equation (6.5) is satisfied automatically. This fact provides the equivalent physical model (Figure 6.2) of EDL and suggests that an inverted EDL from scalp potential is just like a new surface potential when the outer layers are stripped (He et al., 2002b; Yao and He, 2003).

6.1.2.4 Boundary with an Infinite Outside Conductivity

According to equation (6.4), if we have

$$\Phi'(\vec{r}') = 0 \tag{6.7}$$

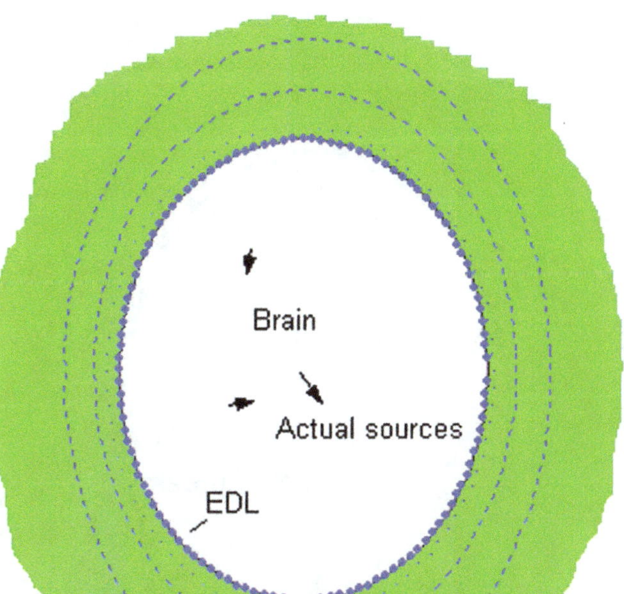

FIGURE 6.2 The equivalent physical model of the EDL. When the ESL in Figure 6.1 is an EDL, the strength of the EDL is proportional to the surface potential of a physical model with the same actual sources and conductivity inside the layer (the white region) but with an absolute insulated medium with zero conductivity filled outside (the gray region). (From Yao and He (2003).)

Then

$$\Phi(\vec{r}) = \oint_{S} G(\vec{r},\vec{r}')\left(\sigma\frac{\partial\Phi'(\vec{r}')}{\partial n'}\right)ds', \quad \text{for } \vec{r}' \text{ on } S \tag{6.8}$$

Equations (6.7) and (6.8) show that if the surface potential of the boundary is zero, then the general integral solution reduces to the continuous ECL form, whose discrete counterpart is already shown in equation (6.1). The strength density of ECL is the normal surface current density produced by the actual sources inside the surface with an infinite outside conductivity, where the boundary condition equation (6.7) is satisfied automatically (Figure 6.3). This fact provides the equivalent physical model of ECL and means that an inverted ECL from the scalp potential is just like the normal current density when the outside medium is replaced by a perfect conductor (Figure 6.3).

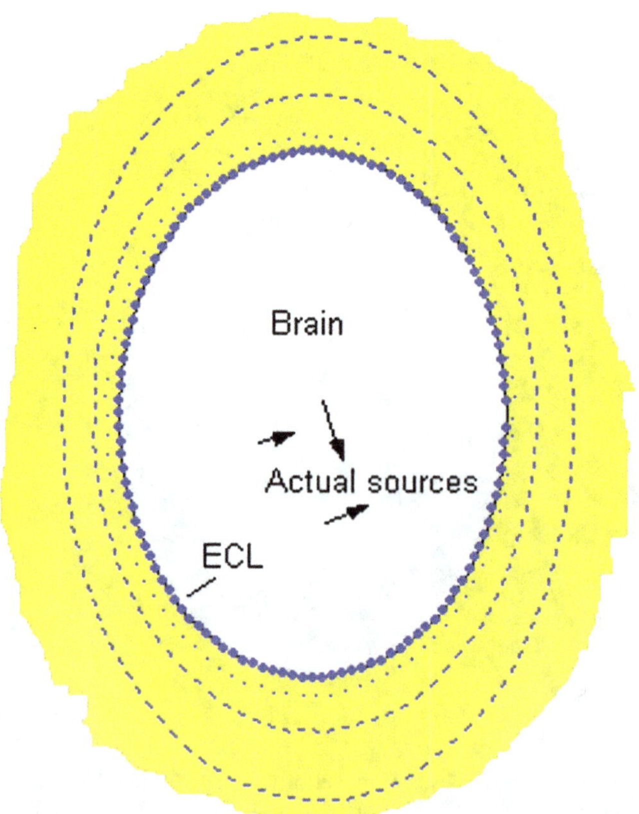

FIGURE 6.3 The equivalent physical model of the ECL. When the ESL in Figure 6.1 is an ECL, the strength of the ECL is equal to the normal current density on the surface of a physical model with the same actual sources and conductivity inside the layer (the white region) but with an absolute conductive medium filled outside. (From Yao and He (2003).)

With the determined physical model and equations (6.6) and (6.8), the forward EDL and ECL can be calculated by the boundary element method (BEM) for layered medium or the volume element method (VEM) such as FEM for an arbitrary volume conductor model. For example, for the EDL, we may get the EDL density, $-\delta\Phi'(\vec{r}')$, by solving a Poisson problem with the actual sources and the boundary condition equation (6.5) for an arbitrary boundary shape. This fact means that the ESL technique can be evaluated even for an arbitrary layered head model and an arbitrary equivalent source layer. They provide the theoretical basis of charge (Yao, 1996) or dipole layer imaging (Sidman et al., 1990).

6.1.2.5 Closed Solution of a Spherical ESL Surface

As noted above, the general solutions of EDL and ECL for a layer with an arbitrary shape and a volume conductor with an arbitrary conductivity distribution require a numeric algorithm such as FEM with much computation, while for a spherical model, such as multi-layer concentric and eccentric spheres and multi-layer spheroidal models, the existing forward series solutions of potential produced by a dipole are just the forward EDL (Chapter 4), and we may get the series solution of the forward ECL, too. Here, we consider the simplest case, in which the dipole is enclosed by a single spherical equivalent layer as used in the early EDL and ECL practice in EEG (Sidman et al., 1992; Yao, 1996; Wang and He, 1998), and the closed solutions can be used to make the calculation of the theoretical EDL and ECL strength very easy.

6.1.2.5.1 Closed-Solution of a Spherical EDL

As shown above, the equivalent physical model of EDL is a boundary between the volume conductor and the outside air, such as the head-air physical boundary. For a spherical surface and a dipole source inside, the closed solution (4.16) has been derived in Chapter 4 (Yao, 2000b).

6.1.2.5.2 Closed Solution of a Spherical ECL

In general, the potential produced by neural electric sources in an infinite medium may be assumed to be (Yao, 2000a)

$$\Phi_0(r,\theta,\phi)=\sum_{l,m}\left(a_l^m\cos m\phi+b_l^m\sin m\phi\right)\frac{P_l^m(\cos\theta)}{r^{l+1}},\quad \text{for}\quad r>r_0 \tag{6.9}$$

where P_l^m is the associated Legendre function of degree l and order m, a_l^m and b_l^m are the spherical harmonic spectra (SHS) of the potential, r_0 is the largest radius of the neural electric source positions, (r,θ,ϕ) are the spherical coordinates of a field point $\vec{r}$. According to the equivalent physical model and the uniqueness theory of the boundary-value problem in Chapter 2, we need to add to equation (6.9), a solution Φ_1 of Laplace's equation which has no singular point in a region with $r\le r_T$ and makes $\Phi'=\Phi_1+\Phi_0$ satisfy the Dirichlet boundary condition $\Phi'=0$ on S, i.e., equation (6.7). Then we may get the equivalent charge density strength by calculating the normal current density on the surface, where the surface with a radius r_T is the surface of the spherical equivalent charge layer. $\Phi_1(r,\theta,\phi)$ can be obtained by reference to equation (6.9) and the Dirichlet boundary condition, and the total potential becomes

$$\Phi'(r,\theta,\phi) = \Phi_1(r,\theta,\phi) + \Phi_0(r,\theta,\phi)$$

$$= \sum_{l,m} \left(a_l^m \cos m\phi + b_l^m \sin m\phi\right)\left(\frac{1}{r^{l+1}} - \frac{r^l}{r_T^{2l+1}}\right) P_l^m(\cos\theta), \quad r_T \geq r > r_0 \tag{6.10}$$

Apparently, $\Phi'(r,\theta,\phi)\,|_s = \Phi'(r,\theta,\phi)\,|_{r=r_T} = 0.0$, and $\Phi_1(r,\theta,\phi)$ is a solution of Laplace's equation in a region that $r \leq r_T$ and so $\Phi'(r,\theta,\varphi)$ is the unique solution in a region $r_T \geq r > r_0$ of Poisson's equation with a Dirichlet boundary condition on $S(r = r_T)$. According to equation (6.10), the strength of the spherical equivalent charge layer is

$$f_q = \sigma \left.\frac{\partial \Phi'}{\partial n}\right|_{r=r_T} = -\sigma \left.\frac{\partial \Phi'}{\partial r}\right|_{r=r_T}$$
$$= \sigma \sum_{l,m} Q \frac{1}{r_T^{l+1}} \left(a_l^m \cos m\phi + b_l^m \sin m\phi\right) P_l^m(\cos\theta) \tag{6.11}$$

With

$$Q = (2l+1)/r_T \tag{6.12}$$

Equations (6.11) and (6.12) show that the charge layer strength density is a result of the original potential processed by the filter Q. This conclusion is the same as obtained in a direct comparison between the outside potential produced by actual sources and the potential produced by a spherical equivalent charge layer (Yao, 2000a). Here, to get a closed solution, we let

$$f_q = \sigma \sum_{l,m} \frac{2l+1}{r_T^{l+2}} \left(a_l^m \cos m\phi + b_l^m \sin m\phi\right) P_l^m(\cos\theta)$$

$$= \sigma \sum_{l,m} \left(2\left(-\frac{\partial}{\partial r}\left(\frac{1}{r^{l+1}}\right)\right)\Big|_{r=r_T} - \frac{1}{r_T^{l+2}}\right)\left(a_l^m \cos m\phi + b_l^m \sin m\phi\right) P_l^m(\cos\theta)$$

$$= -2\sigma \left.\frac{\partial \Phi_0}{\partial r}\right|_{r=r_T} - \sigma \frac{1}{r_T} \Phi_0\Big|_{r=r_T} \tag{6.13}$$

where Φ_0 is the potential produced by actual sources in an infinite homogeneous medium.

For a dipole in an infinite homogeneous medium, its corresponding Φ_0 is

$$\Phi_0\big|_{r=r_T} = \frac{1}{4\pi\sigma} \frac{\vec{p}\cdot(\vec{r}_T - \vec{r}_0)}{\left|\vec{r}_T - \vec{r}_0\right|^3} \tag{6.14}$$

and so

$$\left.\frac{\partial \Phi_0}{\partial r}\right|_{r=r_T} = \frac{1}{4\pi\sigma}\left(\frac{\vec{p}\cdot\vec{r}_T}{r_T^4} - 3\frac{\vec{p}\cdot(\vec{r}_T-\vec{r}_0)}{r_T^5}\frac{r_T^{\,2}-\vec{r}_0\cdot\vec{r}_T}{r_T}\right) \tag{6.15}$$

Substituting equations (6.14) and (6.15) into (6.13), we get the strength density of the equivalent charge layer of a dipole with moment $\vec{p}$ located at $\vec{r}_0$ (Yao and He, 2003)

$$f_q = -\frac{1}{2\pi}\left(\frac{\vec{p}\cdot\vec{r}_T}{r_T^4} - 3\frac{\vec{p}\cdot(\vec{r}_T-\vec{r}_0)}{r_T^5}\frac{r_T^{\,2}-\vec{r}_0\cdot\vec{r}_T}{r_T}\right) - \frac{1}{4\pi r_T}\frac{\vec{p}\cdot(\vec{r}_T-\vec{r}_0)}{|\vec{r}_T-\vec{r}_0|^3} \tag{6.16}$$

where vector $\vec{r}_0$ is the dipole location and vector $\vec{r}_T$ is a point on S. If r_T goes to infinite, the sphere decays to a plane, and the formula (6.16) is simplified to contain only the first term, which was used in an equivalent body surface charge model (He et al., 1995).

Equations (6.13)–(6.16) correspond to the case of a dipole in V_0. Similarly, we can derive a formula for a charge or a quadrupole source. This fact indicates that any source can be equivalently represented by a charge layer, and the physical basis is that any source model is formed by charge or charge combination (Plonsey, 1969; Yao, 1996; Gulrajani, 1998).

6.1.2.6 Summary

In this section, the equivalent physical model and forward formulation of the equivalent source layer utilized in high-resolution EEG imaging are presented. The result of EDL indicates that it can be considered the "potential" on a new "scalp" surface when the outer medium is replaced by an insulator, just like sometimes we get after the skull is open (Yao and He, 2003). While for ECL, its strength density is the normal current density, equation (6.11), $j_r = -\sigma\left.\frac{\partial \Phi}{\partial r}\right|_{r=r_T}$, when the outer medium is replaced by a perfect conducting medium, which guarantees a zero potential on the boundary (Yao and He, 2003). These results provide the theoretical base of the ESL application in practice, and the physical models provide a way to get the theoretical value of the equivalent source layer even for an arbitrary layer shape.

The equivalent source layer imaging is suggested not only because of their practical efficiency as shown by the literature but also because of their clear physical meaning. The above forward theory affirms them being unique representations of the actual sources underneath the layer and being independent of the complexity of the outer medium of the actual head model. Besides, if we approximately assume the sources of our interests are on the cortical surface, the EDL and ECL may be considered an approximate representation of the actual sources.

The numeric results in the literature (Yao, 2000b; He et al., 2002a; Yao et al., 2004) confirmed that ECL and EDL are of higher spatial resolution than the scalp potential map does and therefore of practical significance. In addition, EDL has been realized in a real head model (Babiloni et al., 1997). ECL can be extended to a real head shape, too, so ECL and EDL can be expediently reconstructed from scalp recordings. When comparing EDL and ECL, we can find that ECL is of higher spatial resolution in theory because it is proportional to a derivative of the potential, see equation (6.8), and the strength of EDL is proportional to the potential directly, see equation (6.6). However, the practical inverse of them is an ill-posed problem, and due to their different illnesses that need different regularization levels, the current available practical resolution is similar to each other.

As a historical note, we'd like to point out that EDL was actually utilized by Sidman et al. (1990), but he did not explicitly note the concept of the equivalent source principle (Section 2.8) and also did not prove or discuss the physical rationale of ESL, but took it as a mathematical skill. Later, from the Kirchhoff integration, Amir (1994) made a preliminary discussion about the theory of equivalently distributed sources, and even concluded that the distributed dipoles are generally not perpendicular to the spherical surface adopted to place the equivalent sources. However, our theoretical research above demonstrated that the dipoles should be perpendicular to the surface of the equivalent source distribution.

Finally, we would like to point out that equations (6.6) and (6.8) were quite different from the conventional Green function method, which modifies the Green function in equation (6.2) to satisfy certain boundary conditions (Jackson, 1975; Yamashita, 1982) to remove one of the two terms in equation (6.2). Superficially, the resulting formulas were similar. However, in equations (6.6) and (6.8), we kept the Green function to be that of an actual point source in the head model, which is physically direct and clear and has been adopted in CCIT (Yao and Luo, 1996) and DCIT (Sidman et al., 1990). We removed one of the two terms in equation (6.2) by modifying the coefficients to have one of them being zero for each case, while for the final formula of the conventional Green function method, the modified Green function is a special one satisfying the assumed boundary condition of the assumed layer, which makes it a complex problem and unfeasible to be utilized in CCIT and DCIT.

6.2 SPHERICAL DISTRIBUTED EQUIVALENT CHARGE SOURCE

Before the equivalent physical model of the equivalent distributed source introduced in the above Section 6.1 (Yao and He, 2003), a spherical series solution isshown in 1996 (Yao and Luo, 1996) and is presented here as Sections 6.2 and 6.3.

Starting from equation (6.2), i.e., the Kirchhoff integral of the electrostatic field in unbounded space, the spherical harmonic analysis method may be used to prove that the potential of the real source could be equivalently obtained from the distributed point charge source and thus to provide a theoretical basis for the CCIT (Yao and Luo, 1996).

Assuming that the small spherical surface with a radius r_T of the distributed equivalent source surrounds all the real sources (Figure 6.4), the Kirchhoff integral formula (C.17) of the electrostatic field could then be expressed as:

$$\Phi = \frac{1}{4\pi\sigma} \oiint_{S_T} \left[\frac{1}{R} \sigma \frac{\partial \Phi}{\partial n} - \sigma \Phi \frac{\partial}{\partial n} \left(\frac{1}{R} \right) \right] ds \tag{6.17}$$

where Φ is the potential generated by the real source within the sphere S_T and R is the distance from a point on S_T to any field point between the spherical scalp surface and S_T.

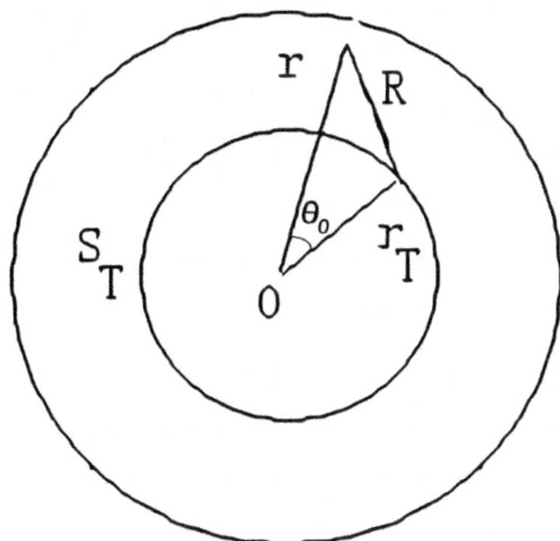

FIGURE 6.4 Schematic diagram of the equivalent distributed source method. r_T is the sphere radius of the equivalent distributed source, S_T is the spherical surface of the equivalent distributed source, and R is the distance from the field point r to a point on S_T (Yao and Luo, 1996).

Equation (6.17) shows that the external field generated by the internal real source within S_T can be equivalently produced by the surface charge (equation (6.11), density $\delta = \sigma \frac{\partial \Phi}{\partial n'}$, where σ is the conductivity) and the surface dipole moment (equation (6.6), density $\tau = \sigma \Phi$) along the normal line of the surface. If S_T is physically an equipotential surface, only the point charge density term is left. Of course, for CIT, S_T is generally not an equipotential surface. As shown in Figure 6.4, $r(r,\theta,\phi)$ is a field point between the surface S_T and the head surface, $r_T(r_T,\theta_T,\phi_T)$ is a point on S_T. Let's investigate the first item in formula (6.17):

$$\Phi_1 = \frac{1}{4\pi} \oiint_{S_T} \frac{1}{R} \frac{\partial \Phi}{\partial n'} ds \tag{6.18}$$

Based on the spherical harmonic theory of the potential (Chapter 4), Φ can be expressed as:

$$\Phi = \sum_{l=0}^{\infty} \sum_{m=0}^{l} \left(a_l^m \cos m\phi + b_l^m \sin m\phi\right) \frac{P_l^m(\cos\theta)}{r^{l+1}} \tag{6.19}$$

where P_l^m is the associated Legendre function and a_l^m and b_l^m are the expansion coefficients, which are a constant for a certain source distribution and coordinate system.

From equation (6.19), it can be got:

$$\frac{\partial \Phi}{\partial n'} = -\frac{\partial \Phi}{\partial r} = \sum_{l=0}^{\infty} \sum_{m=0}^{l} \left(a_l^m \cos m\phi + b_l^m \sin m\phi\right)(1+l) \frac{P_l^m(\cos\theta)}{r^{l+2}} \tag{6.20}$$

According to the basic property of Legendre function, we have:

$$\frac{1}{R}=\frac{1}{\sqrt{r^2+r_T^2-2rr_T\cos\theta_0}}=\sum_{l=0}^{\infty}\frac{r_T^l}{r^{l+1}}P_l(\cos\theta_0)$$

$$=\sum_{l=0}^{\infty}\frac{r_T^l}{r^{l+1}}\sum_{m=0}^{l}\frac{2(l-m)!}{\mu_m(l+m)!}(\cos m\phi\cos m\phi_T+\sin m\phi\sin m\phi_T)\,p_l^m(\cos\theta)P_l^m(\cos\theta_T)$$

(6.21)

where $\mu_m=2(m=0)$ or $1(m\neq 0)$ and θ_0 is the angle between r and r_T. Based on equations (6.20) and (6.21), (6.18) can be transformed as:

$$\Phi_1=\frac{1}{4\pi}\sum_{l=0}^{\infty}\sum_{m=0}^{l}\sum_{l'=0}^{\infty}\sum_{m'=0}^{l'}\int_0^{2\pi}\int_0^{\pi}(l+1)\frac{p_l^m(\cos\theta_T)}{r_T^{2+l}}$$

$$\left(a_l^m\cos m\phi_T+b_l^m\sin m\phi_T\right)\frac{r_T^{l'}}{r^{l'+1}}\frac{2(l'-m')!}{(l'+m')!\,\mu_{m'}}$$

$$\left(\cos m'\phi\cos m'\phi_T+\sin m'\phi\sin m'\phi_T\right)p_{l'}^{m'}(\cos\theta)\,p_{l'}^{m'}(\cos\theta_T)\,r_T^2\sin\theta_T d\theta_T d\phi_T$$

$$=\sum_{l=0}^{\infty}\sum_{m=0}^{l}\frac{l+1}{2l+1}\left(a_l^m\cos m\phi+b_l^m\sin m\phi\right)\frac{p_l^m(\cos\theta)}{r^{l+1}}$$

(6.22)

The orthogonal property of the Legendre function is applied in the derivation of the formula (6.22).

To compare the formulas of equations (6.19) and (6.22), it is obvious that if Φ in formula (6.18) can be modified as:

$$\Phi_q=\sum_{l=0}^{\infty}\sum_{m=0}^{l}\frac{2l+1}{l+1}\left(a_l^m\cos m\phi_T+b_l^m\sin m\phi_T\right)\frac{p_l^m(\cos\theta_T)}{r_T^{l+1}} \qquad (6.23)$$

Then Φ_1 in formula (6.22) will be the same with Φ.

$$\Phi=\Phi_1=\frac{1}{4\pi}\oiint_{S_T}\frac{1}{R}\frac{\partial\Phi_q}{\partial n}ds=\frac{1}{4\pi\sigma}\oiint_{S_T}\frac{1}{R}\sigma\frac{\partial\Phi_q}{\partial n}ds=\frac{1}{4\pi\sigma}\oiint_{S_T}d\frac{1}{R}ds$$

(6.24)

Equation (6.24) shows that the real potential Φ can be the same as the potential generated by a layer of a point charge. Therefore, equation (6.24) provides a theoretical basis for CCIT, among which the theoretical source density of the equivalent charge source is:

$$d = \sigma \frac{\partial \Phi_q}{\partial n'} = -\sigma \frac{\partial \Phi_q}{\partial r}$$

$$= \sum_{l=0}^{\infty} \sum_{m=0}^{l} \{\sigma(2l+1)/r_T\} \left(a_l^m \cos m\phi_T + b_l^m \sin m\phi_T\right) \frac{p_l^m(\cos\theta_T)}{r_T^{l+1}} \quad (6.25)$$

$$= \sum_{l=0}^{\infty} \sum_{m=0}^{l} K_l(q) \left(a_l^m \cos m\phi_T + b_l^m \sin m\phi_T\right) \frac{p_l^m(\cos\theta_T)}{r_T^{l+1}}$$

To compare with the original potential expression (6.19), the difference is the following spatial filter operator:

$$K_l(q) = \sigma(2l+1)/r_T \quad (6.26)$$

Numerical experiments have shown that the calculated value d is consistent with the inversion result obtained by CCIT from the scalp recordings, suggesting the feasibility of the CCIT algorithm.

According to the multipole expansion theory, Φ_q is also the potential corresponding to a certain source distribution. Comparing equations (6.23) and (6.19), it can be seen that in the two multipole expansions, the term $l = 0$ (i.e., the first one) is the same. This shows that the total charge corresponding to the two cases is the same. So, Φ_q seems to be understood as the potential generated by the same number of sources but with slightly different distributions.

It should be noted that the above proof is based on the Kirchhoff integral of the electrostatic field in unbounded space. However, as explained in Section 6.1, the equivalence between the real source and the equivalent distributed source is valid for the outside of the equivalent layer and is irrelevant to the complexity of the outside medium. Here, we discuss the case of unbounded space, nothing more than using the simple Green function $1/R$. As long as the region around the equivalent distributed source layer is homogeneous, the potential of the real source in this region can be expanded into the form of equation (6.19), and the general Green function that replaces $1/R$ can be expanded similarly in this region. Therefore, the core result, i.e., the spatial filtering operator (6.26), is the same.

In addition, the strategy adopted here (including the next section) is similar to that of Section (6.1), i.e., modifying the coefficients of the equivalent sources in equations (6.17) and (6.4) so that one of the integrals disappears and the other term produces an equivalent overall effect, while the Green function maintains its original form, and this idea is quite different from the conventional Green's function method in electrodynamics, where the Green function is modified so that its value to be zero directly or its derivative being zero on the boundary.

6.3 SPHERICAL DISTRIBUTED EQUIVALENT DIPOLE SOURCE

Similar to Section (6.2) for the charge layer, in this section, we will derive the forward theoretical formula for the equivalent distributed dipole layer in the concentric three-layer sphere model (Section 4.4).

The normalized radii of the widely used concentric three-layer sphere model are $a = 0.87$, $b = 0.92$, and $c = 1.0$, and the conductivity is $\sigma_{1(r<a)} = \sigma_{3(r>b)} = 1.0, \sigma_{2(a<r<b)} = 0.0125$, respectively. Based on Chapter 4, it can be deduced that the potential generated in this model by a unit dipole at any position with any orientation is as follows (Yao, 2000a).

$$\Phi_p\left(r,\theta,\phi;r_0,\theta_0,\phi_0\right) = \sum_{l,m} \frac{1}{r^{l+1}}\left(G_l^m \cos m\phi + H_l^m \sin m\phi\right) P_l^m\left(\cos\theta\right) \quad (6.27)$$

where

$$G_l^m(i) = K_l(i) g_l^m, \quad H_l^m(i) = K_l(i) h_l^m, \quad i = 1, r \le a; \; i = 2, a \le r \le b; \; i = 3, b \le r \le c \quad (6.28)$$

Here, $\left(g_l^m, h_l^m\right)$ is the electric multipole expansion coefficient of the unit dipole in unbounded space, that is, the intensity coefficient of the multipole series located at the center of the sphere, these multipoles together equivalently generate the electric field of the dipole. For the sake of brevity, these coefficients can be expressed in complex numbers as

$$\begin{aligned} g_l^m + jh_l^m = & \frac{r_0^{l-1}}{4\pi\sigma_1}\left(2-\delta_{m0}\right)\frac{(l-m)!}{(l+m)!}\Big[lP_r P_l^m\left(\cos\theta_0\right) \\ & + \frac{mP_\phi}{\sin\theta_0} P_l^m\left(\cos\theta_0\right) e^{j\pi/2} - \frac{P_\theta}{2}\left\{(l-m+1)(l+m)P_l^{m-1}\left(\cos\theta_0\right)\right. \\ & \left. - P_l^{m+1}\left(\cos\theta_0\right)\right\}\Big] e^{jm\phi_0}, \quad j = \sqrt{-1} \end{aligned} \quad (6.29)$$

where $\left(r_0, \theta_0, \phi_0\right)$ is the spherical coordinate position of the dipole in the brain $\left(r_0 < a\right)$. $\delta_{m0} = 0(m \ne 0)$ and $\delta_{00} = 1$. $\left(P_r, P_\theta, P_\phi\right)$ are the components of the dipole moment in the spherical coordinate system. P_l^m is the associated Legendre function. $K_l(i)$ is a correction coefficient related to the parameters of the head model. For the cortical surface ($i = 1$, $r = a = 0.87$) and scalp surface ($i = 3$, $r = c = 1.0$), we have

$$K_l(1) = \frac{(2l+1)\left(a^{2l+1}+\gamma\right)}{l(1-s)a^{2l+1}+\gamma\left(l+s(l+1)\right)}$$

$$K_l(3) = \frac{(2l+1)\left(b^{2l+1}+\gamma\right)}{l(1-s)a^{2l+1}+\gamma\left(l+s(l+1)\right)} \times \frac{l+1}{(l+1)b^{2l+1}+lc^{2l+1}}\left(c^{2l+1}+\chi\right)$$

where

$$s = \frac{\sigma_2}{\sigma_1} = \frac{\sigma_2}{\sigma_3}, \quad f = {}^{b}\!/_{c}, \quad \chi = \frac{l}{l+1}c^{2l+1}, \quad \gamma = \frac{\alpha}{\beta}$$

$$\alpha = f^{2l+1}(1-s)-1-\frac{ls}{1+l}, \quad \beta = f^{-(2l+1)}(1-s)-1-\frac{l+1}{l}s$$

Regarding the potential field of the unit dipole source in equation (6.27), the corresponding equivalent distributed source theory can be expressed by the following surface integral:

$$\Phi(r,\theta,\phi) = \oiint_{S_T} p_r(r_T,\theta_T,\phi_T;r_0,\theta_0,\phi_0)\Phi_{p_r}(r,\theta,\phi;r_T,\theta_T,\phi_T)ds_T \tag{6.30}$$

Here, S_T is the closed surface where the spherical equivalent distribution source is located, situated at the radius of r_T. The $\Phi_{p_r}(r,\theta,\phi;r_T,\theta_T,\phi_T)$ represents the potential generated at (r,θ,ϕ) by a unit radial dipole $(P_r,P_\theta,P_\phi)=(1.0,\ 0.0,\ 0.0)$ locating at (r_T,θ_T,ϕ_T)on S_T. $p_r(r_T,\theta_T,\phi_T;r_0,\theta_0,\phi_0)$ is the density function of the equivalent distribution source of the real source to be determined here. In the spherical coordinate system, it can be assumed as:

$$p_r(r_T,\theta_T,\phi_T;r_0,\theta_0,\phi_0) = \sum_{l,m}\left(a_l^m \cos m\phi_T + b_l^m \sin m\phi_T\right)P_l^m(\cos\theta_T) \tag{6.31}$$

Substituting equations (6.27) and (6.31) into (6.30), we can get

$$\Phi(r,\theta,\phi) = \sum_{l,m}\sum_{l',m'}\frac{1}{r^{l+1}}\frac{r_T^{l-1}}{4\pi\sigma_1}(2-\delta_{m0})\frac{(l-m)!}{(l+m)!}lK_l(i)P_l^m(\cos\theta)$$

$$\times\int_0^{2\pi}\int_0^{\pi} P_l^m(\cos\theta_T)(\cos m\phi_T\cos m\phi+\sin m\phi_T\sin m\phi)\left(a_{l'}^{m'}\cos m'\phi_T+b_{l'}^{m'}\sin m'\phi_T\right)$$

$$\times P_{l'}^{m'}(\cos\theta_T)r_T^2\sin\theta_T d\theta_T d\phi_T$$

Applying the orthogonality of the trigonometric function and the associated Legendre function, we have

$$\Phi(r,\theta,\phi) = \sum_{l,m}\frac{1}{r^{l+1}}\frac{r_T^{l+1}}{\sigma_1}K_l(i)\frac{l}{2l+1}\left(a_l^m\cos m\phi+b_l^m\sin m\phi\right)P_l^m(\cos\theta) \tag{6.32}$$

Comparing (6.27), (6.28), and (6.32), it can be seen that if

$$a_l^m = g_l^m\Big/\left(\frac{r_T^{l+1}}{\sigma_1}\frac{l}{2l+1}\right), \quad b_l^m = h_l^m\Big/\left(\frac{r_T^{l+1}}{\sigma_1}\frac{l}{2l+1}\right) \tag{6.33}$$

then the expression (6.32) is the same as (6.27). Therefore, substituting (6.33) into (6.31), we can get the forward theoretical formula of the equivalent distributed dipole source.

$$p_r\left(r_T,\theta_T,\phi_T;r_0,\theta_0,\phi_0\right)=\sum_{l,m}\frac{\sigma_1}{r_T^{l+1}}\frac{2l+1}{l}\left(g_l^m\cos m\phi_T+h_l^m\sin m\phi_T\right)P_l^m\left(\cos\theta_T\right) \tag{6.34}$$

For the case of L dipole sources with intensities of $p_i, i=1,\ldots,L$, the total potential is:

$$\Phi(r,\theta,\phi)=\sum_{i=1}^{L}p_i\Phi_p\left(r,\theta,\phi;r_i,\theta_i,\phi_i\right) \tag{6.35}$$

The corresponding equivalent source theoretical formula is

$$\Phi(r,\theta,\phi)=\oiint_{S_T}P_r\left(r_T,\theta_T,\phi_T\right)\Phi_{p_r}\left(r,\theta,\phi;r_T,\theta_T,\phi_T\right)ds_T \tag{6.36}$$

where

$$P_r\left(r_T,\theta_T,\phi_T\right)=\sum_{i=1}^{L}p_ip_r\left(r_T,\theta_T,\phi_T;r_i,\theta_i,\phi_i\right) \tag{6.37}$$

According to expression (6.34), if $g_0^0\neq 0$, then u_T will be infinite $\left(\frac{l+1}{l},l=0\right)$. It means that the field of point charge cannot be equivalent by a dipole layer, or that a combination of dipoles cannot form an equivalent charge, which shows that the DCIT cannot be used for imaging a point charge. Of course, for the EEG source, the sum of the positive and negative current source density (charge) in the brain is zero, i.e., there is no net point charge. So, DCIT can be used for EEG imaging.

Compare the expression (6.34) with the potential expansion formula, expression (6.27) in $K_l(i)=1$, their difference is only a spatial filter (Yao et al., 2001).

$$K_l(d)=\sigma(2l+1)/l \tag{6.38}$$

This factor is consistent with the results derived for a simple spherical model (Figure 6.4 (Yao and Luo, 1996)). This fact just confirmed the fact noted in Section 6.2 that the equivalent distribution source has nothing to do with the medium outside the distribution layer. So, when deriving the equivalent distribution source from the real source, we do not need to consider the influence of the CSF and scalp layer. Therefore, the Green function here can be in the form of a realistic head model or an unbounded space model, as illustrated in Section 6.2. The difference between whether there is a scalp or not is in the Green function adopted. The form of the Green function is the same for the true source and the equivalent source.

REFERENCES

Amir, A. 1994. Uniqueness of the generators of brain evoked potential maps. *IEEE Trans Biomed Eng* 41:1–11.

Babiloni, F., C. Babiloni, F. Carducci, et al. 1997. High resolution EEG: A new model-dependent spatial deblurring method using a realistically-shaped MR-constructed subject's head model. *EEG Clin Neuro* 102:69–80.

Dampney, C.N.G. 1969. The equivalent source techniqe. *Geophysics* 34:539–553.

Gulrajani, R. 1998. *Bioelectricity and Biomagnetism*. New York: John Wiley & Sons Press.

He, B., D. Yao, J. Lian, et al. 2002a. An equivalent current source model and Laplacian weighted minimum norm current estimates of brain electrical activity. *IEEE Trans Biomed Eng* 49(4):277–288.

He, B., D. Yao, J. Lian. 2002b. High-resolution EEG: On the cortical equivalent dipole layer imaging. *Clin Neurophysiol* 113:227–235.

He, B., Y.B. Chernyak, R.J. Cohen. 1995. An equivalent body surface charge model representative three-dimensional bioelectrical activity. *IEEE Trans Biomed Eng* 42:637–646.

Jackson, J.D. 1975. *Classical Electrodynamics*, 2nd ed. New York: John Wiley & Sons Press.

Plonsey, R. 1969. *Bioelectric Phenomena*. New York: Mcgraw Hill Education.

Sidman, R., M.R. Ford, G. Ramsey, et al. 1990. Age-related features of the resting and P300 auditory evoked responses using the dipole localization method and cortical imaging techniques. *J Neurosci Meth* 33(1):23–32.

Sidman, R.D, D.J. Vincent, D.B. Smith, et al. 1992. Experimental tests of the cortical imaging technique-applications to the response to median nerve stimulation and the localization of epileptiform discharges. *IEEE Trans Biomed Eng* 39:437–444.

Wang, Y., B. He. 1998. A computer simulation study of cortical imaging from scalp potentials. *IEEE Trans Biomed Eng* 45(6):724–735.

Yamashita, Y. 1982. Theoretical studies on the inverse problem in electrocardiography and the uniqueness of the solution. *IEEE Trans Biomed Eng* 29(11):719–725.

Yao, D. 2000a. Electric potential produced by a dipole in a homogeneous conducting sphere. *IEEE Trans Biomed Eng* 47(7):964–966.

Yao, D. 2000b. High-resolution EEG mappings: A spherical harmonic spectra theory and simulation results. *Clin Neurophysiol* 111:81–92.

Yao, D. 1995. Study of complex Huygens' principle. *Int J Infrared Millimeter Waves* 16(4):831–838.

Yao, D. 1996. Simplification of the integral solution of wave equation and Huygens' principle. *J UEST China* 25(3):245–250 (in Chinese).

Yao, D., B. He. 2003. Equivalent physical models and formulation of equivalent source layer in high-resolution EEG imaging. *Phys Med Biol* 48(21):3475–3483.

Yao, D., B. Luo. 1996. Theory of the EEG cortical imaging techniques. *Chin J Biomed Eng (English edition)* 5(3):128–136.

Yao, D., L. Wang, K.D. Nielsen, et al. 2004. Cortical mapping of EEG Alpha power using a charge layer model. *Brain Topogr* 17(2):65–71.

Yao, D., Y. Zhou, M. Zeng, et al. 2001. A study of equivalent source techniques for high-resolution EEG imaging. *Phys Med Biol* 46(8):2255–2266.

7 High-Resolution Cortical Imaging

In the history of EEG and ECG imaging, the most simple model of the head and chest is a plane, and the sphere is the second simple approximation model that people most easily think of. For this reason, the spherical harmonic spectra (SHS) method has long been used in bioelectromagnetics, especially in the forward/inversion of EEG. It is valuable to note that SHS methods have been utilized in geophysical exploration for a long time, and some methods there may be adapted to EEG, MEG and ECG. Even though, the work of deriving the cortical surface potential from the scalp surface potential by SHS only started in the 1990s (Yao, 1995, 1996; Srinivasan et al., 1996; Edlinger et al., 1998), here SHS acts as the intermediate role similar to the equivalent distribution source in Chapter 6. Our work learned from the spherical harmonic analysis of geomagnetic potential where the up-and-down continuation of the field is used to understand the origin of the geomagnetic field. Obviously, given the spherical head model like the earth, we can refer to the practice of geomagnetism to use SHS to realize the downward continuation of the scalp potential to obtain the cortical surface potential and to establish a new cortical imaging method.

Besides, except the cortical potential in the brain, we noticed that if we got the actual electric source, it can produce potential not only in the head model, may also have it in homogeneous unbounded space, where the potential will be a naked expression of the electric source, i.e., without disturbance from Cerebrospinal fluid (CSF) and the skull in the head model, and therefore should be a better imaging parameter, we took it as a new imaging modality by using potential in the unbounded space (Yao, 1995), now for the sake of simplicity, we call it the source potential mapping (SPM) technique, because it uses pure source potential, which is completely unrelated to the volumetric conductor head model (Yao, 2001b). In this chapter, based on SHS and the equivalent distributed source method introduced in Chapter 6, we will present a method to realize cortical potential imaging and SPM. In the last section, some simulation results are listed. In addition, following the idea of spherical harmonic analysis, if the distribution range of scalp surface data is limited, we can also use the spherical cap harmonic analysis (Haines, 1985) or rectangular harmonic analysis (Alldredge, 1981) or other similar methods in geophysics.

DOI: 10.1201/9781032639260-7

7.1 SPHERICAL HARMONIC SPECTRA APPROACH

7.1.1 Spherical Harmonic Analysis of the Scalp Potential

The theory of spherical harmonic analysis of brain potential has been introduced in detail in Section 2.4. Here, as an introduction, we choose the simplest single-layer spherical model adopted in the early stage (Yao, 1995), with an emphasis on clarifying the relevant physical concepts and processes.

Assume that the conductivity of the sphere head model is σ, the radius is R, and the outside of the sphere is a complete insulator, so the boundary condition is $\left.\frac{\partial \Phi}{\partial n}\right|_{r=R} = 0$, Φ represents the potential, and $\vec{n}$ is the normal unit vector of the interface. If the origin of the spherical and polar coordinate system is located at the center of the sphere, then the potential of a current source density I at $\vec{b}$ in the sphere is

$$\Phi_\infty = \frac{I}{4\pi\sigma r_b} = K\frac{I}{r_b} = K\sum_{n=0}^{\infty}\frac{b^n}{r^{n+1}}P_n(\cos\beta) \tag{7.1}$$

Where, Φ_∞ represents the potential generated by I in an unbounded space; $r_b = |\vec{r} - \vec{b}|$ is the distance between source point $\vec{b}$ and field point $\vec{r}$; $K = \frac{I}{4\pi\sigma}$; P_n is the first kind of Legendre polynomial; β is the angle between $\vec{r}$ and $\vec{b}$. Referring to Chapter 3, the source here is the divergence of the actual current density (equivalent unipolar source or equivalent charge source or current source density).

For the EEG problem, the total amount of charge in the brain is zero or there is no net charge in the brain, that is, positive and negative current source densities always exist at the same time. If the positive and negative ones are located in $\vec{a}$ and $\vec{b}$ respectively, it can be obtained by the superposition principle

$$\Phi_\infty = K\sum_{n=1}^{\infty}\left[a^n P_n(\cos\alpha) - b^n P_n(\cos\beta)\right]\frac{1}{r^{n+1}} \tag{7.2}$$

Where α is the angle between $\vec{a}$ and $\vec{r}$.

According to the uniqueness theorem of electromagnetic field (Section 2.3), a Φ_i is added to equation (7.2), here Φ_i satisfies the Laplace equation of the passive region and it is of no singularity in the sphere, which makes the total potential satisfies $\left.\frac{\partial(\Phi_i + \Phi_\infty)}{\partial n}\right|_{r=R} = 0$, then $\Phi = \Phi_i + \Phi_\infty$ is the potential solution of the above current source in the conductive sphere head model.

Referring to equation (7.2) and the spherical harmonic series solution theory of the Laplace equation, we have

$$\Phi = K\sum_{n=1}^{\infty}\left(\frac{1}{r^{n+1}} + \frac{n+1}{n}\frac{r^n}{R^{2n+1}}\right)\left(a^n p_n(\cos\alpha) - b^n P_n(\cos\beta)\right) \tag{7.3}$$

Further, let the positions of $\vec{a}, \vec{b}$ and $\vec{r}$ in spherical coordinates be $(a,\theta_a,\phi_a),(b,\theta_b,\phi_b)$ and (r,θ,ϕ), then use the addition theorem of Legendre polynomials, (7.3) change to

$$\Phi = \sum_{n=1}^{\infty}\sum_{m=0}^{n}\left(\frac{n+1}{n}\frac{r^n}{R^{2n+1}} + \frac{1}{r^{n+1}}\right)\left(c_n^m \cos m\phi + d_n^m \sin m\phi\right)P_n^m(\cos\theta) \tag{7.4}$$

Where

$$c_n^m = K(2-\delta_{m0})\frac{(n-m)!}{(n+m)!}\left[a^n P_n^m(\cos\theta_a)\cos m\phi_a - b^n P_n^m(\cos\theta_b)\cos m\phi_b\right]$$

$$d_n^m = K(2-\delta_{m0})\frac{(n-m)!}{(n+m)!}\left[a^n P_n^m(\cos\theta_a)\sin m\phi_a - b^n P_n^m(\cos\theta_b)\sin m\phi_b\right]$$

$\delta_{m0} = 0(m \neq 0)$ or $\delta_{m0} = 1(m = 0)$, P_n^m is the first associated Legendre function. When there are L pairs of current sources presented in the brain, simply change c_n^m and d_n^m to $c_n^m = \sum_{l=1}^{L} c_n^m(l)$, $d_n^m = \sum_{l=1}^{L} d_n^m(l)$, and the form of equation (7.4) remains unchanged.

Equation (7.4) shows that, for any EEG source, potential solution can be expressed by a spherical harmonic series, which means that the EEG source potential can be equivalently generated by a series of 2^n-poles $(n = 1,2,\ldots,L)$ at the origin, including Φ_∞ and their mirror image sources potential Φ_i. Apparently, here the SHS approach can also be taken as an equivalent source approach with the equivalent sources located at the origin instead of a closed surface used in the equivalent distributed source approach introduced in Chapter 6. Obviously, this theoretical model provides a physical basis for imaging techniques based on SHS. Equation (7.3) shows that the potential value of each term is decreased along with the order increase when $r > a$ and $r > b$, so the potential from deep source in the brain can be approximated with fewer terms, while more terms are required for shallow source.

Theoretically, the Legendre function $P_n^m(\cos\theta)$ for $\theta \in (0,\pi)$ interval is the orthogonal and complete function, equation (7.4) can be seen as a potential field Φ in the sense of a generalized Fourier series, then c_n^m and d_n^m accordingly may have corresponding analytic calculation formula. However, the actual recording of the scalp potential is mainly limited to an approximate half sphere $\theta \in \left(0,\frac{\pi}{2}\right)$, rather than $\theta \in (0,\pi)$, therefore, needing to adopt other approximate numerical calculation methods. In practice, the direct inversion method and spherical spline fitting method are adopted. The direct inversion method is introduced here and the spline fitting inversion method will be introduced in Chapter 9.

7.1.2 Scalp Potential Spherical Harmonic Analysis Algorithm

The spherical harmonic analysis (SHA) of brain potential is to solve the unknown coefficients c_n^m and d_n^m in equation (7.4). For the potential observation on the scalp surface, $r=R$, and equation (7.4) is reduced to

$$\Phi = \sum_{n=1}^{\infty}\sum_{m=0}^{n}\left(g_n^m \cos m\phi + h_n^m \sin m\phi\right)P_n^m(\cos\theta) \tag{7.5}$$

In the formula

$$g_n^m = \frac{2n+1}{nR^{n+1}}c_n^m, \quad h_n^m = \frac{2n+1}{nR^{n+1}}d_n^m \tag{7.6}$$

Assuming that the series expressed in equation (7.5) can be approximately truncated at $n=N$, an equation can be formulated for each observation point $\Phi_j\left(R,\theta_j,\varphi_j\right)$

$$\Phi_j\left(R,\theta_j,\phi_j\right) = \sum_{n=1}^{N}\sum_{m=0}^{n}\left(g_n^m \cos m\phi_j + h_n^m \sin m\phi_j\right)P_n^m\left(\cos\theta_j\right) \tag{7.7}$$

Since sin $0=0$, there are $L = N(N+2)$ unknown coefficients g_n^m and h_n^m at the right end of equation (7.7), and a set of equations can be obtained for J observation points

$$[\Phi]_{J\times 1} = A_{J\times L}\begin{bmatrix} g_n^m \\ h_n^m \end{bmatrix}_{L\times 1} \tag{7.8}$$

Where, $A_{J\times L}$ is composed of the coefficients associated with g_n^m and h_n^m in equation (7.7). The equation (7.8) is usually underdetermined ($J<L$) and often has a large condition number of the matrix A, so it is necessary to choose a good solution method. Truncated Singular Value Decomposition (SVD), which has a strong ability to overcome ill condition and does not magnify errors in operation, is generally chosen in current practice.

7.1.3 Cortical Potential Imaging

The 3D mapping of EEG refers to the inward extension of EEG recordings so that the potential distribution on the cortical surface close to the neural electric sources can be used to display the underlying sources. Obviously, on the SHS analysis above, with g_n^m and h_n^m substituted into equation (7.6) to get c_n^m and d_n^m, and then substituted them into equation (7.4) to obtain the N-order approximate potential formula of the potential in the head

$$\Phi_r = \sum_{n=1}^{N}\sum_{m=0}^{n}\left(\frac{n+1}{n}\frac{r^n}{R^{2n+1}} + \frac{1}{r^{n+1}}\right)\left(c_n^m \cos m\phi + d_n^m \sin m\phi\right)P_n^m(\cos\theta) \tag{7.9}$$

Equation (7.9) can be used to calculate the potential distribution on any sphere with radius $r<R$, to realize the inward continuation and imaging of EEG signals.

7.1.4 Potential in Unbounded Space

Since equation (7.4) consists of two parts, one is the potential of the actual neural sources in unbounded space (source potential), and the other is the influence of the boundary of the head model. Therefore, after getting c_n^m and d_n^m, the potential of the actual neural sources in unbounded space (SPM) can be obtained by using the following formula:

$$\Phi_\infty = \sum_{n=1}^{N}\sum_{m=0}^{n}\frac{1}{r^{n+1}}\left(c_n^m \cos m\phi + d_n^m \sin m\phi\right)P_n^m(\cos\theta) \tag{7.10}$$

The potential imaging in unbounded space can be realized by selecting a radius r of the imaging sphere.

7.2 EQUIVALENT DISTRIBUTED DIPOLE APPROACH

The low pass effect of the skull greatly limits the spatial resolution of the original scalp surface EEG (Yao, 2000), and this low resolution cannot be solved by increasing the number of electrodes, just like a low-frequency signal, adding sampling points cannot turn it into a high-frequency signal. The above cortical imaging method based on SHA is an effort to recover a higher spatial resolution, however, it is limited to a spherical head model. Actually, some other efforts werealso investigated in the past two decades, such as the following:

1. Equivalent source technology (EST), such as the equivalent distributed dipole layer technology (Sidman et al., 1990) and the equivalent distributed charge layer technology (Yao, 1996). The dipole layer and the charge layer themselves can also be used for imaging (Freeman, 1980; Yao, 2000), the corresponding image can be called the equivalent dipole layer mapping and the equivalent charge layer mapping (Yao, 1995, 2000; Yao, et al., 2001b).
2. Numerical calculation methods for the realistic head model, including boundary element method (BEM) (Srebro et al., 1993), finite element method (FEM) (Gevins et al., 1993), finite volume element method, finite difference method, and other numerical methods as introduced in Chapter 5.
3. Laplacian imaging. It will be introduced in the next chapter.

In this section, we will analyze and compare CPM and SPM (Yao, 2001b) based on the equivalent distributed source technique. It includes three aspects: physical model and forward simulation, inverse problem, simulation result, and discussion. The head model adopted here is a concentric four-layer spherical model (Figure B.2), and the specific formula derivation is shown in Chapter 4 or Appendix B.

7.2.1 Forward Model Test

7.2.1.1 Physical Model of SPM and CPM

Here we have the head approximated as consisting of four concentric spheres, as shown in Figure B.1, conductivity and radius data were obtained from Cuffin and Cohen (1977). CPM and SPM are usually set at the location of the cortex surface with a radius of 7.9 cm.

Figure 7.1 shows a model of a uniform unbounded volume conductor in which the four layers are the same as those shown in Figure B.1, but are only symbolic here because they do not differ in physical parameters. The conductivity of the whole conductor is 0.461 A/(Vm). The potential generated by the real source in this model is called SPM, because, without the influence of conductivity inhomogeneity, SPM is a direct reflection of the source itself. In practice, the SPM in Figure 7.1 should be recovered from the potential information in the scalp surface recordings ($r = 8.8$ cm) in Figure B.1.

When comparing the physical models of CPM and SPM, the difference is obvious. What CPM attempts to recover is the cortical surface potential generated by the actual neural electric sources in the multilayer head model, which still suffered from the influence of a non-uniform conductor in the extra-cortical region on the cortical surface potential. What SPM attempts to recover is the potential generated by the actual sources in unbounded space, in which the influence of the non-uniform conductor

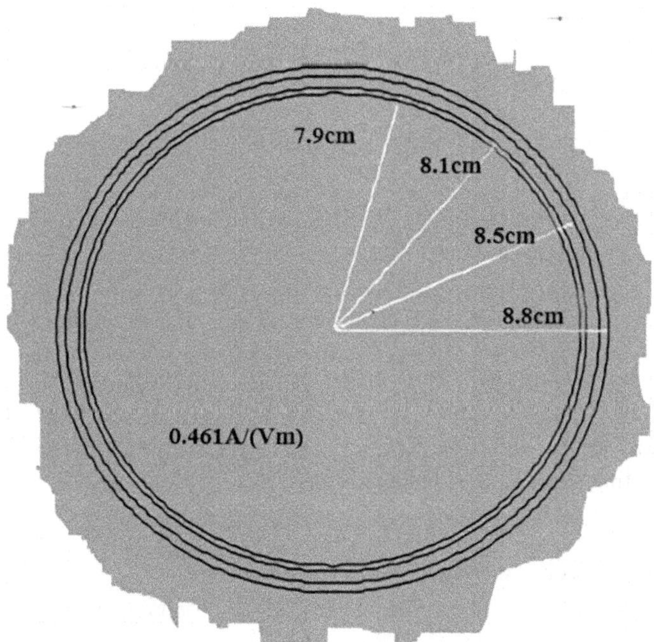

FIGURE 7.1 Model of a uniform unbounded volume conductor. Conductivity is 0.461 A/(VM). The circles in the figure only represent the same locations as the corresponding areas in Figure B.1 (Yao, 2001b).

model on the source potential distribution has been completely eliminated. This difference shows that, compared with CPM, SPM really corresponds to the potential of the actual sources and is a direct reflection of neural activity. To further explain the difference, we present the mathematical forward theory for the four-layer concentric spheres model below, which is actually quite similar to that in the above Section 7.1.

7.2.1.2 SPM Forward Theory – The Forward Theory of Dipoles in Unbounded Media

The potential distribution of a dipole at any position and orientation in unbounded space is (Yao, 2000)

$$\Phi_\infty = \frac{1}{4\pi\sigma_1}\vec{P}\cdot\nabla_{r_0}\left(\frac{1}{r_p}\right) = \sum_{l,m}\frac{1}{r^{l+1}}S_l^m\left(\vec{P},\theta,\phi\right),\ r > r_0 \tag{7.11}$$

Where

$$S_l^m\left(\vec{P},\theta,\phi\right) = \left(g_l^m\cos m\phi + h_l^m\sin m\phi\right)P_l^m\left(\cos\theta\right) \tag{7.12}$$

And

$$\begin{aligned} g_l^m + jh_l^m = {} & \frac{r_0^{l-1}}{4\pi\sigma_1}\left(2-\delta_{m0}\right)\frac{(l-m)!}{(l+m)!}\Big[lP_rP_l^m\left(\cos\theta_0\right) \\ & + \frac{mP_\phi}{\sin\theta_0}P_l^m\left(\cos\theta_0\right)e^{j\pi/2} - \frac{P_\theta}{2}\big\{(l-m+1)(l+m)P_l^{m-1}\left(\cos\theta_0\right) \\ & - P_l^{m+1}\left(\cos\theta_0\right)\big\}\Big]e^{jm\phi_0},\quad j=\sqrt{-1} \end{aligned} \tag{7.13}$$

$\left(r_0,\theta_0,\phi_0\right)$ is the coordinate of the dipole in the inner brain region $\left(r_0 < 0.79\text{ cm}\right)$, $\left(r,\theta,\phi\right)$ is the spherical coordinate of a field point, r_p is the distance between the field point and the dipole position, and P_l^m is the associated Legendre function of degree l and order m. σ_1 is the electrical conductivity of the inner brain layer in Figure B.1 and the unbounded space model in Figure 7.1. δ_{m0} is the Kronecker delta function, $\delta_{m0}=0(m\neq 0)$, $\delta_{00}=1$. This $\left(P_r,P_\theta,P_\phi\right)$ is the moment component of the dipole in the spherical coordinate system. The degree must be truncated in practice, and the truncation position is set as $L=150$ in our work. According to equation (7.11), the forward modeling of SPM is established.

7.2.1.3 CPM Forward Theory – The Forward Theory of Dipoles in a Concentric Four-Layer Sphere Model

The potential solution of the dipole in the four-layer concentric spherical volume conductor model is (see Appendix B for details).

$$\Phi(i) = \sum_{l,m}\frac{W_l(i)}{r^{l+1}}S_l^m = \sum_{l,m}\frac{1}{r^{l+1}}\left(U_l^m(i)\cos m\phi + V_l^m(i)\sin m\phi\right)P_l^m\left(\cos\theta\right) \tag{7.14}$$

$$U_l^m(i) = W_l\ (i)g_l^m,\quad V_l^m(i) = W_l\ (i)h_l^m \tag{7.15}$$

$$7.9 \geq r > r_0, i = 1;\quad 8.1 \geq r \geq 7.9,\ i = 2;\quad 8.5 \geq r \geq 8.1, i = 3;\quad 8.8 \geq r \geq 8.5, i = 4$$

Where $i = 1$, 2, 3, and 4 represent the cerebral region, CSF, skull, and scalp layer, respectively (as shown in Figure B.1).

For the inner surface of the cerebral cortex and the surface of the scalp, we have

$$W_l(1) = A_l r_1^{2l+1} + 1 \qquad W_l(4) = G_l\left(r_4^{2l+1} + \chi\right) \tag{7.16}$$

The specific form of A_l, G_l, χ in the equation can be found in Appendix B. Equations (7.15) and (7.16) tell us that the potential on the surface of the inner cerebral cortex consists of two parts: one is generated by the sources in an unbounded space with an electrical conductivity of 0.461 A/(Vm), namely Φ_∞ in equation (7.11) or $\Phi(1)$ in equation (7.16) when $A_{l=}0$, which are SPM; Another part of the effect comes from the head model, which is represented by non-zero A_l in equation (7.16). This situation clearly expresses the mathematical difference between CPM and SPM, and the relationship between them can be simply expressed as

$$\begin{aligned} \text{CPM} &= \text{SPM} + \text{The effects of the head model} \\ &= \Phi_\infty + \text{The effects of the head model} \end{aligned} \tag{7.17}$$

Obviously, if we can get the potential Φ_∞ (SPM), there is no need to use $\Phi(1)$ (CPM), which still contains the effects of the head model, because one of the main purposes of developing the high-resolution EEG is to eliminate the effects of the head model. When we get CPM from scalp surface potential, we have eliminated the main effect of the head model, but not all of it. When we get SPM, we have eliminated the effect of the head model in theory. Thus, SPM can be regarded as a further cleaned version of CPM. Actually, the phenomenon in equation (7.17) has been displayed in Section 7.1 for the simplest spherical head model.

7.2.1.4 Forward Modeling Method

In general, the scalp surface potential of the four-layer sphere head model can be expressed by the following linear equation

$$\Phi_{4\,\text{scalp}} = G_{4\,\text{scalp}} S \tag{7.18}$$

And the potential on the surface of the cortex can be expressed as

$$\Phi_{4\,\text{brain}} = G_{4\,\text{brain}} S \tag{7.19}$$

Here S is the actual or equivalent neural electrical activity sources, and $\Phi_{4\,\text{scalp}}$ is the potential obtained at the Scalp. $G_{4\,\text{scalp}}$ and $G_{4\,\text{brain}}$ are the leadfield matrices determined by the head model, for four-layer sphere model, they are represented by equation (7.14) with the position information of source and electrode. $\Phi_{4\,\text{brain}}$ is the potential on the surface of the cerebral cortex.

Similarly, for the unbounded space model, we have

$$\Phi_{\text{scalp}} = I_{\text{scalp}} S \tag{7.20}$$

And

$$\Phi_{\text{brain}} = I_{\text{brain}} S \tag{7.21}$$

Here, S is the same as in equations (7.18) and (7.19). Φ_{scalp} and Φ_{brain} are potential generated by the same sources at the same radius in the unbounded space. I_{scalp} and I_{brain} are the leadfield (transfer) matrices derived from equation (7.11). It should be pointed out that for the unbounded space, the complex spherical harmonic function calculation in equation (7.11) is unnecessary because the related calculation can be accomplished quickly and accurately by using the first equation in equation (7.11), i.e

$$\Phi_{\text{scalp(brain)}}(\vec{r}) = \frac{1}{4\pi\sigma_1}\vec{P}\cdot\nabla_{r_0}\left(\frac{1}{r_p}\right) = \frac{1}{4\pi\sigma_1}\frac{\vec{P}\cdot\vec{r}_p}{r_p^3}, \quad \vec{r}_p = \vec{r} - \vec{r}_0 \tag{7.22}$$

7.2.1.5 Forward Simulation Experiment

We assume S to be three dipoles in the model of a four-layer concentric volume conductor sphere.

Figure 7.2 shows the results of the forward modeling. The two above in the figure are the results of the four-layer spherical model obtained from equation (7.14). The location information is seriously blurred by the four-layer sphere model. The map on the right shows the cortical surface potential map (CPM) ($r = 7.9$ cm) from the equation (7.19). The three sources in the map are clearly shown, showing the higher spatial resolution of CPM than EEG, which is the fundamental reason why so much attention has been paid to the study of CPM. The lower left figure shows the potential (SPM) generated on the surface $r = 8.8$ cm in an unbounded space by the same set of sources derived from equations (7.20). The lower right figure shows the potential (SPM) generated on the surface $r = 7.9$ cm in an unbounded space by the same set of sources derived from equations (7.21). The SPM without the influence of the head model can provide a higher spatial resolution image than EEG and CPM on both the cerebral cortex ($r = 7.9$ cm) and the scalp ($r = 8.8$ cm), demonstrating the high-resolution imaging capability of the SPM.

By comparing equations (7.14) and (7.11), it can be seen that their difference lies in spatial filtering. For the four-layer concentric sphere model, $W_l(1) = A_l r_1^{2l+1} + 1$, $W_l(4) = G_l\left(r_4^{2l+1} + \chi\right)$, and for the unbounded space medium $W_l(1) = 1$, $W_l(4) = 1$. Figure 7.3 shows the normalized spatial filter, which clearly shows how the high spatial frequency component (large l) is suppressed by the four-layer head model. Under such suppression, the scalp surface EEG with the low spatial resolution is obtained as shown in the upper left of Figure 7.2.

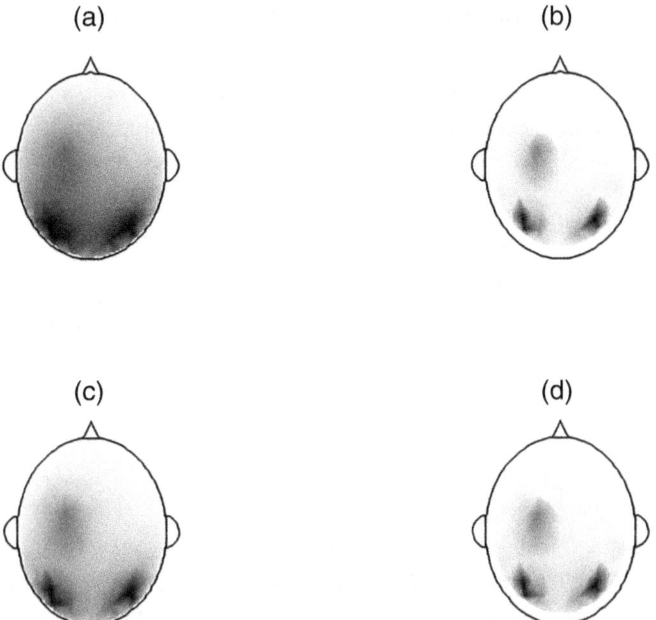

FIGURE 7.2 Normalized topographic map of the three dipole sources. The top row shows the potential topography of the source surrounded by a concentric four-layer volumetric sphere of conductors (Figure B.1), EEG on the scalp surface and CPM at $r=7.9$ cm. The lower row shows the corresponding SPM (Yao, 2001b). (a) Potential at $r=8.8$ cm in the four-sphere model, (b) potential at $r=7.9$ cm in the four-sphere model, (c) potential at $r=8.8$ cm in infinite medium, and (d) potential at $r=7.9$ cm in infinite medium.

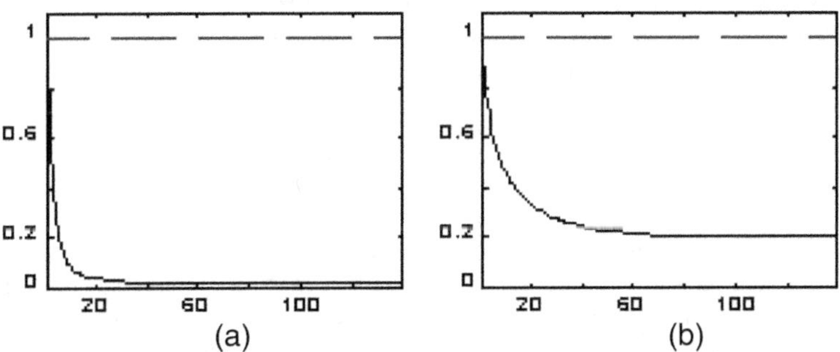

FIGURE 7.3 Normalized amplitude of spatial spectrum. The ordinate is the spatial-spectral amplitude, and the abscissa is the degree l of the spatial filter. The solid line represents CPM, the solid line in panel (a) represents the spatial filter of the scalp surface (8.8 cm) EEG, and the solid line in panel (b) represents the spatial filter of the cortical surface (7.9 cm) CPM. The dotted line indicates SPM, which has a value of 1.0 regardless of the spatial position of 7.9 or 8.8 cm, that is, there is no filtering effect of the conductor model (Yao, 2001b).

7.2.2 Inversion Simulation Test

CPM has almost become a synonym for high resolution EEG. However, the above comparison of SPM and CPM shows that the physical model of SPM is simpler than that of CPM and the calculations are simpler, and the resolution is higher than CPM, which means that SPM is likely to be a strong competitor to CPM. In this part, we provide a technique to recover CPM and SPM from scalp potential information by using the equivalent distributed source method, and then carry out a simulation experiment.

7.2.2.1 Equivalent Source Technique

In Chapter 6, we systematically introduced the equivalent distributed source theory and method, and here we introduce its numerical implementation.

According to equation (7.18), we have

$$S = G^{+}_{4\,\text{scalp}}\Phi_{4\,\text{scalp}} \tag{7.23}$$

Where "+" denotes the generalized inverse. In practice, "+" is derived from SVD. For data with noise, the truncated SVD pseudo-inverse can be used to reduce the impact of noise on the results. Substituting equation (7.23) into equation (7.19–7.21), we obtain $\Phi_{4\,\text{brain}}$, Φ_{scalp} and Φ_{brain}. This procedure shows that if we know the source $\vec{S}$, we can get CPM (7.19) and SPM (7.20 and 7.21). However, it should be noted that if we know the source, we don't need SPM/CPM anymore, because all we want is the source. It is very difficult in practice to obtain the actual sources by solving the inverse problem, because it faces the non-uniqueness of the solution, which is discussed more in Chapter 10 later in this book. We may take the high-resolution imaging as a compromise between low-resolution scalp EEG and actual source information.

Interestingly, the difficulty of the EEG inverse problem arises from the non-uniqueness of the solution, ,however, it is this non-uniqueness that induces the equivalent source technique that performs well in high-resolution imaging techniques and zero reference reconstruction, namely some equivalent sources can produce the same scalp and cortical surface potential as the actual source. Writing it in mathematical terms is

$$\Phi_{*} = H_{*}S' \tag{7.24}$$

Where, S' is the equivalent source vector of the actual source vector S, H_{*} is one of the leadfield (transfer) matrices $G'_{4\,\text{scalp}}$, $G'_{4\,\text{brain}}$, I'_{scalp} and I'_{brain}, and the potential vector Φ_{*} is one of $\Phi_{4\,\text{scalp}}$, $\Phi_{4\,\text{brain}}$, Φ_{scalp} and Φ_{brain} corresponding to the equivalent source S' instead of the actual source vector S. According to equation (7.24), we have

$$S' = G'^{+}_{4\text{scalp}}\Phi_{4\text{scalp}} \tag{7.25}$$

Substitute S' back into equation (7.24)

$$\Phi_{4\text{brain}} = G'_{4\text{brain}}S' \tag{7.26}$$

$$\Phi_{\text{scalp}} = I'_{\text{scalp}} S' \tag{7.27}$$

$$\Phi_{\text{brain}} = I'_{\text{brain}} S' \tag{7.28}$$

By comparing equation (7.25) with equation (7.23), it can be seen that they are the same in form. The essential difference between the two is that the position of the equivalent source in equation (7.25) is pre-assumed, so equation (7.25) is a linear inversion, and the unique solution can be obtained in the sense of minimum norm. The location of the actual source in equation (7.23) is unknown, as it is a nonlinear problem requiring determining the location of the source, which is naturally much more difficult than the linear problem. The equivalent source method and cortical imaging skillfully avoid the nonunique and non-linear inversion of the real sources.

In Chapter 6 and Section 6.1, we have introduced three EST, that is, S' can be either an equivalent multipole source of the actual electrical source at the origin of coordinates (SHS) or a closed dipole layer or a closed charge layer enclosing the real source. In the past, all three ESTs have been used to obtain CPM. In the work of obtaining SPM with EST, the equivalent dipole layer was initially adopted (Yao, 1995), and then the equivalent multipole method was adopted (Yao, 2000).

7.2.2.2 Inverse Results of the Four-Layer Concentric Sphere Model

SPM and CPM are accomplished simultaneously with an equivalent dipole layer. The specific implementation process is as follows:

There is a discrete radial dipole layer at $r = 7.5$ cm. In the simulation, $N = 3{,}000$ dipoles are used, and they are approximately uniformly distributed on the surface of the upper hemisphere with a spherical radius of 7.5 cm.

The moment vector S' of N dipoles is obtained by solving the equation (7.25) with SVD from the potential $\Phi_{4\,\text{scalp}}$.

S' was substituted into the equations (7.27) and (7.28) to obtain SPM at $r = 8.8$ and 7.9 cm, and CPM at $r = 7.9$ cm with equation (7.26).

We evaluated the CPM and SPM reconstructed from the scalp surface potential V shown in the upper left figure in Figure 7.2 (Yao, 2001b). The scalp surface potential here is noise-free, and the SVD is not truncated in the reconstruction. Correlation analysis between the results(figure omitted) with corresponding figures in Figure 7.2, the correlation coefficients are 1.0 (upper left, EEG), 0.9909 (upper right, CPM), 0.9998 (lower left, SPM), and 0.9909 (lower right, SPM), respectively. These results indicate that the equivalent dipole layer technique restores these four images very well. For more details on computation, refer to Yao (2001a).

7.2.2.3 Inverse Results of a Realistic Head Model

Figure 7.4 shows CPM, SPM, and the difference between them (DPM) with the help of the boundary element method based on the realistic head model (Figure 5.3) (Cai and Yao, 2003). The results again show that SPM has a higher resolution.

Source potential and cortical potential imaging maps of two dipoles in a realistic head model. CPM-cortical potential mapping, SPM-source potential mapping, DPM = SPM − CPM.

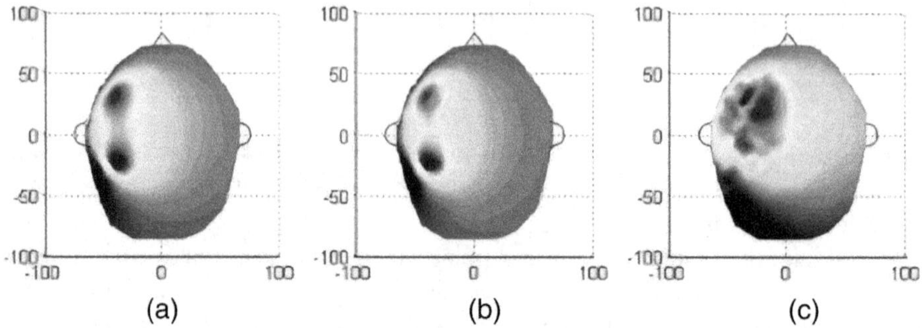

FIGURE 7.4 Source potential and cortical potential imaging maps of two dipoles in a realistic head model. (a) CPM image. (b) SPM image. (c) DPM image.

7.3 COMPARISON BETWEEN DIPOLE AND CHARGE APPROACH

The equivalent distributed dipole source and charge source have been introduced in Chapter 6, and their implementation methods are similar to that in 7.2. Therefore, they will not be repeated in this section. Here we mainly introduce some comparative simulation results, including the results of different calculation methods. It should be noted that these results were done in 1993–1995 on a single-layer spherical model, but the implementation steps on a multi-layer spherical or multi-layer realistic head model are exactly the same. Here the equivalent source has been used as the imaging quantity — ESM, including Equivalent Dipole/charge Layer Mapping (EDL/ECL) .

7.3.1 Equivalent Distributed Sources Inversion

Assume there is a homogeneous sphere model with a radius $r = 1.0$ and a conductivity of 1.0. There are two unit dipoles located at $\left(0.2\sin(\pi/6), 0.0, 0.2\cos(\pi/6)\right)$ and $\left(-0.2\sin(\pi/6), 0.0, 0.2\cos(\pi/6)\right)$. There are 33 electrodes uniformly distributed in the upper hemisphere.

7.3.1.1 Equivalent Charge Layer Inversion

Calculate the scalp potential using the analytical solution described in Section 4.2, only one peak (figure omitted) could be seen on the scalp (Yao and Luo, 1996). Similar to the method in Section 7.2, there are 160 uniformly distributed equivalent point charges on the surface of the upper hemisphere of $r_T = 0.40$, and then use SVD to get the charge distribution. Figure 7.5a is the inversion result. This method is called the Equivalent Charge Source Based Cortical Imaging Technique (CCIT) (Yao, 1996). Obviously, it clearly shows the presence of two sources, with a maximum value of 0.1624 and a minimum value of −0.0502. Figure 7.5b is the equivalent

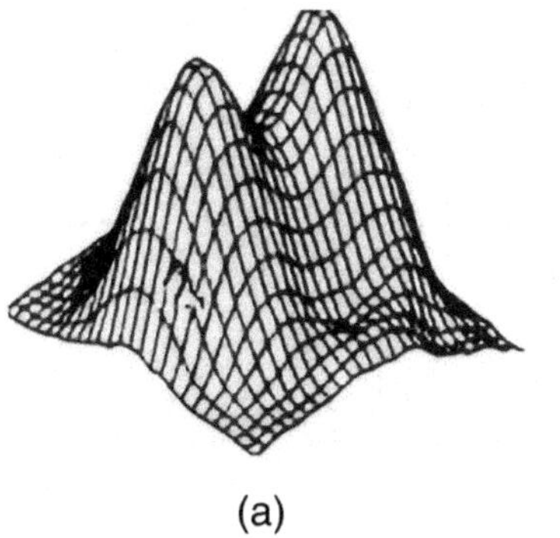

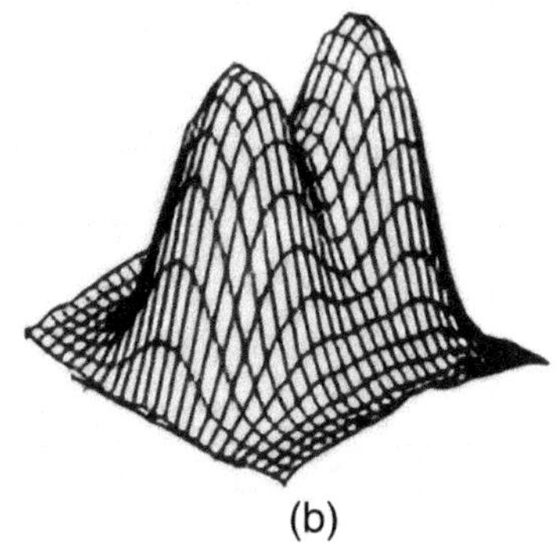

FIGURE 7.5 Inversion of equivalent distributed point charge layer (ECL). (a) Direct SVD inversion; (b) spherical harmonic analysis inversion.

distributed point charge layer calculated by spherical harmonic analysis inversion (refer to Section 7.1) and then using the forward formula of the equivalent distributed point charge source in Section 6.2 (equations 6.13 and 6.14). Its maximum value is 0.1474 and its minimum value is −0.036. The value is a little smaller, but the pattern is almost the same.

7.3.1.2 Equivalent Dipole Layer Inversion

The idea is the same as in Section 7.2, 160 evenly distributed dipoles are set on the surface of the upper sphere of $r_T = 0.28$, and then solved by SVD. The result is similar to Figure 7.5a, thus omitted. This method is called the Equivalent Dipole Source Based Cortical Imaging Technique (DCIT) (Yao, 1996). It also clearly shows the presence of two sources, with a maximum value of 0.0585 and a minimum value of −0.0047. Similar to the above charge layer inversion, first is a spherical harmonic analysis inversion (refer to Section 7.1) and then a spherical harmonic series forward to obtain an equivalent distributed dipole source with equation (6.38), the result figure is similar to Figure 7.5b, thus omitted. Its maximum value is 0.0544 and its minimum value is −0.01048. These results indicate that equivalent distributed sources (EDS) can also be used as the imaging physical quantity of high-resolution EEG, and its physical significance has been introduced in Chapter 6.

7.3.2 Simulation Test for the Charge, Dipole, and Quadrupole Sources Test

In the following experiments, the number of distributed sources is 160 and the electrode is 33. Figure 7.6 is a cortical potential imaging comparison of the two dipole sources. In the experiment, both CCIT and DCIT used $r_T = 0.35$. The two dipoles are located at $(x, y, z) = \left(\pm 0.3 \sin(\pi/6), 0.0, 0.3 \cos(\pi/6)\right)$, and their dipole moments are $(p_x, p_y, p_z) = \left(\pm 0.5 \sin(\pi/6), 0.0, 0.5 \cos(\pi/6)\right)$. Obviously, both methods achieve good cortical potential imaging of the dipoles (Yao, 1996).

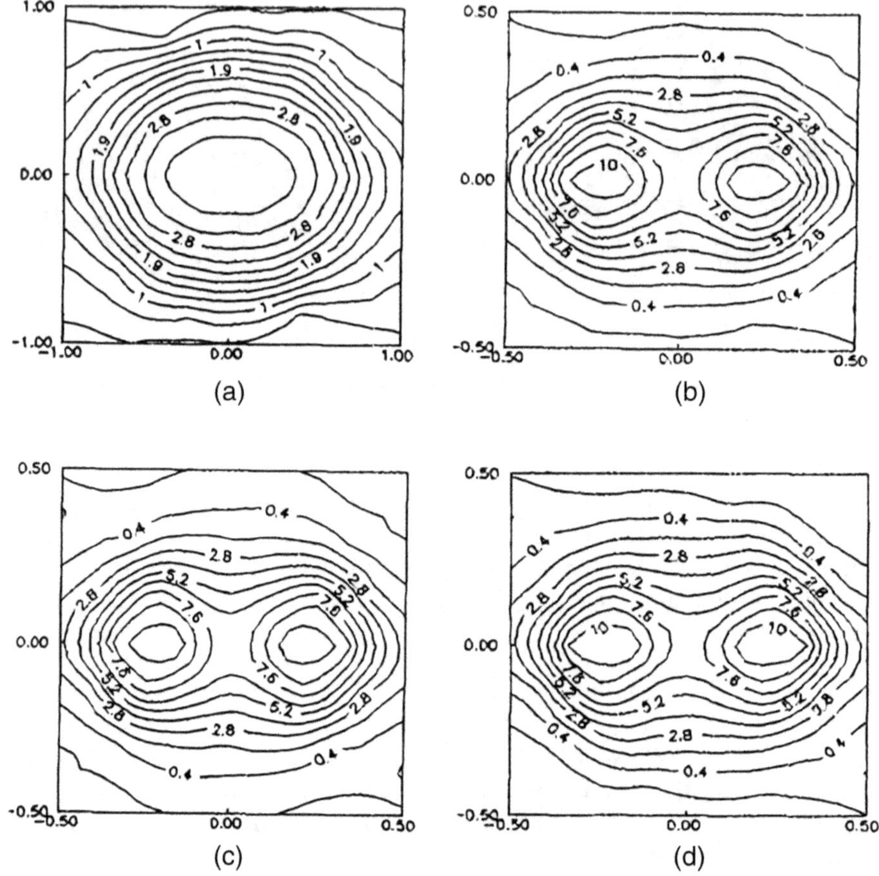

FIGURE 7.6 Potential imaging of dipole sources by DCIT and CCIT. (a) Forward scalp potential ($r=1.0$); (b) forward cortical potential ($r=0.50$); (c) cortical potential of DCIT reconstruction ($r=0.50$); (d) cortical potential of CCIT reconstruction ($r=0.50$) (Yao, 1996).

Figure 7.7 is the cortical potential imaging of two-point charge sources. The CCIT and DCIT parameters used here are the same as those in Figure 7.6. The point charge is at $\left(\pm 0.3\sin\left(\pi/6\right), 0.0, 0.3\cos\left(\pi/6\right)\right)$ and the intensity is ±1 (total charge is zero). The results in Figure 7.7 show that the effect of CCIT is better than that of DCIT (Yao, 1996).

Figure 7.8 is the cortical potential imaging results of a pair of quadrupole sources Q_{zz} (Q_{33} component of the quadrupole moment tensor) located at $\left(\pm 0.3\sin\left(\pi/6\right), 0.0, 0.3\cos\left(\pi/6\right)\right)$ with an intensity of 1. The theoretical record is calculated by the analytical solution formula of the quadrupole source potential in the homogeneous sphere introduced in Section 4.3. The number of distributed sources used is still 160, the number of electrodes is still 33, and the spherical surface of the distributed source is $r_T = 0.35$. The results in Figure 7.8 show that CCIT and DCIT can also accurately image the quadrupole sources (Yao, 1995).

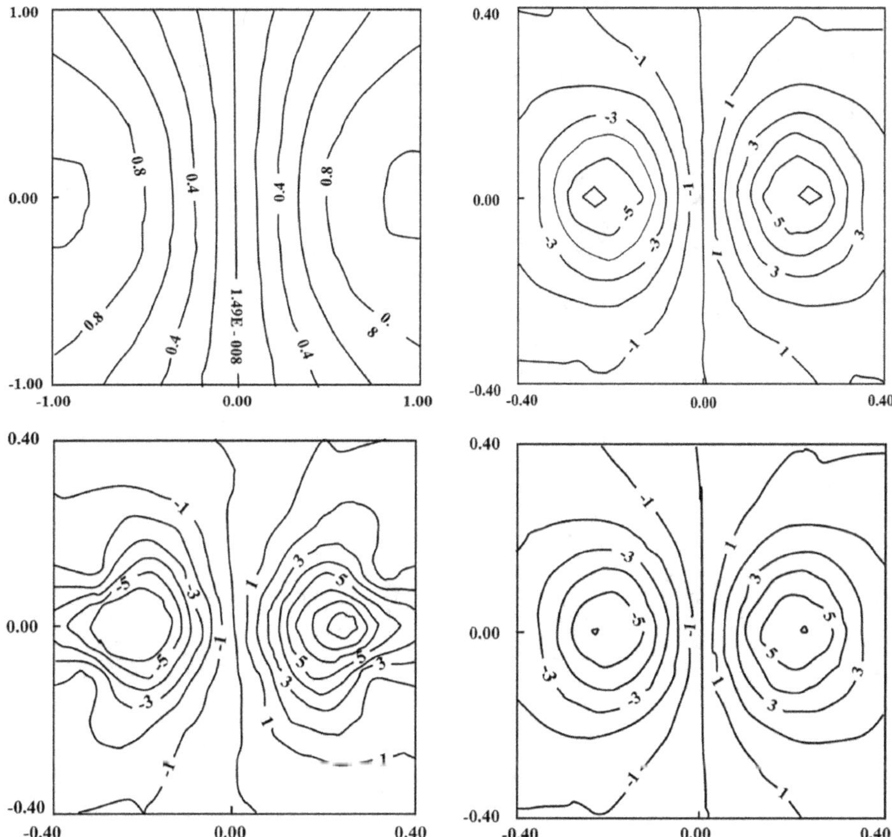

FIGURE 7.7 Potential imaging of point charge sources with DCIT and CCIT. (a) Forward scalp potential ($r = 1.0$); (b) forward cortical potential ($r = 0.40$); (c) the cortical potential of DCIT reconstruction ($r = 0.4$); (d) the cortical potential of CCIT reconstruction ($r = 0.4$) (Yao, 1996).

Based on the above results, it can be seen that the equivalent source and the real source may not be the same kind of source, because the equivalence refers to the potential without the requirement of the source type, which is also the implicit conclusion of the forward theory of equivalent sources in Chapter 6. It should be noted that the actual EEG sources may be neither point charge sources, nor simple dipole sources or quadrupole sources, but some of their combinations. Therefore, this conclusion is an important basis for the practical application of CCIT and DCIT.

In addition, the results in Figures 7.6 and 7.7 also indicate that equivalent sources identical to real sources realize better results, (c) in Figure 7.6 and (d) in Figure 7.7, most notably CCIT imaging of point charges. This fact means that we should probably try cortical imaging techniques based on multiple equivalent sources to achieve optimal results.

Finally, it is also found in our simulation study (Yao, 1995) that the equivalent distributed source does not necessarily surround the real source, which seems to contradict the forward modeling theory in Chapter 6. In fact, for some problems, the

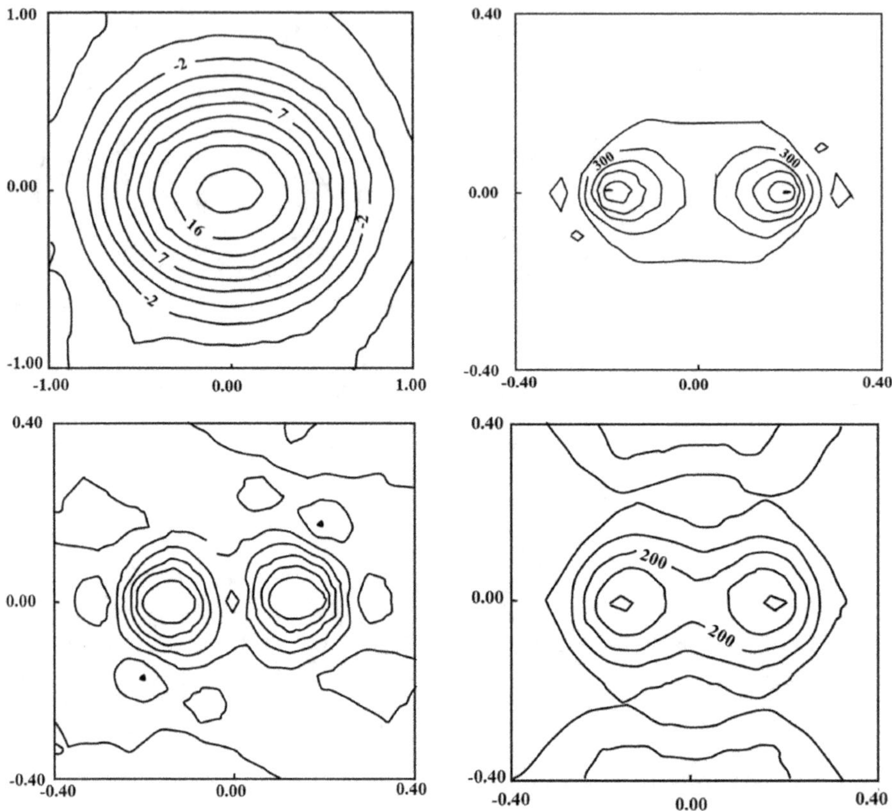

FIGURE 7.8 Quadrupole imaging by DCIT and CCIT. (a) Forward scalp potential ($r=1.0$); (b) forward cortical potential ($r=0.40$); (c) the cortical potential of DCIT reconstruction ($r=0.40$); (d) the cortical potential of CCIT reconstruction ($r=0.40$).

equivalent distributed source which is even slightly deeper than the real source can achieve better results (Yao, 1995). Based on the theory of spherical harmonic analysis, it is not difficult to understand this phenomenon. The equivalence between the equivalent distributed source and the real source can only be realized approximately in practice, specifically to be equivalent mainly in the low-frequency spatial spectrum. In the sense of inverting the low-frequency spatial spectrum, it is possible to reach similar effectiveness by using an equivalent distributed source whose depth varies within a certain range, it can be a multipole series which is located at the center of the sphere, it also can be a 3D distributed source within a volume almost overlapping the actual source, and certainly, it can be a distributed source layer with all the actual sources enclosed inside. Of course, due to the non-uniqueness of the inversion and the singularity of the equation, there will be differences in the accuracy and stability of the solution when different schemes are used to invert the low-frequency spatial spectral components, so that a better performance of a deeper equivalent distribution source is likely to appear than the shallower one. Of course, this situation also illustrates the complex relationship between theory and practice, practice is the most important, theory may suggest possible implementations, and the two help each other.

7.4 SPONTANEOUS ALPHA WAVE IMAGING

7.4.1 METHODS

In Yao et al. (2004a), we recruit 11 healthy right-handed male adult volunteers (age: 18–30 years) from the Aalborg University staff and students. Informed consent was obtained from each subject before the study. The EEG was recorded from 128 surface electrodes. The EEG epochs of 2 minutes were sampled at 512 Hz. The on-line reference was the left mastoid, and the data were re-referenced off-line to average reference for further analysis (Yao etal., 2019).

We took the charge layer (CL) inversion, the head model is a four-concentric spheres volume conductor (Cuffin and Cohen 1977, Appendix B). The parameters are the same as in Figure B.1. The discrete CL with 680 charges was set and uniformly distributed on an inner sphere surface with a radius of 7.5 cm. The CL estimation X was obtained by solving a linear equation $V=GX$ where V is the scalp potential recordings, G is a leadfield matrix determined by the head volume conductor and the CL configuration on the layer. The inverse is computed by truncated SVD pseudo-inverse (minimum norm solution), and the truncation is chosen at 31 where the ratio of the 31st singular value to the first singular value is 0.0178.

The scalp potentials V were segmented to epochs of 1 second, and epochs free of muscle, electrooculogram (EOG), and movement artifacts were visually selected offline. Both the inverted CL and scalp V were transformed by fast Fourier transform (FFT) to get the power (Welch's method in MATLAB). For each subject, an Individual Alpha Peak Frequency (IAPF) of the power curves of the electrodes at O1 and O2 was identified, and the frequency band with IAPF $\pm$ 2 Hz was used to plot the power maps for the scalp and cortex CL, separately.

7.4.2 RESULTS

In this study, the IAPF values at O1 and O2 for both the scalp potential and cortex CL are the same for 10 of the 11 subjects: 11 Hz for four subjects, 10 Hz for five subjects, and 12 Hz for one. Only one subject has an 11 Hz IAPF on the scalp and 12 Hz on the cortex. The results of three representative subjects are shown in Figure 7.11. For the scalp power map, 7 of the 11 subjects show two distinctive occipital activities as Ss-1 and Ss-2 in Figure 7.9. The other four cases show blurred occipital activities as Ss-3. However, for CL, all the maps of the 11 subjects clearly show a distinct location of the main activities at the left and right occipital regions. Seven CL maps of the 11 subjects have mainly two occipital activities as Ss-1, and the other four subjects have clear sources located at the middle occipital region as Ss-2 and Ss-3. A parietal alpha, in addition to the occipital sites, is also visible in the Ss-2.

Normalized spatial positions of the maxima and their spherical coordinates are shown in Figure 7.10. They have searched automatically within each hemisphere. Based on Figure 7.10, the maxima (+) of the cortex CL power are much closer to each other (some of them are overlapped) than the maxima (o) of the scalp potential. The azimuth angles (radians) of the maxima positions are -0.5 ± 0.2 and 0.5 ± 0.2 for the left and right scalp potential power, respectively, and -0.57 ± 0.08

and 0.5 ± 0.1 for the left and right cortex CL power, respectively. The elevation angles (radians) are 0.3 ± 0.1 for both the left and right scalp potential power and 0.27 ± 0.07 and 0.28 ± 0.06 for the left and right cortex CL power, respectively. Paired statistic tests of the 11 subjects were conducted. The results show that the elevation angles between the left and right hemisphere for either the scalp potential or cortex CL, or between the scalp potential and the cortex CL for either the left or the right hemisphere are non-significant while the azimuth angles between the maxima positions of the left and right hemisphere are significantly ($p < 0.001$) different for either the scalp potential or the cortex CL. The azimuth angles between the maxima positions of the scalp potential power and that of the cortex CL power for either the left or right hemisphere are also significantly ($p < 0.05$) different. These results show that the local maximum of the scalp potential power is consistent in the elevation with that of the cortical CL power, and the maxima distributions between the scalp potential and the cortex CL are significantly different in azimuth as shown in Figure 7.10. This difference may be caused by the non-zero average reference for the scalp potential that needs further clarifying by using zero reference(Yao, 2001).

Regarding the frontal Alpha activities (Nunez et al., 2001), all our 11 subjects including the three in Figure 7.9 did not show distinct frontal alpha activities. However, distinct IAPFs can be identified in 10 of the 11 subjects at the normalized power spectra curves of the scalp potential at the electrode positions F3 and F4, such as the three subjects shown in Figure 7.11. This fact means that the frontal does have alpha electric activities (Nunez et al., 2001) though its strength is much smaller than the occipital region. While for the cortex CL, three F3 and five F4 (including Ss-1 and Ss-3 in Figure 7.11) miss the IAPFs. Meanwhile, the IAPF values at F3 and F4 are the same as those at O1 and O2.

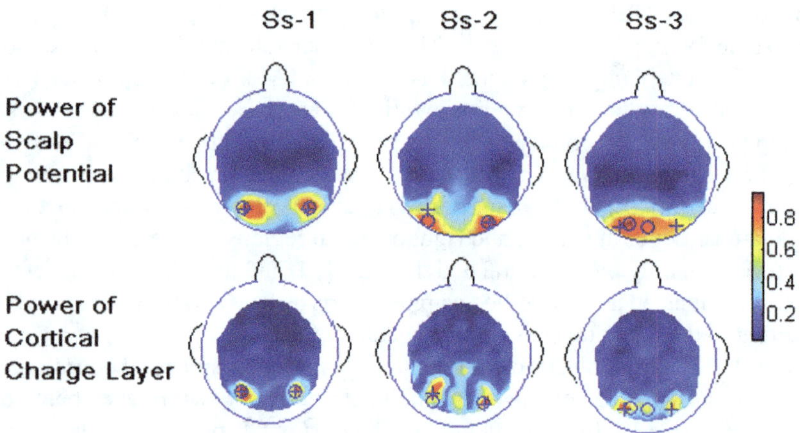

FIGURE 7.9 Normalized power maps of Alpha rhythms in the frequency domain. The "o" and "+" indicate the left and right power maxima of the scalp potential and the cortex CL, respectively. (Modified from Yao et al. (2004a).)

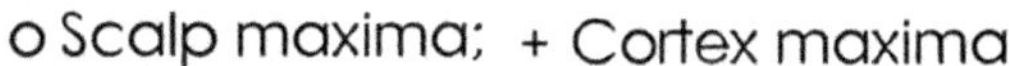

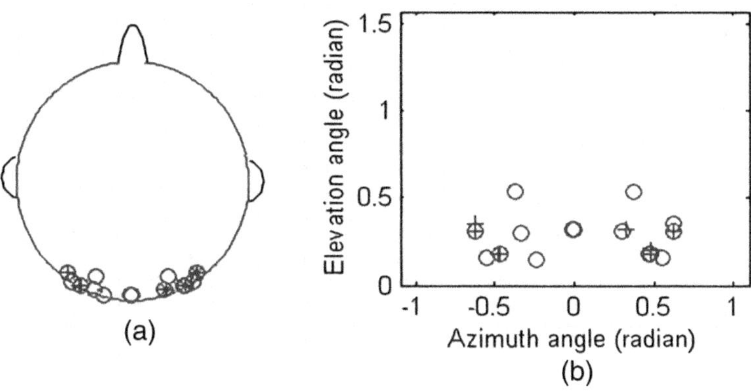

FIGURE 7.10 Normalized spatial positions of the maxima and their spherical coordinates. (a) Normalized spatial positions of both the scalp potential power maxima (o) and the cortex CL power maxima (+). (b) The spherical coordinates of the maxima. For a Cartesian coordinate system with $+x$ axis towards the right ear, y-axis towards the nasion, and +z axis towards the vertex, the azimuth angle is defined as the angle between the projection on the xoy plane of a maxima point and the negative y-axis. A negative azimuth means the projection is in the $-x$ direction. The elevation angle is defined as the angle between the maxima point and its projection on the xoy plane. (From Yao et al. (2004a).)

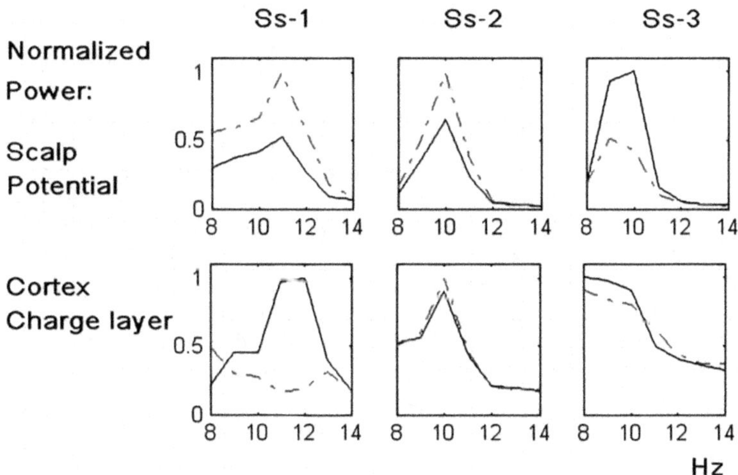

FIGURE 7.11 Normalized power spectra curves of scalp potential and cortex CL. The lateral axis is the frequency (Hz). The spatial positions are F3 (solid line) and F4 (dashed line) on the scalp surface or cortex points with the same spherical coordinates (Yao et al., 2004a).

Specifically, even for the scalp surface, we may have the potential mapping (similar to an equivalent dipole layer (EDL) mapping), power mapping, Scalp Laplacian (SL, Chapter 8), and also may have an ECL mapping (Yao et al., 2004b) and source potential mapping (SPM) (Yao, 2001b). As shown by the theory (Chapter 8) and example in Yao et al. (2004b), SL and ECL would be of higher spatial resolution, and then SPM and EDL, for more details, pls refer to Yao et al. (2004b) and Chapter 8.

Finally, it's important to note that the spheric harmonic analysis method introduced in this chapter can only be used for the multilayer spherical head model. However, the distributed equivalent dipole/charge layer methods have nothing to do with the head model, they can be used for both the spherical head model and the realistic head model. Therefore, the CPM, SPM, and ESM introduced in this chapter also can be implemented in a realistic head model. The basic idea is still to implement ESM inversion on the realistic head model (Zhai and Yao, 2004), and then use ESM as the imaging result, or further calculate the potential of ESM on the quasi-imaging surface in the realistic head model or uniform model to obtain CPM or SPM.

REFERENCES

D Yao, Y Qin, S Hu, L Dong, MLB Vega, PAV Sosa. Which Reference Should We Use for EEG and ERP practice? Brain topography. 2019, 32: 530–549.

Alldredge, L.R. 1981. Rectangular harmonic analysis applied to the geomagnetic field. *J Geophys Res-Sol Ea* 86(B4):3021–3026.

Cai, Y., D. Yao. 2003. A method to calculate source potential in realistic head model and its effectiveness. *J UEST China* 32(2):149–154.

Cuffin, B.N., D. Cohen. 1977. Magnetic fields of a dipole in spherical volume conductor shapes. *IEEE Trans Biomed Eng* 24:372–381.

Edlinger, G., P. Wach, G. Pfurtscheller. 1998. On the realization of an analytic high-resolution EEG. *IEEE Trans Biomed Eng* 45:736–745.

Freeman, W.J. 1980. Use of spatial de-convolution to compensate for distortion of EEG by volume conductor. *IEEE Trans Biomed Eng* 37:421–429.

Gevins, A., J. Le, N.K. Martin, et al. 1993. High resolution EEG: 124-channel recording, spatial enhancement and MRI integration methods. *EEG Clin Neuro* 90:337–358.

Haines, G.V. 1985. Spherical cap harmonic analysis. *J Geophys Res* 90(B3):2583–2591.

Nunez, P.L., B.M. Wingeier, R.B. Silberstein. 2001. Spatial-temporal structures of human alpha rhythms: Theory, microcurrent sources, multiscale measurements, and global binding of local networks. *Hum Brain Mapp* 13:125–164.

Sidman, R., M.R. Ford, G. Ramsey, et al. 1990. Age-related features of the resting and P300 auditory evoked responses using the dipole localization method and cortical imaging techniques. *J Neurosci Meth* 33(1):23–32.

Srebro, R., R.M. Oguz, K. Hughlett, et al. 1993. Estimating regional brain activity from evoked potential field on the scalp. *IEEE Trans Biomed Eng* 40:509–516.

Srinivasan, R., P.L. Nunez, D.M. Tucker, et al. 1996. Spatial sampling and filtering of EEG with spline Laplacians to estimate cortical potentials. *Brain Topogr* 8:355–367.

Yao, D. 1995. Research on 3D imaging technology of EEG by equivalent source method. *J Biomed Eng* 12(4):332–341.

Yao, D. 1996. The equivalent source technique and cortical imaging. *EEG Clin Neuro* 98:478–483. doi: 10.1016/0013-4694(96)94694-5.

Yao, D. 2000. Electric potential produced by a dipole in a homogeneous conducting sphere. *IEEE Trans Biomed Eng* 47(7):964–966.

Yao, D. 2001a. A method to standardize a reference of scalp EEG recordings to a point at infinity. *Physiol Meas* 22(4):693–711.

Yao, D. 2001b. Source potential mapping: A new modality to image neural electric activities. *Phys Med Biol* 46(12):3177–3189.

Yao, D., B. Luo. 1996. Theory of the EEG cortical imaging techniques. *Chin J Biomed Eng (English edition)* 5(3):128–136.

Yao, D., L. Wang, K.D. Nielsen, et al. 2004a. Cortical mapping of EEG Alpha power using a charge layer model. *Brain Topogr* 17(2):65–71.

Yao, D., Y. Zhou, M. Zeng, et al. 2001. A study of equivalent source techniques for high-resolution EEG imaging. *Phys Med Biol* 46(8):2255–2266.

Yao, D., Z. Yin, X. Tang, et al. 2004b. High-resolution electroencephalogram (EEG) mapping: Scalp charge layer. *Phys Med Biol* 49(22):5073–5086.

Zhai, Y., D. Yao. 2004. A study on the reference electrode standardization technique for a realistic head model. *Comput Methods Programs Biomed* 76:229–238.

8 Scalp Laplacian Imaging

The surface Laplacian (SL) technique is a method in EEG and ECG studies for a long time. It emerged from the seminal works of Nicholson (1973), Nicholson and Freeman (1975), and Hjorth (1975). These were followed by efforts to develop better computational methods (Perrin et al., 1989; Gevins et al., 1990; Law et al., 1993; Yao, 2002b; Kayser and Tenke, 2006; Carvalhaes and Suppes, 2011; Carvalhaes and De Barros, 2015), making the technique increasingly popular among the EEG community. The first EEG surface Laplacian (SL) algorithm was implemented by Hjorth (1975) to improve spatial resolution and eliminate the influence of non-zero reference potential distortion. In the field of electrocardiograph (ECG), it was first applied by Kleber et al. (1978). The other pioneering work was known as current source density (CSD) measurement, which was popular with physiologists doing depth recordings in the animal cortex. CSD consists of estimating the second derivatives of recorded field potentials along a single cortical depth coordinate to reveal local synaptic current sources and sinks more sharply (Nunez, 2014). Currently, some neuroscientists and psychologists use the labels "CSD" and "Laplacian" interchangeably, though they are quite different, as shown below (Nunez and Srinivasan, 2006).

This chapter includes three parts: biophysics of Laplacian (Yao, 2002b), a new variable scale Laplacian calculation method for the spherical head (Yao, 2002a) and realistic head model (Zhai and Yao, 2004), and high-order approximation local Laplacian calculation (Lai and Yao, 2009), etc.

8.1 BIOPHYSICS OF LAPLACIAN

8.1.1 Introduction

The detection of current sources and sinks on the scalp surface was the primary impetus for the development of Laplacian topography. The EEG source is assumed to be a layer composed of current dipoles with different dipole moments, originating from parallel arranged pyramidal cells. Based on the boundary condition that no current flows into the air, the instantaneous current snapshot may be like the case, as shown in Figure 8.1. At any given moment, some areas will receive the current from the other areas that the scene consists of small "sources" and "sinks" in the cortex and the scalp. The scalp Laplacian topography considers "sources" and "sinks" in the scalp layer, i.e., the areas where the net current flows through the skull into the scalp layer are called "sources" and the opposite areas are called "sinks". Apparently, if we place the reference electrode far away and the active electrode close to the source or sink region in Figure 8.1, we can get a potential difference reflecting the current flow between the two electrodes. However, if we symmetrically place the active and reference electrodes on both sides of the source region, the potential difference will be zero. Such as that between 2 and 3 in Figure 8.1. It means that the usual potential

DOI: 10.1201/9781032639260-8

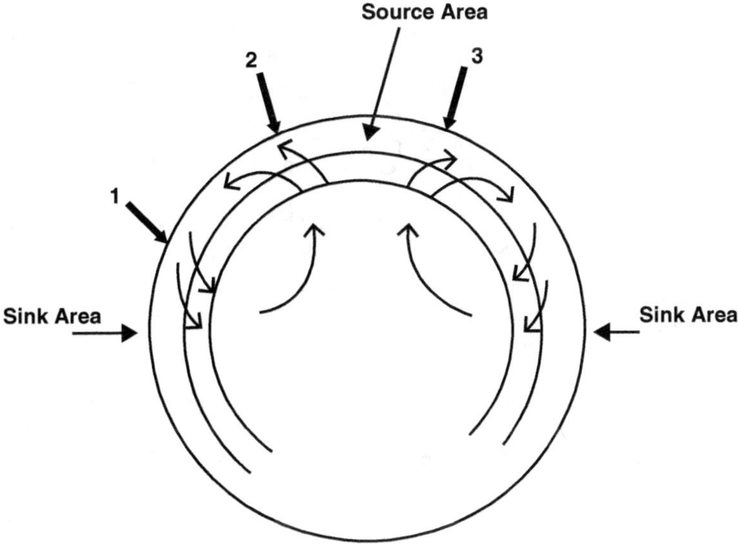

FIGURE 8.1 Schematic diagram of cortical sources and volume conduction currents. The scalp source/sink corresponds to the area where the current flows out/into the skull, respectively (Katznelson, 1981).

difference may not necessarily reflect the "source" or "sink", and the Laplacian is designed to display the sources and sinks directly.

8.1.2 Physics of Scalp Laplacian

Simply, we can take the scalp as a two-dimensional piece of conductor on the poorly conductive skull. As shown in Figure 8.1 or 8.2, electrical currents from sources in the brain pass through the skull and into the scalp and later pass through the skull back into the brain, forming a continuous current line. For an area where there is no source or sink, the current line is parallel to the scalp and will not go in or out of the skull. To determine whether a point is a source or sink of scalp current, we need to estimate the two-dimensional (2D) divergence of the scalp current density, which is defined as

$$\nabla_{xy} \cdot \vec{J} = \frac{\partial J_x}{\partial x} + \frac{\partial J_y}{\partial y} \tag{8.1}$$

where x and y are surface coordinates. The above equation can be expressed by the potential distribution Φ on the scalp. According to Ohm's law: $\vec{J} = \sigma\vec{E}$ and $\vec{E} = -\nabla\Phi$, Formula (8.1) becomes

$$-\frac{1}{\sigma}\nabla \cdot \vec{J} = -\frac{1}{\sigma}\left(\frac{\partial J_x}{\partial x} + \frac{\partial J_y}{\partial y}\right) = \nabla_{xy}{}^2\Phi = \frac{\partial^2\Phi}{\partial x^2} + \frac{\partial^2\Phi}{\partial y^2} \tag{8.2}$$

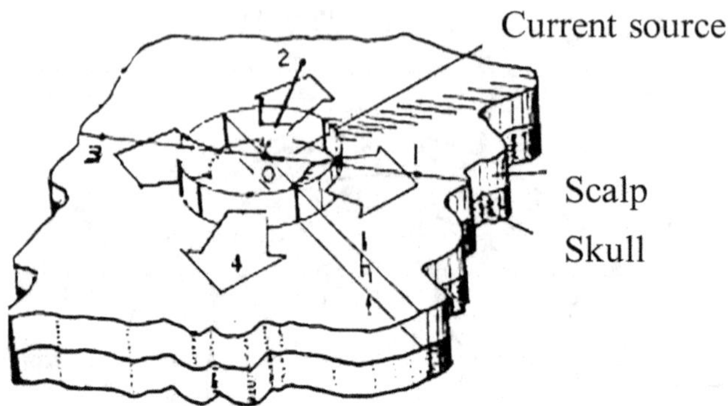

FIGURE 8.2 A small cephalic source area. EEG scalp currents are represented by flat arrows. (From Nunez (1981).) The thickness is h, the radius is d, and the z-axis is normal to the skull from the skull through the scalp.

It should be noted that the electrical potential divergence of equations (8.1) and (8.2) is a 2Ddivergence, which corresponds to the appearance or disappearance of the current in the scalp. The 3D divergence of the current is zero due to the conservation of charge. Equation (8.2) indicates that the 2D Laplacian (second-order derivative calculation) of the potential will obtain a quantity proportional to the 2D divergence of the current source or sink. Obviously, based on equation (8.2) and the fact that the 3D divergence of the current is zero, Laplacian can also be denoted as

$$L_{xy} = \frac{\partial^2 \Phi}{\partial x^2} + \frac{\partial^2 \Phi}{\partial y^2} = -\frac{\partial^2 \Phi}{\partial z^2} \tag{8.3}$$

Thus, Laplacian can also be achieved by the depth-oriented second-order derivative recording, which has been used in invasive implanted electrophysiological studies for many years. Another approach is to use a computer to solve the 2D second-order derivative of the potential in equation (8.3), based on the scalp potential records. The original approach (Hjorth, 1975) was a difference approximation of equation (8.3)

$$L_{xy} \approx (\Phi_1 + \Phi_2 + \Phi_3 + \Phi_4 - 4\Phi_0)/(2d^2) \tag{8.4}$$

Here, the Laplacian is calculated according to the potentials of four points around Φ_0, d is the distance between two electrodes, so the above equation is the second-order approximation for a rectangular electrode array. If the observation array is non-uniform and asymmetrical, more points can be used, and the corresponding accuracy can be estimated by the different theories. Over the past decades, the development of Laplacian technology has focused on the calculation method of different head models with different complexities and the direct measurement of the scalp SL. They are described separately in the later sections of this chapter. Here, to further illustrate the physical meaning of Laplacian, we introduce Laplacian from the point of view of integration rather than differentiation, which is described separately in the later sections of this chapter. Considering the cylindrical disc of the scalp surface shown in Figure 8.2, assume that the total current entering the cylinder is I, the electrode spacing is d, the scalp layer thickness is h, and the electrical conductivity is σ, which is much larger than the electrical conductivity of the skull. Thus, the total current across the side of the cylinder is I, and it can be expressed as

$$I = \iint_S \vec{J} \cdot d\vec{s} = \sigma \iint_S \vec{E} \cdot d\vec{s} \tag{8.5}$$

Now divide the surface of the cylinder into four equal parts (like the four sides of a square cylinder), and use the potential gradient of the center points as an estimate of the electric field at the center points of each side

$$\begin{aligned} I &= \sigma \iint_S \vec{E} \cdot d\vec{s} \approx \sigma \left[\left(\frac{\Phi_0 - \Phi_1}{d} \right) \frac{S}{4} + \left(\frac{\Phi_0 - \Phi_2}{d} \right) \frac{S}{4} + \left(\frac{\Phi_0 - \Phi_3}{d} \right) \frac{S}{4} + \left(\frac{\Phi_0 - \Phi_1}{d} \right) \frac{S}{4} \right] \\ &= \frac{\pi \sigma h}{2} \left(4\Phi_0 - \Phi_1 - \Phi_2 - \Phi_3 - \Phi_4 \right) \end{aligned} \tag{8.6}$$

where $S = \pi(2d)h$. Now if dividing this current by the base area of the cylinder, we can get an estimate of the current density entering the scalp vertically (Nunez, 1981)

$$J_{\uparrow} \approx \frac{I}{\pi(d)^2} = \frac{\sigma h}{2d^2} \left(4\Phi_0 - \Phi_1 - \Phi_2 - \Phi_3 - \Phi_4 \right) = -\sigma h L_{xy} \tag{8.7}$$

This equation has the same calculation operation as that of Laplacian in equation (8.4), with only the difference of a constant σh, where h is the thickness of the scalp or a thickness to which the "first-order" difference approximation in equation (8.6) is valid. This formula can also be extended to the case of multi-electrode recording, with the result being proportional to the current source/sink density too. The difference is that the former estimate, equation (8.4), is the 2D divergence of the current density, which can be regarded as the current flow out/in from the side wall of the cylinder with unit thickness, while equation (8.7) is the current flow out/in from the side wall of a cylinder with a thickness of h. The electric field in equation (8.7) is an average of those in the range of thickness h, while the former is the electric field in the very thin layer of the realistic scalp layer. Therefore, both can be considered the same quantity in a realistic sense, although they are different from each other in details. Equations (8.4) and (8.7) correlate Laplacian and vertical current density, which entered the scalp layer of thickness h, in the sense of a 2D flat piece, they provide a physical explanation for Laplacian. Appendix D shows the relation between the Laplacian and the potential of the skull surface. Furthermore, Nunez and Srinivasan (2006) proposed the assumption that the skull is very thin and most current flows in the radial direction through the skull into the scalp, they proved that the radial current is approximately proportional to the inner surface potential of the skull, so the Laplacian can be applied as an approximate cortical imaging method.

For more details, one can refer to the book of Nunez and Srinivasan (2006) on pages 267–268 and 317–322, where the following relationship was derived

$$L_{xy} = \frac{1}{d_{\text{skull}} d_{\text{scalp}}} \frac{r_{\text{skull-inner}} \sigma_{\text{skull}}}{r_{\text{skull-outer}} \sigma_{\text{scalp}}} \left(\Phi_{\text{skull-inner}} - \Phi_{\text{skull-outer}} \right) \tag{8.8}$$

Here, $\Phi_{\text{cortex}} \approx \Phi_{\text{skull-inner}}$, $\Phi_{\text{skull-outer}} \approx \Phi_{\text{sclap}}$ are the potentials at the inner and outer surfaces of the skull, $r_{\text{skull-inner}}$ and $r_{\text{skull-outer}}$ are the radii of the skull innter/outer surface, σ_{skull} and σ_{scalp} are the conductivities of the skull and the scalp, d_{skull} and d_{sclap} are the thicknesses of the skull and scalp, and the scalp Laplacian $L_{xy} = L_{\text{scalp}} = \nabla^2 \Phi_{\text{scalp}}$. From a large number of simulation studies, it can be found that $\Phi_{\text{cortex}} \gg \Phi_{\text{skull-outer}}$ (Nunez et al., 2019). So, it can be concluded that

$$L_{\text{scalp}} \propto \Phi_{\text{cortex}} \tag{8.9}$$

Therefore, the scalp SL can be used as an estimate of cortical surface potential. In terms of relative values, this approximation does not require information on the skull conductivity but is held only when the skull conductivity is significantly weaker than that of the other layers (Nunez and Srinivasan, 2006).

The following is a further analysis from the perspective of a spherical model that is closer to a realistic head shape than a planar model (Yao, 2002a).

Assume the surface of the skull and scalp is composed of two concentric spheres. Then the potential is

$$\Phi(r,\theta,\phi) = \sum_{l,m} K_l(r) G_l S_l^m \left(\vec{P},\theta,\phi\right) \tag{8.10}$$

G_l is determined by the conductor model. For three- and four-layer concentric sphere models, refer to the formula in Chapter 4. $S_l^m\left(\vec{P},\theta,\phi\right)$ depends on the spherical polar coordinates of the dipole and the field points. $K_l(r)$ is a quantity that is only related to the radius in the scalp layer

$$K_l(r) = r^l + \chi \frac{1}{r^{l+1}}, \quad \chi = \frac{l}{l+1} c^{2l+1} \tag{8.11}$$

c is the radius of the scalp. Accordingly, the scalp Laplacian is (Yao, 2000).

$$\begin{aligned} Ls &= -\frac{1}{r^2} \frac{\partial}{\partial r} \left(r^2 \frac{\partial \Phi}{\partial r} \right) \Bigg|_{r=c} = -\sum_{l,m} \left[l(l+1) r^l + (l+1) l \chi \frac{1}{r^{l+1}} \right]_{r=c} \frac{1}{c^2} G_l S_l^m \\ &= \sum_{l,m} \left(\frac{l(l+1)}{-c^2} \right) K_l(c) G_l S_l^m = -\sum_{l,m} l(2l+1) c^{l-2} G_l S_l^m \end{aligned} \tag{8.12}$$

Thus, the Laplacian factor is

$$L_l = -(l+1)l/c^2 \tag{8.13}$$

Compared with the scalp surface potential, the extra factor (8.13) indicates that Laplacian is the result of high-pass filtering on the potential in the spatial domain.

Then let's study the current density from the skull to the scalp

$$\begin{aligned} J_{\uparrow} &= -\sigma \frac{\partial \Phi}{\partial r}\bigg|_{r=r_k} = -\sigma \sum_{l,m} \left[l r^{l-1} - (l+1)\chi \frac{1}{r^{l+2}} \right]_{r=r_k} G_l S_l^m \\ &= -\sigma \sum_{l,m} \left[l r_k^{\,l-1} - l c^{2l+1} \frac{1}{r_k^{\,l+2}} \right] G_l S_l^m \end{aligned} \tag{8.14}$$

This formula indicates that we can calculate the true current density from the skull to the scalp layer if the results of spherical harmonic analysis are known.

Suppose the thickness of the scalp is h, there is $r_k = c - h$, the outer radius of the skull, then

$$\begin{aligned} J_{\uparrow} &= -\sigma \sum_{l,m} \left[l(c-h)^{l-1} - l c^{2l+1}(c-h)^{-(l+2)} \right] G_l S_l^m \\ &= -\sigma \sum_{l,m} \left[\left(1-\frac{h}{c}\right)^{l-1} - \left(1-\frac{h}{c}\right)^{-(l+2)} \right] l c^{l-1} G_l S_l^m \\ &= \sum_{l} F(c,h,l)\sigma L_s \end{aligned} \tag{8.15}$$

And

$$F(c,h,l) = \frac{c}{2l+1} \left[\left(1-\frac{h}{c}\right)^{l-1} - \left(1-\frac{h}{c}\right)^{-(l+2)} \right] \tag{8.16}$$

where h/c is a small quantity. For example, for the four concentric spheres model, its radius is 7.9, 8.1, 8.5, and 8.8 cm, respectively, then $h/c = 0.3/8.8 = 0.0341$. For the normalized concentric three spheres model, the radii are 0.87, 0.92, and 1.0, respectively, then $h/c = 0.08/1.0 = 0.08$. Therefore, the following series expansion on $x = h/c$ converges.

$$(1+x)^p = 1 + px + \frac{p(p-1)}{2!}x^2 + \frac{p(p-1)(p-2)}{3!}x^3 + \cdots \tag{8.17}$$

Using equation (8.17), (8.15) can be approximated as:

$$J_{\uparrow} = -\sigma\sum_{l,m}\left[\left(1-\frac{h}{c}\right)^{l-1} - \left(1-\frac{h}{c}\right)^{-(l+2)}\right] lc^{l-1}G_l S_l^m$$

$$= -\sigma\sum_{l,m} lc^{l-1}G_l S_l^m\left\{-(2l+1)\frac{h}{c} + \frac{-8l-4}{2}\left(\frac{h}{c}\right)^2 - \frac{2l^3+3l^2+37l+18}{6}\left(\frac{h}{c}\right)^3 + \cdots\right\} \tag{8.18}$$

Compare (8.18) with (8.12), it shows

$$J_{\uparrow} = -\sigma\sum_{l,m} lc^{l-1}G_l S_l^m\left\{-(2l+1)\frac{h}{c} + \frac{-8l-4}{2}\left(\frac{h}{c}\right)^2 - \frac{2l^3+3l^2+37l+18}{6}\left(\frac{h}{c}\right)^3 + \cdots\right\}$$

$$= \sum_{l} L_s(l)\sigma\left\{f_1(h,c) + f_2(h,c) + f_3(l,h,c)\right\} \tag{8.19}$$

Where the three filters are

$$f_1(h,c) = -\frac{h}{c}$$

$$f_2(h,c) = -\left(\frac{h}{c}\right)^2 \tag{8.20}$$

$$f_3(l,h,c) = -\frac{2l^3+3l^2+37l+18}{6(2l+1)}\left(\frac{h}{c}\right)^3$$

If just keeping the first two filters and omitting the third one, we have

$$J_{\uparrow} = -\sigma\sum_{l,m} lc^{l-1}G_l S_l^m\left\{-(2l+1)\frac{h}{c} + \frac{-8l-4}{2}\left(\frac{h}{c}\right)^2\right\}$$

$$= \sigma\sum_{l,m} l(2l+1)c^{l-2}G_l S_l^m\left(h + 2\frac{h^2}{c}\right) = -\sigma\left(h + 2\frac{h^2}{c}\right)L_s \tag{8.21}$$

If we just keep the first filter, then we have $J_{\uparrow} = -\sigma h L_s$. It can be seen that in the case of first-order approximation, the result here degenerates to that of equation (8.7). However, if the quadratic approximation is retained, (8.22) can be obtained from equation (8.21)

$$L_s = J_{\uparrow}\Big/\left\{\sigma\left(h + 2\frac{h^2}{c}\right)\right\} \tag{8.22}$$

This equation shows that up to the second-order approximation of h/c, there is a linear relationship between Laplacian and the current density entering the scalp vertically, which is independent of the spatial frequency (i.e., l), but does not hold when the approximation of h/c is higher than the second order. Figure 8.3 shows the form of the three filtering functions for (8.16) and (8.20). Set $c = 1.0$ and $h = 0.08$, then the relative value of the second order term to the first one is 2.0*0.08*0.08/0.08 = 0.16 = 16%, implicating a very important role in objectively estimating the current density.

Since current density is a real physical quantity and Laplacian is only the result of a mathematical operation, which is physically a first- or second-order approximation of "current density entering the scalp vertically", we believe that current density should be calculated directly when it is possible to do so. Due to the boundary condition of the scalp, the vertical current density at the scalp is zero. Therefore, the calculation of the vertical current density should be based on the skull surface and the result is a scalar quantity. Although a current is a vector, when fixing its direction to be perpendicular to the skull surface, a scalar can fully express the current density. In principle, $J_{\uparrow}$ can be calculated by the numerical method with any shape head model, but it requires a large amount of calculation. For the spherical model, the problem is relatively simple. After the spherical harmonic coefficient is obtained by inversion, $J_{\uparrow}$ can be easily estimated, just as other imaging physical quantities do.

The practical value of equation (8.22) is that for a realistic head model, the calculation of scalp SL is easier than the numerical solution of a 3D EEG problem; Laplacian does not need to know the conductivities and inner boundary interfaces of the head model. Laplacian can use (8.22) to approximate estimate the $J_{\uparrow}$, with the local electric conductivity, the local scalp thickness h, and the local scalp curvature radius c. In current Laplacian practice, the combination of these three terms is taken as a constant neglectable, and the Laplacian is then directly assumed to be $J_{\uparrow}$.

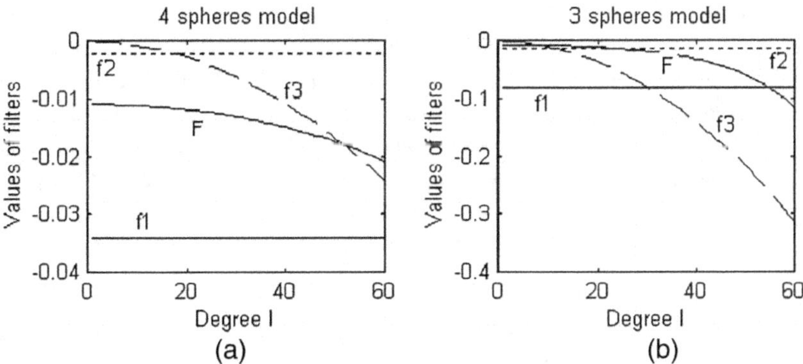

FIGURE 8.3 Values of filters versus degree l. Panel (a) is for the four spheres head model with a scalp radius $c = 8.8$ and a scalp thickness $h = 0.3$. Panel (b) is for the normalized three spheres head model with $c = 1.0$ and $h = 0.08$. F is the filter in equation (8.16) and f_1, f_2, and f_3 are the filters in equation (8.20) (Yao, 2002a).

According to equations (8.21) and (8.22), the condition for the Laplacian to be interpreted as the current density vertically into the scalp is whether a second-order approximation of formula (8.22) is reasonable, and the key is h/c. For the scalp EEG problem, the approximation should be taken as a rational choice. But when it comes to the ECG problem, Laplacian is widely adopted in ECG, where h has no definite corresponding physical layer, how do we understand Laplacian? Perhaps we can only think of Laplacian as the current density in a sufficiently thin but non-zero "layer" vertically into the body surface. This thickness cannot be equal to zero, otherwise, the current density entering the body surface vertically will be zero. Formally, the second derivative of the body surface potential, Laplacian, is a quantity independent of thickness; it thus indicates that there is indeed a physical difference between Laplacian and current density. In other words, Laplacian is the result after a second-order mathematical filter of the surface potential, it can be applied to any potential distribution; however, the relation between such a Laplacian and the radial (normal) current density is model-dependent.

8.1.3 Direct Measurement of Scalp Laplacian

The application of SL is thought to be due to its ability to provide different and perhaps more localized source information. In the past decades, many methods have been developed to estimate Laplacian from its surface potential. However, on the one hand, there are errors in such an estimation. On the other hand, such a result is a deduced quantity and does not add any more original information than the potential. Therefore, it is valuable to understand the feasibility and superiority of direct measurement of Laplacian.

According to the formula, the Laplacian calculation requires a body surface model, such as a spherical model for the head and a plane model for the chest surface. Although a realistic model can be established, it is not a scheme that can be easily popularized and applied. Meanwhile, as potential measurements are point-by-point convenient, they would also be rendered meaningless if Laplacian measurements are too cumbersome.

Here introduced is the study of Geselowitz and Ferrara (1999). First, they assumed that a plane model can be used on the chest surface for ECG, and the Laplacian approximation can be estimated with a five-point difference formula, with the calculation results not changing with the rotation of the four points surrounding the center, that is, it should have rotation invariance. The second issue they investigated was the signal-to-noise ratio (SNR), and they assumed the SNR should not be too low, otherwise, it would render Laplacian meaningless.

8.1.3.1 Measurement Method

SL was recorded using a five-electrode array, with a grid spacing of 2 cm, of 10 normal male subjects, aged 20–32. Two locations were recorded, one in the left thoracic region, 4 cm to the left, and 10 cm above the xiphoid process, and another in the left shoulder region, ~8 cm to the left, and 20 cm above the xiphoid process. As the body size changes, the position changes slightly. These two locations are chosen because, based on the simulation study, these two regions exactly correspond to regions of large and small values. Each position was recorded twice, with the electrode array of the second round rotated 45° relative to the first. The time of each recording is about 1 minute. Each set of data includes four potential differences and their average, which is proportional to Laplacian. Measurements were made using four differential amplifiers and a multi-channel A/D conversion (Biopac MP100A). The center electrode is connected to the cathode of each amplifier, while each of the other four is connected to the anode of each amplifier, which has a gain of 10,000, a low cut of 0.05 Hz, and a high cut of 1 kHz. The MP100A has a resolution of 16 bits, a dynamic range of ±10 V, and a sampling rate of 3237.29 Hz. The effect of 60 Hz AC was treated with a Notch filter. To analyze the noise level, a signal-averaging technique is used. For each channel of each subject, select a window that exactly contains QRS, and then use this window to average the signals over 30 beats. The corresponding mean signal is subtracted from each signal as an estimate of the noise, and then the root mean square (RMS) value of the noise is calculated.

8.1.3.2 Results

Ten subjects had 40 different channels in two positions, the mean RMS potential of the different channels in the chest was 83 ± 42 μV, and that in the shoulder was 44 ± 33 μV. The mean SNR of Laplacian was 3.25 ± 1.26 on the chest and 2.46 ± 1.97 on the shoulder. In 3 of the 20 shoulder records, their SNRs were less than one.

Among the 20 Laplacian records of ten subjects, a comparison of the results of the two electrode arrays showed that 9 pairs had a normalized cross-correlation >0.92, while the other 11 pairs had a normalized cross-correlation between 0.42 and 0.88. Although the planar surface Laplacian is theoretically independent of the rotation of the electrode array, the above data indicate that there is a significant rotation-dependent on the actual recording. In a simulation study where 23 dipole sources were adopted, rotational independence was found in the sphere model, while in the realistic body model, some rotational dependence was shown, but not as severe as in the measurements. These results raise a big question mark to the feasibility of actually measuring Laplacian on the body surface. Therefore, the Laplacian study at present can only be used as a subsequent processing of potential recordings, and its value as the original data remains to be determined. For example, in recent years, some researchers have carried out inverse problem studies based on Laplacian. Obviously, if there is no original Laplacian observation data but only potential data, then the inverse problem method based on Laplacian is equivalent to a Laplacian weighted inverse problem method in measuring space (also refer to Section 12.1). In addition, when conducting inverse problem studies based on Laplacian, it is necessary to take into account the fact that the observed SNR of Laplacian is lower than the SNR of potential.

Besides, Besio et al. (2006a,b) and Makeyev and Besio (2016) have extensively studied the effectiveness of tripolar concentric ring electrodes and their superiority over bipolar concentric ring electrodes for direct measurements of Laplacian. Later in this chapter, we will describe our development of higher-order difference methods for plane and spherical cases.

8.2 SCALP LAPLACIAN BASED ON TUNABLE BASIS FUNCTIONS

8.2.1 Introduction

Laplacian mapping (LM) is the result of a 2D Laplacian operator applied to the scalp surface potential, so it can provide a higher spatial resolution than the scalp surface potential. LM needs to know neither the details of the volume conductor model nor the neural activity source model, which is its advantage over cortical potential imaging.

Since Hjorth (1975), a variety of Laplacian estimation techniques have been developed, and can usually be divided into two categories: local and global approximation techniques. The local approximation technique uses a numeric operator to estimate the LM of each electrode, and a typical example is Hjorth's second-order finite difference scheme, which adopts an approximate local plane model and a rectangular electrode array to simplify the calculation of second-order finite difference (Hjorth, 1975). Following Hjorth's pioneering work, several improvements have been developed, such as for a non-rectangular electrode array on a flat surface (Wallin and Stalberg, 1980; Nunez, 1981) or a nonplanar scalp surface model. The local approximation technique has an advantage in the easier calculation, but also has some disadvantages where (1) its imaging may be discontinuous and the extreme points are all located at the electrode position (Lo, 1999); (2) the estimation of LM around the boundary of the electrode array is inaccurate or unavailable as there is not enough information surrounding the electrode.

The global approximation technique first constructs a global interpolation function (Babiloni et al., 2001; He et al., 2001; Zhai and Yao, 2004), and then analytically applies the second-order and 2D differential operators to this function. Till now, many interpolation functions have been adopted, such as the 2D thin-plate spline function in the interpolation of EEG (Perrin et al., 1987), the SSF in the LM estimation of the spherical head model (Perrin et al., 1989; Yao, 2000; Kayser and Tenke, 2015), the 3D thin-plate spline function in the LM estimation of spherical and ellipsoid head models (Law et al., 1993), and the realistic head model (Babiloni et al., 1996). The advantage of the global approximation technique is that it can estimate the potential of both internal and boundary electrode points, and the performance of spline/global LM is better than that of local LM, according to some comparative studies. Its main drawback is sometimes a distorted Laplacian estimate from the global mathematical interpolation scheme itself. At this point, one view is that the coherence between the electrodes is bound to increase if the LM at an electrode position is estimated with an algorithm in which a great weight is applied to electrodes far from this point.

To sum up, it can be seen that the global or local approximation techniques have their advantages and disadvantages, and the reason is the global or local interpolation functions they are based on. In this section, we will introduce another approximation method, which is based on the RBF to interpolate potential functions and Laplacian. Different from the above schemes, it adopts an adjustable spatial support range of the basis function, so as to be called the Laplacian technique based on the adjustable basis function. The spatial support range is generally wider than the local five-neighborhood regions, but smaller than the entire global electrode array to better integrate the advantages of the existing global and local approximation techniques. The details of the algorithm are described as follows. Then the numerical comparison between this method and the SSF-based method is carried out for the spherical model. Since the SSF-based approach is considered to be the best in the case of the spherical model, this comparison helps to evaluate the effectiveness of the RBF approach. In Section 8.4, the RBF approach for a real head model is discussed.

8.2.2 Radial-Basis Function Laplacian

8.2.2.1 The RBF Interpolation

RBF was first used to solve the problem of multivariable interpolation and, more recently, in neural networks for interpolation and classification (Haykin, 1999). The RBF technique begins by selecting a function with the following form (Powell, 1988)

$$\Phi(\vec{x}) = \sum_{i=1}^{N} w_i \psi\left(\|\vec{x} - \vec{x}_i\|\right) \tag{8.23}$$

Here, $\left\{\psi\left(\|\vec{x} - \vec{x}_i\|\right) \quad i = 1,2,\ldots,N\right\}$, a group of N nonlinear functions, is known as the radial basis function, where $\|\cdot\|$ represents a module, usually refers to the Euclidean module. Given the data $\vec{x}_i \in R^k, \;\; i = 1,2,\ldots,N$, they are selected as the center of the RBF, where k is the dimension of the position vector. $\{w_i\}, \; i = 1,\ldots,N$ is the unknown weight. The interpolation problem can be expressed in a strict sense as

$$\Phi(\vec{x}_i) = d_i, \quad i = 1,2,\ldots,N \tag{8.24}$$

where $d_i \in R^1 \left(i = 1,2,\ldots,N\right)$ is the known data point, which in this case is the observed potential value at the electrode position $\vec{x}_i \in R^k \;\; \left(i = 1,2,\ldots,N\right)$ on the scalp surface. Substituting equation (8.24) into (8.23), we can obtain the following linear equation in matrix form

$$\Phi = \Psi w \tag{8.25}$$

For this interpolation problem, the following theorem holds (Micchelli, 1986):

Let $\vec{x}_i \in R^k, \;\; i = 1,2,\ldots,N$ be a set of discrete points in R^k, then the interpolating square matrix ψ of size N whose ji-th element is $\psi_{ij}\left(\|\vec{x}_j - \vec{x}_i\|\right)$ is nonsingular.

There is a big class of RBFs that satisfy Micchelli's theorem, such as multi-quadrics, inverse multi-quadric functions (Hardy, 1971), and the Gaussian function, which is chosen here. Recently, Janin Jager et al. (2016) compared multi-quadrics with SSF, and the results showed that the RBF function is more effective.

8.2.2.2 Sphere Scalp Surface Potential Interpolation and LM Based on RBF

The Gaussian function is defined as (Haykin, 1999).

$$\psi(\vec{x}, \vec{x}_i) = \exp\left(-\frac{1}{2\sigma^2} S_i^2\right) \tag{8.26}$$

where $\sigma \neq 0$ is a constant that varies from problem to problem and S_i is the distance between $\vec{x}$ and $\vec{x}_i$ For spherical scalp surface interpolation, we specifically define S_i as the arc length on a large circle passing through $\vec{x}$ and $\vec{x}_i$.

$$S_i = r\theta_i \tag{8.27}$$

where θ_i is the arc angle of $\vec{x}$ and $\vec{x}_i$ from the center of the sphere and r is the radius of the sphere. Based on this assumption, $k=1$, greatly reduced the complexity.

In practice, the output of the interpolation function often contains a constant independent of the data. Therefore, an offset b (Haykin, 1999) needs to be introduced into the interpolation matrix equation. Correspondingly, the equation (8.25) becomes

$$\Phi = \Psi w + Tb \tag{8.28}$$

where T is a column vector, and each of its entries is 1. Let's define $\Psi' = \begin{bmatrix} \Psi & T \end{bmatrix}$ and $W' = \begin{bmatrix} w \\ b \end{bmatrix}$, so we have

$$\Phi = \Psi' W' \tag{8.29}$$

By substituting the equations (8.26) and (8.27) into (8.29), we can obtain the weighted coefficients represented by the generalized linear inversion

$$W' = \Psi'^{+}\Phi \tag{8.30}$$

Here, the generalized inverse "+" can be accomplished by the truncated SVD algorithm.

Substituting equations (8.26) and (8.27) into (8.23) and taking into account the offset b given in equation (8.28), we obtain

$$\Phi(\vec{x}) = \sum_{i=1}^{N} w_i \exp\left(\beta\theta_i^{\,2}\right) + b \tag{8.31}$$

where $\beta = -\dfrac{r^2}{2\sigma^2}$ is a constant that needs to be selected according to the problem. According to equation (8.31), in spherical coordinates (r,θ,ϕ), LM of the surface potential of the sphere head is

$$L_S(\vec{x}) = \left\{ \frac{1}{r^2 \sin\theta} \frac{\partial}{\partial\theta}\left(\sin\theta \frac{\partial}{\partial\theta}\right) + \frac{1}{r^2 \sin^2\theta} \frac{\partial^2}{\partial\phi^2} \right\}\Phi = \nabla^2_{\theta\phi}\Phi \tag{8.32}$$

where $\nabla^2_{\theta\phi}$ is the Laplacian operator. Substitute (8.31) into (8.32) to get

$$L_S(\vec{x}) = \nabla^2_{\theta\phi}\Phi = \sum_{i=1}^{N} w_i \nabla^2_{\theta\phi}\left\{\exp\left(\beta\theta_i^{\,2}\right)\right\} = \sum_{i=1}^{N} w_i L_S(\vec{x},\vec{x}_i) \tag{8.33}$$

The Laplacian interpolation function based on RBF is

$$Ls(\vec{x},\vec{x}_i) = \nabla^2_{\theta\phi} \exp\left(\beta\theta_i(\vec{x},\vec{x}_i)^2\right) \tag{8.34}$$

Equation (8.33) shows that the Laplacian at a point $\vec{x}$ consists of a linear superposition of the contributions of all measured positions $\vec{x}_i,\ i = 1,\ldots,N$. Equation (8.34) shows that the contribution of a measuring point $\vec{x}_i$ to a field point $\vec{x}$ is determined by the relative arc angle $\theta_i(\vec{x},\vec{x}_i)$ between them. Obviously, based on the weight coefficient $\vec{W}$ obtained by equation (8.30), if we get $Ls(\vec{x},\vec{x}_i)$ located at all $\vec{x}_i,\ (i = 1,\ldots,N)$, then we obtain $Ls(\vec{x})$ by simple linear summation according to equation (8.33). So, the problem becomes finding the $Ls(\vec{x},\vec{x}_i)$ that corresponds to any pair of $\vec{x}_i$ and $\vec{x}$.

Since the operator $\nabla^2_{\theta\phi}$ in equation (8.34) is independent of the coordinate system, we can choose a specific coordinate system for each point $\vec{x}_i\,(i = 1,\ldots,N)$ to simplify the analytical expression of the operator $\nabla^2_{\theta\phi}$. Obviously, if we are aiming at every point $\vec{x}_i\,(i = 1,\ldots,N)$, choose a particular spherical coordinate system, make the z-axis just pass through the point $\vec{x}_i$, the $\theta_i(\vec{x},\vec{x}_i)$ in equation (8.34) comes to the latitude θ of the spherical coordinates (r,θ,ϕ), it also means that the exponential function in equation (8.34) has nothing to do with the angle of longitude ϕ, so that we get $\nabla^2_{\theta\phi} = \frac{1}{r^2\sin\theta}\frac{\partial}{\partial\theta}\left(\sin\theta\frac{\partial}{\partial\theta}\right)$ in the local coordinate system as a simplified Laplacian operator, finally, we have

$$\begin{aligned} Ls(\vec{x},\vec{x}_i) &= \frac{1}{r^2\sin\theta}\frac{\partial}{\partial\theta}\left(\sin\theta\frac{\partial}{\partial\theta}\exp\left(\beta\theta^2\right)\right) \\ &= 2\beta\exp\left(\beta\theta^2\right)\left\{\frac{\theta}{\sin\theta}\cos\theta + 1 + 2\beta\theta^2\right\} \end{aligned} \tag{8.35}$$

Equation (8.35) has no singularity as long as $\lim\limits_{\theta\to 0}\frac{\theta}{\sin\theta} = 1$ is considered here.

Now, by substituting the equation (8.35) into (8.33), the following spherical surface Laplacian is obtained, which is very easy to calculate

$$Ls(\vec{x}) = \sum_{i=1}^{N} 2\beta w_i \exp\left(\beta\theta_i(\vec{x},\vec{x}_i)^2\right)\left\{\frac{\theta_i(\vec{x},\vec{x}_i)}{\sin\theta_i(\vec{x},\vec{x}_i)}\cos\theta_i(\vec{x},\vec{x}_i) + 1 + 2\beta\theta_i(\vec{x},\vec{x}_i)^2\right\} \tag{8.36}$$

According to equation (8.36), once we have the weighting coefficient from equation (8.30), the Laplacian of the field point $\vec{x}$ can easily be obtained by substituting into the arc angle $\theta_i(\vec{x},\vec{x}_i)$ between $\vec{x}$ and $\vec{x}_i,\ i = 1,\ldots,N$ into equation (8.36).

In some practices of neural networks, the parameter $\beta = -\frac{r^2}{2\sigma^2}$ is recommended to be $\beta = -\frac{m}{d^2_{\max}}$ (Haykin, 1999), where m is the number of interpolation centers, in this case, the number of electrodes N, and $d_{\max}$ is the maximum distance between these electrode pairs, in this case $d_{max} = \pi$, which is the maximum relative radian on the hemisphere electrode array. In this work, we first select $\beta = -m/\pi^2$ and leave only m as a tunable parameter. Then, the parameter m is selected through experiments to make the algorithm conform to EEG potential interpolation and LM. Obviously, in such an arrangement, the parameter m plays the role of the control parameter in the RBF approximation.

8.2.2.3 Scalp Surface Potential Interpolation and LM Based on SSF

For a spherical surface, some studies have shown that order 2 or 3 spherical splines provide the best estimation of Laplacian. Therefore, the choice of the spherical spline as a comparison is a good starting point.

The SSF interpolation formula is (Wahba, 1981; Perrin et al., 1987, 1989; Yao, 2000)

$$\Phi(\vec{x}) = \sum_{i=1}^{N} G(\vec{x}, \vec{x}_i) w(\vec{x}_i) + b \tag{8.37}$$

where SSF is

$$G(\vec{x}, \vec{x}_i) = \sum_{l=1}^{L\max} \frac{1}{4\pi} \frac{2l+1}{\left(l(l+1)\right)^n} P_l\left(\cos\theta_i\left(\vec{x}, \vec{x}_i\right)\right) \tag{8.38}$$

The control parameter n in equation (8.38) can be selected as 2 (Yao, 2000), 4 (Perrin et al., 1989), or 3. In this work, we tested different n, and the results showed that 2 was the best. Where $p_l(\cos\theta)$ is the Legendre function, and the value of L_{max} is 150 (Yao, 2000).

The surface Laplacian based on spherical splines is the result of the surface potential passing through the following spatial filter equation (8.13)

$$K_l = -l(l+1)/r^2 \tag{8.39}$$

Apply the filter shown in equations (8.39) to (8.37) - (8.38), and finally get

$$Ls(\vec{x}) = \sum_{i=1}^{N} w_i Ls(\vec{x}, \vec{x}_i) \tag{8.40}$$

The Laplacian interpolation function based on SSF is

$$Ls(\vec{x}, \vec{x}_i) = \sum_{l=1}^{L\max} \frac{1}{4\pi} \frac{2l+1}{(l(l+1))^n} \left\{-\frac{l(l+1)}{r^2}\right\} P_l\left(\cos\theta_i\left(\vec{x}, \vec{x}_i\right)\right) \tag{8.41}$$

A comparison of potential and Laplacian interpolation function formulas based on RBF and SSF shows that they are completely consistent in form. Just because of this similarity, we use the same SVD algorithm to solve the weighting coefficient for the two methods. However, we still need to point out that there are two essential differences between these two approximations: (1) the interpolation function is different and (2) the tunable parameters in the interpolation function are different. For RBF, the tunable parameter is m in β, which controls the spatial support range of the interpolation function. As will be seen later, adjusting m smoothly adjusts the spatial support range of the RBF-based potential (equation 8.26) and Laplacian (equation 8.35) interpolation functions. For SSF, the control parameter is n in the low-pass filter $1/\left(l(l+1)\right)^n$. The function of the low-pass filter is to suppress the high spatial frequency components to control the spatial false frequency interference caused by insufficient sampling (Perrin et al., 1989; Srinivasan et al., 1998).

8.2.2.4 Comparison of the RBF and SSF Interpolation Functions

In Figure 8.4, the exponential function $\exp\left(\beta\theta_i^{\,2}\right)$ (equation 8.31) in the RBF interpolation function and $G(\vec{x},\vec{x}_i) = G\left(\theta_i(\vec{x},\vec{x}_i)\right)$ (equation 8.38) in the SSF interpolation function are plotted. The RBF interpolation function is shown as a focus adjustable function with an adjustment parameter m. The larger m is, the smaller the spatial support range of the interpolation function is. The SSF interpolation function is a global function, which has little difference between $n=2$–4, and they are obviously different from $n=1$. This situation should be the reason why $n=2$–4 has been used in the past, but $n=1$ has never been used.

Figure 8.5 illustrates two Laplacian interpolation functions. The difference between RBF-based Laplacian interpolation and RBF-based potential interpolation is their different spatial support ranges. The Laplacian interpolation function (Figure 8.5a) is more local than the potential interpolation function (Figure 8.4a). The SSF-based Laplacian interpolation functions (equations 8.38 and 8.41) are too local at $n=1$ ($n=1$, equation 8.41, Figure 8.5b), and the curves of the Laplacian interpolation function at $n=2$ are also significantly different from those at $n=3$ and 4 (Figure 8.5b).

Assuming that the 120 electrodes are uniformly distributed on the upper hemisphere, the number of electrodes covered by the spherical caps with different latitude angles can be calculated. Figure 8.6b shows the relationship between the latitude

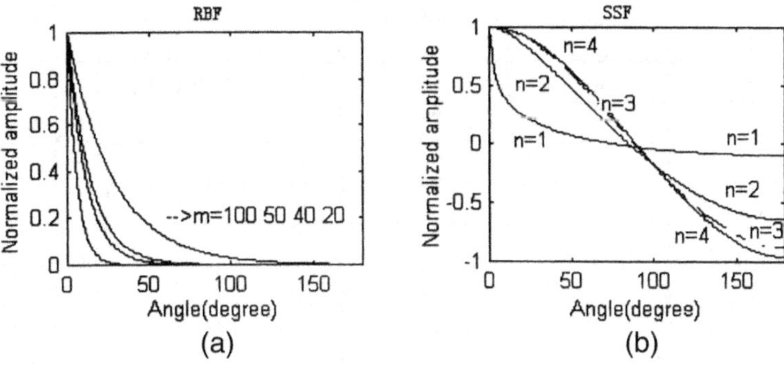

FIGURE 8.4 Potential interpolation functions for RBF (a) and SSF (b). The horizontal axis is the latitude angle and the vertical axis is the normalized amplitude. The parameters m and n are the control parameters in the two potential interpolation methods. The curves in the left figure are from left to right for $m=100$, 50, 40, and 20, respectively (Yao, 2002b).

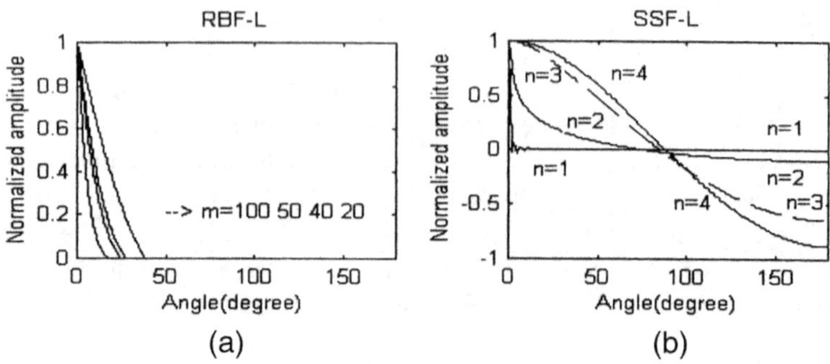

FIGURE 8.5 Laplacian interpolation functions for RBF (a) and SSF (b). The meaning of the parameters in the figure is the same as in Figure 8.4 (Yao, 2002b).

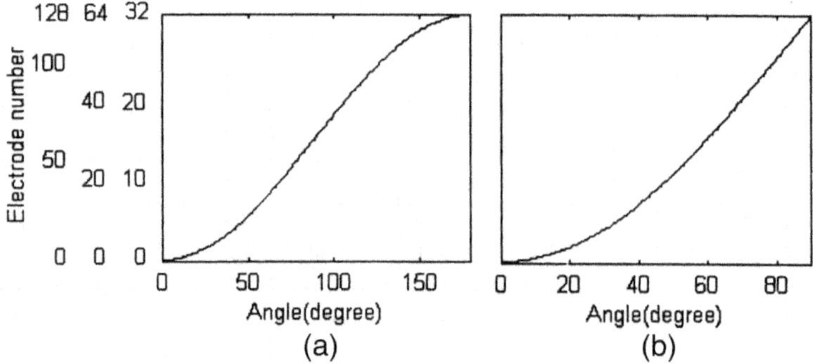

FIGURE 8.6 Electrode number versus relative latitude angle. The vertical axis shows the number of electrodes covered by the area corresponding to the relative latitude angle shown by the horizontal axis. Panel (a) corresponds to a border point of the hemisphere electrode array, the angle differences between such a border point to other points range from zero (itself) to 180 (its opposite point). Panel (b) corresponds to the vertex point, the largest angle is 90, which is between the vertex and a border point (Yao, 2002b).

angle and the number of electrodes when the vertex of the sphere is taken as the reference point. At this point, the maximum latitude angle is 90°. Figure 8.6a shows the relationship between the latitude angle and the number of electrodes when an edge point of the hemisphere is taken as the reference point, and the maximum latitude angle is 180°. According to Figures 8.4–8.6, we see the effective ranges are quite different for RBF with different m and for SSF with different n.

To sum up, since m plays a role in regulating the contribution of the variance in the Gaussian function, both the potential and Laplacian interpolation functions based on RBF are adjustable according to m. The larger m is, the smaller the spatial support range of the corresponding interpolation function will be. For potential and Laplacian interpolation functions based on SSF, they are basically global functions, especially when $n = 3$–4. Among the various SSF-based interpolation functions shown in Figures 8.4 and 8.5, except for the cases of n = 1 in Figure 8.4b and $n = 1$–2 in Figure 8.5b, the functions are not only global but also give a large weight to the distant points, such as the 180° case. For the numerical simulation effect of this method and the effectiviness of actual data, please refer to the literature (Yao, 2002a).

8.2.3 Discussion

The scalp Laplacian has the following properties (Hjorth, 1975; Nunez, 1981; Nunez et al., 1994). (1) It is a scalar quantity, so its topographic distribution is easy to display, just like the scalp surface potential; (2) it is independent of the reference electrode, and its estimation does not require the internal details of the head model; (3) compared with magnetic fields, EEG Laplacian is associated with both the radial and tangential components of neural activity currents; (4) compared with potential distribution, Laplacian distribution emphasizes shallow sources (high spatial frequency component); (5) surface Laplacian can be approximately interpreted as the current source density from the scalp to the skull, which is why it is called the current density map in some literature; and (6) simulation experiments show that it is approximately proportional to the potential of the cortical surface, so some literature also calls LM a cortical imaging technology. Due to these properties, surface Laplacian has attracted the attention of scholars in the fields of technology and application in recent years.

From a technical point of view, Laplacian is considered to have two basic approximation methods: the local method and the global method. As the literature has commented, they each have advantages and disadvantages. The local approximation technique is affected by the local noise and cannot deal with the points on the edge of the electrode array. The reason is that the local approximation technique only uses the information of the nearest electrode to estimate the Laplacian of a certain point. Due to the limited information used, it is difficult to distinguish the local noise and signal, and for the points on the edge of the electrode array, there are not enough nearest electrodes available. For global approximation techniques, which are usually based on global spline functions, the potential record of the entire electrode array will be utilized when estimating the Laplacian at a point on the scalp. Experiments show that the global approximation technique is usually better than the local approximation technique because it can smooth out the noise caused by eye movement in the global sense by introducing proper regularization or adjusting its low pass filter. Also, in a global sense, it can use global spline functions to predict the Laplacian on the boundary. However, the global spline function is not the real potential function after all, it is only a fitness function based on the limited potential observation data. In general, this fitting is based on the least squares criterion, and the fitting based on this criterion is dominated by relatively large noises (Menke, 1989), so that in the presence of noise, especially non-uniform noise, the estimation result is a distorted global Laplacian estimate. In fact, it has been argued that LM should "strictly follow the surface Laplacian definition, using only local properties of potential and scalp shape" (Le et al., 1994).

In this section, we present a relatively new algorithm to estimate surface Laplacian, which is based on the Gaussian RBF. Such RBF is a function with a tunable support range, which can be viewed as an intermediate approximation technique compared with existing global and local approximation techniques. The spatial support range is wider than that of the local approximation technique, which makes it capable of denoising to a certain extent, while the spatial support range is narrower than that of the global approximation technique, which makes it avoid interference between electrodes that are far apart from each other. In addition, the spatial support range of the interpolation function of the new method is adjustable, which is not possible in the previous global and local approximation techniques.

The simulation results show that the control parameter m in the Gaussian RBF should vary according to the depth of the source. For deep sources, a smaller m is better, while for shallow sources, a larger m is better. For shallow sources, the RBF-based approach is obviously superior to the SSF-based approach. In addition, for very shallow sources, we need to increase the number of electrodes to improve the RBF-based method.

As a new attempt to estimate Laplacian based on RBF, the algorithm in this section is only implemented in a relatively ideal four-layer concentric sphere model. However, the approach introduced here can be further extended in principle to realistic head models, in the next section 8.3. We believe that the main contribution of the RBF-based method introduced in this section is that it introduces the concept of a new interpolation approximation, fills the gap between local approximation and global approximation, and provides a new choice for related theoretical and experimental research. Definitely, the scalizable idea introduced here can be realized by using other interpolation functions such as multi-quadric (Jager et al., 2016), and the regularization method can also be updated (Bortel and Sovka, 2013), and then a better result may be obtained.

8.3 SCALP LAPLACIAN FOR REALISTIC HEAD MODEL

8.3.1 Introduction

In this section, a global Laplacian based on spline function, a mesoscale Laplacian based on spline function, and a Laplacian based on support range tunable function RBF are introduced for realistic head models (Zhai and Yao, 2004; Zhai, 2005). Compared with global Laplacian, mesoscale Laplacian based on spline function only utilized the scalp surface position and potential data within a limited region, thus avoiding the disadvantage that global spline Laplacian suffered from the global disturbance. The RBF-based scheme directly utilizes the tunable support range to control the region involved.

From a historical perspective, global Laplacian began with the development of the spherical spline function Laplacian (Perrin et al., 1989), followed by the development of the ellipsoidal spline function Laplacian (Law et al., 1993), and then the realistic head spline Laplacian (Babiloni et al., 1996). For the real head model, the previous spline function Laplacian is based on the position information and potential at all electrodes, which can be called global spline Laplacian (Global Realistic Geometry Spline Laplacian – GSL) (He et al., 2001). In this section, we introduced a Mesoscale/Moderate Scale Realistic Geometric Spline Laplacian (MSL) based on spline function, which essentially confined the idea of spline function fitting to a limited area (Zhai and Yao, 2004). The global Laplacian method based on the spline function is introduced first, followed by the mesoscale Laplacian based on the spline function. Finally, Laplacian based on RBF and its verification are introduced. Simulation evaluations on the effects of potential noise and electrode position noise can be found in Zhai and Yao (2004).

8.3.2 Global Laplacian Based on Spline Functions

For a scalp surface Ω, it can be expressed in terms of $u = x$, $v = y$, and $z = f(x,y)$, and the continuous second derivative of the function $f(x,y)$ exists. If $\Phi(u,v)$ as the scalp potential function over Ω with continuous second derivative exists, then the scalp Laplacian is as follows (Babiloni et al., 1996)

$$\nabla^2{}_s\Phi = \frac{1}{\sqrt{g}}\left\{\frac{\partial}{\partial u}\left[\sqrt{g}\left(g_{11}\frac{\partial\Phi}{\partial u} + g_{12}\frac{\partial\Phi}{\partial v}\right)\right] + \frac{\partial}{\partial v}\left[\sqrt{g}\left(g_{21}\frac{\partial\Phi}{\partial u} + g_{22}\frac{\partial\Phi}{\partial v}\right)\right]\right. \tag{8.42}$$

Among them,

$$g = 1 + \left(\partial f/\partial u\right)^2 + \left(\partial f/\partial v\right)^2, \quad g_{11} = \left(1 + \left(\partial f/\partial v\right)^2\right)\Big/g$$

$$g_{12} = g_{21} = -\left(\partial f/\partial u\right)\left(\partial f/\partial v\right)/g, \quad g_{22} = \left(1 + \left(\partial f/\partial u\right)^2\right)\Big/g$$

It can be seen from equation (8.42) that to calculate the Laplacian of the Scalp, both the scalp geometry $f(x,y)$ and potential distribution $\Phi(u,v)$ functions must be obtained. The spline function is the most widely used for $f(x,y)$ and $\Phi(u,v)$. For M sample points (electrodes) on the scalp surface $(x_i, y_i, z_i), i = 1\ldots M$, the mathematical model of the scalp can be expressed by the 2D thin disk spline function (Babiloni et al., 1996; Harder and Desmarais, 1972; He et al., 2001)

$$z = f(x,y) = \sum_{i=1}^{M} p_i K_{m-1} + Q_{m-1} = \sum_{i=1}^{M} p_i d_i^2 \log\left(d_i^2 + w^2\right) + \sum_{d=0}^{m-1}\sum_{k=0}^{d} q_{dk} x^{d-k} y^k \tag{8.43}$$

where $m = 2$, $d_i^2 = (x - x_i)^2 + (y - y_i)^2$, w is a constant, K_{m-1} and Q_{m-1} are the basic function and osculating function, respectively, and the coefficients p_i and q_{dk} are the solutions of the following matrix equation

$$(K + \omega I)P + EQ = Z \tag{8.44}$$

$$E^T P = 0 \tag{8.45}$$

P, Q, and Z are the vectors, including p_i, q_{dk}, and z_i, respectively. I is the identity matrix, K and E are composed of basis functions and sampling coordinate elements, and ω is a regularization parameter to improve the stability of the system.

Similarly, based on the potential Φ_i, $i = 1 \ldots M$, of M sampling points (x_i, y_i, z_i) on the given scalp surface, the potential distribution of the scalp can be expressed by a 3D spline function in 3D space (He et al., 2001)

$$\Phi(x,y,z) = \sum_{i=1}^{M} t_i H_{m-1} + R_{m-1} = \sum_{i=1}^{M} t_i r_i^{(2m-3)/2} + \sum_{d=0}^{m-1} \sum_{k=0}^{d} \sum_{g=0}^{k} r_{dkg} x^{d-k} y^{k-g} z^{g} \tag{8.46}$$

where $m = 3$, $r_i^2 = (x - x_i)^2 + (y - y_i)^2 + (z - z_i)^2$, H_{m-1} and R_{m-1} are the basis function and the osculating function, respectively, and the coefficients t_i and r_{dkg} are the solutions of the following matrix equation

$$(H + \lambda I)T + FR = \Phi \tag{8.47}$$

$$F^T T = 0 \tag{8.48}$$

Among them, T, R, and V are the vector containing t_i, r_{dkg}, and Φ_i, H and F are composed of basis functions and sampling coordinate elements, and λ is a regularization parameter to improve the stability of the system.

Equations (8.44), (8.45), (8.47), and (8.48) are reorganized into one linear system equation (He et al., 2001)

$$AX = B \tag{8.49}$$

$$\text{where,} \quad A = \begin{bmatrix} K + \omega I & E & 0 & 0 \\ E^T & 0 & 0 & 0 \\ 0 & 0 & H + \lambda I & F \\ 0 & 0 & F^T & 0 \end{bmatrix}, \quad X = \begin{bmatrix} P \\ Q \\ T \\ R \end{bmatrix}, \quad B = \begin{bmatrix} Z \\ 0 \\ \Phi \\ 0 \end{bmatrix}.$$

To simplify the inverse calculation, set the two regularization parameters ω and λ equal to zero and then solve equation (8.49) by a unified regularized method, such an approach also avoids searching on a 2D (ω, λ) plane. Here, the truncated singular value decomposition method (TSVD) is used to find the inverse of matrix A:

$$A^{+} = V\sum\nolimits_{k}^{-1} U^{T} \tag{8.50}$$

U and V are matrices whose columns are orthogonal eigenvectors of AA^T and A^TA, respectively. $\sum_{k}^{-1}$ is a diagonal matrix, whose elements are the reciprocal of non-zero singular values arranged from large to small. Only the largest k singular values are retained, and k is a truncation parameter.

It can be seen that using equation (8.50), only one parameter k needs to be determined when solving A^{+}, instead of two parameters, ω and λ (He et al., 2001).

8.3.3 Mesoscale Laplacian Based on Spline Function

The above is a global Laplacian based on a spline function. Here we propose a mesoscale Laplacian (MSL) based on a spline function. Suppose 128 electrodes are used. For each electrode, only those electrodes whose spatial distance from the electrode is less than a set value are searched and used for the above spline approach, and the number of the searched electrodes is the *M*. In this work, we use a search radius of 34 mm, and the number of adjacent electrodes of each electrode obtained by the search is 6–13 (in the case of 128 total electrodes). Then the spatial position and potential of each electrode and its adjacent electrodes were adopted in the global Laplacian algorithm based on the spline function above. Then do the same calculation for each electrode to get the MSL for all the electrodes.

8.3.4 Tunable Laplacian Based on Radial Basis Functions

The general idea is the same as that introduced in Section 8.2. Here, the Gauss function is still chosen as the basis function, which is defined as

$$\psi(\vec{p},\vec{p}_i) = \exp\left(-\frac{1}{2\sigma^2}s_i^2\right) \tag{8.51}$$

where $\sigma \neq 0$ is a constant that varies from problem to problem and s_i is the distance between $\vec{p}$ and $\vec{p}_i$. For the realistic head model, s_i is specially defined as the Euclidean distance between $\vec{p}$ and $\vec{p}_i$, i.e.,

$$s_i = \|\vec{p} - \vec{p}_i\| \tag{8.52}$$

Assume that the scalp geometry of the head model is $f(x,y)$ and the potential distribution is $\Phi(u,v)$. The Gauss basis function is used to replace the spline function used in GSL to interpolate the potential distribution and the geometry of the scalp. For the sample points in the scalp surface (x_i,y_i,z_i), $i=1, \ldots, N$, the mathematical model of scalp geometry $f(x,y)$ can be expressed in the following interpolation form of RBF

$$z = f(x,y) = \sum_{i=1}^{N} w_i\psi\left(\|\vec{p}-\vec{p}_i\|\right) + b = \sum_{i=1}^{N} w_i \exp\left(-\frac{1}{2\sigma^2}s_i^2\right) + b \tag{8.53}$$

where $s_i^2 = (x - x_i)^2 + (y - y_i)^2$, ψ is the Gaussian function, and b is the offset. Similar to equation (8.44), the coefficient w_i can be calculated from the following equation by using the known values of the Cartesian coordinates of the electrodes.

$$Z = Kw + Tb \tag{8.54}$$

wherein Z is composed of electrode position coordinates z_i, where $i = 1,\dots,N$, matrix K is composed of RBFs, and T is a column vector with each element equal to 1.

Let the potential at the scalp surface (x_i, y_i, z_i) be Φ_i, where $i = 1,\dots,N$ and N is the number of scalp surface electrodes. Therefore, the potential distribution of the scalp surface in 3D space can be expressed by the following RBF model

$$\Phi(x, y, z) = \sum_{i=1}^{N} \omega_i \psi\left(\|\vec{p} - \vec{p}_i\|\right) + c = \sum_{i=1}^{N} \omega_i \exp\left(-\frac{1}{2\sigma^2} s_i^2\right) + c \tag{8.55}$$

where $s_i^2 = (x - x_i)^2 + (y - y_i)^2 + (z - z_i)^2$, c is the offset, and ω_i can be obtained from the following matrix (8.56) using the known scalp surface potential

$$\Phi = H\omega + Tc \tag{8.56}$$

wherein Φ is a vector composed of scalp recordings Φ_i, where $i = 1,\dots,N$, H is composed of RBF elements, and T is a column vector whose elements are all 1.

Equations (8.54) and (8.56) can be solved separately (Babiloni et al., 1996) or together (He et al., 2001), which combine the equations (8.54) and (5.56) as follows

$$AX = B \tag{8.57}$$

Among them $A = \begin{bmatrix} L & 0 \\ 0 & M \end{bmatrix}$, $X = \begin{bmatrix} w \\ b \\ \omega \\ c \end{bmatrix}$, $B = \begin{bmatrix} Z \\ 0 \\ \Phi \\ 0 \end{bmatrix}$, $L = [K \quad T]$, $M = [H \quad T]$.

The solution to equation (8.57) can be given by the following equation

$$X = A^{+}B \tag{8.58}$$

where A^{+} is the pseudo-inverse of A because A is ill-posed, the TSVD is used to improve the stability of the solution, as shown in equation (8.50). After we get the scalp shape function f and the potential function Φ, then we may go back to equation (8.42) to get the Laplacian.

For the spherical model, the parameter σ in equations (8.53) and (8.55) can be found in Section 8.2. For the realistic head model here, redefine parameter $-s_i^2/2\sigma^2 = s_i^2/R^2\left(-R^2/2\sigma^2\right) = \left(s_i^2/R^2\right)\beta = s_i^2/R^2\left(-m/\pi^2\right)$, where R is the average radius of the surface of the realistic head model and $\beta = -m/\pi^2$.

The adjustable parameter m can make the algorithm more suitable for the model shape and potential distribution. Therefore, m is the control parameter. When 128 or 120 electrodes were selected, 60 was chosen for m according to practical experience.

8.3.5 Simulation Test

It is generally believed that global Laplacian can obtain better results than local Laplacian (Babiloni et al., 1995). We compared the mesoscale Laplacian based on spline function and the scale adjustable RBF Laplacian with the global Laplacian based on spline function. Our results show that the performance of the former two Laplacian is similar, and better than the global Laplacian (Zhai and Yao, 2004).

8.3.6 Discussion

The RBFL/MSL approach presented in this section, to some extent, overcomes the shortcomings of both local and global Laplacian. Compared with the original global and common local scalp surface Laplacian, RBFL and MSL are mesoscale Laplacian between local and global Laplacian (Yao, 2002a; Zhai and Yao, 2004); they have a wider spatial support range than earlier local-approximation techniques, which enables potential and electrode position noise removal. Moreover, their spatial support ranges are flexible and narrower than those of the global approximation technique, so they can better deal with the boundary points and avoid the interference between the electrodes, which are far away from each other, to avoid the effect of a distant singular noise. More importantly, by adjusting the adjustable parameter m, or the search radius for MSL, their space support range can be adjusted to meet different actual needs. In the future, a RBF and spline function combined approach may be evaluated to see if they are complementary and to work together to have an even better result.

8.4 LOCAL LAPLACIAN BASED ON HIGH-ORDER DIFFERENCE APPROXIMATION

8.4.1 Introduction

As described in the previous sections, three types of methods can be used for surface Laplacian calculations. One is local Laplacian, which only calculates the difference of potential of some local electrodes. The second one is global Laplacian. This kind of method usually uses the basis function, such as the spline function, to carry out the interpolation fitting of potential and then carries out the difference calculation of the fitting function. The third one is the mesoscale Laplacian method, which is an intermediate between local and global Laplacian (Yao, 2002a; Zhai and Yao, 2004). The experimental results of Tandonnet et al. (2005) show that local Laplacian is more suitable for research aimed at local electrode areas, especially in clinical applications, where there are often only a small number of electrodes.

A typical local Laplacian algorithm is a second-order difference approximation of planar Laplacian that adopts five symmetric points. In this section, the theoretical calculation formula of planar Laplacian is first given, then the second-order difference approximation formula of planar Laplacian is derived, and then the fourth and sixth-order difference approximation formulas are, respectively, given (Zhai, 2005). The fourth-order difference approximation uses 21 symmetrical grid points, and the sixth-order difference approximation uses 53 symmetrical grid points. To save space, the detailed simulation test is not shown, and those interested can refer to the literature (Zhai, 2005). Conceptually, 21 or 53 electrodes make the difference approach closer and closer to the mesoscale approach.

8.4.2 Difference Approximation Laplacian

8.4.2.1 Theoretical Potential on the Plane and the Theory of Laplacian

Suppose the model is a semi-unbounded uniform conductor model with conductivity σ, such a simple plane model may be used to represent a small area of the brain's epidermis or anterior chest wall. In this model, the potential on the plane can be considered as the sum of the potential generated by a current dipole and the potential generated by its mirror dipole (which together meet the potential boundary condition – the surface normal current is zero). The current dipole and its mirror dipole are symmetrical about the plane in position and are assumed to be in a uniform conductor, as shown in Figure 8.7.

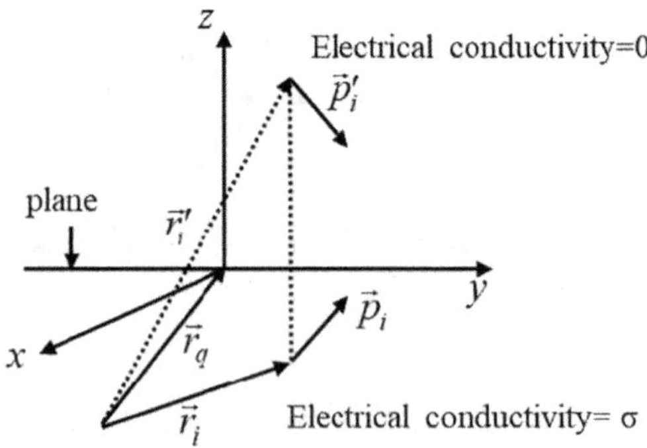

FIGURE 8.7 Semi-unbounded volume conductor model. The Cartesian coordinate is located on the plane $z=0$, and the z-axis is perpendicular to the plane, pointing into the air from the volume conductor (He and Cohen, 1992).

The potential generated by the N current dipoles on the plane is

$$\Phi=\sum_{i=1}^{N}\Phi(i)=\sum_{i=1}^{N}\left\{(1/4\pi\sigma)(\vec{r}_q-\vec{r}_i)\cdot\vec{p}_i/|\vec{r}_q-\vec{r}_i|^3+(1/4\pi\sigma)(\vec{r}_q-\vec{r}_i')\cdot\vec{p}_i'/|\vec{r}_q-\vec{r}_i'|^3\right\}$$

$$=\sum_{i=1}^{N}(1/2\pi\sigma)(\vec{r}_q-\vec{r}_i)\cdot\vec{p}_i/|\vec{r}_q-\vec{r}_i|^3$$

(8.59)

where $\Phi(i)$ represents the potential generated by the i-th current dipole and its mirror dipole, their values are the same on the plane surface, $\vec{r}_q$ is the observation point on the plane, $\vec{p}_i$ and $\vec{r}_i$, respectively, represent the vectors of the i-th current dipole moment and position, and $\vec{p}_i'$ and $\vec{r}_i'$ represent the vectors of the i-th mirror current dipole moment and position.

The theoretical Laplacian formula of surface potential (He and Cohen, 1992) is

$$L_{xy}=\partial^2\Phi/\partial x^2+\partial^2\Phi/\partial y^2=\Delta\Phi$$

$$=\sum_{i=1}^{N}(3/2\pi\sigma)\left\{5(z_q-z_i)^2(\vec{r}_q-\vec{r}_i)\cdot\vec{p}_i/|\vec{r}_q-\vec{r}_i|^7-\left[(\vec{r}_q-\vec{r}_i)\cdot\vec{p}_i+2(z_q-z_i)p_{z,i}\right]/|\vec{r}_q-\vec{r}_i|^5\right\}$$

(8.60)

Where $\vec{p}_i=(p_{x,i},p_{y,i},p_{z,i})$, $\vec{r}_i=(x_i,y_i,z_i)$, and $\vec{r}_q=(x_q,y_q,z_q)$.

8.4.2.2 Finite Difference Approximation Formula

Here, we derive Laplacian's finite-difference approximation formula from the Taylor expansion of a function. We give only the derivation of the second-order difference approximation formula, and the derivations of the fourth- and sixth-order difference approximation formulas are similar. Let the grid sizes of the X and Y axes on the 2D plane be equal, and they are both h.

Define $\xi=h\dfrac{\partial}{\partial x}$, $\eta=h\dfrac{\partial}{\partial y}$, there is $\xi^2+\eta^2=h^2\Delta$, and the Taylor expansion formula is

$$\Phi(x+h)=\left(1+h\frac{d}{dx}+\cdots+\frac{h^n}{n!}\frac{d^n}{dx^n}+\cdots\right)\Phi(x)=\left(e^{h\frac{d}{dx}}\right)\Phi(x)$$

And so, we can get

$$\Phi_1=e^{\xi}\Phi_0,\ \Phi_2=e^{\eta}\Phi_0,\ \Phi_3=e^{-\xi}\Phi_0,\ \Phi_4=e^{-\eta}\Phi_0,\ \Phi_5=e^{\xi+\eta}\Phi_0,\ldots$$

Define

$$S_0 = \Phi_0,$$

$$\begin{aligned} S_1 &= \Phi_1 + \Phi_2 + \Phi_3 + \Phi_4 = e^{\xi}\Phi_0 + e^{\eta}\Phi_0 + e^{-\xi}\Phi_0 + e^{-\eta}\Phi_0 \\ &= \left(1 + \xi + \frac{1}{2}\xi^2 + 1 + \eta + \frac{1}{2}\eta^2 + 1 - \xi + \frac{1}{2}\xi^2 + 1 - \eta + \frac{1}{2}\eta^2\right)\Phi_0 + O\left(h^4\right) \\ &= \left(4 + \xi^2 + \eta^2\right)\Phi_0 + O\left(h^4\right) \\ &= 4\Phi_0 + h^2\Delta\Phi_0 + O\left(h^4\right) \end{aligned}$$

So, there is

$$\Delta\Phi_0 = \frac{S_1 - 4S_0}{h^2} + O\left(h^2\right) \tag{8.61}$$

If the mesh step size is defined as $h = 1$, then the Laplacian second-order difference approximation formula on the plane is obtained

$$L_{2xy} = S_1 - 4S_0 \tag{8.62}$$

The five symmetric grid points adopted by the second-order finite-difference approximation are shown in Figure 8.8a. By a similar method, the fourth-order difference approximation formula with 21 grid points and the sixth-order difference approximation formula with 53 grid points can be derived. Figure 8.8b and c shows the 21 grid points and 53 grid points, respectively.

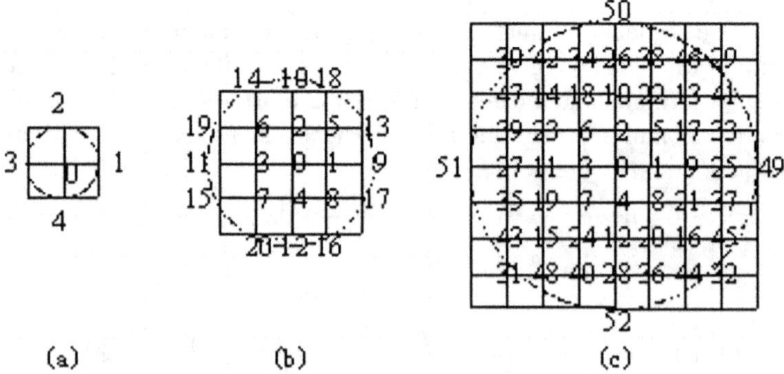

FIGURE 8.8 Mesh planes and mesh points for order 2, 4, and 6 finite difference approximations. (a) Five grid points used in the second-order difference approximation; (b) 21 grid points used in the fourth-order difference approximation; (c) 53 grid points used in the sixth-order difference approximation.

Define $S_0 = \Phi_0$, $S_1 = \Phi_1 + \Phi_2 + \Phi_3 + \Phi_4$, ..., $S_5 = \Phi_{17} + \Phi_{18} + \Phi_{19} + \Phi_{20}$, the Laplacian difference approximation formula of the fourth order is obtained

$$L_{4xy} = 64S_1/69 + 16S_2/69 - 7S_3/276 - 2S_4/69 - 2S_5/69 - 99S_0/23 \quad (8.63)$$

Further defining S_6 to S_{13}, with $S_{13} = \Phi_{49} + \Phi_{50} + \Phi_{51} + \Phi_{52}$, the Laplacian difference approximation formula of sixth-order is obtained

$$\begin{aligned} L_{6xy} = {} & 163264S_1/137155 + 20408S_2/137155 - 10204S_3/137155 \\ & -2551S_4/274310 - 2646S_5/137155 - 2646S_6/137155 - 17836S_7/3703185 \\ & -1372S_8/3703185 + 1462S_9/411465 + 1462S_{10}/411465 + 688S_{11}/411465 \\ & +688S_{12}/411465 + 5117S_{13}/6583440 - 72458941S_0/14812740 \end{aligned} \quad (8.64)$$

8.4.2.3 Simulation Research

The Laplacian finite difference approximations of orders 2, 4, and 6 are compared by simulation experiments. The results show that, for the noiseless case, the higher the difference precision, the resulted Laplacian is closer to the theoretical value, while for the noise case, the results depend on the noise feature as the higher difference (favor to high frequency noise) is more sensitive to noise.

8.4.3 Discussion

In practice, the measured potential distribution of EEG or ECG is a blurred version of the cortical or pericardium potential by the volume conductor. The surface Laplacian technique is developed to obtain a spatial distribution with a slightly better spatial resolution than EEG or ECG potential and eliminate the uncertainty introduced by the reference electrode. The five-point local difference approximation algorithm is the simplest method to obtain the planar surface Laplacian. In this section, we show the higher-order finite difference approximations of Laplacian. The positions of these grid points are shown in Figure 8.8, in which three circles drawn with dotted lines are used to approximate the range of 5, 21, and 53 grid points on the plane. It can be seen that more grid points in a larger range are needed for higher-order difference approximation. When there was no noise, the simulation results again demonstrated that Laplacian was more effective at shallow sources, i.e., neural activity close to the anterior chest wall or cortical surface. For the potential generated by sources at the same depth, the higher-order Laplacian difference approximation obtained better results. However, when global noise and peak noise are added, the high-order difference approximation results are better than the low-order difference approximation only when the noise is relatively small, including no noise. When the noise is relatively high, the low-order difference approximation gets a better result. Higher-order difference approximation requires more grid points, which means that more spatial information must be included and more noise, especially high frequency noise, will be kept or enlarged in the calculation. Therefore, when the noise is high, the low-order difference approximation is better because it will introduce less noise. In EEG and ECG practice, surface Laplacian should be obtained by using the differential approximation of appropriate order according to different noise levels. Furthermore, Laplacian is not only of better spatial resolution but also of more accurate time information by removing the reference effect to obtain reliable time domain process information.

8.5 LOCAL LAPLACIAN ESTIMATE ON SPHERICAL SCALP SURFACE

8.5.1 Introduction

As shown in Section 8.4, the earliest and simplest local method, the difference approximation, was still based on a planar surface model. In this section, the local difference approximation approach was extended to a spherical surface model (Lai and Yao, 2009). The second- and fourth-order approximations were derived, and a proportional coefficient difference was revealed between the planar and spherical surface models. This work provided the theoretical basis of the local difference approach for Laplacian on a spherical scalp surface.

As the Laplacian estimation should strictly comply with the definition of the surface Laplacian and should only be implemented with local information about potential and scalp surface shape (Gevins et al., 1991; Le et al., 1994), the local method is the most reasonable one as it usually estimates the Laplacian locally at an electrode site by a numerical difference estimator. Furthermore, it is also the most convenient and popular one in practical applications because of its simple computation and implementation. Till now, several local approaches have been proposed in the last 30 years. Hjorth (1975) first reported a rectangular uni-polar electrode array to estimate the second-order finite difference of surface potential, and in Section 8.4, we introduced fourth- and sixth-order approximations. Based on the fact that a coaxial electrode was more sensitive to local electrical events than a unipolar electrode, He and Cohen (1992) adopted a bi-polar concentric ring electrode sensor to approximate the analytical Laplacian. Based on a nine-point numerical approximation, Besio et al. (2006b) recently suggested a tri-polar concentric ring electrode for estimating the surface Laplacian with a precision of fourth-order residual error, higher than the bi-polar/ring and five-point difference electrode systems.

However, all the above three local approaches were developed on the assumption of a planar surface, which is inconsistent with the actual situation of the realistic head shape and may result in an inaccurate calculation. So, in this section, a spherical surface was used for the tri-polar concentric ring electrode to get a more realistic Laplacian estimation for scalp potentials.

8.5.2 Local Difference Approximation on a Planar Surface

8.5.2.1 Second-Order Planar Surface Laplacian

As shown in Figure 8.9a, the planar five-point electrode method first measures the local potentials at v_0, v_1, v_2, v_3, v_4. Then the laplacian at the point v_0 can be given by a numerical difference as:

$$\nabla^2\Phi = \frac{4}{r^2}\left(\frac{1}{4}\sum_{i=1}^{4}\Phi_i - \Phi_0\right) + O\left(r^2\right) \tag{8.65}$$

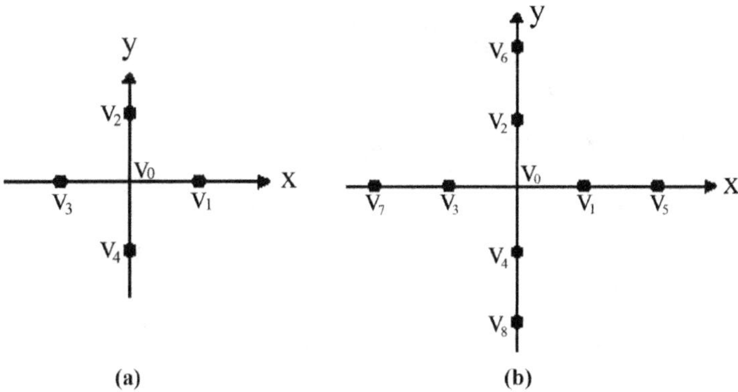

FIGURE 8.9 Schematic diagram of the planar five-point (a) and nine-point (b) electrode configurations (Lai and Yao, 2009).

where Φ_i is the potential at the point v_i, r is the distance between the central point v_0 and its neighboring points v_1, v_2, v_3 and v_4. The second-order residual error is:

$$O\left(r^2\right) = -r^2 \sum_{n=4}^{\infty} \frac{2r^{n-4}}{n!}\left(\frac{\partial^n \Phi}{\partial x^n} + \frac{\partial^n \Phi}{\partial y^n}\right)(n = 4,6,8,\ldots,\infty) \tag{8.66}$$

8.5.2.2 Fourth Order Planar Surface Laplacian

As shown in Figure 8.9b, the nine points can be viewed as two five-point systems: points v_0, v_1, v_2, v_3, v_4 and v_0 form a five-point arrangement with a distance of r, points v_5, v_6, v_7, v_8 and v_0 form another five-point arrangement with a distance of $2r$. Then the second-order term in equation (8.66) can be canceled by a combination of these two five-point arrangements, and a fourth order higher precision Laplacian at point v_0 can be written as:

$$\nabla^2 \Phi = \frac{16}{3r^2}\left[\left(\frac{1}{4}\sum_{i=1}^{4} \Phi_i - \Phi_0\right) - \left(\frac{1}{4}\sum_{i=5}^{8} \Phi_i - \Phi_0\right)\right] + O\left(r^4\right) \tag{8.67}$$

where the fourth-order residual error is:

$$O\left(r^4\right) = r^4 \sum_{n=6}^{\infty} \frac{2\left(2^{n-2} - 4\right)r^{n-6}}{3 \times n!}\left(\frac{\partial^n \Phi}{\partial x^n} + \frac{\partial^n \Phi}{\partial y^n}\right)(n = 6,8,10,\ldots,\infty) \tag{8.68}$$

8.5.2.3 Laplacian Measurement by Coaxial Electrodes on a Planar Surface

According to equation (8.67), the fourth-order derivative of the potential at point v_0 is approximately related to the average potentials for v_1, v_2, v_3, v_4 and v_5, v_6, v_7, v_8.

By replacing the average potential with the average estimate over the whole ring, a concentric tri-polar electrode system, as shown in Figure 8.10, can be used to facilitate a direct measure of the local Laplacian with better sensitivity for local electrical

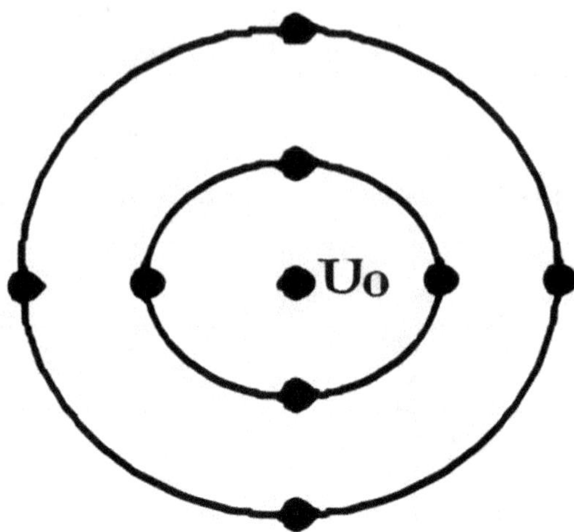

FIGURE 8.10 Schematic diagram of the circular tri-polar electrode configuration on a plane (Lai and Yao, 2009).

events than the rectangular unipolar electrode array. So, the equation (8.67) can be applied to a concentric disc and rings as follows (Besio et al., 2006a,b):

$$\nabla^2\Phi = \frac{16}{3r^2}\left[\left(\frac{1}{2\pi}\int_0^{2\pi}\Phi_r\, d\theta - \Phi_0\right) - \left(\frac{1}{2\pi}\int_0^{2\pi}\Phi_{2r}\, d\theta - \Phi_0\right)\right] + O\left(r^4\right) \quad (8.69)$$

Similarly, the second-order approximation of the ring-electrodes can be written as:

$$\nabla^2\Phi = \frac{4}{r^2}\left(\frac{1}{2\pi}\int_0^{2\pi}\Phi_r\, d\theta - \Phi_0\right) + O\left(r^2\right) \quad (8.70)$$

8.5.3 Local Difference Approximation on a Spherical Surface

The above local approaches are useful and reasonable for various surface Laplacian that strictly complies with the definition of the surface Laplacian and can be implemented only with the local information of potential and scalp surface shape. But all of the approaches are based on a planar surface assumption. That's maybe an effective approximation of the chest surface, but not the case of the scalp surface, where it is more like a sphere, especially in a local sense. So, in this section, the above functions are to be extended to a more general spherical surface. Before deducing the spherical local Laplacian, two crucial approximating formulas, which are important for the following work, are to be put forward first.

8.5.3.1 Two Lemmas

The first lemma is about the relationship between $\ln(\cos 2\theta)$ and $\ln(\cos\theta)$. By applying the least-squares method, a directly proportional coefficient of 4.0601 was found in the duration of $\theta \in [1°,10°]$. Approximately, we take 4.0 for simplicity and get the formula:

$$\ln(\cos 2\theta) \approx 4\ln(\cos\theta) \quad (\theta \in [1°,10°]) \tag{8.71}$$

As shown in Figure 8.11a, there is little deviation between the curves of $\ln(\cos 2\theta)$ and $4\times\ln(\cos\theta)$. A detailed computation showed that the maximum error between $\ln(\cos 2\theta)$ and $4\times\ln(\cos\theta)$ is only 0.0009671. So, it is reasonable to conclude that $\ln(\cos 2\theta)$ is about four times to ln (cos θ) in the duration of $\theta \in [1°,10°]$.

Another one is about the relationship between ln (cos θ) and θ^2. Similarly, a negative correlation coefficient of −0.5024 between ln (cos θ) and θ^2 was found by applying the least-squares method in the duration of $\theta \in [1°,10°]$. For convenience, an approximation value of −0.5 was taken to get:

$$\ln(\cos(\theta)) \approx -0.5\theta^2 \quad (\theta \in [1°,10°]) \tag{8.72}$$

As shown in Figure 8.11b, the two curves of ln (cos θ) and $-0.5\theta^2$ were almost overlapped in the whole range of $\theta \in [1°,10°]$. Detailed computation showed that the maximum error between ln (cos θ) and $-0.5\theta^2$ is only 0.00007796. So, it is rational to assume that ln (cos θ) is about −0.5 times to θ^2 in the duration of $\theta \in [1°,10°]$.

8.5.3.2 Local Laplacian by Coaxial Electrodes on a Spherical Surface

Due to the clear interrelation between equations (8.67) and (8.69), here we will directly derive the formula related to equation (8.69).

The Laplacian of the scalp potential function Φ in spherical coordinates (r, θ, φ) is:

$$\nabla^2\Phi = \frac{1}{r^2\sin\theta}\frac{\partial}{\partial\theta}\left(\sin\theta\frac{\partial\Phi}{\partial\theta}\right) + \frac{1}{r^2\sin^2\theta}\frac{\partial^2\Phi}{\partial\phi^2} \tag{8.73}$$

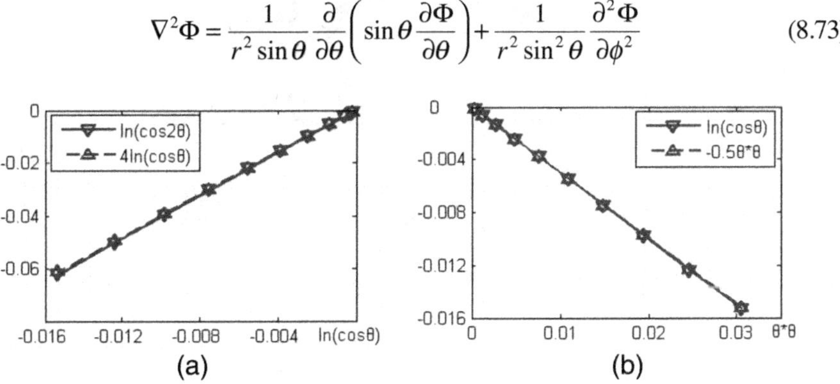

FIGURE 8.11 Two approximate lemmas. (a) The linear relationship between $\ln(\cos 2\theta)$ and $\ln(\cos\theta)$. The lateral axis shows $x = \ln(\cos\theta)$, the vertical axis shows $Y_1 = \ln(\cos 2\theta)$, $Y_2 = 4\ln(\cos\theta)$, and $\theta \in [1°,10°]$. The tested angle range $(\theta \in [1°,10°])$ is assumed to be larger than the possible range of the inter distance of the electrodes utilized for the actual calculation of Laplacian at a point (Lai and Yao, 2009); (b) The proportional relationship between $\ln(\cos\theta)$ and θ_2. The lateral axis shows $x = \theta_2$, the vertical axis shows $Y_1 = \ln(\cos\theta)$, $Y_2 = -0.5\theta_2$, and $\theta \in [1°,10°]$ (Lai and Yao, 2009).

Since the Laplacian operator is independent of the coordinate system, a specific spherical coordinate system may be chosen for each point, with the z-axis passing through the point and its origin at the center of the sphere. So, the second-order term in equation (8.73) will tend to zero, and a simplified Laplacian operator can be approximated as:

$$\begin{aligned}\nabla^2\Phi &= \frac{1}{r^2\sin\theta}\frac{\partial}{\partial\theta}\left(\sin\theta\frac{\partial\Phi}{\partial\theta}\right)=\frac{1}{r^2\sin\theta}\left[\sin\theta\frac{\partial^2\Phi}{\partial\theta^2}+\cos\theta\frac{\partial\Phi}{\partial\theta}\right]\\ &=\frac{1}{r^2}\left[\frac{\partial^2\Phi}{\partial\theta^2}-\frac{\partial\Phi}{\partial(\ln\cos\theta)}\right]\end{aligned} \tag{8.74}$$

Considering a set of four symmetric points on the ring (Figure 8.12), the nine-point numerical difference method can be used to estimate the second-order derivative of Φ in function (8.74). Taking the average approximation on the two rings, we have the following function as the counterpart of function (8.69):

$$\frac{\partial^2\Phi}{\partial\theta^2}=\frac{16}{3(\Delta\theta)^2}\left[\left(\frac{1}{2\pi}\int_0^{2\pi}\Phi_{\Delta\theta}\,d\theta-\Phi_0\right)-\left(\frac{1}{2\pi}\int_0^{2\pi}\Phi_{2\Delta\theta}\,d\theta-\Phi_0\right)\right]+O\left((\Delta\theta)^4\right) \tag{8.75}$$

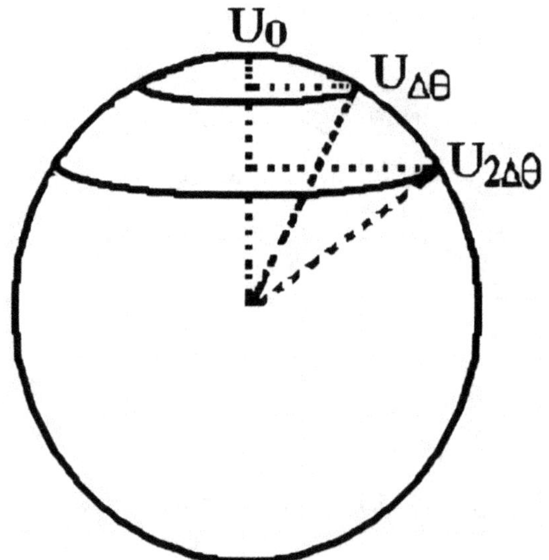

FIGURE 8.12 Schematic diagram of the circular tri-polar electrode structure on a spherical surface (Lai and Yao, 2009).

where $\Delta\theta$ is the latitude angle between the z-axis and the inner ring, and $2\Delta\theta$ is the outer ring. The fourth-order residual error is:

$$O\left((\Delta\theta)^4\right)=(\Delta\theta)^4\sum_{n=6}^{\infty}\frac{2\left(2^{n-2}-4\right)(\Delta\theta)^{n-6}}{3\times n!}\frac{\partial^n\Phi}{\partial\theta^n} \tag{8.76}$$

To consider the difference approximation of the first-order term in equation (8.74), the potential for an arbitrary point on the rings can be expressed by applying the Taylor expansions:

$$\Phi_{\Delta\theta}=\Phi_0+\sum_{n=1}^{\infty}\frac{\left(\ln\cos(\Delta\theta)\right)^n}{n!}\frac{\partial^n\Phi}{\partial(\ln\cos\theta)^n} \tag{8.77}$$

$$\Phi_{2\Delta\theta}=\Phi_0+\sum_{n=1}^{\infty}\frac{\left(\ln\cos(2\Delta\theta)\right)^n}{n!}\frac{\partial^n\Phi}{\partial(\ln\cos\theta)^n} \tag{8.78}$$

According to formula (8.71), $\ln\left(\cos(2\Delta\theta)\right)$ is about 4 times to $\ln\left(\cos(\Delta\theta)\right)$ if $\Delta\theta$ is only in a small duration of [1°, 10°], as the situation for a local Laplacian approximation. Then the above two functions can be combined as $16\times(8.77)$ minus (8.78). So, the first-order derivative of Φ in equation (8.74) can be written as:

$$\begin{aligned}\frac{\partial\Phi}{\partial(\ln\cos\theta)}=\frac{16}{12\ln\cos(\Delta\theta)}&\left[\left(\frac{1}{2\pi}\int_0^{2\pi}\Phi_{\Delta\theta}\,d\theta-\Phi_0\right)\right.\\&\left.-\left(\frac{1}{2\pi}\int_0^{2\pi}\Phi_{2\Delta\theta}\,d\theta-\Phi_0\right)\right]+O\left[\left(\ln\cos(\Delta\theta)\right)^2\right]\end{aligned} \tag{8.79}$$

where the residual error is

$$\begin{aligned}&O((\ln\cos(\Delta\theta))^2)=\\&(\ln\cos(\Delta\theta))^2\times\sum_{m=3}^{\infty}\frac{\left(4^m-16\right)\left(\ln\cos(\Delta\theta)\right)^{m-3}}{12\times m!}\frac{\partial^m\Phi}{\partial(\ln\cos\theta)^m}\end{aligned} \tag{8.80}$$

By a combination of equations (8.75), (8.79), and (8.74), the spherical local Laplacian can be estimated as:

$$\begin{aligned}\nabla^2\Phi=\frac{1}{12r^2}&\left[16\left(\frac{1}{2\pi}\int_0^{2\pi}\Phi_{\Delta\theta}\,d\theta-\Phi_0\right)-\left(\frac{1}{2\pi}\int_0^{2\pi}\Phi_{2\Delta\theta}\,d\theta-\Phi_0\right)\right]\\&\times\left(\frac{4}{(\Delta\theta)^2}-\frac{1}{\ln\cos(\Delta\theta)}\right)+O\left((\Delta\theta)^4\right)\end{aligned} \tag{8.81}$$

and the fourth-order residual error is:

$$O\left((\Delta\theta)^4\right)=\frac{1}{r^2}\left[\sum_{n=6}^{\infty}\frac{2\left(2^{n-2}-4\right)\times(\Delta\theta)^{n-2}}{3\times n!}\frac{\partial^n\Phi}{\partial\theta^n}+\sum_{m=3}^{\infty}\frac{\left(16-4^m\right)\left(\ln\cos(\Delta\theta)\right)^{m-1}}{12\times m!}\frac{\partial^m\Phi}{\partial(\ln\cos\theta)^m}\right]$$

$$\left(n=6,8,10,\ldots,\infty;\quad m=3,4,5,\ldots,\infty\right)\tag{8.82}$$

According to formula (8.72), $\ln\left(\cos(\Delta\theta)\right)$ is about -0.5 times to $(\Delta\theta)^2$. So, the function (8.81) can be further simplified as:

$$\nabla^2\Phi\approx\frac{1}{2r^2(\Delta\theta)^2}\left[16\times\left(\frac{1}{2\pi}\int_0^{2\pi}\Phi_{\Delta\theta}\,d\theta-\Phi_0\right)-\left(\frac{1}{2\pi}\int_0^{2\pi}\Phi_{2\Delta\theta}\,d\theta-\Phi_0\right)\right]+O\left((\Delta\theta)^4\right)\tag{8.83}$$

where the fourth-order residual error is:

$$O\left((\Delta\theta)^4\right)=\frac{(\Delta\theta)^4}{r^2}\left[\sum_{n=6}^{\infty}\frac{2\left(2^{n-2}-4\right)\times(\Delta\theta)^{n-6}}{3\times n!}\frac{\partial^n\Phi}{\partial\theta^n}+\sum_{m=3}^{\infty}\frac{\left(16-4^m\right)(\Delta\theta)^{2m-6}}{12\times(-2)^{m-1}\times m!}\frac{\partial^m\Phi}{\partial(\ln\cos\theta)^m}\right]$$

$$\left(n=6,8,10,\ldots,\infty;\quad m=3,4,5,\ldots,\infty\right)\tag{8.84}$$

To compare the function (8.83) with (8.69), it is evident that the spherical local Laplacian has a similar form as that of the planar local Laplacian (Besio et al., 2006a,b), except for a proportional coefficient difference.

Similarly, we may have the second-order approximation as:

$$\nabla^2\Phi\approx\frac{6}{r^2(\Delta\theta)^2}\left(\frac{1}{2\pi}\int_0^{2\pi}\Phi_{\Delta\theta}d\theta-\Phi_0\right)+O\left((\Delta\theta)^2\right)\tag{8.85}$$

where the second-order residual error is:

$$O\left((\Delta\theta)^2\right)=\frac{(\Delta\theta)^2}{r^2}\left[\sum_{n=4}^{\infty}\frac{2(\Delta\theta)^{n-4}}{n!}\frac{\partial^n\Phi}{\partial\theta^n}+\sum_{m=2}^{\infty}\frac{(-0.5)^{m-1}(\Delta\theta)^{2m-4}}{m!}\frac{\partial^m\Phi}{\partial(\ln\cos\theta)^m}\right]$$

$$\left(n=4,6,8,\ldots,\infty;\ m=2,3,4,\ldots,\infty\right)\tag{8.86}$$

For the actual measurement of the scalp surface Laplacian, the spherical tri-electrode system, as shown in Figure 8.12, can be used to get the local Laplacian. It consists of three parts: a conductive disk at the center and two concentric conductive rings surrounding the central disk. The surface area of the central disk should be set to the same as that of the inner and outer rings. The latitude angles between the z-axis and the rings are $\Delta\theta$ and $2\Delta\theta$, respectively.

8.5.3.3 Local Laplacian by Discrete Electrode Net on a Spherical Surface

For numeric calculation of the spherical scalp Laplacian at a concern point similar to that in Figure 8.10, the fourth-order approximation can be expressed as:

$$\nabla^2\Phi \approx \frac{1}{2r^2(\Delta\theta)^2}\left[16\times\left(\frac{1}{4}\sum_{i=1}^{4}\Phi_{\Delta\theta}(i)-\Phi_0\right)-\left(\frac{1}{4}\sum_{i=5}^{8}\Phi_{2\Delta\theta}(i)-\Phi_0\right)\right]+O\left((\Delta\theta)^4\right) \tag{8.87}$$

where $\Phi_{\Delta\theta}(i)$ $(i=1,2,3,4)$ is the neighboring discrete symmetric measurement points on the first ring, and $\Phi_{2\Delta\theta}(i)$ $(i=5,6,7,8)$ is the one on the second ring. Similarly, we can also get the second-order approximation as:

$$\nabla^2\Phi = \frac{6}{r^2(\Delta\theta)^2}\left(\frac{1}{4}\sum_{i=1}^{4}\Phi_{\Delta\theta}(i)-\Phi_0\right)+O\left((\Delta\theta)^2\right) \tag{8.88}$$

In practice, if the electrode net is not designed symmetrically, we may have a local interpolation step at first to get the potential at the symmetric points of each concern point. Then the above formula can be utilized.

8.5.4 Discussion

No matter what the approach, local, moderate, or global, each was designed to reduce the blurring effect of volume conduction on scalp EEG and to avoid the effect of an arbitrary non-zero reference electrode choice. The local approach is alive for its mathematical convenience and physical advantage to locating the current sources more properly than the other approaches since its approximation strictly complies with the definition of the surface Laplacian and is only related to the local information of neighboring potentials. However, in the last 30 years, all local approaches were based on a planar surface model, which was not fit for the realistic head situation. In this part, the local difference approach was extended to a spherical head model. Based on two approximating lemmas, we obtained the formula for a spherical surface with a form quite similar to the counterparts of a planar surface model. Thus, the current techniques developed for the planar model can be directly applied to the spherical case. And that's why we did not present any simulation results but referred to the literatures (Oostendrop and Oosterom, 1996; Besio et al., 2006a, b).

Though the scalp surface for a realistic head model cannot be taken as a regular sphere, a spherical approximation for each local area will still be more reasonable than the planar assumption. So, the spherical surface model can be used to preserve a more accurate representation of the scalp's potential distribution. As for the irregular surface, the local Laplacian can be approximated by an average of the directional derivatives from each surrounding electrode to the given electrode or by a method named nearest neighbor Surface Laplacian Derivative (Le et al., 1994), which included three main procedures: first, finding neighboring electrodes within a chosen search radius, then determining a local optimal sphere to fit this set of electrodes (including the given electrode) by using the Taylor expansion and the least-squares technique, and lastly estimating the Laplacian on this local optimal sphere by using measured potential values at the given electrode and its neighbors.

Additionally, for a discrete situation where the electrodes are not symmetric, it is unfit to apply the function (8.83) directly to estimate the spherical local Laplacian. There is a prerequisite that the spherical local Laplacian estimation must be based on a spherical symmetric four-point system to use the nine-point numerical difference to approximate the second-order derivative of Φ in equation (8.73). But this problem can be solved by choosing a suitable interpolation function to construct a local interpolation for each point, as in Le et al. (1994). Then a spherically symmetric system can be obtained by applying the numerical difference estimator in equation (8.83).

In this chapter, we discuss the different estimation methods or measurement methods of Laplacian at the framework level, for methods that need to solve the equation, the different regularization methods can affect the effectiveness of the method. A method with good practical results must be an overall optimization of these aspects. The mesoscale method is the optimal method only in terms of physical concepts, so we can use it as a starting point and then optimize other aspects to establish the optimal practical method. In theory, we give a complex relationship between Laplacian and current density in the spherical model. In any case, the core of Laplacian is a spatial high-pass filtering output of the scalp surface EEG, aiming to display the electrical activities of the head with higher spatial resolution. Nunez, under certain assumptions, proved that the Scalp Laplacian was proportional to the cortical potential, as shown in equation (8.9). Therefore, approximate Scalp Laplacian is a kind of high-resolution cortical imaging method as a result of high-pass filtering.

REFERENCES

Babiloni, F., C. Babiloni, L. Fattorini, et al. 1995. Performances of surface Laplacian estimators: A study of simulated and real scalp potential distributions. *Brain Topogr* 8(1):35–45.

Babiloni, F., C. Babiloni, L. Fattorini, et al. 1996. Spline Laplacian estimate of EEG potentials over a realistic magnetic resonance-constructed scalp surface model. *EEG Clin Neurosci* 98:363–373.

Babiloni, F., C. Babiloni, L. Fattorini, et al. 2001. Spatial enhancement of EEG data by surface Laplacian estimation: The use of magnetic resonance imaging-based head models. *Clin Neurophysiol* 112:724–727.

Besio, G., K. Koka, R. Aakula, et al. 2006a. Tri-polar concentric ring electrode development for Laplacian electroencephalography. *IEEE Trans Biomed Eng* 53(5):926–933.

Besio, W., R. Aakula, K. Koka, et al. 2006b. Development of a tri-polar concentric ring electrode for acquiring accurate Laplacian body surface potentials. *Ann Biomed Eng* 34(3):426–435.

Bortel, R., P. Sovka. 2013. Potential approximation in realistic Laplacian computation. *Clin Neurophysiol* 124:462–473.

Carvalhaes, C., J.A. De Barros. 2015. The surface Laplacian technique in EEG: Theory and methods. *Int J Psychophysiol* 97:174–188.

Carvalhaes, C., P. Suppes. 2011. A spline framework for estimating the EEG surface Laplacian using the Euclidean metric. *Neural Comput* 23:2974–3000.

Geselowitz, D.B., J.E. Ferrara. 1999. Is accuracy recording of the ECG surface Laplacian feasible? *IEEE Trans Biomed Eng* 46:377–381.

Gevins, A., J. Le, P. Brickett, et al. 1991. Seeing through the skull: Advanced EEGs use MIRs to accurately measure cortical activity from the scalp. *Brain Topogr* 4(2):125–131.

Gevins, A., P. Brickett, B. Costales, et al. 1990. Beyond topographic mapping: Towards functional-anatomical imaging with 124-channel EEGs and 3-D MRIs. *Brain Topogr* 3(1):53–64.

Harder, R.L, R.N. Desmarais. 1972. Interpolation using surface splines. *J Aircraft* 9(2):189–191.

Hardy, R.L. 1971. Multiquadric equations of topography and other irregular surfaces. *J Geophys Res* 76(8):1905–1915.

Haykin, S. 1999. *Neural Networks: A Comprehensive Foundation*. Hoboken, NJ: Prentice Hall Press.

He, B., J. Lian, G. Li. 2001. High-resolution EEG: A new realistic geometry spline Laplacian estimation technique. *Clin Neurophysiol* 112:845–852.

He, B., R.J. Cohen. 1992. Body surface Laplacian mapping in man. *IEEE Trans Biomed Eng* 39(11):1179–1191.

Hjorth, B. 1975. An on-line transformation of EEG scalp potentials into orthogonal source derivations. *EEG Clin Neurosci* 39:526–530.

Jager, J., A. Klein, M. Buhmann, et al. 2016. Reconstruction of electroencephalographic data using radial basis functions. *Clin Neurophysiol*. doi: 10.1016/j.clinph.2016.01.003.

Katznelson, R.D. 1981. Normal modes of the brain: Neuroanatomical basis and a physiological theoretical model. In Nunez, P.L. (ed.), *Electric Fields of the Brain: The Neurophysics of EEG*. New York: Oxford University Press, 401–442.

Kayser, J., C.E. Tenke. 2006. Principal components analysis of Laplacian waveforms as a generic method for identifying ERP generator patterns: I. *Clin Neurophysiol* 117(2):348–368.

Kayser, J., C.E. Tenke. 2015. Issues and considerations for using the scalp surface Laplacian in EEG/ERP research: A tutorial review. *Int J Psychophysiol* 97(3):189–209.

Kleber, A.G., M.J. Janse, F.J. Van Capelle, et al. 1978. Mechanism and time course of ST and TQ segment changes during acute regional myocardial ischemia in the pig heart determined by extracellular and intracellular recordings. *Circ Res* 42:603–613.

Lai, Y., D. Yao. 2009. Local Laplacian estimate on spherical scalp surface. *Chin J Electron* 18(4):681–685.

Law, S.K., P.L. Nunez, R.S. Wijesinghe. 1993. High resolution EEG using spline generated surface Laplacians on spherical and ellipsoidal surfaces. *IEEE Trans Biomed Eng* 40:145–153.

Le, J., V. Menon, A. Gevins. 1994. Local estimate of surface Laplacian derivation on a realistically shaped scalp surface and its performance on noisy data. *EEG Clin Neurosci* 92:433–441.

Lo, P.C. 1999. Three-dimensional filtering approach to brain potential interpolation. *IEEE Trans Biomed Eng* 46(5):574–583.

Makeyev, O., W.G. Besio. 2016. Improving the accuracy of Laplacian estimation with novel variable inter-ring distances concentric ring electrodes. *Sensors* 16(6):858. doi: 10.3390/s16060858.

Menke, W. 1989. Geophysical data analysis: Discrete inverse theory. Orlando, FL: Academic Press.

Micchelli, C.A. 1986. Interpolation of scattered data: Distance matrices and conditionally positive definite functions. *Constr Approx* 2:11–12.

Nicholson, C. 1973. Theoretical analysis of field potentials in anisotropic ensembles of neuronal elements. *IEEE Trans Biomed Eng* 20:278–288.

Nicholson, C., J.A. Freeman. 1975. Theory of current source-density analysis and determination of conductivity tensor for anuran cerebellum. *J Neurophysiol* 38(2):356–368.

Nunez, P.L. 1981. *Electric Fields of the Brain.* New York: Oxford University Press.

Nunez, P.L. 2014. A historical context of the EEG surface Laplacian. https://ssltool.sourceforge.net/history.html.

Nunez, P.L., M.D. Nunez, R. Srinivasan. 2019. Multi-scale neural sources of EEG: Genuine, equivalent, and representative. *Brain Topogr* 32:193–214.

Nunez, P.L., R. Srinivasan. 2006. *Electric Fields of the Brain: The Neurophysics of EEG.* New York: Oxford University Press.

Nunez, P.L, R.B. Silberstein, P.J. Cadusch, et al. 1994. A theoretical and experimential study of high-resolution EEG based on surface laplacian and cortical imaging. *EEG Clin Neurosci* 90:40–57.

Oostendrop, T.F., A.V. Oosterom. 1996. The surface Laplacian of the potentials: Theory and application. *IEEE Trans Biomed Eng* 43(4):394–405.

Perrin, F., J. Pernier, O. Bertrand, et al. 1987. Mapping of scalp potentials by surface spline interpolation. *EEG Clin Neurosci* 66:75–78.

Perrin, F., J. Pernier, O. Bertrand, et al. 1989. Spherical splines for scalp potential and current density mapping. *EEG Clin Neurosci* 72:184–187.

Powell, M.J.D. 1988. Radial basis function approximations to polynomials. In *Proceedings of 12th Biennial Numerical Analysis Conference*, Dundee, pp. 223–241.

Srinivasan, R., P.L. Nunez, R.B. Silberstein. 1998. Spatialfiltering and neocortical dynamics: Estimates of EEG coherence. *IEEE Trans Biomed Eng* 45(7):814–826.

Tandonnet, C., B. Burle, T. Hasbroucq, et al. 2005. Spatial enhancement of EEG traces by surface Laplacian estimation: Comparison between local and global methods. *Clin Neurophysiol* 116:18–24.

Wahba, G. 1981. Spline interpolation and smoothing on the sphere. *SIAM J Sci Stat Comput* 2(1):5–16.

Wallin, G., E. Stalberg. 1980. Source derivation in clinical routine EEG. *EEG Clin Neurosci* 50:282–292.

Yao, D. 2000. High-resolution EEG mappings: A spherical harmonic spectra theory and simulation results. *Clin Neurophysiol* 111:81–92.

Yao, D. 2002a. The theoretic relation of scalp Laplacian and scalp current density of a spherical head model. *Phys Med Biol* 47(12):2179–2185.

Yao, D. 2002b. High-resolution EEG mapping: A radial-basis function based approach to the scalp Laplacian estimate. *Clin Neurophysiol* 113(6):956–967.

Zhai, Y. 2005. Research on reference electrode independent technology (in Chinese). PhD Thesis, University of Electronic Science and Technology of China.

Zhai, Y., D. Yao. 2004. A radial-basis function based surface Laplacian estimate for a realistic head model. *Brain Topogr* 17(1):55–62.

9 A Unified Framework for High-Resolution EEG

In previous chapters, five mappings were introduced: Laplacian mapping (LM), equivalent distributed dipole layer mapping (EDM), equivalent distributed charge layer mapping (ECM), cortical potential mapping (CPM), and source potential mapping (SPM). People naturally ask which one is the best and what is the relation among them. In this chapter, as a summary of these high-resolution EEG techniques, we will discuss the uniqueness of the cortical potential and establish a unified theoretical framework for the five high-resolution EEG mapping based on spherical harmonic spectrum analysis.

In this chapter, we no longer care about a specific algorithm as we have introduced a variety of methods in previous chapters, and in some literatures, there are also other methods to obtain cortical surface potential such as boundary element, finite element, and other numerical methods. More importantly, it can be expected that with the development of computer resources and computing technology, even better algorithms will be developed in the future.

In Chapter 2, we investigated the multipole expansion (ME) theory of EEG sources in an unbounded infinite conductor. It will be shown below that it is this ME theory that provides a unified theoretical framework for the above five high-resolution EEG techniques (Yao, 2000).

9.1 SPHERICAL HARMONIC SPECTRUM-BASED FRAMEWORK

9.1.1 ME and Spherical Harmonic Spectrum of Dipole Source

According to the deductive analysis in Chapters 2 and 4, the potential of a dipole at any position and orientation in an infinite medium can be expressed as

$$\Phi_\infty = \sum_{l=1}^{\infty}\sum_{m=0}^{l}\frac{1}{r^{l+1}}\left(g_l^m \cos m\phi + h_l^m \sin m\phi\right)P_l^m(\cos\theta) \tag{9.1}$$

where g_l^m, h_l^m is completely determined by the position of the dipole (r_0,θ_0,ϕ_0) and its moment. The items in equation (9.1) are corresponding to various multipole models in physics (Section 4.1); (r,θ,ϕ) is the spherical coordinate of a field point. In another word, equation (9.1) can also be regarded as the result of the orthogonal expansion of potential function Φ_∞ according to Legendre function, where g_l^m, h_l^m can therefore be called the SHS (spherical harmonic spectra) of the potential. It should be noted that for any potential function, there is a corresponding SHS, so the SHS here is directly

DOI: 10.1201/9781032639260-9

related to potential. ME, by contrast, is directly defined by the source distribution. Certainly, the potential is generated by the source; thus ME and SHS are intrinsically connected to each other with the influence of the volume conductor model, or they are the two aspects of a coin, MEs are the equivalent sources at the origin, and SHS are the strengths of the equivalent sources modified by the conductor.

9.1.2 Potential SHS in a Three-Layer Concentric Sphere

Among various volume conductor head models, the three-layer spherical model shown in Figure 9.1A is the most widely used. The parameters can be found in Section 4.4. By using the related boundary conditions, the dipole potential formula in the concentric three-layer sphere model (Figure 9.1A) can be obtained (Yao, 2000) (Chapter 4).

$$\Phi(i) = \sum_{l,m} \frac{1}{r^{l+1}} \left(G_l^m(i) \cos m\phi + H_l^m(i) \sin m\phi \right) P_l^m(\cos\theta) \tag{9.2}$$

where

$$G_l^m(i) = K_l(i) g_l^m, \qquad H_l^m(i) = K_l(i) h_l^m \tag{9.3}$$

$$a \geq r > r_0,\ i = 1; \quad b \geq r \geq a,\ i = 2; \quad c \geq r \geq b,\ i = 3$$

The $i = 1, 2, 3$ correspond to the brain, skull, and scalp regions, respectively. In the formula,

$$K_l(1) = A_l r^{2l+1} + 1, \quad K_l(2) = B_l \left(r^{2l+1} + \gamma \right), \quad K_l(3) = E_l \left(r^{2l+1} + \chi \right)$$

The parameters in equations (9.2) and (9.3) are the same as in formulas (4.88)–(4.94). The G_l^m and H_l^m in the formula are the SHS of the dipole potential Φ in a concentric three-layer spherical model. In particular, when $r = a$, $\Phi(1)$ = cortical surface potential, $r = c$, $\Phi(3)$ = scalp surface potential.

Here $K_l(i)$, $(i = 1,2,3)$ can be interpreted as three spatial filters, and the SHS can be regarded as the result of the ME passing through different spatial filters. For an infinite spatially conductive medium, we assume that its corresponding spatial filter $K_l(0) = 1.0$ is an all-pass spatial filter. Similarly, the cortical potential imaging (CPM) corresponds to a situation after ME passing $K_l(1)$; it is the potential on the surface of $r = a$ in Figure 9.1A, and the unbounded source potential mapping (SPM) is that ME passes through $K_l(0) = 1$, it is based on the potential on the surface $r \geq$ ia of Figure 9.1B. Of course, CPM can also be regarded as the result of the scalp surface SHS filtered by $K_l^{-1}(3)K_l(1)$, while SPM is the result of the scalp surface SHS filtered by $K_l^{-1}(3)K_l(0)$, that is, removing the influence of the concentric three-layer sphere model by the inverse filtering $K_l^{-1}(3)$, and then obtain the desired output through the corresponding forward spatial filter.

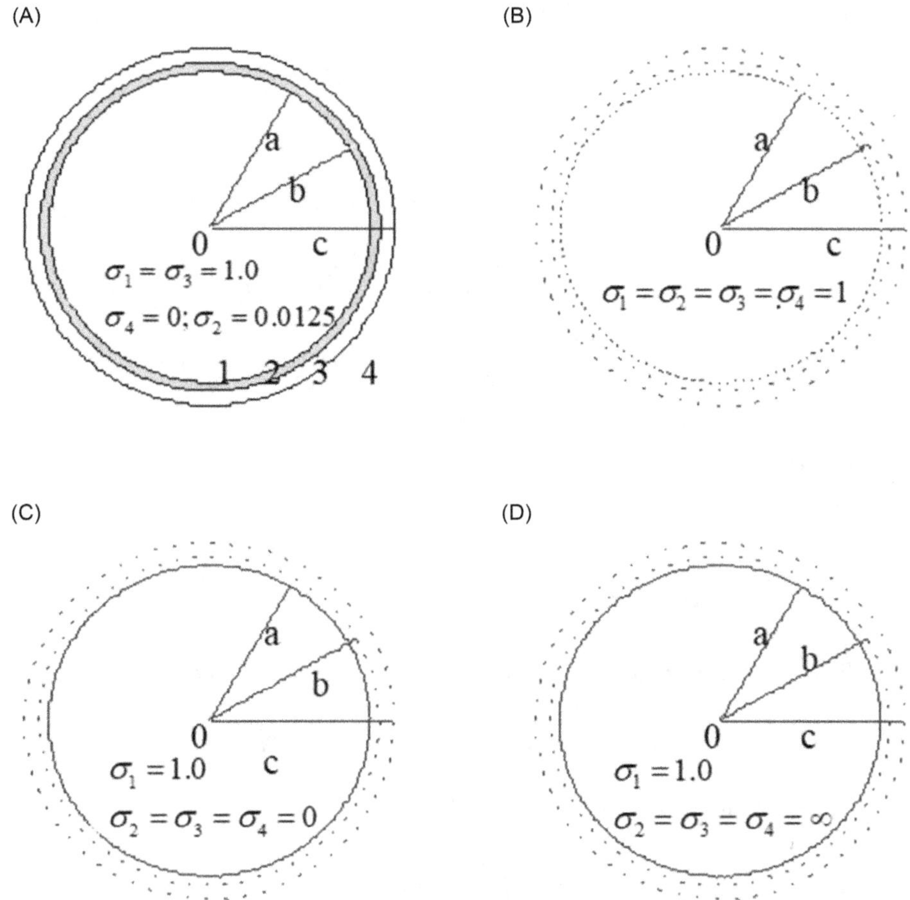

FIGURE 9.1 Various volume conductor models. (k) A concentric three-layer spherical conductor model; (l) unbounded space conductor model; (m) the equivalent distributed dipole layer ($r=0.87$); (n4) equivalent distribution charge layer ($r=0.87$). The a, b, and c in the figure are the radii of the three layers; σ_i is the corresponding regional conductance, $i=1, 2, 3, 4$ (Yao, 2000).

9.1.3 SHS of Equivalent Distributed Charge Layer Mapping on the Cortical Surface

According to the results in Chapters 6 and 7, for the spherical equivalent distributed point current source density/equivalent distributed charge, its intensity can be expressed as

$$S_q = \sum_{l,m}^{\infty} \frac{1}{r^{l+1}} \left(G_l^m(q) \cos m\phi + H_l^m(q) \sin m\phi \right) p_l^m(\cos\theta), \quad r \leq a \tag{9.4}$$

The SHS of the corresponding equivalent distribution point current source density is

$$G_l^m(q) = K_l(q)g_l^m, \quad H_l^m(q) = K_l(q)h_l^m \tag{9.5}$$

where the spatial filter (equation 6.12) is

$$K_l(q) = \sigma_1(2l+1)/r, \quad r \le a \tag{9.6}$$

Obviously, the key factor $(2l + 1)$ is a high spatial frequency amplification factor. The equivalent physical model confirmed in Section 6.1 is shown in Figure 9.1D.

9.1.4 SHS of Equivalent Distributed Dipole Layer Mapping on the Cortical Surface

According to the results in Chapters 6 and 7, for the spherical radial equivalent distributed dipole source, the intensity can be expressed as

$$S_d = \sum_{l=0}^{\infty}\sum_{m=0}^{l} \frac{1}{r^{l+1}}\left(G_l^m(d)\cos m\phi + H_l^m(d)\sin m\phi\right) p_l^m(\cos\theta) \tag{9.7}$$

The SHS of the corresponding equivalent distributed dipole source is

$$G_l^m(d) = K_l(d)g_l^m, \quad H_l^m(d) = K_l(d)h_l^m \tag{9.8}$$

The corresponding spatial filter (equation 6.38) is

$$K_l(d) = \sigma_1(2l+1)/l = 2 + 1/l \tag{9.9}$$

Obviously, the key factor $K_l(d) \sim 1/l$ is a relatively low spatial frequency amplification factor. The equivalent physical model confirmed in Section 6.1 is shown in Figure 9.1C.

9.1.5 SHS of Laplacian Mapping

According to Chapter 8, the scalp surface Laplacian is defined as

$$\text{Lscalp} = -\frac{1}{r^2}\frac{\partial}{\partial r}\left(r^2\frac{\partial\Phi(3)}{\partial r}\right)\Bigg|_{r=c} = \sum_{l,m}\frac{1}{c^{l+1}}\left(G_l^m(L)\cos m\phi + H_l^m(L)\sin m\phi\right)P_l^m(\cos\theta) \tag{9.10}$$

The corresponding Laplacian SHS is

$$G_l^m(L) = K_l(L)g_l^m, \quad H_l^m(L) = K_l(L)h_l^m \tag{9.11}$$

The corresponding spatial filter is

$$K_l\left(L\right)=-E_l\left(c^{2l+1}+\chi\right)(l+1)l/c^2=-K_l(3)(l+1)l/c^2 \tag{9.12}$$

It is easy to show that the spatial filter of Laplacian in other regions of the concentric three-layer sphere model is

$$K_l\left(L,i\right)=-K_l(i)(l+1)l/r^2 \tag{9.13}$$

$a\geq r>r_0,\ i=1;\ b\geq r\geq a,\ i=2;\ c\geq r\geq b, i=3$. For simplicity, in the following discussion, the minus sign and the constant r^2 in equations (9.12) and (9.13) are omitted and abbreviated to $K_l\left(L,0\right)=l(l+1)$ and $K_l\left(L,i\right)=K_l\left(L,0\right)K_l(i),\ i=1,2,3$.

In summary, we obtained five SHS and the corresponding spatial filters (Yao, 2000; Yao and He, 2003). With the help of these expressions, it is possible to uniquely relate any high-resolution EEG to ME of the EEG generator through these spatial filters and thus to correlate the five high-resolution EEG pairwise. Thus, these formulas essentially establish a unified theoretical framework for various high-resolution EEG technologies; it lays a foundation for the objective evaluation of various high-resolution EEG techniques, that is to say, the best one is a method that can recover the true SHS and so ME of the EEG.

9.2 RELATIONS AMONG FIVE HIGH-RESOLUTION EEG

9.2.1 Spatial Filters

Section 9.1 shows that the spatial filter K_l acts as a connection between the ME of the EEG source and the SHS of the various imaging. So the spatial filter K_l actually plays a crucial role in determining the relative effectiveness of various imaging techniques.

Figure 9.2 shows the normalized spectral characteristics of the above spatial filters K_l. From these curves, we may check the normalized value K_{10}/K_1 for $l=10$ and $l=1$. For the spatial filter K_l of the three concentric sphere head model, the relative values are 0.0879 (Figure 9.2d) for the scalp surface ($r=1.0$), 0.7532 (Figure 9.2b) and 0.5040 (Figure 9.2a) for the cortical surface with $r=0.87$ and $r=0.80$, respectively. For LM, EDM, and ECM, the relative values are 4.836 (Figure 9.2h), 0.70 (Figure 9.2g), and 7.0 (Figure 9.2f), respectively.

Compared with SPM $K_l(0)$ (Figure 9.2e), it is obvious that LM and ECM will produce higher spatial resolution maps (Figure 9.2f,h), while CPM and EDM will produce lower spatial resolution maps than SPM does (Figure 8.2a,c,g). At the same time, it is clear that EDM (Figure 9.2g) and CPM ($r=0.87$) (b) will have almost equivalent resolutions. The minimum value of Laplacian $K_l\left(L,3\right)$ on the scalp surface is 0.4174 when $l=8$, and the relative value is 0.4252 when $l=10$ (in Figure 9.2l). The relative value 0.4252 of the Laplacian filter at $l=10$ and $r=1.0$ is smaller than the relative value 0.7532 of the cortical potential at $l=10$ and $r=0.87$ (Figure 9.2l versus Figure 9.2b). In practice, the spatial frequency component that can be reliably reconstructed from

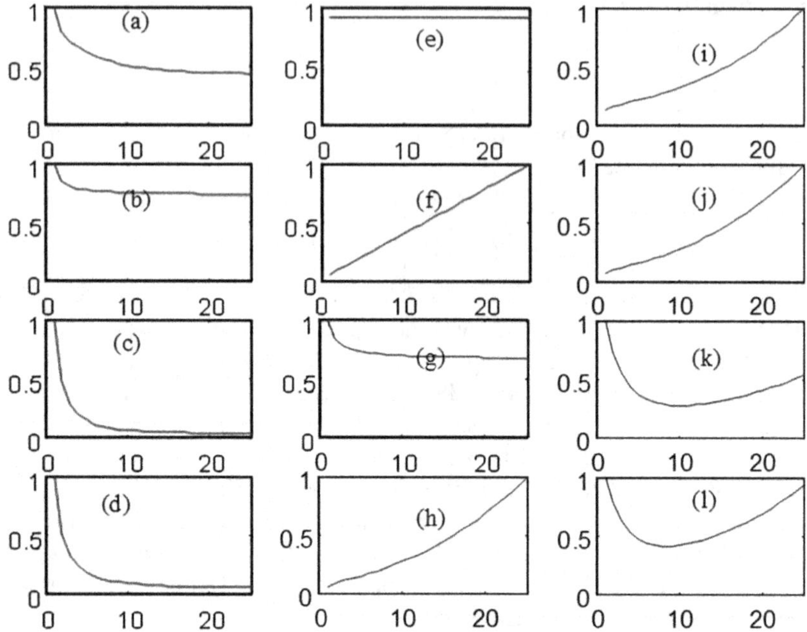

FIGURE 9.2 Normalized spectral curve of spatial filter K_l. The horizontal axis is l, and the vertical axis is the normalized spectral amplitude. (a) and (b) are $K_l(1)$ with r=0.80 and 0.87 (CPM); (c) $K_l(2)$ at r=0.92, the skull surface; (d) $K_l(3)$ at r=1.0, the scalp surface. They are filters of the concentric three-layer sphere model; (e) $K_l(0)$ at r=1.0, the infinite conductor model (SPM); (f) $K_l(q)$, the spherical equivalent charge layer(ECM); (g) $K_l(d)$, the spherical equivalent dipole layer (EDM); (h) $K_l(L,0)=l(l+1)$, the filter of Laplacian operator; (I)–(l) are $K_l(L,i)=K_l(i)K_l(L,0), i=1,2,3$ at r=0.80 (i=1), 0.87 (i=1), 0.92 (i=2), and 1.0 (i=3), respectively (Yao, 2000).

scalp data containing noise is around l= 10 (Edlinger et al., 1998); this fact provides support for the results in Nunez et al. (1994) that CPM at r=0.87 is better than LM at r= 1.0.

9.2.2 Forward Imaging Maps

A simulation example was given in Yao (2000), where a concentric three-layer spherical model was adopted, and three radial dipoles were used.

The results (Yao, 2000) clearly showed that the three sources are difficult to be distinguished on the surface of the three-layer concentric spherical head model, but they can be clearly distinguished on the cortical surface or deeper and on the derived graphs SPM, EDM, ECM, and LM, and that is why they are called high-resolution EEG mapping. In fact, the spherical harmonic spectra calculated simultaneously also clearly demonstrate the differences in the spectral distribution among different imaging quantities.

9.2.3 Inverse Imaging Maps

Based on the theoretical framework in Section 9.1, the multipolar expansion (ME) of EEG sources is not only an objective description of the EEG source but also a common basis for evaluating various high-resolution mapping techniques. As the different high-resolution maps are defined on different physical quantities, it is difficult to compare them directly, but they all are tightly related to the same ME of the actual sources, so they can be compared objectively and directly in the sense of ME.

For the multilayer sphere head model, as mentioned in the previous chapters, we can use different approaches to get high-resolution EEG, such as the equivalent distributed source method (Chapter 6), spheric harmonic analysis method (Chapter 7), and the numerical calculation method (Chapter 5), however, based on the above-unified framework, no matter which approach adopted, the eternal goal is to get the ME of the actual source as accurately as possible, and then, we can get various maps by the corresponding filter and the following high-resolution EEG map.

In Yao (2000), we made a simple comparison between the equivalent distributed dipole source method and the spherical spline fitting method (Edlinger et al., 1998; Yao, 2000; Perrin et al., 1989), in terms of cortical potential, the equivalent distributed dipole source intensity, and the spherical harmonic spectrum. The specific algorithm of the equivalent distributed dipole source method has been introduced in detail in Section 7.2.

Table 9.1 shows the correlation coefficients and relative errors between the three kinds of data calculated by the two methods and the corresponding theoretical values. The results also show that the equivalent distributed dipole source method is superior to the spherical spline method.

Considering that the equivalent distributed source method is easy to be applied under the realistic head model; it is of greater practical significance to further develop and improve the algorithm of the equivalent distributed source method including both dipole and charge source approaches.

TABLE 9.1
Correlation and Relative Error between Theoretical Values and Inversion Results (Yao et al., 2001)

Correlation Coefficient / Relative Error	Equivalent Source Distribution	Cortical Potential Distribution	Spheric Harmonic Spectrum
Sphere spline	0.8118	0.8959	0.9281
	0.5733	0.4350	0.3794
Equivalent distributed	0.9624	0.9904	0.9783
dipole source	0.3495	0.1342	0.2096

9.3 UNIQUENESS AND FUTURE TREND

9.3.1 Uniqueness of High-Resolution EEG Imaging

Let us review the uniqueness of various high-resolution EEG physical quantities. Laplacian, as a direct derivation of the scalp surface potential, is undoubtedly unique in theory. It has a higher spatial resolution than the scalp surface potential, which is similar to the cortical surface potential. The previous theoretical analysis also shows that it is approximately proportional to the cortical surface potential under certain approximate assumption (Section 8.1). The challenge for Laplacian is to make a robust estimate or develop reliable actual measurement technique. The equivalent distributed source theory shows that the distributed equivalent dipole source or point charge source is a forward solution of the realistic source in a specific physical model, respectively. In the case that the equivalent source distribution is spherical, we further presented the analytic solution of equivalent distributed source intensity (Chapter 6); these results indicate that the equivalent distributed source is determined in theory, and has a clear physical model, including a definite integral formula from the equivalent distributed source to scalp surface potential, thus they are unique in theory. The problem is that the known potential is on the limited upper scalp surface; we need to find a good method to reach enough precision in inverting the equivalent distributed source.

Under the framework of the Green Function Method, Yamashita (1982) studied the uniqueness of determining epicardial potential from the body surface, and he deduced a Fredholm integral equation relating epicardial to body surface potential and further proved that the epicardial potential can be determined uniquely from the body surface map with known geometry and conductivity of the torso volume conductor. In addition, Martin and Pilkington (1972) discussed the problem of uniqueness from the perspective of spherical harmonic series expansion, and obtained similar result. Obviously, these conclusions are also applicable to EEG in theory. For cortical potential imaging or unbounded space potential imaging, they can be regarded as the extension of the inversion result of the equivalent distributed source, which is the result of a further forward calculation after the distribution source is obtained. In theory, the uniqueness of cortical potential and unbounded space potential is determined by the uniqueness of the equivalent distributed source. Of course, the calculation of cortical potential can also be obtained by using the boundary element method (He et al., 2002), finite element method, etc., to solve the discretized Laplace equation in the passive region from the cortical surface to the scalp surface with boundary conditions, thus actually consist of the scalp and the cortical surface potential as the variable and the solution is uniqueness in theory.

9.3.2 Future Trend of High-Resolution EEG

Based on the previous three Chapters 6–8, a unified theoretical framework of the five kinds of high-resolution EEGs in sense of ME and SHS is presented in this chapter. It should be noted that although we have introduced the method of inversion, the core of this chapter is not to introduce one or two methods of inversion. Any other new and more efficient inversion method is acceptable. What we emphasize is the inherent relationship of various physical approximation methods. In this chapter, it is shown that each high-resolution map can be related to ME of EEG generator through its spatial filter, thus the theoretical relations among various high-resolution EEG are established, which lays a foundation for the direct comparison of various high-resolution maps and their reconstruction techniques.

Based on the forward theory presented above, ECM, LM, and SPM have higher spatial resolution than CPM, while EDM has equivalent spatial resolution to CPM. Therefore, if we can get a good reconstruction of SHS, ECM, LM, and SPM will be better than CPM and EDM. However, due to many practical limitations of EEG experiments and reconstruction algorithms, the reliable high spatial resolution component in reconstructed SHS is limited. In this case, the actual effect may be that CPM/EDM is better than ECM/LM/SPM. If we choose ECM/LM, and cannot correctly reconstruct the required high spatial frequency component, the wrong high-frequency component may lead to a strong interference in the result, or the strong regularization smoothing technology is adopted to smooth out the high spatial frequency component, and the final obtained frequency component might be equivalent to that of CPM/EDM. Thus, among the various high-resolution EEG techniques, it is meaningless to simply assume that one type is superior to another but should be evaluated on a common scale: ME. We believe that ME is the basic basis for objectively evaluating existing methods and guiding the development of new methods which can improve the ME inversion that should be developed in the future.

From the perspective of the physical quantity of imaging, LM only has an approximate physical interpretation (see Section 8.1), CPM suffers from the effect of the complex model, while the EDM, ECM, and SPM are only related to the source and imaging position (imaging surface), and has a clear physical meaning, particularly, the physical significance of the SPM is the most simple and clear, and as a result, we think that SPM could become a widely adopted standard high-resolution imaging physical quantity in the future.

REFERENCES

Edlinger, G., P. Wach, G. Pfurtscheller. 1998. On the realization of an analytic high-resolution EEG. *IEEE Trans Biomed Eng* 45:736–745.

He, B., X. Zhang, J. Lian, et al. 2002. Boundary element method-based cortical potential imaging of somatosensory evoked potentials using subjects' magnetic resonance images. *NeuroImage* 16(3):564–576.

Martin, R.O., T.C. Pilkington. 1972. Unconstrained inverse electrocardiography: Epicardial potentials. *IEEE Trans Biomed Eng* 19:276–285.

Nunez, P.L, R.B. Silberstein, P.J. Cadusch, et al. 1994. A theoretical and experimential study of high resolution EEG based on surface laplacian and cortical imaging. *EEG Clin Neuro* 90:40–57.

Perrin, F., J. Pernier, O. Bertrand, et al. 1989. Spherical splines for scalp potential and current density mapping. *EEG Clin Neuro* 72:184–187.

Yamashita, Y. 1982. Theoretical studies on the inverse problem in electrocardiography and the uniqueness of the solution. *IEEE Trans Biomed Eng* 29(11):719–725.

Yao, D. 2000. High-resolution EEG mappings: A spherical harmonic spectra theory and simulation results. *Clin Neurophysiol* 111:81–92.

Yao, D., B. He. 2003. Equivalent physical models and formulation of equivalent source layer in high-resolution EEG imaging. *Phys Med Biol* 48(21):3475–3483.

Yao, D., Y. Zhou, M. Zeng, et al. 2001. A study of equivalent source techniques for high-resolution EEG imaging. *Phys Med Biol* 46(8):2255–2266.

10 Basic Theory of EEG Inverse Problem

The theory of equivalent distributed sources described in Chapter 6 above is an illustration of non-uniqueness in EEG because the potential generated by real sources on the scalp surface can be equivalently generated by different distributed sources, so for the same EEG, the corresponding source is non-unique. Directly related to uniqueness is the sampling problem of scalp surface EEG, which involves many factors. This chapter includes four sections that cover the scalp EEG sampling problem, the uniqueness of EEG source inverse and potential reconstruction, the dipole localization problem, and the reciprocity theorem.

10.1 EEG SAMPLING PROBLEM

10.1.1 Introduction

According to the International Federation of Clinical Neurophysiology (IFCN) Guidelines on EEG electrode array, the main points are (Seeck et al., 2017): using at least 25 electrodes; for source localization purposes use of the entire or parts of the 10–10 system or high-density systems with 64–128 or more electrodes. Apparently, these rules come from long-term clinic experience and thus would be quite valuable for practice.

At present, most EEG signal recording systems are designed based on rules such as 10–20 or 10–10 or 10–5. In practice, the number of electrodes such as 8, 16, 20, 32, 64, 96, 128, and 256 is used for different purposes. So what system should be used for EEG inverse problems?

From the view of equation solving, as long as the number of independent equations is more than the unknown variables, the problem is overdetermined, and it is possible to resolve the parameters from the equation group (Appendix E), for example, if the supposed source is one dipole, then the number of the unknown variables is six: three coordinates (x, y, z), and three components of the dipole moment (P_x, P_y, P_z), and at least six sampling electrodes are needed. Similarly, if there are two dipoles to be determined, at least 12 sampling electrodes are needed, and so on. This is the theoretical basis of the dipole source localization algorithm. The question is who knows the number of dipoles in the brain? Actually, the actual electrical activity in the brain may be a distribution source of varying intensity across the whole brain, and whether it can be approximately equivalent to a few dipoles is a problem. Apparently, if the prior information is strong enough to support an assumption of limited and spatially separable dipoles, combined with sufficient independent EEG recordings, then the solution can be retrieved. Of course, if the sources are relatively

DOI: 10.1201/9781032639260-10

deep and close to each other, it will also lead to the "ill-condition" of the equation. In short, such an effort looks like a "knowen the answer to find the answer", and the key is whether you "know the answer" in some way beforehand.

From the sampling theory in information science, the spatial samplings should be substantially higher than the spatial information content of the EEG signals to avoid the loss of the high frequency and the distortion of the low frequencies due to aliasing. Thus, the choice of a given electrode montage depends upon assumptions about the information content in a problem to explore. Vaidyanathan and Buckley (1997) studied the sampling problem of EEG based on this idea as shown below.

10.1.2 Sampling Theorem

Vaidyanathan and Buckley (1997) assumed that the head model is a hemisphere (Figure 10.1), and the potential function on its surface can be expressed as

$$\Phi(\theta,\phi) = 2\sum_{n=0}^{\infty}\{a_{n0}\,Y_{n0}(\theta) + \sum_{m=1}^{n} a_{nm}Y_{nm}^{0}(\theta,\phi)\}, (0<\theta<\pi), (0<\phi<\pi) \quad (10.1)$$

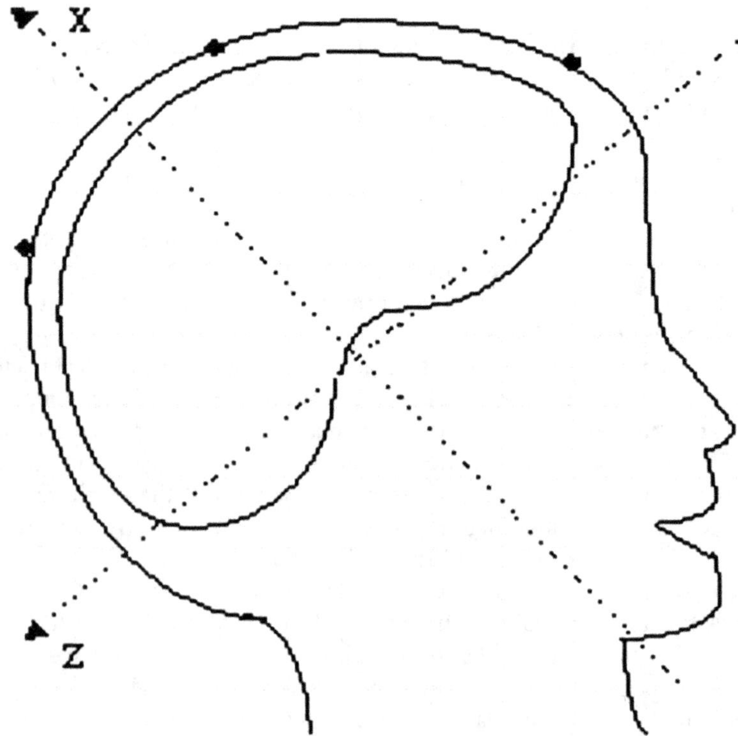

FIGURE 10.1 Head model and coordinate system. (From Vaidyanathan and Buckley (1997).) The area where $x > 0$ in the figure is the observation aperture.

Where Y_{n0} and $Y_{nm}^{0}(\theta,\phi)$ are n degree spherical harmonic functions, it is further assumed that the potential is an even symmetric periodic extension in $-\pi < \phi < \pi$, so only the cosine function is required in equation (10.1). The functions Y_{n0} and $Y_{nm}^{0}(\theta,\phi)$ have $n+1$ cases for each n, $n = 0, 1, 2, \ldots$, they form a complete orthogonal function system on the surface of the hemisphere.

Since the source is inside the head, the spherical harmonic series in equation (10.1) is attenuated (Chapters 2, 4, and Appendix B), so a finite term truncation (N) can be used to approximate the potential function. Based on an actual measure η_N (the mean square error [MSE]), the sampling requirement problem becomes finding the truncation N that satisfies the following equation

$$\left\| \Phi(\theta,\phi) - \sum_{n=0}^{N} \sum_{m} a_{nm} Y_{nm}^{0}(\theta,\phi) \right\|^2 \leq \eta_N \tag{10.2}$$

As can be seen from equation (10.1), for each n, $(n + 1)$ coefficients are needed, so the number of spatial points required is

$$L = \sum_{n=0}^{N} (n+1) = \frac{(N+1)(N+2)}{2} \tag{10.3}$$

This equation gives the number of sensors required to reconstruct the potential function for a given approximation requirement η_N. Of course, if the assumption of even symmetry is not made in equation (10.1), such as the general spherical harmonic function, then $(2n + 1)$ coefficients are needed for each n, and more space points are needed for the same η_N. If you use other functions to fit the scalp potential, the situation will be different, too.

For a practical EEG problem, the scalp surface recordings are

$$\Phi_{\text{meas}}(\theta,\phi) = \Phi(\theta,\phi) + N_b(\theta,\phi) + N_s(\theta,\phi) \tag{10.4}$$

Where the sensor noise covariance may be assumed to be $\varepsilon\left[N_s N_s^{\,T}\right] = \sigma_s^2 I$, here I is the identity matrix and σ_s^2 is the covariance value, and assuming $\varepsilon\left[N_b N_b^{\,H}\right] = \sigma_b^2 \Sigma$, where Σ is the covariance matrix related to the part of the dipole in the volume corresponding to the background neural noise. Comparatively speaking, σ_b is a more basic interference source, so it determines the selection of electrode number L. MSE_{η_N} should be chosen above the background noise σ_b because the signal submerged in the background noise is indistinguishable. For evoked potentials, the background EEG can be considered random and constitutes the major source of background noise σ_b.

The method for estimating the number of electrodes introduced by Vaidyanathan and Buckley (1997) is directly aimed at a proper potential fitting by spherical harmonic functions. For theoretical analysis, they further set up a hypothetical EEG generation model to generate scalp surface EEG, and then theoretically explore the number of electrodes needed. This process also shows that the number of electrodes is related to the EEG problem to solve and the model assumed.

In the magnetoencephalography study, Ahonen et al. (1993) studied the spatial spectrum sampling problem, more directly to the information theory on the concept of sampling theory, namely the sampling interval is determined by the spatial information of the valuable highest frequency, similar to Vaidyanathan and Buckley (1997), the decision of the highest spatial frequency is based on the truncation of the spherical harmonic spectrum.

10.2 UNIQUENESS OF EEG INVERSE

10.2.1 Introduction

EEG inverse problem especially evoked potential inverse problem, has been paid much attention in current studies. The most direct purpose is to localize the sources, including their locations, orientations, and intensities of the assumed dipoles though the dipole model is only an approximation of the realistic neural sources. The inverse problem is different from the forward problem in that it often has no unique solution, which has been proved theoretically a long time ago (Helmholtz, 1853), that there is no unique solution for the most general, unconstrained case. In fact, the previous equivalent distributed source theory (Chapter 6) is based on this fact or provides another proof that the EEG inverse is not unique. However, due to the importance of EEG source information, the EEG community has been developing methods to realize source localization with the help of some additional priors or constraints. The following is a summary of the main efforts. To sum up, there are two main evidences for the non-unique EEG inverse problem. One is focused on observation of the scalp EEG, the existence of a silent source which makes it impossible to determine the source based on the scalp EEG data (no information in the data) (Section 3.3). The second is the existence of the equivalent source principle which means that the same scalp EEG data can be produced by the true source or by its equivalent source, and it is impossible to distinguish which one is the true source and which one is the equivalent source except some priors hinting us something (Section 2.8). In practice, another important factor is the low SNR or not enough independent observations, they may lead to uncertainty or singularity in the algorithm realization, which is not discussed here but partly in Chapter 12, where a lot of EEG inverse algorithms are working to tackle these problems by regularization and priors.

10.2.2 Uniqueness Problem When the Scalp Surface Potential is Completely Known

Amir (1994) argued that any method for source reconstruction essentially implied a field interpolation method. Therefore, the continuous potential distribution could be used to discuss the physical uniqueness of the solution. To simplify the problem, the non-uniqueness caused by the dimension of the measured data being smaller than the dimension of the source parameter is not considered, because it is obvious. The main conclusions drawn by Amir (1994) are as follows:

1. Two spatially distinct point generators cannot create the same scalp potential map. In particular, it applies to dipole generators. This fact provides support for limited sparse source localization.

2. Two nonoverlapping distributed generators cannot create the same scalp potential map.
3. In most cases, zero-volume sources such as line and open surface sources are uniquely defined by the scalp potential map.
4. Distributed generators that create the same scalp potential map must intersect. The intersection of all such generators defines a unique dipole layer distribution having a minimum volume or zero volume generator which creates that map.

These results may be helpful for inverse algorithm design in practice. This work is within the non-unique framework caused by the equivalent source principle, but some special cases are identified where a unique result can be obtained. For more details please look at (Amir, 1994).

10.2.3 Silent Source-Induced Non-Uniqueness

Due to silent sources defined by the null space of the forward matrix (Riera et al., 2006), the inverse solution definitely cannot recover them. For example, Magnetoencephalography (MEG) inverse solution is non-unique, owing to the existence of so-called 'magnetically silent' current sources; an example of a magnetically silent source is a radial current dipole in a spherically symmetric conductor, which does not produce a magnetic field outside the head, thus such a current source can always be added to a solution of the inverse problem without affecting the field outside (Sarvas, 1987). Definitely, since the human head is not exactly spherically symmetric, an approximately radial source in the brain is not necessarily silent in practical MEG (Cuffin, 1990), but the contribution may be quite small. For EEG, the contribution of the two dipoles on the opposite walls of a gyrus may cancel each other, that's why the scalp EEG is assumed mainly from the activities of the gyrus instead of the sulcus. Furthermore, the rotational component of the current distribution also is a silent component of EEG (Grave de Peralta Menendez et al., 2000; Dassios and Fokas, 2013). The concrete sensitivity of the source orientation can be evaluated by simulation (Ahlfors et al., 2010).

10.2.4 Uniqueness of the Cortical Potential Imaging

Continuing the previous idea, we still assume that the scalp surface potential is completely known, without considering the measurement noise and other recording problems, and simply consider the uniqueness of cortical imaging from the physical perspective.

Firstly, the cortical equivalent distributed source imaging, based on the studies on the sphere model (Yao and Luo, 1996) and arbitrary realistic head model (Yao and He, 2003) as shown in Chapter 6, both the equivalent distributed dipole source (current source) and the charge source (current source density) have clear equivalent physical model, and according to the uniqueness theorem of electrostatic (Chapter 2), these equivalent source distributions are unique(the layer position is pre-defined). Of course, according to the forward uniqueness theorem of electrostatics, the scalp potential data itself is unique. In practice, the scalp potential from the true source (except the silent source) or the equivalent source has a clear and unique mathematical relationship. In the opposite case, if we have enough scalp observation data, the singularity from the scalp to the cortical surface is controllable. Therefore, the equivalent source information can be approximately calculated from the scalp potential data.

About cortical potential imaging: at present, there are two implementations, the first is the equivalent source inversion approach if the equivalent sources are properly reconstructed as noted above, the next step from the equivalent sources to the cortical surface potential is uniqueness ensured by the uniqueness of the forward calculation of the EEG sources in a head model. (Chapters 4 and 5); The second implementation method is the boundary element method (He et al., 1999) or other numerical methods where the mathematic relation between the cortical potential and scalp potential is established according to the theory of static field where the uniqueness is guaranteed by the electromagnetic theory (Chapters 4 and 5).

10.2.5 Discussion

From the above summary, it can be seen that the localization of the EEG source is not unique in general, but the reconstruction of the cortical imaging is unique, because a prior hypothesis has been introduced in cortical imaging, that is, the position of the imaging surface. From the point of view of the equivalent source, the scalp potential of the real source can be generated by the equivalent source, and the equivalent source is not unique, for example, the layer position may be different that determining the "source" from the scalp potential is non-uniqueness. However, the equivalent distributed source on a known curved surface is unique (Yao and Luo, 1996; Yao and He, 2003), and then the cortical potential from the equivalent distributed source is determined. Here we see that the uniqueness of cortical imaging potential is mainly due to the pre-determined cortical imaging surface. This also gives us a revelation that the non-uniqueness of the EEG inverse problem can only be alleviated by introducing appropriate prior information.

As Nunez and Srinivasan (2006) noted: "source localization makes use of constraints based on physiological assumptions to overcome the fundamental nonuniqueness of the inverse problem. These inverse solutions depend critically on the accuracy of the chosen constraints, regardless of algorithmic sophistication or beautiful computer graphics. In other words, fancy mathematics can never trump fundamental physics." The constraints "have included plausible conjectures based on physiology, independent anatomical information perhaps obtained from MRI, fMRI, or PET data, hopeful assumptions, or combinations of these constraints." In the following chapters and currently ongoing publications, various assumptions of constraints are shown, and some different performances are obtained.

10.3 EEG DIPOLE LOCALIZATION

Dipole localization is the basic dream of EEG inverse problems. It assumes that the scalp records are generated by a finite number of dipoles and hopes to identify these dipoles efficiently from the scalp EEG.

If the locations of the sources are known in advance, only the dipole moments are estimated, such a case is called the fixed dipole model. If both the position and the dipole moment are unknown and therefore need to be estimated at the same time, the model is called a moving dipole model. The fixed dipole model has fewer parameters than the moving dipole model, thus allowing more dipoles to be processed simultaneously. When the model parameters are more than the number of scalp electrodes, it is necessary to have some prior information available to obtain a reasonable solution to the inverse problem.

10.3.1 Dipole Localization

The neural current sources are approximated by several current dipoles whose positions and moments are estimated by the least mean square error, or other *lp* ($p<=2$) norm measure on the error, of the predicted potential and the measured potential. Assuming the number of electrodes on the scalp surface is M, the potential recorded by them is expressed as a vector Φ_{meas}, which is generated by a few dipoles in the brain.

$$\Phi_{\text{meas}} = GX \tag{10.5}$$

If each dipole has three components, then for *N* dipoles, the dimension of *X* is 3*N*. If the direction of the dipole is known, suppose it is along the normal direction of the cortex, and the normal direction of the cortex can be determined by a high-resolution image, the dimension of *X* is reduced to *N*. *G* is a leadfield matrix composed of the scalp surface potential generated by each dipole component with unit strength, so *G* is a function of the position of the dipole and is independent of the strength of the dipole. The EEG inverse problem is to solve:

$$\min_{\vec{r},X} O(\vec{r},X) = \min_{\vec{r},X} \left\| \Phi_{\text{meas}} - G(\vec{r})X \right\|^{p}, p \in [0,2] \tag{10.6}$$

So we need to determine the position of the dipole, $\vec{r}$, the moment components of the dipole, *X*, and the number of dipoles, *N*. If the positions are given, $p=2$, then we have

$$X = G(\vec{r})^{+}\Phi_{\text{meas}} \tag{10.6a}$$

and

$$\min_{\vec{r},X} O(\vec{r},X) = \min_{\vec{r},X} \left\| \Phi_{\text{meas}} - G(\vec{r})X \right\|^{p} = \min_{\vec{r}} \left\| \left(I - G(\vec{r})G(\vec{r})^{+}\right)\Phi_{\text{meas}} \right\|^{p} \tag{10.7}$$

Equation (10.7) shows that the crucial unknown parameters are the position coordinates of the dipoles since the optimal dipole moment can be determined by equation (10.6a) when the scalp potential and the dipole position are known (Musha and Okamoto, 1999).

Another problem is the number of dipoles. Due to the non-uniqueness of the EEG inverse problem described in Section 10.2, different combinations of dipoles may produce the same scalp surface potential. Therefore, the number of dipoles is also a problem, which often requires reasonable assumptions based on the physiological, psychological, and pathological background of the data. In general, more dipoles can better fit the observed potential and even overfit the data, meanwhile, more dipoles can also make the inverse problem more illness in a sense of separability and noise sensitivity. Therefore, it is advisable to use one or at most two dipoles to approximate the EEG sources without knowing the location of the source in advance (Musha and Okamoto, 1999).

In general, for $p=2$, unless the positions of the dipoles are very close together, G^TG is invertible, and the Moore-Penrose inverse of G can be simply expressed as $G^+ = \left(G^TG\right)^{-1} G^T$. According to this formula, the inverse problem optimization process starts with a hypothetical dipole or a combination of dipoles, then minimizes the error function O, equation (10.7), through repeated iteration. The idea of the steepest descent method, which is widely used, is that the error function can be minimized at each step by choosing the direction and step size of the modified parameter. The method needs to calculate the gradient of the error function in each iteration to find the steepest descent direction. The simplex rule, another optimization method, solves this problem well, too. In the iterative process of any optimization method, there are usually many local minima, although they do not correspond to the global minimum error function, but may make the iterative process stop. To avoid this error condition, iterate from multiple different initial values, such as six initial positions, then compare the minimum values reached by these initial values and choose the smallest as the true minimum. If multiple dipoles are localized at the same time, the corresponding solution will have a higher degree of mathematical ill-condition and therefore need the help of prior knowledge, but the method and idea are the same.

10.3.2 Discussion

A measure that indicates whether a dipole model is good or bad is dipolarity (Musha and Okamoto, 1999). The equivalence of an equivalent dipole to the true source is that it produces a similar scalp potential. The equivalent dipole of a simple distributed source will have a larger dipole moment and is usually deeper than that of the distributed source. In other words, an equivalent dipole can fit the observed potential well, but this does not mean that its location is consistent with the location of the real active region, unless the real active source is very local. Therefore, the commonly referred source localization refers to the situation of sparse sources in a physiological sense.

10.3.2.1 Dipolarity

The "equivalence" of the dipole to real brain activity is based on the fact that the potential produced by the equivalent dipole at the electrode position is closest to the observed potential. In this case, "closest" means that the mean of the squared difference between the predicted potential and the measured potential is smaller than any other possible dipole. Since the true source is not a point, the equivalent dipole cannot fully represent the true source, thus the so-called dipolarity problem exists.

Musha and Okamoto (1999) define dipolarity as the fitting degree from which the dipole potential to the actual potential.

$$D = \sqrt{1 - \frac{\left\|\Phi_{\text{meas}} - \Phi_{\text{dipole}}\right\|^2}{\left\|\Phi_{\text{meas}}\right\|^2}} \tag{10.8}$$

Where Φ_{meas} and Φ_{dipole} are the measured potential and the potential generated by the equivalent dipole, respectively. The second term under the square root sign is sometimes referred to as the normalized residual. A dipole is a point source, so we should pay attention to the D parameter when discussing the electrical activity of the brain in terms of equivalent dipoles. Then, what kind of source is most likely to reduce the dipolarity? The simplest answer is "a spatially extended source decreases D because the dipole is a point source, after all".

10.3.2.2 The Optimal Number of Equivalent Dipoles

When the active areas of the cortex are simply connected, one dipole can approximate the observed surface potential distribution very well. When cortical activity is divided into two active parts, two dipoles are the best approximation. In general, increasing the number of equivalent dipoles increases the polarity, regardless of the actual source conditions, because more model parameters better fit the measured potential distribution. On the other hand, increasing the number of electrodes decreases the polarity, because it becomes more difficult for the potential produced by the dipole to match the actually measured potential. Therefore, it is not useful to simply compare dipolarity for different head models and different numbers of electrodes.

So, for a given number of electrodes, what is the optimal number of equivalent dipoles? There is an empirical rule summarized by Musha and Okamoto (1999). Suppose there are 21 electrodes according to the international 10–20 standard, the single dipole model may be used when the dipolarity is >98% for one dipole and <98% for a couple of dipoles. In contrast, the double dipole model may be used when the dipolarity <98% and that of a couple of dipoles >99%.

10.3.2.3 Experiment on Artificial Dipole Source

To test the effectiveness of a dipole localization realization, the artificial dipole can be used as the gold standard test as it is really known. Artificial dipoles are formed by feeding pulses of current into electrodes implanted in the human head (Cuffin, 1991; Homma et al., 1994). Then, based on the collected scalp data, the dipole localization method is used to locate and compare with the true source location.

10.3.2.4 Deep Learning and Other Methods for Dipole Localization

All the difficulties in EEG inverse are related to the general inverse in equation (10.6a), and various methods including regularization and priors are being explored. Now, the artificial neural network (ANN) was also introduced to tackle this problem, it uses a neural network to realize the function of the leadfield matrix G (EEG forward) and its inverse matrix (EEG inverse). As EEG forward is determined and well established, ANN is unnecessary. However, the EEG inverse is non-uniqueness, the fighting is always on the way. Specifically, ANN is trained by paired data of the sources and scalp recordings, with the scalp recordings as input and sources as output, thus avoiding an inappropriate reverse process. The big data for all conceivable situations may be generated by EEG forward model, then efficient ANN (Van Hoey et al., 2000) or deep learning (DL) network (Adler and Öktem, 2017; Daoud and Bayoumi, 2020; Bore et al., 2021; Sun et al., 2022) can be obtained with good performance. Here the training process with various data and noise completes both inverse and regularization in the traditional algorithm. However, the full story of DL application in EEG is just beginning, its advantages and disadvantages are still waiting for evaluation. The fundamental problem is the explainable of the current DL algorithm.

For EEG inverse, various optimization methods can be adopted in principle (Awan et al., 2018), except the widely adopted gradient descent method, simplex method, some global optimization methods are also used, such as the simulated annealing method (Haneishi et al., 1994) and genetic algorithm (Uutela et al.,1998).

10.3.2.5 Other Sources Model

The above is all based on the dipole source model, actually, we may have other choices as noted in Section 3.3. Unfortunately, there is not much work on the others. It is a big space and a chance to explore.

10.4 RECIPROCITY THEOREM AND ITS APPLICATIONS

To ensure accuracy in estimating the equivalent current dipole, researchers would naturally place scalp electrodes near the supposed EEG source. Clearly, the best electrode array depends not only on the position of the source but also on the orientation of its moment. So how do we determine whether an electrode array is good or not? Reciprocity Theorem (Appendix C) provides an idea for this problem (Plonsey, 1963; Rush and Driscoll, 1969; Malmivuo and Plonsey, 1995). We are interested in using the reciprocity theorem to derive the relationship between the scalp surface EEG and the source. A dipole with a current density of $\vec{J}_s$ in the Ω region generates a potential of Φ at the boundary Γ.

$$\nabla \cdot (\sigma \nabla \Phi) = \nabla \cdot \vec{J}_s, \text{ within } \Omega \tag{10.9}$$

$$\begin{cases} \Phi = \bar{\Phi} \\ \partial_n \Phi = 0 \end{cases} \tag{10.10}$$

Boundary condition (10.10) means that there is no current moving in and out of the scalp.

Now let us suppose that a unit current flows from the electrode at $\vec{r}_a$ on the scalp and is received by the electrode at $\vec{r}_b$ on the scalp, thus generating a potential Ψ, which satisfies the following equation

$$\begin{cases} \nabla\cdot\sigma(\vec{r})\nabla\Psi(\vec{r})=0, & \text{within}\,\Omega \\ \sigma(\vec{r})\partial_n\Psi(\vec{r})=\delta(\vec{r}-\vec{r}_a)-\delta(\vec{r}-\vec{r}_b), & \text{on the boundary}\,\Gamma \end{cases} \tag{10.11}$$

Where $\delta(r)$ is the Dirac delta function. Applying Green's theorem to Φ and Ψ, we get (Appendix C)

$$\begin{aligned} &\int_\Omega dv\left(\Phi\nabla\cdot\sigma\nabla\Psi-\Psi\nabla\cdot\sigma\nabla\Phi\right)=-\int_\Omega dv\Psi\nabla\cdot\vec{J}_s \\ &=\int_\Omega dv(\nabla\Psi)\cdot\vec{J}_s=\int_\Gamma ds\left(\Phi\sigma\,\partial_n\Psi-\Psi\sigma\,\partial_n\Phi\right) \\ &=\int_\Gamma ds\Phi(\vec{r})\left\{\delta(\vec{r}-\vec{r}_a)-\delta(\vec{r}-\vec{r}_b)\right\}=\Phi(\vec{r}_a)-\Phi(\vec{r}_b) \end{aligned} \tag{10.12}$$

therefore

$$\Phi_{ab}\equiv\Phi(\vec{r}_a)-\Phi(\vec{r}_b)=-\int_\Omega dv\vec{E}_L(\vec{r})\cdot\vec{J}_s(\vec{r});\quad \vec{E}_L(\vec{r})\equiv-\nabla\Psi(\vec{r}) \tag{10.13}$$

This equation is the reciprocity theorem: The potential difference, between position $\vec{r}_a$ and $\vec{r}_b$, generated by the current source $\vec{J}_s$ inside the brain is equal to the volume integral of the scalar obtained after the dot product of $\vec{J}_s$ and $\vec{E}_L$. The field strength $\vec{E}_L$ is generated by the unit current fed in at $\vec{r}_a$ and left at $\vec{r}_b$. Equation (10.13) is for electric field and current, a similar formula for charge and potential has been given by equation (C.45) (Appendix C). When the current $\vec{J}_s$ is a current dipole located at $\vec{r}_p$, and the moment of the dipole is $\vec{J}_s(\vec{r})=\vec{p}\delta(\vec{r}-\vec{r}_p)$, we get

$$\Phi_{ab}=-\int_\Omega dv\vec{E}_L(\vec{r})\cdot\vec{p}\delta(\vec{r}-\vec{r}_p)=-\vec{E}_L(\vec{r}_p)\cdot\vec{p} \tag{10.14}$$

The potential difference between $\vec{r}_a$ and $\vec{r}_b$, Φ_{ab}, is equal to $-p\cdot\vec{E}_L(\vec{r}_p)$, and $\vec{E}_L$ is called the leadfield vector. Obviously, when the current orientation is parallel to the electric field $\vec{E}_L$ at the location, the potential difference generated at a and b is the largest. When its orientation is perpendicular to the electric field $\vec{E}_L$, the potential difference generated is zero. For example, a radial dipole in a spherical model has zero potential difference between the two symmetric points on both sides of the projection point on the scalp surface of the radial arrow (Figure 8.1).

We can imagine a steady current field, which is formed from a unit current through electrodes a, b to the brain region. We can see the inflow of current at point "a", and the outflow of current at point "b", and also see a current field at "$\vec{r}_p$". Such a stable field can occur by $\vec{J}_s$ at "$\vec{r}_p$" and be maintained, or it can occur by unit current at $\vec{r}_a$ and $\vec{r}_b$, and be maintained, and thus linking the two cases.

In the more general case, suppose that a distributed current density $K(\vec{r})$ instead of unit strength is applied to two points on the scalp, which produces a lead field $\vec{E}_L$ in the brain, then we have a general form of equation (10.12) as

$$\Phi_K \equiv \int_\Gamma dsK\Phi = -\int_\Omega dv\vec{E}_L \cdot \vec{J}_{\text{ext}} \tag{10.15}$$

Here the law of charge conservation can be described as

$$\int_\Gamma dsK = 0 \tag{10.16}$$

Using the reciprocity theorem, we can find the most appropriate electrode array (Rush and Driscoll, 1969) according to the principle of being most sensitive to the source in the brain.

The reciprocity theorem can be used to study the sensitivity of electrodes (Rush and Driscoll, 1969), the forward and inverse problems of EEG, and trans-cranial electrical stimulation (Wagner et al., 2016).

REFERENCES

Adler, J., O. Öktem. 2017. Solving ill-posed inverse problems using iterative deep neural networks. *Inverse Probl* 33(12):1–24.

Ahlfors, S.P., J. Han, J.W. Belliveau, et al. 2010. Sensitivity of MEG and EEG to source orientation. *Brain Topogr* 23:227–232.

Ahonen, A.I., M.S. Hamalainen, R.T. Ilmoniemi, et al. 1993. Sampling theory for neuromagnetic detecor arrays. *IEEE Trans Biomed Eng* 40(9):859–869.

Amir, A. 1994. Uniqueness of the generators of brain evoked potential maps. *IEEE Trans Biomed Eng* 41:1–11.

Awan, F.G., O. Saleem, A. Kiran. 2018. Recent trends and advances in solving the inverse problem for EEG source localization. *Inverse Probl Sci Eng* 27(11):1–16. doi: 10.1080/17415977.2018.1490279.

Bore, J.C., P. Li, L. Jiang, et al. 2021. A long short-term memory network for sparse spatiotemporal EEG source imaging. *IEEE Trans Med Imaging* 40(12):3787–3800.

Cuffin, B.N. 1990. Effects of head shape on EEG's and MEG's. *IEEE Trans Biomed Eng* 37:44–52.

Cuffin, B.N. 1991. Eccentric spheres models of the head. *IEEE Trans Biomed Eng* 38(9):871–878.

Daoud, H., M. Bayoumi. 2020. Deep learning approach for epileptic focus localization. *IEEE Trans Biomed Circuits Syst* 14(2):209–220.

Dassios, G., A.S. Fokas. 2013. The definite non-uniqueness results for deterministic EEG and MEG data. *Inverse Probl* 29(6):065012.

Grave de Peralta Menendez, R., S.L. Gonzalez Andino, S. Morand, et al. 2000. Imaging the electrical activity of the brain: ELECTRA. *Hum Brain Mapp* 9:1–12.

Haneishi, H., N. Ohyama, K. Sekihara, et al. 1994. Multiple current dipole estimation using simulated annealing. *IEEE Trans Biomed Eng* 41(11):1004–1009.

He, B., Y. Wang, D. Wu. 1999. Estimating cortical potentials from scalp EEGs in a realistically shaped inhomogeneous head model by means of the boundary element method. *IEEE Trans Biomed Eng* 46(10):1264–1268.

Helmholtz, H. 1853. Ueber einige gesetze der vertheilung elektrischer strome in korperlichen leitern, mit anwendung auf die thierischelektrischen versuche. *Annalen der Physik* 165(7):353–377.

Homma, S., T. Musha, Y. Nakajima, et al. 1994. Location of electric current sources in the human brain estimated by the dipole tracing method of the scalp-skull-brain (SSB) head model. *EEG Clin Neuro* 91:374–382.

Malmivuo, J., R. Plonsey. 1995. *Bioelectromagnetism: Principles and Applications of Bioelectric and Biomagnetic Fields*. New York: Oxford University Press.

Musha, T., Y. Okamoto. 1999. Forward and inverse problems of EEG dipole localization. *Crit Rev Biomed Eng* 27(3–5):189–239.

Nunez, P.L., R. Srinivasan. 2006. *Electric Fields of the Brain: The Neurophysics of EEG*. New York: Oxford University Press.

Plonsey, R. 1963. Reciprocity applied to volume conductors and the ECG. *IEEE Trans Biomed Electron* 10(1):9–12.

Riera, J.J., P.A. Valdés, K. Tanabe. 2006. A theoretical formulation of the electrophysiological inverse problem on the sphere. *Phys Med Biol* 51(7):1737–1758.

Rush, S., D.A. Driscoll. 1969. EEG electrode sensitivity: An application of reciprocity. *IEEE Trans Biomed Eng* 16:15–22.

Sarvas, J. 1987. Basic mathematical and electromagnetic concepts of the biomagnetic inverse problem. *Phys Med Biol* 32(1):11–22.

Seeck, M., L. Koessler, T. Bast, et al. 2017. The standardized EEG electrode array of the IFCN. *Clin Neurophysiol* 128:2070–2077.

Sun, R., A. Sohrabpour, G.A. Worrell, et al. 2022. Deep neural networks constrained by neural mass models improve electrophysiological source imaging of spatiotemporal brain dynamics. *PNAS* 119(31):e2201128119.

Uutela, K., M. Hamalainen, R. Salmelin. 1998. Global optimization in the localization of neuromagnetic sources. *IEEE Trans Biomed Eng* 45(6):716–723.

Vaidyanathan, C., K.M. Buckley. 1997. A sampling theorem for EEG electrode configuration. *IEEE Trans Biomed Eng* 44:94–97.

Van Hoey, G., J. De Clercq, B. Vanrumste, et al. 2000. EEG dipole source localization using artificial neural networks. *Phys Med Biol* 45(4):997.

Wagner, S., F. Lucka, J. Vorwerk. 2016. Using reciprocity for relating the simulation of transcranial current stimulation to the EEG forward problem. *NeuroImage* 140(15):163–173.

Yao, D., B. He. 2003. Equivalent physical models and formulation of equivalent source layer in high-resolution EEG imaging. *Phys Med Biol* 48(21):3475–3483.

Yao, D., B. Luo. 1996. Theory of the EEG cortical imaging techniques. *Chin J Biomed Eng (English edition)* 5(3):128–136.

11 Signal Space-Based EEG Inverse Solution

The EEG inverse problem can be roughly divided into two categories: one is the imaging method, including cortical potential imaging, unbounded spatial potential imaging, scalp Laplacian, equivalent distributed source imaging, etc., the details of these methods can be found in Chapters 6–9. The other is the source localization method, which consists of three subcategories: the nonlinear inverse algorithm introduced in Section 10.3, the beamformer/signal subspace decomposition/multi-signal classification algorithm to be introduced in this chapter, and the iterative linear inverse method approach to be introduced in Chapter 12.

The methods presented in Chapters 10 and 12 usually take the observed potential value at a certain moment as the known information, while the multiple signal classification (MUSIC)-based signal subspace decomposition utilizes the temporal process, and electrode spatial distribution simultaneously (Mosher et al., 1992). The MUSIC algorithm is different from the multi-dimensional search method, which is a multi-dimensional nonlinear optimization problem, with heavy calculation and is easy to fall into local minimum points (Chapter 10). The MUSIC algorithm has a computational complexity significantly lower than that of a multi-dimensional nonlinear optimization algorithm. This chapter will introduce the state of the art and some of our efforts.

It should be noted that there are some other closely related methods not included in this book. (1) *The beamformer*: The goal is to design a bank of spatial filters where each filter passes signals originating from a specified location within the brain while attenuating signals from other locations. A display of the variance or power at the output of each filter as a function of the filter's spatial "passband" location provides an estimate of the distribution of activity within the brain (Van Veen et al., 1997; Ilmoniemi and Sarvas, 2019); (2) *Independent component analysis* (*ICA*): ICA is a blind source separation method, which can be used to analyze the time series of EEG measurements. ICA does not require either a head model or a forward field solver, and so it is a solely data-driven method. ICA is often used to separate artifacts from brain signals. ICA assumes that the temporal activities of sources are non-Gaussian random signals, each source is statistically independent of the others (Ilmoniemi and Sarvas, 2019). (3) *Blind source separation* (*BSS*) *method*: Each BSS is based on the specific assumptions of the sources, and ICA is a special BSS method. BSS by joint diagonalization (JD) is a widely used method. In practice, an exact JD is possible only in a few special cases, and so the various numerical algorithms for JD only perform an approximate JD (AJD). A popular BSS method using AJD is second order blind identification (SOBI) (Belouchrani et al., 1997), where AJD is applied to autocorrelation matrices. It was confirmed that SOBI is able to recover known noise

DOI: 10.1201/9781032639260-11

sources that were either spontaneously occurring or artificially induced; it is also able to recover neuronal sources activated by median nerves.

11.1 SIGNAL SUBSPACE SCAN ALGORITHM

The MUSIC algorithm, originally designed in radar signal processing (Schmidt, 1986), was introduced to the EEG community by Mosher et al. (1992). This section is a primary introduction to MUSIC. We proposed two parameters, the normalized blurring index (NBI) and the singular value ratio (SVR), to describe the performance of MUSIC, and the time-domain process reconstruction of the EEG source after the position of the source is determined by MUSIC (Yao and Fu, 2002).

11.1.1 Multi-Signal Classification Algorithm

The EEG inverse problem can be summarized as the following model

$$\Phi(t) = A(L)X(t) + E(t),\ \ t = 1,\ldots,N \tag{11.1}$$

where $\Phi(t) \in R^{M\times 1}$ is the observation record of M electrodes on the scalp surface, $X(t) \in R^{L\times 1}$ is the source function of L independent sources inside the brain, and $E(t) \in R^{M\times}$ is the noise at M electrodes. $A(L)$ is the source-to-electrode leadfield matrix, which is related to the head model. In this work, we adopted a concentric three-layer spherical head model for $A(L)$.

The correlation matrix of $\Phi(t)$ is

$$R = \left\langle \Phi(t)\Phi^T(t) \right\rangle = APA^T + W = APA^T + \sigma^2 I \tag{11.2}$$

Here, $P = \left\langle X(t)X^T(t) \right\rangle$, $W = \left\langle E(t)E^T(t) \right\rangle$, and T represent the transposition of the matrix. For the model expressed in equation (11.1), assuming it meets the following three conditions:

Condition 1: $M > L$ means that the number of observed electrodes is more than the number of unknown independent sources, and the electrode positions do not overlap with each other. $A(L) = [a_1,\ldots,a_L]$, When $i \neq j$, a_i and a_j are independent of each other, so rank(A) = L.

Condition 2: Noise is uniform white noise and uncorrelated with the EEG signal, $W = \left\langle E(t)E^T(t) \right\rangle = \sigma^2 I$

Condition 3: The sources are incoherent, that is, the matrix $P = \left\langle X(t)X^T(t) \right\rangle$ is nonsingular, so, rank(P) = L and $\text{rank}\left(\text{APA}^T\right) = L$.

In addition, $N > M$ to ensure the calculation accuracy of the autocorrelation matrix. If the above conditions are satisfied, with the singular value decomposition of R, then

$$R = U\Sigma V^T \tag{11.3}$$

where $\Sigma = \text{diag}[\lambda_1,\ldots,\lambda_M]$, λ_i is the eigenvalue of the matrix R, and $\lambda_i > \sigma^2$, $i = 1,\ldots,L$; $\lambda_i \approx \sigma^2$, $i = L+1,\ldots,M$, Orthogonal matrices U and V are composed of corresponding eigen vectors. According to the distribution of eigen values, orthogonal matrices U or V decompose into signal and noise subspaces, such as, $U = [S \mid G] = [u_1,\ldots,u_L \mid u_{L+1},\ldots,u_M]$ where $S = [u_1,\ldots,u_L]$ and $G = [u_{L+1},\ldots,u_M]$ are, respectively, called signal subspace and noise subspace.

Since σ^2 and G are eigenvalues and eigenvectors of R, respectively, there is

$$\text{RG} = \sigma^2 G \tag{11.4}$$

On the other hand, by right multiplication of G (11.2), we get

$$\text{RG} = \text{APA}^T G + \sigma^2 G \tag{11.5}$$

By combining (11.4) and (11.5), it can be concluded that

$$\text{APA}^T G = 0 \tag{11.6}$$

So, we have $G^T \text{APA}^T G = (A^T G)^T P (A^T G) = 0$, and since P is nonsingular, we have

$$A^T G = 0, \quad G^T A = 0, \quad A^T G G^T A = 0 \tag{11.7}$$

Equation (11.7) show that

$$a_i^T G G^T a_i = 0, \ i = 1,\ldots,L. \tag{11.8}$$

Equation (11.8) shows that if position i is the location of a real source, it will make the value at the left end of equation (11.8) close to zero, thus obtaining an objective function to find the location of the real source. Accordingly, the specific MUSIC implementation is as follows: scan every point i in the solution space to determine the minimum of the L spatial positions of the following equation

$$f(i) = a_i^T G G^T a_i = \min \tag{11.9}$$

For the EEG inverse problem, the orientation of the dipole moment at a certain position may be fixed or rotatable during the whole observation period. For the dipole moment rotatable case, the EEG inverse problem can be summed up as finding the distribution of the following positioning operator (Mosher et al., 1992)

$$J(i) = 1 \Big/ \left(\frac{\|GG^T A_i\|_F^2}{\|A_i\|_F^2} \right) \tag{11.10}$$

Here, F stands for the Frobenious norm of a matrix. A_i is the total transfer function matrix consisting of the three components of a dipole along the three coordinates.

For the case that the orientation of the dipole moment is fixed, let $a(i) = A_i\hat{M}_i$ and $\hat{M}_i$ be the unit orientation vector of the dipole with a fixed orientation, and then the positioning operator can be defined as (Mosher et al., 1992; Sekilhara et al., 1997)

$$J(i) = 1\Bigg/\left(\frac{\left\|G^T\left(A_i\hat{M}_i\right)\right\|_2^2}{\left\|A_i\hat{M}_i\right\|_2^2}\right) = 1/\lambda_{\min}\left\{A_i^T G G^T A_i, A_i^T A_i\right\} \tag{11.11}$$

The denominator of equation (11.11) denotes the minimum generalized eigenvalue of the matrix pair in braces.

To sum up, the MUSIC algorithm of the EEG inverse problem consists of the following steps:

1. Calculate the correlation matrix $R = E\left\{\Phi\Phi^T\right\}$ of the recordings;
2. Singular value decomposition (SVD) of R $R = U\Sigma V^T$;

   According to the singular value distribution and some prior physiological knowledge, the signal and noise subspaces are determined; that is, the matrix U is divided. In the simulation study, L can be considered to be known. For practical application, how to effectively determine L remains to be further studied. Possible methods include the application of information criteria such as AIC for selection (Knosche, 1997). Let $U = [S \mid G]$, where S is composed of eigenvectors corresponding to L large singular values in Σ, and G is composed of M-L eigenvectors following. Calculate $J(i)$ for every possible source position, and obtain the source position information according to the $J(i)$ distribution. In practice, it is to interpret the first L large values of the $J(i)$ distribution as the possible source positions. In this way, MUSIC is an indirect method of source localization.
3. According to the obtained source position information, the linear inverse of equation (11.1) is solved to obtain the time-domain process of the source.

In practice, the corresponding signal and noise space can also be obtained directly by the singular value decomposition of the observation data matrix.

11.1.2 Evaluation Parameters

11.1.2.1 Normalized Blurring Exponent

The blurring degree of the $J(i)$ distribution is directly related to the spatial resolution of the source. For this reason, we propose the following NBI

$$\mathrm{NBI}_i = \sqrt{\sum_k \|\vec{r}_k - \vec{r}_i\|^2 X_k{}^2 \Big/ \sum_k X_k{}^2} \Bigg/ \sqrt{\sum_k \|\vec{r}_k - \vec{r}_i\|^2 \Big/ \sum_k 1} \tag{11.12}$$

In the formula, the serial number i corresponds to the grid point of the 3D discrete solution space and $\vec{r}_k$ is the grid point vector. We calculate NBI in a sub-sphere with such a radius $r_s = 0.3$ around each point. Obviously, if the distribution of $J(i)$ is uniform within the sub-sphere, then NBI will approach 1.0. However, if $J(i)$ shows a sharp distribution similar to the Dirc function, then NBI will approach 0. Obviously, a small NBI is what we would expect.

11.1.2.1.1 Temporal Process Reconstruction Relative Error and Similarity Coefficient

After the source spatial location is obtained, the corresponding transfer matrix is determined, and thus the time domain process $X(t)$ can be obtained by solving the linear inverse of equation (11.1). In the literature, the Moore-Penrose generalized inverse is generally adopted. Here we propose to use the newly emerged total least squares (TLS) (Markovskya and Van Huffel, 2007), which can better reconstruct the time-domain evolution process of the EEG source

$$X(t) = \mathrm{TLS}(\Phi(t) = AX(t)), \quad t = 1,\ldots,N \tag{11.13}$$

The relative error is defined as the error between the reconstructed source time-domain process and the real source time-domain process. Specific for

$$\mathrm{RE} = \|X_c - X\|_F / \|X_c\|_F \tag{11.14}$$

In the formula, $X_c(t)$ and $X(t)$ correspond to the time-domain process of the real source and the inversion source, respectively.

The correlation coefficient (CC) is defined as

$$\mathrm{CC} = \frac{\left\|(X - \langle X\rangle)^T (X_c - \langle X_c\rangle)\right\|_F}{\|X - \langle X\rangle\|_F \|X_c - \langle X_c\rangle\|_F} \tag{11.14a}$$

where $\langle * \rangle$ is mean. CC characterizes the similarity between the two arrays, and its variation trend is usually opposite to that of RE. It should be pointed out that the application of equations (11.14) and (11.14a) implies that the positions of the real source and the inversion source are the same.

11.1.2.2 Singular Value Ratio (SVR)

The MUSIC algorithm depends on the degree of separation of the signal-noise space, which in turn is related to the distribution of the SVD. The result of various interference effects makes a fuzzy signal-noise demarcation point of the SVD. Therefore, we define the singular value ratio (SVR) to reflect this problem.

$$\mathrm{SVR} = \sum_{k=L+1}^{M} \lambda_k \Big/ \sum_{k=1}^{L} \lambda_k \tag{11.15}$$

In the formula, $\lambda_k\ (k = 1,\ldots,M)$ is the element on the diagonal of the matrix Σ, obviously, the smaller the SVR, the clearer the cut-off point. To sum up, we define three types of indicators to describe the performance of the algorithm from three aspects, namely, the blurring degree (NBI) of spatial distribution in the spatial domain, the reconstruction accuracy in the time domain process (RE and CC), and the separation degree of signal-noise (SVR). Obviously, these indexes are also of general significance for evaluating other inversion methods.

11.1.3 Simulation Test

The normalized parameters of the three-layer concentric spherical head model adopted in this study were: scalp radius 1, skull radius 0.92, and brain radius 0.87. The time-domain process of the EEG activity is simulated by a sinusoidally modulated Gaussian function. By adjusting the parameters of the function, similar but not identical time-domain processes can be obtained, so that source combinations with different correlations can be conveniently constructed as a discrete cosine function

$$f(t) = \exp\left[-\left(2\pi f_m (t-t_0)/\gamma\right)^2\right]\cos\left(2\pi f_m (t-t_0)+v\right)$$
$$i = 1,\ldots,150,\ dt = 2\text{ ms},\ v = \frac{\pi}{2},\ \gamma = 3,\ \pi = 3.14159 \tag{11.16}$$

as shown in Figure 11.1a. The number of scalp surface electrodes is 120, according to the NEUROSCAN system. The potential forward is conducted by the formula in Chapter 4. The noise is independent, uniform Gaussian white noise. The noise corresponding to 120 records was obtained by running the white noise program 120 times, respectively. The SNR (N/S) is defined as the ratio of the standard deviation of all noise data to the standard deviation of all signals. Therefore, the SNR of each channel is different. The channel with a strong signal has a high SNR, which is similar to the actual situation.

Figure 11.1b is the result of the simulation calculation. With the increase of noise, the SVR becomes larger and larger, reflecting the influence of noise on the separability of signal-noise. Accordingly, the imaging results become more and more blurry, that is, NBI gradually increases, and the reconstruction error of the time-domain process correspondingly increases, that is, RE increases and CC decreases. It should be pointed out that we have carefully checked the $J(i)$ distribution of various cases in Figure 11.1b, and the two points with the largest $J(i)$ value are the same as the positions of the assumed two sources, thus ensuring the premise of the RE and CC calculations.

Figure 11.1c is the time-domain process of the reconstructed source. Due to the difference between the orientation of the reconstructed dipole and the actual situation, the comparison between the reconstructed results and the real source is made by decomposing them into three coordinate components. The figure shows the x-axis component P_x of the dipole moment of the real dipole and the corresponding reconstructed dipole. The results in Figure 11.1c again show that the MUSIC algorithm is stable to Gaussian white noise from the perspective of the time-domain process.

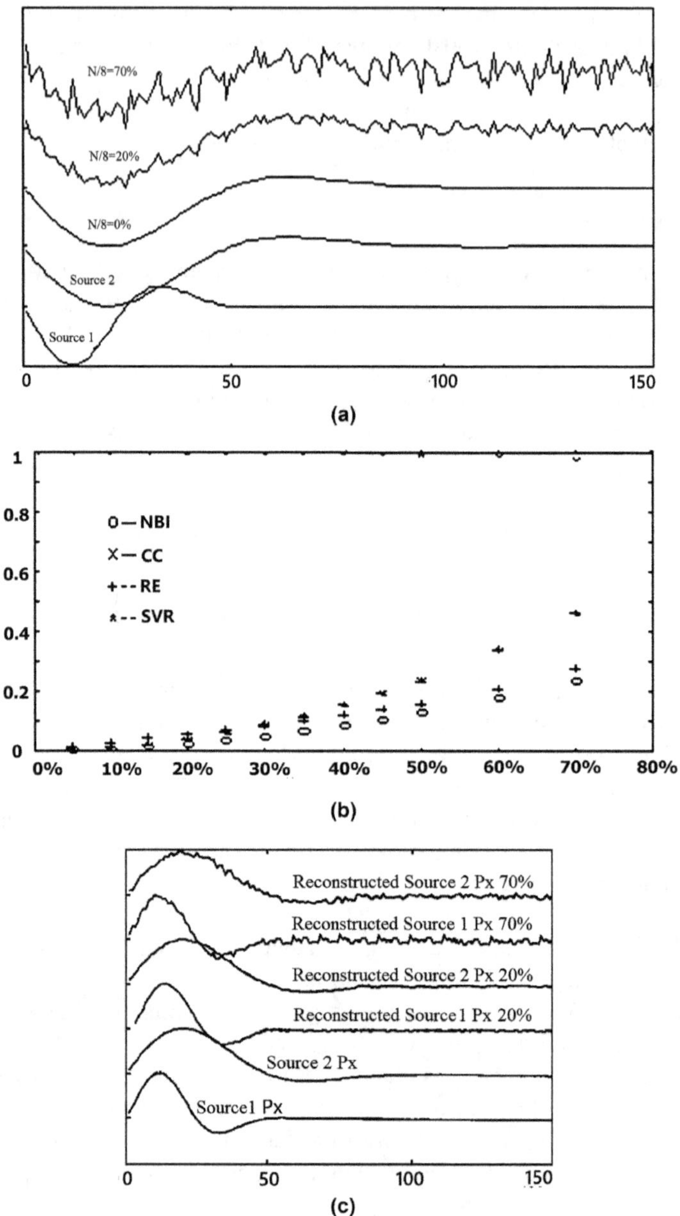

FIGURE 11.1 Assumed source signals and reconstructed signals. (a) The time domain sequence of the assumed sources and the forward scalp surface recording of an electrode with different SNR. The horizontal axis is the time-sequence number, and the vertical axis is the normalized amplitude. (b) The evaluation index values for different noise levels. The horizontal axis is the noise level and the vertical axis is the evaluation index value (NBI, CC, RE, SVR). (c) The time-domain sequence of the P_X component of the assumed sources and reconstructed sources at different *N/S* conditions. The horizontal axis is the time-sequence number and the vertical axis is the normalized amplitude.

11.2 DELAY-CORRELATION-BASED SOLUTION

In Section 11.1, we systematically evaluate the performance of the classic MUSIC established by Mosher et al. (1992). Since 1992, various extensions of MUSIC have been reported. For example, Sekihara et al. (1997) reported a method when a "task experiment" and a "control experiment" were carried out simultaneously, and Koles (1998) combined it with common-spatial-patterns (CSP) decomposition to study the localization of epileptic waves in spontaneous EEG. These studies show that the typical MUSIC method makes it difficult to distinguish coherent sources and is sensitive to colored noise (Yao and Fu, 2002). An existing method for suppressing colored noise is the pre-whitening method based on the known noise covariance matrix (Sekilhara et al., 1997). However, in many cases, the noise covariance matrix is unknown or difficult to estimate. In this section, we put forward a delay-correlation method aimed to use the delay-correlation difference between the signal and noise, to suppress the spatial color noise, and improve the imaging ability of the MUSIC algorithm (Yao et al., 2000).

Continuing with equation (11.1) in Section 11.1, the delay-correlation matrix of the $\Phi(t)$ is

$$R(\beta) = \langle \Phi(t)\Phi^T(t+\beta) \rangle = AP(\beta)A^T + W(\beta) \tag{11.17}$$

In the formula, $P(\beta) = \langle X(t)X^T(t+\beta) \rangle$, $W(\beta) = \langle E(t)E^T(t+\beta) \rangle$, T represents the transposition of the matrix. Here, we list the following three conditions:

Condition 1: $M > L$, that is, the number of observed electrodes is more than the number of unknown independent sources, and the electrode positions do not overlap with each other. $A(L) = [a_1 \dots a_L]$. When $i \neq j$, a_i and a_j are independent of each other, thus, rank(A)$=L$.

Condition 2: With evoked EEG as the application background, noise is spontaneous EEG, so the correlation with the delay length β of the signal will be much larger than that of noise. Therefore, we can choose β as the delay corresponding to the first zero of the delay autocorrelation function of the noise, so there is

$$R = \mathrm{APA}^T, \ W(\beta) = 0 \tag{11.18}$$

Condition 3: The sources are irrelevant, but the autocorrelation delay time of each source is much larger than β. Therefore, rank (P)$=L$, rank$\left(\mathrm{APA}^T\right) = L$.

In addition, assume $N > M$ to ensure the calculation accuracy of the autocorrelation matrix. Decompose R,

$$R = U\Sigma V^T \tag{11.19}$$

where $\Sigma = \mathrm{diag}\left[\sigma_1^2,\dots,\sigma_M^2\right] = \mathrm{diag}[\lambda_1,\dots,\lambda_M]$, $\lambda_i = \sigma_i^2$ is the eigenvalue of the matrix R, and $\lambda_i > 0,\ i = 1,\dots,L;\ \lambda_i \approx 0,\ i = L+1,\dots,M$. It is assumed that in the time-dependent matrix, the eigenvalue corresponding to the noise is zero, which may be an infinitesimal for the actual case. The orthogonal matrices U and V are composed of corresponding eigenvectors. According to the distribution of eigenvalues, the orthogonal matrices U or V can be decomposed into signal and noise subspaces. For example, $U = [S \mid G] = [u_1,\ldots,u_L \mid u_{L+1},\ldots,u_M]$ where $S = [u_1,\ldots,u_L]$ and $G = [u_{L+1},\ldots,u_M]$ are called signal subspaces and noise subspaces, respectively. Since the eigenvalue corresponding to the noise subspace is zero, therefore

$$\mathrm{RG} = \mathrm{APA}^T G = 0 \tag{11.20}$$

The steps below are the same as in Section 11.1.

Simulation results (Yao, 2008) showed that the delay-correlation MUSIC can suppress spatially dependent colored noise to a certain extent, and its effectiveness depends on the difference in the autocorrelation time between the signal and the noise.

11.3 HIGH-ORDER CUMULANT-BASED SOLUTION

The main body of various MUSIC algorithms is based on the second-order statistics of the records, and their effectiveness is significantly affected by spatially colored noise. An important feature of higher-order statistics is their insensitivity to any Gaussian process. Therefore, we developed a MUSIC method based on the fourth-order cumulant (Yao, 2000).

11.3.1 Algorithm

11.3.1.1 Fourth-Order Cumulants

EEG recordings are typical multi-sensor signals. Assume M signals form the vector Y

$$Y^T = [y_1(t), y_2(t),\ldots, y_M(t)] \tag{11.21}$$

$\forall k_1,k_2,k_3,k_4 \in \{1,2,3,\ldots,M\}$, The first-, second-, third-, and fourth-order spatial moments are

$$\begin{aligned} \mu_1(k_1) &= E\{y_{k1}(t)\},\\ \mu_4(k_1,k_2,k_3,k_4) &= E\{y_{k1}(t)y_{k2}(t)y_{k3}(t)y_{k4}(t)\},\\ \mu_2(k_1,k_2) &= E\{y_{k1}(t)y_{k2}(t)\},\\ \mu_3(k_1,k_2,k_3) &= E\{y_{k1}(t)y_{k2}(t)y_{k3}(t)\}. \end{aligned} \tag{11.22}$$

Obviously, the spatial second-order moment is the spatial cross-correlation function, which is adopted in the above classical and its variant MUSIC, and the spatial first-order moment is the mathematical expectation.

The spatial fourth-order cumulant is

$$\begin{aligned}\kappa_4(k_1,k_2,k_3,k_4) = {} & \mu_4(k_1,k_2,k_3,k_4) - \big[\mu_3(k_2,k_3,k_4)\mu_1(k_1) \\ & +\mu_3(k_1,k_3,k_4)\mu_1(k_2) + \mu_3(k_1,k_2,k_4)\mu_1(k_3) \\ & +\mu_3(k_1,k_2,k_3)\mu_1(k_4)\big] - \big[\mu_2(k_1,k_4)\mu_2(k_2,k_3) \\ & +\mu_2(k_1,k_3)\mu_2(k_2,k_4) + \mu_2(k_1,k_2)\mu_2(k_3,k_4)\big] \\ & +2\big[\mu_2(k_3,k_4)\mu_1(k_1)\mu_1(k_2) + \mu_2(k_2,k_4)\mu_1(k_1)\mu_1(k_3) \\ & +\mu_2(k_2,k_3)\mu_1(k_1)\mu_1(k_4) + \mu_2(k_1,k_4)\mu_1(k_2)\mu_1(k_3) \\ & +\mu_2(k_1,k_3)\mu_1(k_2)\mu_1(k_4) + \mu_2(k_1,k_2)\mu_1(k_3)\mu_1(k_4)\big] \\ & -6\mu_1(k_1)\mu_1(k_2)\mu_1(k_3)\mu_1(k_4)\end{aligned} \tag{11.23}$$

For a zero mean signal, the above equation is simplified to

$$\begin{aligned}\kappa_4(k_1,k_2,k_3,k_4) = {} & \mu_4(k_1,k_2,k_3,k_4) - \mu_2(k_1,k_4)\mu_2(k_2,k_3) \\ & - \mu_2(k_1,k_3)\mu_2(k_2,k_4) - \mu_2(k_1,k_2)\mu_2(k_3,k_4)\end{aligned} \tag{11.24}$$

This is a commonly used fourth-order cumulant expression in spatial spectrum estimation. For EEG signals, the zero mean hypotheses may not be practical, especially for evoked EEG signals. For this reason, the general form of equation (11.23) above will be used as the starting point in this work.

There are many good properties of cumulants that are valuable to be adopted, three of which are listed below:

1. If $\{\alpha_i\}_{i=1}^{n}$ is a constant variable and $\{y_i\}_{i=1}^{n}$ is a random variable, then the cumulant satisfies the following relation

$$\text{cum}(a_1 y_1,\ldots,\alpha_n y_n) = \prod_{i=1}^{n} \alpha_i \,\text{cum}(y_1,\ldots,y_n) \tag{11.25}$$

2. If the random variable $\{y_i\}_{i=1}^{n}$ and the random variable $\{z_i\}_{i=1}^{n}$ are statistically independent, then

$$\text{cum}(z_1 + y_1,\ldots,z_n + y_n) = \text{cum}(z_1,\ldots,z_n) + \text{cum}(y_1,\ldots,y_n) \tag{11.26}$$

   Here, cum (.,.,.,.) stands for cumulative quantity.
3. The fourth-order cumulant completely suppresses the Gaussian noise. If $\{y_i\}_{i=1}^{n}$ is a Gaussian random variable, then

$$\text{cum}(y_{k1}, y_{k2}, y_{k3}, y_{k4}) = 0.0 \tag{11.27}$$

11.3.1.2 Inverse Problem Model of EEG

The EEG forward problem equation (11.1) can be summarized as the following model

$$y_i(t) = \sum_{j=1}^{L} a_{ij} x_j(t) + e_i(t), \quad t = 1,\ldots,N; \quad i = 1,\ldots,M \tag{11.28}$$

where $Y(t) \in R^{M\times 1}$ is the observation record of M electrodes in the scalp, $X(t) \in R^{L\times 1}$ is the L independent source function in the head, and $E(t) \in R^{M\times 1}$ is the noise at M electrodes. $A\left(a_1,\ldots,a_j,\ldots,a_L\right)$ is the source-to-electrode transfer function matrix, the leadfield matix, which is related to the head model. A concentric three-layer spherical head model or other in chapters 4 and 5 can be used.

In general, it can be assumed that the noise in the EEG model is Gaussian noise. Previous MUSIC based on second-order statistics are for the independent noise between channels, and the performance of such algorithms will greatly degenerate in the case of spatially dependent noise (spatial color noise). Based on the properties of the fourth-order cumulant shown by equation (11.27), the algorithm based on the fourth order can completely suppress Gaussian noise. Therefore, as long as the time process of the noise is Gaussian, whether it is independent or correlated among channels, it will not affect the algorithm based on the fourth order.

According to the properties of the fourth-order cumulant and the above EEG model, we have

$$\begin{aligned} k_4\left(k_1,k_2,k_3,k_4\right) &= \operatorname{cum}\left(y_{k1},y_{k2},y_{k3},y_{k4}\right) \\ &= \sum_{j=1}^{L} a_{k_1 j} a_{k_2 j} a_{k_3 j} a_{k_4 j} \operatorname{cum}\left(x_j,x_j,x_j,x_j\right) \end{aligned} \tag{11.29}$$

It shows that the fourth-order cumulant can completely suppress the Gaussian noise. Assume that $k_4\left(k_1,k_2,k_3,k_4\right)$ is in the row $\left(k_1-1\right)M+k_2$ and column $\left(k_3-1\right)M+k_4$ of a $M^2 \times M^2$ matrix C, then equation (11.29) can be expressed by a matrix as

$$C = \mathrm{BSB}^T \tag{11.30}$$

In this formula

$$\begin{aligned} &B = \left[a_1 \otimes a_1,\ldots,a_L \otimes a_L\right] \\ &S = \operatorname{diag}\left\{s_1,s_2,s_3,\ldots,s_L\right\}, \quad s_j = \operatorname{cum}\left(x_j,x_j,x_j,x_j\right) \end{aligned} \tag{11.31}$$

where $\otimes$ is the Kronecker product and T indicates transpose. Diag{} represents a diagonal matrix with the elements in parentheses as diagonal elements.

For the model expressed in equation (11.28), we list the following two restrictions:

Condition 1: $M > L$, that is, the number of observed electrodes is more than the number of unknown independent sources, and the electrode positions do not overlap each other. Therefore, in the transfer matrix $A(a_1,\ldots,a_j,\ldots,a_L)$, when i is not equal to j, a_j and a_i are independent of each other, thus rank $(A) = L$ and rank $(B) = L$.

Condition 2: The time process of the signal source is non-Gaussian, so rank $(S) = L$ and rank$\left(\mathrm{BSB}^T\right) = L$.

The singular value decomposition of C is

$$C = U\Sigma V^T \tag{11.32}$$

$\Sigma = \mathrm{diag}\left[\sigma_1^2,\ldots,\sigma_{M^2}^2\right] = \mathrm{diag}\left[\lambda_1,\ldots,\lambda_{M^2}\right]$, $\lambda_i = \sigma_i^2$ is the eigenvalue of matrix C, and $\lambda_i > 0, i = 1,\ldots,L; \lambda_i \approx 0, i = L+1,\ldots,M^2$. It is assumed that the eigenvalue corresponding to the noise is zero, which may be an infinitesimal for the actual case. Similar to the above two Sections (11.1 and 11.2), orthogonal matrices U and V are composed of corresponding eigenvectors. According to the distribution of eigenvalues, orthogonal matrices U or V can be decomposed as signal and noise subspaces, as shown in $U = [S \mid G] = \left[u_1,\ldots,u_L \mid u_{L+1},\ldots,u_{M^2}\right]$, where $S = [u_1,\ldots,u_L]$ and $G = \left[u_{L+1},\ldots,u_{M^2}\right]$ span the signal subspace and noise subspace, respectively. Clearly

$$\mathrm{CG} = \mathrm{BSB}^T G = 0 \tag{11.33}$$

Therefore

$$G^T \mathrm{BSB}^T G = \left(B^T G\right)^T S\left(B^T G\right) = 0 \tag{11.34}$$

And since S is nonsingular, there is (Zhang, 1995a,b)

$$B^T G = 0,\ \ G^T B = 0,\ \ B^T G G^T B = 0 \tag{11.35}$$

Equation (11.35) shows that

$$\left(a_j \otimes a_j\right)^T GG^T \left(a_j \otimes a_j\right) = 0,\ \ j = 1,\ldots,L \tag{11.36}$$

According to equation (11.36), the actual implementation is as follows: scan every point i in the solution space and calculate the following objective function or its reciprocal

$$f(j) = \left(a_j \otimes a_j\right)^T GG^T \left(a_j \otimes a_j\right) \tag{11.37}$$

The objective function will go to zero where the source actually exists and not anywhere else. Therefore, the extreme point on the 3D distribution of the objective function can be used as the location of the real source position.

For the EEG inverse problem, the orientation of the dipole moment of the dipole at a certain position may be fixed or rotated during the whole observation period. Referring to the second-order subspace decomposition method (Section 11.1), the objective function or location operator can be defined as if the dipole moment is rotatable

$$J(i) = 1 \Big/ \left(\frac{\left\| GG^T (A_i \otimes A_i) \right\|_F^2}{\left\| (A_i \otimes A_i) \right\|_F^2} \right) \tag{11.38}$$

where F represents the Frobenious norm of the matrix. A_i is the transfer function matrix consisting of the three components of a dipole along the three coordinates. For the case that the orientation of the dipole moment is fixed, let $a(i) = A_i\hat{M}_i$ and $\hat{M}_i$ be the unit orientation vectors of the fixed orientation of the dipole, and the positioning operator can be defined as

$$\begin{aligned} J(i) &= 1 \Big/ \left(\frac{\left\| G^T \left[\left(A_i\hat{M}_i\right) \otimes \left(A_i\hat{M}\right) \right] \right\|_2^2}{\left\| \left(A_i\hat{M}_i\right) \otimes \left(A_i\hat{M}\right) \right\|_2^2} \right) = 1 \Big/ \left(\frac{\left\| G^T (A_i \otimes A_i)\left(\hat{M}_i \otimes \hat{M}\right) \right\|_2^2}{\left\| (A_i \otimes A_i)\left(\hat{M}_i \otimes \hat{M}\right) \right\|_2^2} \right) \\ &= 1 \Big/ \left(\frac{\left\| G^T \left(A_i'\hat{M}_i'\right) \right\|_2^2}{\left\| \left(A_i'\hat{M}_i'\right) \right\|_2^2} \right) = 1 \big/ \lambda_{\min} \left\{ A_i'^T GG^T A_i', A_i'^T A_i' \right\} \end{aligned} \tag{11.39}$$

The denominator of the last equation, equation (11.39), denotes the minimum generalized eigenvalue of the matrix pair in braces.

In summary, the fourth-order cumulant MUSIC algorithm for the EEG inverse problem consists of the following steps:

Calculate the fourth-order cumulant matrix C of the observed records.
$C = U\Sigma V^T$ is obtained by SVD.

According to the distribution of singular value Σ and some physiological knowledge, the decomposition of the signal and noise space is completed, that is, the matrix U is divided. In this simulation study, we believe that L is known. For practical applications, how to effectively determine L remains to be further studied (Knosche, 1997). Let $U = [S \mid G]$, where S is composed of eigenvectors corresponding to L large singular values in Σ and G is composed of M^2-L eigenvectors in the rear. Calculate $J(i)$ for every possible source location. According to the $J(i)$ distribution, the source location information is obtained. In practice, the first L large values of the $J(i)$ distribution are interpreted as the real source location.

11.3.2 Simulation Test

11.3.2.1 Description of the Simulation Method

The three-layer concentric spherical head model used in this study is the same as that used in the previous two sections. The EEG activity time course is simulated by a sine function. By adjusting the frequency and initial phase of the sine function, similar but different time processes can be obtained, so that non-Gaussian and mutually independent source combinations can be conveniently constructed. The sine function is of the form

$$f(t) = \sin\left(2\pi f_m t + v\right) = \sin\left(2\pi/N_S i + j\right), \quad i = 1,\ldots,N \tag{11.40}$$

If $dt = 4$ ms, the corresponding frequencies of $N_{S=}$10, 13, and 14 are 25.0, 19.2, and 17.8 Hz. In this work, we take $j = 0$.

Since the matrix size of the fourth-order method is $M^2 \times M^2$, too large M will lead to difficulties in the calculation. In this work, we take $M = 11$, and they are approximately evenly distributed in the upper half of the scalp. The noise used in this study is independent, uniform Gaussian white noise. The white noise corresponding to the M channels is obtained by running the white noise program M times, respectively, then they are statistically independent between channels. If the same segment of white noise data is used for all the M channels, they are spatially correlated noises.

In this study, the SNR is defined as the ratio of the standard deviations of all noise data to the standard deviations of all signals. Therefore, the SNR of each channel is different. The channel with a strong signal has a high SNR, which is similar to the actual situation.

The following calculations are carried out by comparing the second- and fourth-order methods. For the details of the second-order method, please refer to Section 11.1 above.

11.3.2.2 The Calculation Results

11.3.2.2.1 Influence of Spatially Unrelated Noise

Suppose there are dipole sources 1 and 2 located in the spherical head model (−0.7, 0.4, and 0.3) and (−0.7, −0.4, and 0.3) with a radial coordinate of 0.86, so they are at the approximately cortical surface ($r = 0.87$). The dipole moment is in the radial direction. *SNR* is 40%, $M = 11$, $N_S = 13$, 14, and $N = 500$. The noise is different for each channel. The results show that the fourth-order and second-order methods have similar performance in the case of spatially uncorrelated noises.

11.3.2.2.2 The Influence of Spatially Correlated Noise

Except the noise added to each channel is the same noise data, thus it is spatially dependent noise, all other conditions being the same as in (11.3.2.2.1). Figure 11.2 shows the results of the fourth-order and second-order methods, respectively. They show that the second-order method leads to the loss of one source in the case of spatially correlated noise, while the fourth-order method still achieves the imaging of two source positions well.

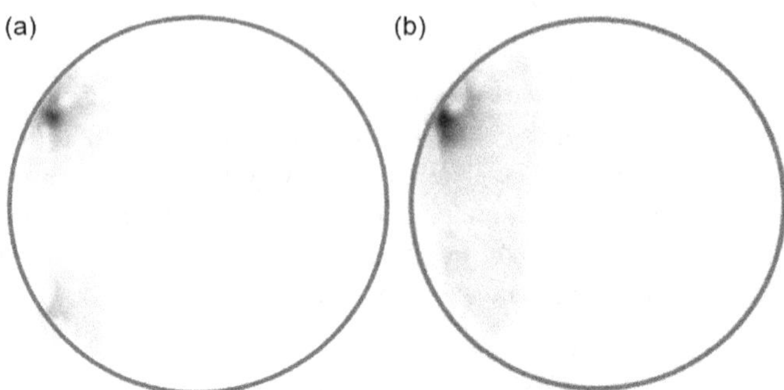

FIGURE 11.2 Fourth-order (a) and second-order (b) methods result in spatially coherent noise.

The result shows that the fourth-order MUSIC can realize the imaging of two sources by using 11 electrode records, so it has considerable practical significance. Compared with the existing second-order methods, the method based on fourth-order cumulants has made some progress in overcoming spatially dependent noise. But there is still a lot of work to be done on how to bring this approach into practice. For example, in this simulation study, we assume that the signal is sinusoidal, so it is not a Gaussian signal, and the noise is a Gaussian white noise, they are the preferred combination to magnify the advantages of the fourth-order method. But when applying this method in practice, we need to confirm the degree of non-Gaussian for actual EEG signals, etc.

In summary, the MUSIC method has two problems: sensitive to colored noise and cannot handle coherent sources. However, color noise and coherent sources are common in EEG and MEG practices. The above time-delay correlation and fourth-order cumulant methods are two of the efforts in tackling these problems. Meanwhile, there are also some other efforts, such as the recursive MUSIC method (R-MUSIC) (Mosher and Leahy, 1998), the recursive multidimensional MUSIC method (Yao, 2008), which is an integration of delay-MUSIC and R-MUSIC, and the methods shown below.

11.4 TRANSFORMED SPACE-BASED SOLUTION

In MUSIC implementation, if the sources are coherent, the source correlation matrix will be rank defect, which means that all the coherent sources correspond to one eigenvector in the signal space, in this case, with subspace scanning, only an equivalent source can be estimated somewhere different from any one of the coherent sources (Ma et al., 1998). In this section, especially for two coherent sources, suppose that the positions of two coherent sources can be estimated roughly in advance, then a digital filter can be designed to suppress one of the two coherent sources, and then another coherent source can be well reconstructed (Zhang et al., 2010).

For the observation equation, $\Phi = AX$, assumed a transformation matrix T which changes the data to

$$Y = T\Phi = TAX \tag{11.41}$$

here, $T = I - \vec{a}_i\left(\vec{a}_i^T\vec{a}_i\right)^{-1}\vec{a}_i^T$, $\vec{a}_i$ is the leadfield vector of source x_i, in this way, $T\vec{a}_i = \left(I - \vec{a}_i\left(\vec{a}_i^T\vec{a}_i\right)^{-1}\vec{a}_i^T\right)\vec{a}_i = 0$, thus in the new data space y, the contribution from x_i is totally suppressed, which means source x_i being a silent source for y. By the way, for a vector $\vec{a}_i$, this T is a unitary transformation that preserves length and satisfies $T^T = T^{-1}$ (Menke, 2012).

If what you want to suppress is not a voxel in the solution space but a region Ω with k_n voxels, we may have

$$T = I - E\left(E^TE\right)^{-1}E^T \tag{11.41a}$$

where $E = \left[\vec{a}_{k1},\ldots,\vec{a}_{kn}\right],\quad k_1,\ldots,k_n \in \Omega$

In the new space after transformation, the signals in the coherent source region Ω will be suppressed and silent. Then classical MUSIC (Mosher et al., 1992) and the Beamformer algorithm (Van Veen and Buckley, 1988; Van Veen et al., 1997; Zhang et al., 2010), etc. can be applied to the transformed data to achieve coherent source localization. It should be noted here that the leadfield vector used in the MUSIC algorithm also should be transformed $T\vec{a}_i$ rather than the original $\vec{a}_i$, the location of the real coherent interference source is often unknown, and the approximate location of the coherent source needs to be roughly determined with the help of physiological knowledge or other imaging methods. For example, the approximate position of the source is obtained by the LORETTA method, and then a suppression region is defined to accurately locate the other coherent sources.

The following are the results of simulated MEG data (Zhang et al., 2010). For two synchronous sinusoidal time processes (frequency of 10Hz), the sources are placed at (0, 30, and 40) mm and (0, –30, and 40) mm, respectively (coordinate definition is shown in Figure 11.3a), and a single-layer spherical volume conduction model is used to compute the leadfield vector. The spacing of mesh points was 5mm. Gaussian noise is added to the simulated MEG, and the parameters are adjusted to SNR=2. The results are shown in Figure 11.3, where 11.3b is the result of the classical MUSIC, and 11.3c and 11.3d are results of the MUSIC with the transform (11.41a) region suppression region. The suppression method is obviously superior to the classical MUSIC method. For the case of n coherent sources, the method may be adopted to sequentially suppress $(n-1)$ and image the remaining one in turn. This is similar to RAP-MUSIC (Mosher and Leahy, 1999; Yao, 2008), in which the source with the strongest signal is located first, then the second source is located after a projection operator is designed to suppress the first one, and the process is recursive. The disadvantage of RAP-MUSIC is that its first source may have been influenced by other associated sources. More details can be found in Zhang et al. (2010) for such as the calculation of $\left(E^TE\right)^{-1}$ and the design of T.

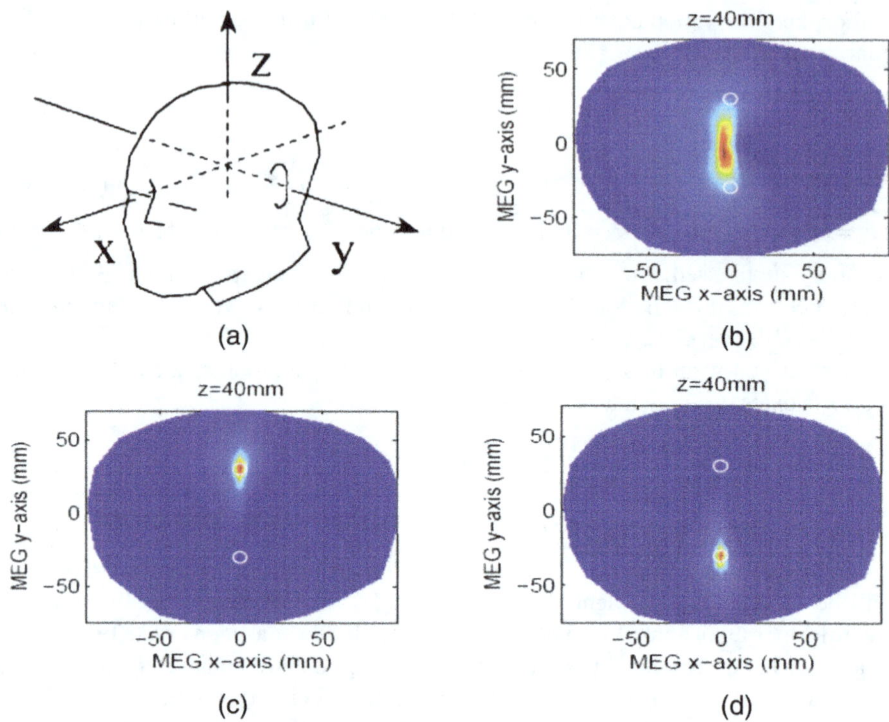

FIGURE 11.3 Results of two simulated coherent sources. (a) The coordinate system. (b) Classical MUSIC. (c) Imaging result after suppressing a source, approximately estimated in (b). (d) Imaging result after suppressing the source estimated in (c) (Zhang et al., 2010).

11.5 NOISE SPACE INVARIANT-BASED SOLUTION

In the above MUSIC method, the signal subspace may be affected by the intensities of the sources. While the intensity change of the source does not change the noise subspace. Therefore, a new MUSIC based on the noise space invariant,INvariance of Noise (INN) algorithm, was recently proposed (Zhang et al., 2013).

Theoretically, if we increase the source strengths, the variance of the signal subspace will increase, whereas the distribution of the noise subspace remains the same. In other words, the noise subspace is invariant with respect to the strengths of the sources. INN is designed according to this property. Let us define a new matrix

$$D^{\theta} = R + h\vec{a}(\theta)\vec{a}(\theta)^{T} \tag{11.42}$$

where R is the data correlation matrix, $\vec{a}(\theta)$ is the leadfield vector corresponding to a unit source at location θ, and h is a positive scalar constant. To balance the magnitude between R and $\vec{a}(\theta)\vec{a}(\theta)^{T}$, we further rewrite equation (11.42) as

$$D^{\theta} = R + h\frac{\text{trace}(R)}{\text{trace}\left(\vec{a}(\theta)\vec{a}(\theta)^{T}\right)}\vec{a}(\theta)\vec{a}(\theta)^{T} \tag{11.42a}$$

SVD of D^θ generates the ordered singular values $\{\mu_i^\theta,\ \ i=1,\ldots,k,\ \ \mu_i^\theta \geq \mu_{i+1}^\theta\}$ and the corresponding singular vectors $V^\theta = \{v_1^\theta,\ldots,v_k^\theta\}$, where k is the channel number. Equation (11.42a) is the formula that we adopted in the following computation. As h is a positive scalar, it can be proved that (Horn and Johnson, 1985)

$$\mu_i^\theta \geq \lambda_i, \quad i=1,\ldots,k \tag{11.43}$$

Where λ_i is the eigenvalue of R. Importantly, when θ matches one of the source locations, the last $(k-p)$ eigenvalues of D^θ and R would be the same, i.e.,

$$\mu_i^\theta = \lambda_i,\ i = p+1,\ldots,k \tag{11.44}$$

where a priori p is the assumed known number of dipoles.. In other words, the energy of the noise subspace obtained from D^θ is the same as that obtained from R.

Note that the property stated in equation (11.44) does not depend on the value of scalar h explicitly and will hold for any positive h. In numerical calculations, because R is obtained from a finite number of time instants, the variance of the original and new noise subspaces may not be exactly the same, and equation (11.44) does not exactly hold. Thus, in practice, we search the sources in a manner where μ_i^θ in equation (11.44), not exactly but as closely as possible equals to λ_i. Therefore, an appropriate objective function can be taken as:

$$O(\theta) = \frac{1}{\sum_{i=p+1}^{k}\left(\mu_i^\theta - \lambda_i\right)} \tag{11.45}$$

The values of the objective function $O(\theta)$ can be used as the imaging index at each grid points θ_i within the whole brain volume to generate an image. The positions where peaks are found are regarded as the locations of the sources.

In summary, the INN algorithm consists of the following steps:

1. Compute R from the EEG/MEG data.
2. Compute the SVD of R.
3. Choose a positive h and compute D^θ and its SVD for each possible source location θ_i.
4. Compute $O(\theta)$.
5. Repeat steps (3) and (4) for all putative source locations.
6. The positions of sources are those where $O(\theta)$ has a local maximum.

Compared to MUSIC, one important advantage of INN is that it is insensitive to strength differences and correlations between sources, even for closely spaced sources. This property is especially important when a weak source is present in the vicinity of a strong one.

In the simulation tests, the spherically symmetric MEG forward model was employed (Sarvas, 1987). The source space had a spherical shape (radius = 90 mm).

The simulated sensor array comprised 272 magnetometers arranged in a hemispheric array on a sphere with a 100 mm radius. The average distance between the sensors was 22 mm. The SNR was defined as the ratio of the Frobenius norm of the data matrix to that of the noise matrix. We used the correlation coefficient to measure the degree of linear correlation between two current source waveforms. Two equally strong tangential sources were simulated: current dipole 1 was located at (25, 45, and 40 mm) with orientation (20.5797, 20.1380, and 20.6621), and current dipole 2 was at (5, 45, and 40 mm) with orientation (20.9341, 0.1489, and 20.3246). The distance of the sources was thus 10 mm. The waveforms of the two sources were 10 Hz sine functions with different phases and a 500 ms duration. The sampling frequency was 1,000 Hz. The correlation coefficient (r^2) between the two sources was set to 0.99, 0.7, and 0.5 by adjusting the phase difference between the waveforms. Uncorrelated white Gaussian noise was added to all data points scaled such that SNR was 3. Figure 11.4 shows the localization results of the four methods with varying degrees of correlation between the sources and with different SNRs. As r^2 decreases, INN resolved the two sources accurately with less spatial blurring. However, beamformer (Dalal et al., 2006) could not resolve the two sources except for $r^2 = 0$ (not shown). As r^2 was decreased to 0.7, MUSIC (Mosher et al., 1992) started to resolve the two sources, with an increasingly clear separation with smaller r^2 values. RAP-MUSIC (Mosher and Leahy, 1999) improved similarly as MUSIC since its performance depended on the MUSIC results in the first iteration.

In summary, comparing INN with MUSIC, conventional BEAMFORMER, and RAP-MUSIC on simulation MEG data shows that INN is promising. For

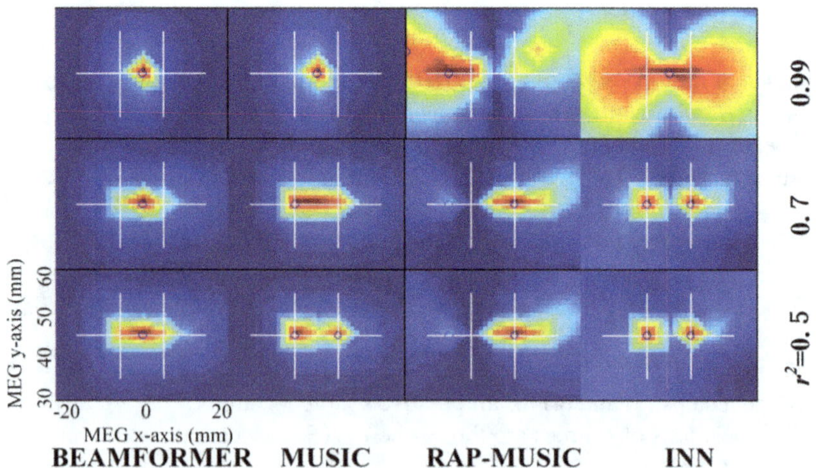

FIGURE 11.4 Comparison of BEAMFORMER, MUSIC, RAP-MUSIC, and INN with two closely spaced (10 mm apart) simulated sources at different levels of correlation ($r^2 = 0.5$, 0.7, or 0.99). The true source locations are (25, 45, and 40) mm and (5, 45, and 40) mm. SNR is 3. The white crosshairs indicate the true source locations and the blue circles are the local maxima found by the methods. The left bottom corner shows the x- and y-axis scales (Zhang et al., 2013).

closely spaced correlated sources, INN offers better performance than MUSIC, RAP-MUSIC, and BEAMFORMER. One shortcoming of INN, similar to other subspace-based methods, is that it cannot directly recover the amplitudes of the sources, and the parameter p cannot be effectively determined in practice. Therefore, another approach, such as a conventional linear least squares fit, with fixed source locations and orientations, is necessary to estimate the time courses of the sources.

This chapter introduces the delay-dependent MUSIC method, higher order MUSIC method, spatial region suppression MUSIC method, and noise space energy invariant method (INN), they have achieved good results in simulation and real data experiments. In addition, the time-frequency MUSIC method has been developed in the literature, that is, the time-frequency analysis of ERP/EEG records is performed first, and then the MUSIC method is applied to locate specific time-frequency components (Sekilhara et al., 1999, 2000). Recently, Makela et al. (2018), based on RAP-MUSIC (Mosher and Leahy, 1999), suggested that each step of the RAP-MUSIC iteration identified a source, and therefore the dimension of signal-subspace should be reduced by 1, thus proposing applying a sequential dimension reduction to RAP-MUSIC, called truncated recursively applied - and - projected MUSIC (TRAP - MUSIC), and achieved a good effect on the simulation and real data. At the same time, Dinh et al. (2017) introduced a fast RAP-MUSIC method that can be used for real-time needs by using the leadfield clustering method. These conditions indicate that the MUSIC method for the inverse EEG problem has been recognized by the field and is still under continuous development and improvement. Based on these developments, MUSIC-type localization could become more reliable and suitable for various online and offline MEG and EEG applications, which have a large development space and a good development prospect.

REFERENCES

Belouchrani, A., A. Belouchrani, K. Abed-Meraim, et al. 1997. A blind source separation technique using second-order statistics. *IEEE Trans Signal Proces* 45(2):434–444.

Dalal, S.S., K. Sekihara, S.S. Nagarajan. 2006. Modified beamformers for coherent source region suppression. *IEEE Trans Biomed Eng* 53:1357–1363.

Dinh, C., L. Esch, J. Rühle, et al. 2017. Real-time clustered multiple signal classification (RTC-MUSIC). *Brain Topogr*. doi: 10.1007/s10548-017-0586-7.

Horn, R., C.R. Johnson. 1985. *Matrix Analysis*. Cambridge: Cambridge University Press.

Ilmoniemi, R.J., J. Sarvas. 2019. *Brain Signals: Physics and Mathematics of MEG and EEG*. Boston, MA: MIT Press.

Knosche, T.R. 1997. Solutions of the neuroelectromagnetic inverse problem: An evaluation study. PhD Dissertation, University Twente.

Koles, Z.J. 1998. Trends in EEG source localization. *EEG Clin Neurosci* 106:127–137.

Ma, C., Y. Peng, L. Tian. 1998. Blind estimation of DOA's of coherent sources. *The Fourth International Conference on Signal Processing*, Beijing, China, pp. 327–330.

Makela, N., M. Stenroos, J. Sarvas, et al. 2018. Truncated RAP-MUSIC (TRAP-MUSIC) for MEG and EEG source localization. *NeuroImage* 167:73–83.

Markovskya, I., S. Van Huffel. 2007. Overview of total least-squares methods. *Signal Process* 87:2283–2302.

Menke, W. 2012. *Geophysical Data Analysis: Discrete Inverse Theory*. London: Academic Press.

Mosher, J.C., P.S. Lewis, R.M. Leahy. 1992. Multiple dipole modeling and localization from spatio-temporal MEG data. *IEEE Trans Biomed Eng* 39:541–557.

Mosher, J.C., R.M. Leahy. 1998. Recursive MUSIC: A framework for EEG and MEG source localization. *IEEE Trans Biomed Eng* 45(11):1342–1354.

Mosher, J.C., R.M. Leahy. 1999. Source localization using recursively applied and projected (RAP) MUSIC. *IEEE Trans Signal Proces* 47(2):332–340.

Sarvas, J. 1987. Basic mathematical and electromagnetic concepts of the biomagnetic inverse problem. *Phys Med Biol* 32(1):11–22.

Schmidt, R. 1986. Multiple emitter location and signal parameter estimation. *IEEE Trans Antenn Propag* 34:276–280.

Sekilhara, K., D. Poeppel, A. Marantz, et al. 1997. Noise covariance incorporated MEG-MUSIC algorithm: A method for multiple-dipole estimation tolerant of the influence of background brain activity. *IEEE Trans Biomed Eng* 44(9):839–847.

Sekilhara, K., S. Nagarajan, D. Poeppel, et al. 1999. Time-frequency MEG-MUSIC algorithm. *IEEE Trans Med Imaging* 18:92–97.

Sekilhara, K., S.S. Nagarajan, D. Poeppel. 2000. Estimating neural sources from each time-frequency component of magnetoencephalographic data. *IEEE Trans Biomed Eng* 47(5):624–653.

Van Veen, B.D., K.M. Buckley. 1988. Beamforming: A versatile approach to spatial filtering. *IEEE ASSP Mag* 5(2):4–24.

Van Veen, B.D., W. Van Drongelen, M. Yuchtman, et al. 1997. Localization of brain electrical activity via linearly constrained minimum variance spatial filtering. *IEEE Trans Biomed Eng* 44:867–880.

Yao, D. 2000. Algorithm for inverse problem of EEG based on subspace decomposition of fourth-order cumulant matrix. *J Biomed Eng* 17(2):174–178 (in Chinese).

Yao, D. 2008. Brain electric source imaging by an iterative correlation matrix based multiple signal classification approach. In Lars N. Bakker, ed., *Brain Mapping Research Developments*, chapter IX, Nova Science Publishers Inc, 211–230.

Yao, D., N. Rao, S. Fu, et al. 2000. Delay-dependent matrix space decomposition algorithm for inverse EEG problems. *Chin J Electron* 28(4):135–138 (in Chinese).

Yao, D., S. Fu. 2002. Computer simulation study of multi-signal classification algorithm for EEG inverse problem. *Chin J Biomed Eng* 21(1):53–58 (in Chinese).

Zhang, J, T. Raij, M. Hämäläinen, et al. 2013. MEG source localization using invariance of noise space. *PLoS One* 8(3):e58408.

Zhang, J., S.S. Dalal, S.S. Nagarajan, et al. 2010. Coherent MEG/EEG source localization in transformed data space. *Biomed Eng App Basis Commun* 22:351–365.

Zhang, X.D. 1995a. *Modern Signal Processing*. Beijing: Tsinghua University Press.

Zhang, Z. 1995b. A fast method to compute surface potentials generated by dipoles within multilayer anisotropic spheres. *Phys Med Biol* 40(3):335–349.

12 Iterative Minimum Norm Solution

In addition to the subspace decomposition algorithm introduced in Chapter 11 and the multi-dimensional nonlinear optimization method in Chapter 10, another kind of method that has attracted wide attention in the electroencephalogram (EEG) community is the low-resolution tomography method based on the generalized minimum norm solution (LORETA) (Pascual-Marqui et al., 1994). In this kind of method, all the points in a discrete three-dimensional brain space are considered as the potential locations of the sources, and a weighted minimum norm solution is found to represent the locations of the sources; then, the solution is optimized by iteratively updating the weights. The characteristic of this type of method is using a linear algorithm to iteratively approximate the nonlinear problem. Here, the initial weight may come from various priors or other modality images, such as functional magnetic resonance imaging (fMRI). This case is related to the fusion of EEG and fMRI, which can be roughly divided into three categories (Lei et al., 2012): (1) fMRI guided by EEG information (Dong et al., 2015; Qin et al., 2020); (2) fMRI used as prior to reducing the non-uniqueness of the EEG inverse problem (Lei et al., 2011); (3) symmetric fusion by a neural computation model which predicts both fMRI and EEG by, such as the dynamic causal model (DCM) (Jafarian et al., 2020) and digital twins brain (DTB) (Zhang et al., 2021). Now, there is special free software, the neuroscience information toolbox (NIT), which provided the main approaches (Dong et al., 2018). However, in this book, the fusion of EEG and fMRI is not included as it is not an EEG problem alone but an integration of the two modalities; there are some monographs that can be found.

This chapter will cover four aspects: the basic theory of linear EEG inversion; low-resolution tomography, and its derivative methods; sparse sources imaging method; and, finally, the coherent weight iterative method.

12.1 BASIC THEORY OF LINEAR EEG INVERSION

12.1.1 Introduction

The EEG inverse problem is essentially a nonlinear optimization problem as the relation between the source position variables and the scalp EEG recordings is nonlinear. Due to the difficulty and complexity to solve a nonlinear problem, researchers often try to linearize the nonlinear EEG inverse problem, including iteratively using a linear method to indirectly approximate the nonlinearity.. In fact, the linearization approach for EEG Inverse problem is now the mainstream of the current EEG inverse problem study. It is now far more adopted than the optimizing localization method introduced in section 10.3. This chapter is to provide a comprehensive summary of this kind of method.

DOI: 10.1201/9781032639260-12

In theory, determining the source of the measured potential on the scalp is a problem without a unique solution, that is, the potential distribution on the outer surface of a volume conductor can be generated by an infinite number of equivalent sources within the conductor (Section 2.8 or 10.2). At the same time, the number of independent measurements that can be obtained in practice is limited, and the existence of noise makes the non-uniqueness of the solution to the inverse problem more serious. These theoretical and practical limitations mean that the bioelectric inverse problem is inherently indeterminate, and it will be difficult to obtain a unique solution for the source localization method unless additional constraints or priors are available. The additional information may be the potential locations of the sources, such as restricted to the grey matter, or assumed that the solution is of minimum energy (minimum norm solution), sparse in space (sparse solution), smooth and non-singular (regularization solution), etc. This chapter will systematically summarize the basic theoretical methods of linear inversion based on certain constraints or priors (He et al., 2000; Appendix E). The typical approach in EEG research is to linearize the inverse problem by fixing the potential source location. The following two typical scenes have effectively implemented this strategy.

(1) High-resolution EEG imaging. After noticing the low spatial resolution of EEG activity on the topographic map of the scalp surface and the non-uniqueness of the inverse problem of EEG, the EEG community has carried out great effort to realize relatively high-resolution EEG mapping as shown in Chapters 6–9. Compared with early 16 channels or 32 channels EEGs, high-resolution EEG technologies usually adopt 64 channels or more, such as 128 or 256 channels, naturally leading to a topographic map with higher resolution. However, too many electrodes mean that the actual operation is not convenient. Meanwhile, the low-pass filtering of the skull plays a key role in limiting the resolution of the topographic map on the scalp surface, and it cannot be removed by adding more channels according to the signal sampling theory (Section 10.1). For this reason, researchers have developed a variety of methods to counteract the effect of low-pass filtering of the skull, such as the various high-resolution EEG techniques in Chapters 6–9. The main calculation in high-resolution EEG is the solution of a linear equation group, which is based on the assumption that the equivalent distributed sources is located at the pre-determined surface, or equivalent multipole sources is located at the coordinate origin. The number of the potential sources is usually much larger than that of the scalp channels, the equation group is underdetermined.

(2) 3D EEG tomography. The famous LORETA method, similar to the equivalent distributed source method of high-resolution EEG technique, linearizes the problem by assuming that the source is in 3D volume distribution and thus all the source positions are fixed. This is a bold move, because the number of 3D unknowns in LORETA is much larger than that of the cortical imaging unknowns, and the underdetermination is more serious. Therefore, the smart inventor especially introduced a low-resolution constraint, the Laplacian weight (Pascual-Marqui et al., 1994). Of course, LORETA's resolution is very low and so is the spatial resolution of the cortical surface distribution source map obtained by the equivalent distributed source method (Chapter 6). Luckily, for many brain science problems, the most important requirement is the potential position information of the sources, and what LORETA provided is an intensity distribution where the relatively prominent points provide

us the possible location of the sources. EEG LORETA profoundly interprets that "simple is good". Certainly, since LORETA appeared in 1994, some new methods have been developed to further improve its spatial resolution, such as FOCUSS introduced in this chapter.

In short, the linearization of a nonlinear problem is of great importance in many scientific and engineering problems. The linear inverse problem can be reduced to the solution of a linear equation group as follows:

$$\Phi = GX \tag{12.1}$$

Where Φ is known, such as scalp potential records; X is unknown variables, such as the intensity of EEG source; and G is the leadfield matrix, each column of which represents the contribution of the source of unit strength at a certain position to the scalp recordings Φ. The so-called inverse problem is to solve X according to the known Φ. When the dimension of X is greater than the dimension of Φ, the problem is called underdetermined.

12.1.2 Generalized Inverse and Singular Value Decomposition

12.1.2.1 Generalized Inverse

Due to noise in scalp potential measurement, equation (12.1) should be rewritten as

$$\Phi = GX + N \tag{12.2}$$

Where N is the noise vector, whose dimension is equal to the number of scalp-measuring electrodes. According to Equation (12.2), by minimizing the following objective function:

$$O(X) = \|\Phi - GX\|^2 \to \min \tag{12.3}$$

A normal equation can be obtained.

$$G^T GX - G^T \Phi = 0 \tag{12.4}$$

If G is a full-rank square matrix, then

$$X = \left(G^T G\right)^{-} G^T \Phi = G^{-} G^{T-} G^T \Phi = G^{-} \Phi \tag{12.5}$$

For a general matrix G, define

$$X = G^{+} \Phi \tag{12.6}$$

Where G^{+} represents the generalized inverse (Appendix E). The generalized inverse solution is one of the infinite numbers of solutions satisfying the measured potential of the scalp (Golub and Van Loan, 1989). The generalized inverse solution has been widely used in the study of EEG, MEG, and ECG.

If $\left(G^T G\right)$ is invertible, then

$$X = \left(G^T G\right)^{-} G^T \Phi = G^{+}\Phi, G^{+} = \left(G^T G\right)^{-} G^T \tag{12.6a}$$

This situation corresponds to the overdetermined or deterministic case, which means that the number of independent measurements is more or equal to the number of independent variables and that G is a full-column rank matrix or full-rank square matrix, and the solution is the least square solution (LS) of equation (12.3) (Appendix E.15).

If $\left(GG^T\right)$ is invertible, then

$$X = G^T\left(GG^T\right)^{-}\Phi = G^{+}\Phi, \quad G^{+} = G^T\left(GG^T\right)^{-} \tag{12.6b}$$

This situation corresponds to the underdetermined case, that is, the number of independent measurements is less than the number of independent unknown variables, but G is a full-row rank matrix or full-rank square matrix, and this situation is the minimum norm solution (MN) under the constraint $\Phi = GX$ (Appendix E). This situation is mainly discussed in this chapter.

Since the ordinary inverse can be regarded as a special case of the generalized inverse, it is only necessary to study the algorithm of equation (12.6). In the literature, there are many methods of computing the generalized inverse G^{+} of matrix G; here, we mainly introduce the method of singular value decomposition (SVD), which can be directly applied to the over- or under-determined problem, and its detailed algorithm and software can be found in many books and open software. In the classic book on the inverse problem by Menke (2012), this generalized inverse is called the natural generalized inverse (Appendix E).

12.1.2.2 SVD Algorithm

For a matrix G of $m \times n$, the SVD is (Golub and Van Loan, 1989)

$$G = U\Sigma V^T \tag{12.7}$$

Among them

$$U = \left[u_1, u_2, \ldots, u_m\right], \quad V = \left[v_1, v_2, \ldots, v_n\right],$$

$$\Sigma = \operatorname{diag}\left(\lambda_1, \lambda_2, \ldots, \lambda_p\right), \quad \lambda_1 > \lambda_2 > \cdots > \lambda_p, \quad p = \min(m, n).$$

The vectors u_i and v_i are the ith left and right singular vectors of the matrix U and V of $m \times 1$ and $n \times 1$, respectively. λ_i is the singular value of G, and Σ is a diagonal matrix with singular values. In addition, if rank $(G) = r$, then

$$\lambda_1 > \cdots > \lambda_r > \lambda_{r+1} = \cdots = \lambda_p = 0 \tag{12.8}$$

Then, the singular value expansion can be expressed as

$$G = U_r \Sigma_r V_r^T = \sum^{r} \lambda_i u_i v_i^T \tag{12.9}$$

Where U_r and V_r are $m\times r$ and $n\times r$ matrices, composed of the corresponding first r singular values, and Σ_r is the $r\times r$ diagonal matrix, which consists of the first r singular values located on the main diagonal. Based on the SVD of matrix G, the Moore-Penrose inverse or "pseudo-inverse" of G is

$$G^{+} = V_r\Sigma^{-1}{}_r U_r^T = \sum_{i=1}^{r}\frac{1}{\lambda_i}v_i u_i^T \tag{12.10}$$

For a linear system like equation (12.1), the solution X of equation (12.6) is

$$X = G^{+}\Phi = \sum_{i=1}^{r}\frac{1}{\lambda_i}v_i u_i^T\Phi = \sum_{i=1}^{r}\frac{1}{\lambda_i}\left(u_i^T\Phi\right)v_i = \sum_{i=1}^{r}w(i)v_i \tag{12.11}$$

12.1.3 Weighted Generalized Inverse

The general method described above is based on the assumption that the generalized inverse of the leadfield matrix G can be properly calculated. In this case, the only problem that needs to be solved in practical application is to make a reasonable truncation choice or an appropriate correction for the singular value by modifying $\frac{1}{\lambda_i}$ to $\frac{1}{\lambda_i+\varepsilon}$, where ε is an infinitesimal (Appendix E). However, because the leadfield matrix G of a practical problem is often seriously ill-conditioned, the inverse calculation is very difficult, or very inaccurate. Therefore, to obtain a better result, the inverse can often be improved by modifying the leadfield matrix G properly before the inverse calculation. Some of the ways to decorate the leadfield matrix are described below. The following introduction is true for both overdetermined and sub-determined cases.

12.1.3.1 Spatial Weighting Method in Measurement Space

According to equation (12.3), the least squares solution (LS) in the overdetermined case is also the solution of the minimum L2 norm of the objective function of residual error (Gencer and Williamson, 1998). Therefore, the larger the residual error is, the more important the corresponding record channel will be in the fitting. And this may be at odds with the desired effect for each measure to work equally. At this point, we hope to take appropriate measures to balance the role of each measurement point in the fitting. Sometimes, because of some special needs, and want to highlight the role of some measurement points, we also hope to be able to realize through a certain way. All such problems can be summarized as introducing a weighting factor to the left of both sides of the equation (12.1), that is, weighting the measured data in the measurement space.

$$W_d\Phi = W_dGX \tag{12.12}$$

or

$$\Phi' = G_dX \tag{12.12a}$$

Thus,

$$X = G_d^{+}\Phi' \tag{12.12b}$$

Where $\Phi' = W_d\Phi,\quad G_d = W_dG.$

Here are a few examples.

12.1.3.1.1 Normal Equation Method

The normal equation (12.4) introduced above, where $G^T\Phi = G^TGX,\ W_d = G^T$.

12.1.3.1.2 Laplacian Inverse

In the study of the ECG inverse problem, an inverse algorithm based on body surface Laplacian was proposed (He and Wu, 1997), which essentially introduced a Laplacian operator weight in the left end of the linear equation.

$$L_s\Phi = L_sGX \tag{12.13}$$

Obviously, $W_d = Ls$ is a Laplacian operator used in measurement space. The result of this operator can be understood as making those records with high-frequency spatial information plays a greater role in inversion.

12.1.3.1.3 Deep and Shallow Source Separation Method

In the study of the EEG inverse problem, due to the different spatial frequency contents of deep and shallow sources, spatial filtering is possible to achieve the separation of deep and shallow sources. This method can be understood as

$$F\Phi = FGX = G_FX \tag{12.14}$$

In essence, a spatial high-pass operator, such as a Laplacian operator, is used to separate the high-frequency components corresponding to the shallow source and the rest corresponding to the deep source. Such a spatial filter can be realized by a combination of different measurement techniques (Samuelsson et al., 2019).

12.1.3.1.4 Noise Covariance Weighting Algorithm

In the objective function defined in equation (12.3), the measurement points with large residuals have large weights. To balance the importance of each measurement value, the covariance weight can be added to the objective function.

$$O(X) = (\Phi - GX)^T C^- (\Phi - GX) \to \min$$

From this, we can get

$$X = \left(G^TC^-G\right)^- G^TC^-\Phi \tag{12.15}$$

Combining (12.12), $W_d^TW_d = C^- = \mathrm{cov}(N)^-$ is obtained. Noise covariance weighting is a commonly used method in EEG studies (Van Veen et al., 1997). In practical use, we can also introduce other forms of weighting according to the needs and characteristics of the problem, or a further combination of the above weighting.

12.1.3.2 Spatial Weighting Method in Solution Space

The function of the above measurement space weighting is to standardize or adjust the role of different measurement points. But in practice, because of the minimum norm feature of the solution and the properties of matrices, there are often other problems. For example, many numerical experiments show that the source in EEG inverse solution is often located at a shallower depth than the actual source. This is because the column of the transfer function for a shallower point is larger than that for a deep point, then to fit the same scalp potential, the norm of a shallower source may be smaller than that of a deeper one, further in the sense of an MN solution, the shallower point is selected by the MN algorithm. In the case of the LS solution, the shallower source is of more high spatial frequencies in the scalp recordings, which may better fit the high spatial frequencies in the actual scalp data, but the noises in actual recordings are also more commonly hidden in high-frequency components, thus the solution not only be biased to the scalp (up-drift) but also be misled by the high-frequency noises. To overcome or mitigate this kind of problem related to solution space, additional weights need to be introduced to overcome the relative importance of different positions in solution space, which is the solution space weighting method described in detail below.

The solution space weighting method can be summarized as

$$\Phi = GW_x^- W_x X = G_x X' \tag{12.16}$$

Where W_x is the weighting factor acting on solution space, $G_x = GW_x^-$, $X' = W_x X$, and then we have

$$X = W_x^- X' = W_x^- G_x^+ \Phi \tag{12.16a}$$

The following are some solution space weighting factors that have emerged in the study of the biological electromagnetic inverse problems.

12.1.3.2.1 Normalized MN Solution

EEG linear inversion is usually an underdetermined problem. To overcome the up-drift problem, a method is to introduce a weight to balance the column norm of the leadfield matrix. Specifically, the column norms of the leadfield matrix are used as weights in the corresponding positions in the solution space, thus normalizing the column norms (Gorodnitsky et al., 1995). Assuming that $G = (g_1, g_2, \ldots, g_n)$, then the weight may take

$$W_x = \text{diag}\left(\|g_1\|, \ldots, \|g_n\|\right) \tag{12.17}$$

12.1.3.2.2 Laplacian Operator for Low-Resolution Solution

Pascual-Marqui et al. (1994) proposed the low-resolution brain electromagnetic tomography (LORETA); the core idea is a weight factor composed of the weight in equation (12.17) and a Laplacian operator.

$$W_x = B\,\mathrm{diag}\left(\|g_1\|,\|g_2\|,\ldots,\|g_n\|\right) \tag{12.18}$$

Where B is a Laplacian operator acting on the solution space. For a three-dimensional solution space, B is a discrete three-dimensional Laplacian operator. Similarly, for a two-dimensional solution space, for example, on a cortical surface, B is a discrete two-dimensional Laplacian operator. Why is it a low-resolution solution? The Laplacian operator is a second-order spatial derivative that acts as a spatial high-pass filter (Chapter 8), thus the "minimum norm" of "WxX" is a "minimum norm" of the "spatial high-frequency components" of X, and the high-frequency minimizing corresponds to an indirect low-frequency enlarging comparatively, thus inducing a final low-resolution (Pascual-Marqui et al., 1994; Yao and He, 1998); for more details, pls refers to Section 12.3.

12.1.3.2.3 Local Weighting Method

In the absence of additional information about the source, a natural idea is to find a locally smooth (average) solution (Grave de Peralta Menendez et al., 1998).

$$X_{\text{ave}} = AX \tag{12.19}$$

Where A is a local average operator, so $Wx = A$. In practice, many different averaging operators can be constructed according to reasonable assumption, and the inverse solution depends on the form of A. For example, we can define the elements of the matrix A as (Grave de Peralta Menendez et al., 1998)

$$A_{ij} = e^{-d_{ij}^2} \Big/ K_i. \tag{12.19a}$$

Where d_{ij} represents the distance between the ith and jth grid points. The coefficient K_i is a correction factor. Another strategy is for each point, the maximum value (or median value) in its neighborhood in previous primary solution is selected as the weight (Xu et al., 2008).

12.1.3.2.4 Recursive Weighting Factor Method

The solution obtained above may be used to form a recursive weight factor. For example,

$$W_x = X_{k-1}^{-}\ \mathrm{diag}\left(\|g_1\|,\|g_2\|,\ldots,\|g_n\|\right) \tag{12.20}$$

In this method, the previous solution X_{k-1} at step $k-1$ is used recursively as the weight of the kth iteration, and of course, X_{k-1} can be changed to a specific function of $X_{k-1}, X_{k-2}, \ldots, X_0$. The result of this weight function will play a role in strengthening the local solution, so it is named "FOCUSS" by the proposer (Gorodnitsky et al., 1995).

12.1.3.2.5 Equivalent Source Method

Recalling the previous formula (12.16), we may have another expansion.

$$\Phi = GX = G_x X' \tag{12.21}$$

Different from equation (2.16), where G_x and X' are the weighted versions of G and X, here X' is an equivalent source in producing the same scalp potential as the actual source X, X' can be two or three-dimensional source distribution, and G_x is the leadfield matrix of the equivalent source. This case provides the theoretical algorithm basis for the equivalent source method (see Chapters 6 and 7). The advantage of the equivalent source method is that the nonlinear problem of finding the real source is changed into the linear problem of finding the equivalent source strength, and then, the equivalent source can be used to realize high-resolution imaging (Chapter 6) or the standardization of the reference electrode (Chapters 13 and 14). The MN or the LORETA distributed source solution can be taken as a specific equivalent solution under a specific constraint too.

12.1.3.2.6 Magnetic Field Tomography

In the study of brain magnetic problems, Ioannides et al. (1990) developed a method called magnetic field tomography (MFT). The basic idea is to assume that source x can be expanded based on g_i in terms of $G = (g_1, \ldots, g_N)$.

$$x(\vec{r}) = \sum_{k=1}^{M} a_k g_k(\vec{r}) w(\vec{r}) \tag{12.22}$$

where w is the prior distribution density of the source in the solution space; w can be a constant throughout the source space with dimension N, but it also can be any a priori assumption about source location; $\vec{r}$ is any point in the solution space; g_i is a column vector consisted of the contribution of the unit source at each source position to the sensor i among the M sensors. Accordingly, the scalp recordings

$$\Phi_{M\times 1} = G_{M\times N} X_{N\times 1} = P_{M\times M} A_{M\times 1} \quad \text{with} \quad P_{ij} = \sum_{k=1}^{N} g_{ik} g_{jk} w(k) \tag{12.22a}$$

If we get a_k from equation (12.22a), then we get the solution from equation (12.22). Strictly speaking, this method is not a direct solution space weight, but a solution reconstruction, from X to A. The weight factors in equation (12.22) consist of both solution space weight w and measurement space weight a.

12.1.3.2.7 ERP+fMRI

Another relatively new field of solution space weighting is the combination of Electroencephalography/Magnetoencephalography (EEG/MEG), event-related potential/event-related field (ERP/ERF), and other means. For example, the weight of solution space can be selected by the source intensity or network distribution provided by fMRI (Lei et al., 2011) to obtain the solution affected by fMRI.

12.1.3.2.8 Other Weights

Other weights may be the Maximum Neighbor Weight in CMOSS in the following Section 12.2 (Xu et al., 2008), and the Gaussian Weight in the Gaussian distributed source model (Lei et al., 2009).

Obviously, in practice, we can construct various weights, such as one-, two-, or three-order derivatives of the solution, and the combination of the above weights, including the combination of the measurement space and source space weights together (update simultaneously or alternately). The most important thing is that we need to know the problem and the value of a specific weight, thus designing a valuable weight for the problem.

12.1.4 Constrained Inverse

In the previous sections, our main efforts were to modify the leadfield matrix from the sources to the recordings. Similarly, we can also try to modify the objective function, and then, we can get some different formulas to reach the goal, which of course gives us additional choices and latent results in solving the inverse problem.

In the LS solution, the objective function is only related to the residuals. In the MN solution, the objective function is the minimum norm of the solution. More generally, we can additionally associate the objective function with the time-space property of the solution; then, we have

$$O(X) = \|\Phi - GX\|^2 + \beta_x \|W_x X\|^2 + \beta_t \|W_t X\|^2 \rightarrow \min \tag{12.23}$$

It consists of the residual error and the spatial and temporal constraints of the solution, and it can be used in mixed-determined situations (Appendix E). According to the equation (12.23), we have

$$\left[G^T G + \beta_x W_x^T W_x + \beta_t W_t^T W_t\right] X = G^T \Phi \tag{12.23a}$$

Where the $\beta_x \|W_x X\|^2$ term is a spatial constraint on the solution and W_x can be one of the various spatial weights introduced above or any other special form. For example, it can be a function of the previous estimate, for example, Baillet and Garnero (1997) adopted an objective function of the following form:

$$O(X) = \|\Phi - GX\|^2 + \beta_x \left\{\varphi(\nabla_x X) + \varphi(\nabla_y X)\right\} + \beta_t \|W_t X\|^2 \rightarrow \min \tag{12.24}$$

Among them

$\varphi(u) = \dfrac{u^2}{1 + \left(u/K\right)^2}$ is a Lorentzian function and K is a constant. The corresponding solution can be derived as

$$\left[G^T G + \beta_x \Delta + \beta_t W_t^T W_t\right] X = G^T \Phi \tag{12.25}$$

Where Δ is a function of X, it means that equation (12.25) is a nonlinear problem; in practice, an iterative implementation can be realized, where Δ is estimated by the result of the previous iteration. If we take equation (12.23) as a special case of equation (12.25), then $W_x^T W_x = \Delta$, which means that Wx is a function of the last estimated result.

A direct example of a spatial constraint is to define the latent spatial location of the source according to the MRI (magnetic resonance imaging)/CT (computerized tomography) image.

In equation (12.23), the $\beta_t \|W_t X\|^2$ term is the temporal constraint on the solution. For example, we can require that the solution changes slowly over time. The following are two temporal constraints reported in the literature.

12.1.4.1 Temporal Regularization

In this temporal regularization strategy (Twomey, 1963), instead of imposing constraints on the solution or its function, it attempts to minimize the difference between solution X and some initial estimate X_0, which may be an estimate of the solution based on information known at a previous moment or an estimate obtained by other means. If $\beta_x = 0.0, \quad W_t = I$, the equation (12.24) changes to

$$O(X) = \|\Phi - GX\|^2 + \beta_t \|X - X_0\|^2 \rightarrow \min \tag{12.26}$$

This expression can be reformulated to

$$O(X) = \|(\Phi - GX_0) - G(X - X_0)\|^2 + \beta_t \|(X - X_0)\|^2 \rightarrow \min \tag{12.27}$$

So, we have

$$\begin{gathered} (X - X_0) = \left[G^T G + \beta_t I\right]^{-} (\Phi - GX_0) \\ X = \left(G^T G + \beta_t I\right)^{-} \left(G^T \Phi + \beta_t X_0\right) \end{gathered} \tag{12.28}$$

Equation (12.28) has been used in the study of ECG problems (Oster and Rudy, 1992).

12.1.4.2 Temporal Constraint

Let X_{k-1} be the unique time neighbor of X and assume that X_{k-1} and X are very close to each other, so the orthogonal projection of X on the hyperplane $E^{\perp}_{X_{k-1}}$, perpendicular to X_{k-1}, will be small; then, a temporal weight in (12.23) can be defined accordingly (Baillet and Garnero, 1997).

$$W_t = I - \frac{X_{K-1} X_{K-1}^T}{\|X_{K-1}\|^2} \tag{12.29}$$

Where I is the unit matrix.

In general, we can assume more terms in equation (12.24) to get a more general expression. However, the more terms there are, the more undetermined parameters β_*, and determining them is another big problem. In practice, the most common way is to retain one constraint term and the corresponding equation (12.25) becomes

$$\left[G^T G + \beta R\right] X = G^T \Phi \tag{12.30}$$

and

$$X = \left[G^T G + \beta R\right]^{-} G^T \Phi \tag{12.31}$$

This equation is often referred to as the Tikhonov regularization framework (Tikhonov and Arsenin, 1977), where the ordinary inverse form is adopted as it assumes that regularization makes the problem without normal inverse become a problem with normal inverse, and it is the essential point of "regularization". In practice, however, we are not necessarily limited to this. In fact, it is only necessary to change the "–" in equation (12.31) to "+", and "–" is a special case of "+". The β in equation (12.31) is called the regularization parameter, where R usually takes one of the following forms: (1) $R=I$ -- identity matrix (Tikhonov zero-order regularization); (2) $R=G$ -- gradient operator (Tikhonov first-order regularization); (3) $R=L$ -- Laplacian operator (Tikhonov second-order regularization). Obviously, as a natural extension, the various weights and operators described above can also be used here. However, some experiments showed that the effectiveness of the zero-order Tikhonov regularization is almost the same as the first-order or even second-order regularization. Therefore, in practice, it is not that the more complex the formula is, the better the effect will be. A good algorithm often needs to consider various influences, and finally, quite possibly, the simpler, the better.

12.1.5 Regularization

For the SVD method, we need to determine the truncation level of SVD. For the constrained inverse, we need to determine the regularization parameters β. Both of these problems are called regularization problems. It should be noted that some of the following methods apply to both the truncation and regularization approaches, while others apply to only one of them.

12.1.5.1 SVD Truncated Pseudo-Inverse

If the measurement noise n is additive, then the generalized inverse solution will be (Shim and Cho, 1981; Sullivan and Liu, 1984)

$$X = G^{+}(\Phi + N) = \sum_{i=1}^{r} w(i) v_i + \sum_{i=1}^{r} \frac{1}{\lambda_i} n_w(i) v_i, \quad n_w(i) = u_i^T N \tag{12.32}$$

Where $G = U\Sigma V^T$, $w(i) = u_i^T \Phi$, λ is the singular value, and r is the number of singular values. In this case, the condition number of the leadfield matrix G (the ratio of the maximum singular value to the minimum singular value) and the noise level in the measurement become important. Equation (12.32) clearly shows that the effect of noise is amplified by small singular values. The effect of the second term may even crush the contribution of the first term. The effect of the second term can be ignored only when the noise is small enough to cancel out the effect of the minimum singular value. When this is not satisfied, the effect of the second term needs to be suppressed by using a truncated pseudo-inverse scheme (Shim and Cho, 1981). This method is simple to implement, i.e., to truncate the summation at a large singular value ($p < r$) when estimating G^+; certainly, the parameter p needs to be selected carefully as shown below. This method reduces the effect of measurement noise; it also loses some useful information, the high-frequency detail information of the solution, so the result will be a smooth solution.

Some studies show that the optimal truncation order can be determined by the following equation (Shim and Cho, 1981; Gencer and Williamson, 1998):

$$p_{\text{opt}} = \max_k \left\{ k \middle| \frac{\lambda_k^2}{\lambda_1^2} \geq \frac{E\left(\|N\|^2\right)}{E\left(\|\Phi\|^2\right)} \right\} \tag{12.33}$$

It is deduced on the white assumption of both X and N. Operator $E\left(\|X\|\right)$ represents the mathematical expectation of the Euclidean norm.

The truncated SVD method has been widely used in various EEG problems, including high-resolution EEG (Sidman et al., 1992; Yao, 1996).

12.1.5.2 L-Curve Method

For the constrained inverse equation (12.23), in current practice, there still is no appropriate method to determine two or more parameters simultaneously; they are often determined empirically, such as an alternating one-by-one determination. For one parameter case, equation (12.31), there are many methods to determine its regularization parameter. By careful comparison of these methods, it can be seen that their essence is to compromise the norm $\|\Phi - GX\|$ of the residual and the solution norm $\|X\|$ in some way.

The L-Curve method, proposed by Miller (1970) and Lawson and Hanson (1974), has been widely used through the work of Hansen (1992). The L-Curve method involves a graph, usually on a log-log scale, on which a curve is drawn by varying the regularization parameter or the truncation level, with the norm $\|X\|$ of the solution as one coordinate and the norm $\|\Phi - GX\|$ of the residual as another. In most cases, a curve of the shape "L" will appear on this curve, and the parameter β or truncation level corresponding to the corner of "L" is considered to be the best regularization point. The reason is that at the corner point, $\|\Phi - GX\|$ and $\|X\|$ both reach a minimum point of each other, intuitively suggesting an optimal solution.

Hansen (1992) gave three conditions for the appearance of such a corner point (Johnston and Gulrajani, 1997): the first is the discrete Picard condition. It means that for an unperturbed problem with no geometric noise in G and no measurement noise in Φ, in the average sense, the coefficient $w(i)$ decays to zero faster than the non-zero singular value λ_i. This condition is usually assumed to be true for the bioelectricity inverse problem. Hansen's second condition is that the noise vector N in Φ is a random vector with zero mean and has a covariance matrix proportional to the identity matrix. If the uncorrelated Gaussian measurement noise in Φ is much stronger than the highly correlated geometric noise in G, it can be assumed that this condition is satisfied. The last condition is $\|N\| < \|\Phi - N\|$, that is, the norm of the noise vector is smaller than that of the noiseless measurement data, and the purpose of this condition is to ensure a certain signal/noise ratio. Hansen and O'Leary (1993) also put forward a numerical algorithm to determine the corner point of the L-Curve to determine the corresponding β_L. Of course, it is assumed that the corner point exists here. In practice, it is also possible to draw the L curve first and then determine it artificially. Of course, we should also note that there are cases of non-convergence for L-Curve (Vogel, 1996).

12.1.5.3 CRESO Algorithm

Franzone (1985) proposed an empirical method for determining the regularization parameter β, which is called the "composite residual and smoothing operator (CRESO)" method. In this method, $\beta(>0)$ is set as the smallest β value that makes the following function achieve a relative maximum:

$$f(\beta) = \|X\|^2 + 2\beta\frac{d\|X\|^2}{d\beta} \tag{12.34}$$

Based on the singular value decomposition of G and $R=I$ (Tikhonov zero-order regularization), this function can be transformed into (Johnston and Gulrajani, 1997)

$$f(\beta) = \sum_{i=1}^{p}\left[\frac{\lambda_i w(i)}{\lambda_i^2+\beta}\right]^2\left[1-\frac{4\beta}{\lambda_i^2+\beta}\right] \tag{12.35}$$

Obviously, the equation (12.35) is much more convenient than (12.34) to be realized by numerical method. Here, the parameter β is explicitly used in the formula; it is suitable for constrained inverse problems.

12.1.5.4 Discrepancy-Principle Regularization Technique

This regularization technique adopts adjusting the regularization parameter β or truncating the singular value series when solving an inverse problem, so that the residual is exactly equal to the error energy of the scalp surface data (Morozov, 1984; Kirsch, 1996), namely

$$\|GX-\Phi\| = \|N\| = E \tag{12.36}$$

It should be noted that such a β choice in the above equations (12.33)–(12.35) can ensure no over-fitting of the scalp data, thus avoiding the erroneous situation that the better the fitting is, the further the solution away from the true solution. The condition that this regularization has a unique solution is $\left\|\Phi^E-\Phi\right\| \le E < \left\|\Phi^E\right\|$, where $\Phi^E = G\tilde{X}$ is the scalp surface data corresponding to the regularization solution because it can be proven that

$$\lim_{\beta\to\infty}\|GX-\Phi\| = \left\|\Phi^E\right\| > E \tag{12.37}$$

And

$$\lim_{\beta\to 0}\|GX-\Phi\| = 0 < E \tag{12.38}$$

Suppose $O(X) = \|\Phi - GX\|^2 + \beta\|X\|^2$, as β tends to infinity, $\|X\|$ tends to zero, and as β tends to zero, $\|\Phi - GX\|$ tends to zero (overfitting).

Furthermore, as $\beta \to \|GX - \Phi\|$ is continuous and increasing, so the determination of β is the same thing as finding the zero of the monotone function $f(\beta) = \|GX - \Phi\|^2 - E^2$ with fixed $E > 0$.

In addition, it is not necessary to strictly satisfy $\|GX - \Phi\| = E$ in practice, as long as it meets equation (12.39) the above assertion valid.

$$c_1 E \le \|GX - \Phi\| \le c_2 E \tag{12.39}$$

The earliest application of this discrepancy-principle regularization method in EEG research was seen in Yao and He (1998), and it has been successfully used later in high-resolution EEG techniques, such as spherical spline, boundary element, and equivalent distributed source methods (Yao, 2000; Yao and He, 2001). The notable feature of this discrepancy-principle regularization is that it is particularly easy to be implemented by computer.

12.1.5.5 Regional Regularization

Regional regularization is the idea that different target regions in solution space should be regularized differently due to the different signal amplitude and noise levels (Oster and Rudy, 1997), or other regional concerns, but in practice, how to implement such an idea is still to be developed.

12.1.5.6 Discussion on Regularization

In machine learning, all the strategies for reducing test error, possibly accompanied with increase in training error, are called regularization, which may be a modification of the learning algorithm to reduce the generalization error rather than the training error. Specific strategies include parameter norm penalty, where the norm may be L_2, L_1, $L_{0.5}$, etc.; other strategies include training data set enhancement, that is, creating false data and adding it to the training set, also including adding various noisy data. What's important is that, if we compare the regularization methods in machine learning (including deep learning), we can see that many of them are consistent with the methods already used in EEG inversion, and some methods have not yet been introduced into EEG, this fact means the importance of interdisciplinary communication (Goodfellow et al., 2016).

12.2 3D LOW-RESOLUTION EEG TOMOGRAPHY

12.2.1 LORETA

As more and more 3D medical imaging technology like CT, positron Emission Tomography (PET), and MRI are developed and popularized, the EEG community also hopes to obtain a 3D tomography from EEG and Magnetoencephalography (MEG). However, due to the limited observation data and the non-uniqueness of EEG inverse problem, the brain electromagnetic three-dimensional tomographic technique has not been developed for a long time, and this situation changed with the emergence of the LORETA (Pascual-Marqui et al., 1994).

Before introducing the LORETA algorithm, we first clarify the physical quantities for 3D tomography and then the algorithm realization. In Section 3.3, we have presented three equivalent sources, namely current source density (equivalent charge), current density (equivalent dipole), and potential. Therefore, the EEG 3D tomography can be constructed on any one of them, and they have been done. They are the equivalent distributed dipole-based LORETA (Pascual-Marqui et al., 1994; Pascual-Marqui, 1995), the equivalent distributed charge source-based CLORETA (Yao and He, 1998; He et al., 2002), and the potential "source"-based ELECTRA (Grave de Peralta Menendez et al., 2000).

The essence of 3D tomography is the instantaneous and 3D discrete linear inversion of the EEG inverse problem. The main difficulty in developing EEG tomography is the physical fact: the measurement data does not provide enough information about the source, leading to the non-uniqueness of the inverse problem (Chapter 10). Therefore, it can be intuitively said that perfect tomography does not exist; what we are going to do is to find an approximate tomography that is valuable for displaying the true source at an acceptable blurring level.

From a historical point of view, the first tomographic solution in the field of electroencephalography was the minimum norm solution (MNS) published in 1984 (Hamalainen, 1984; Hamalainen and Ilmoniemi, 1994). The performance is good when the source is confined to a plane or on a spherical surface parallel to the measured surface, i.e., when the problem is a two-dimensional problem. For an ideal point source, a 2D fuzzy image with the maximum value corresponding to the real point source position can be obtained. However, this method failed when the solution space was extended to 3D (Pascual-Marqui, 1995).

One of the biggest challenges in developing EEG source 3D tomography is how to generalize MNS, which performs well in 2D, to 3D. The low-resolution brain electromagnetic tomography (LORETA) introduced a Laplacian operator constraint, thus replacing the minimum norm of the solution X with a minimum norm of the "high frequency component" filtered out by the Laplacian operator of the solution X, and the minimum norm of high-frequency components induced a solution with enlarged low-frequency components, that's the low-resolution tomography(Pascual-Marqui, 1995). The details of LORETA and its improved versions, standardized low-resolution brain electromagnetic tomography (sLORETA) and exact low resolution brain electromagnetic tomography (eLORETA), can be found at http://www.uzh.ch/keyinst/loreta.htm. Parallel to this dipole source-based tomography, the 2D equivalent charge layer mapping (Yao, 1996; Yao et al., 2004a,b) and the 3D CLORETA (Yao and He, 1998; He et al., 2002) were developed later. The advantage of CLORETA is that its number of unknown variables is only one-third of the dipole LORETA. Meanwhile, based on the initial low-resolution solution, various iterative algorithms were developed for realizing sparse solutions, such as the dipole source-based FOCUSS (Gorodnitsky and Rao, 1997) and the charge source-based CMOSS (Xu et al., 2008). In the following sections, introduced are CLORETA, CMOSS, and a few sparse solutions.

12.2.2 Charge Source LORETA and Maximum Neighbor Weight Sparse Solution

12.2.2.1 Charge Source Model

The popularly used source model in EEG inverse problem is the dipole model, and the charge model is a relatively lesser-known one. In fact, as shown in Chapter 3, the potential Φ generated by neural electrical activities in an infinite homogeneous head model can be stated as (Yao et al., 2004a),

$$4\pi\sigma\Phi = -\int_V \left(\frac{1}{r}\right)\nabla\cdot\vec{J}_S dv = \int_V \nabla\left(\frac{1}{r}\right)\cdot\vec{J}_S dv$$

$$\nabla\cdot\vec{J}_S(\vec{r}) = I_F(\vec{r}) \tag{12.40}$$

$$4\pi\sigma\Phi = -\int_V \left(\frac{1}{r}\right) I_F dv$$

Where σ is the conductivity, $\vec{J}_S$ is the primary current density (CD) distribution, andI_F is the divergence of the current density usually termed as current source density (CSD). The first equation shows that $\vec{J}_S$ behaves like a dipole, and the third equation shows that I_F behaves like a charge; thus equation (12.40) shows us that both charge and dipole can generate the same potential, and they are theoretically related to each other. In essence, both of them are equivalent source models approximating the complex actual physiological neural activities (Yao et al., 2004a,b), and between them, a dipole may be considered as an equivalent representation of a pair of closely neighboring positive and negative charges. As compared with the dipole, the number of unknown variables to be estimated when using the charge model is only one-third of that when using the dipole model, which can lower the computation complexity and avails for the stable solving of the inverse problem. Some further explanations of the difference between the charge model and the dipole model can be found in Chapters 2 and 3 (Yao, 1996; Yao et al., 2004a,b).

12.2.2.2 CLORETA

As shown in Section 12.1 and equation (12.16), the linear problem can be shaped into a source space weight problem as

$$\Phi = GX = GWW^{+}X \tag{12.41}$$

The weighted minimum-norm solution of the inverse problem is

$$\hat{X} = \left(W^T W\right)^{-1} A^T \left[A\left(W^T W\right)^{-1} A^T\right]^{+} Y \tag{12.42}$$

The widely adopted Laplacian weighted matrix W in EEG inverse problem has the form as equation (12.18).

$$W = LH, \quad \text{with } H = \text{diag}\left(\|g_1\|, \|g_2\|, \ldots, \|g_N\|\right) \tag{12.43}$$

where L denotes the discrete spatial Laplacian operator; $\|g_i\|$ is the ith column norm of the leadfield matrix G. The solution with such a weight strategy is the LORETA which means that the solution is extensive and blurring. The corresponding LORETA based on the charge source model was noted as CLORETA (Yao and He, 1998; He et al., 2002). Figure 12.1 shows an example of CLORETA with a comparison to

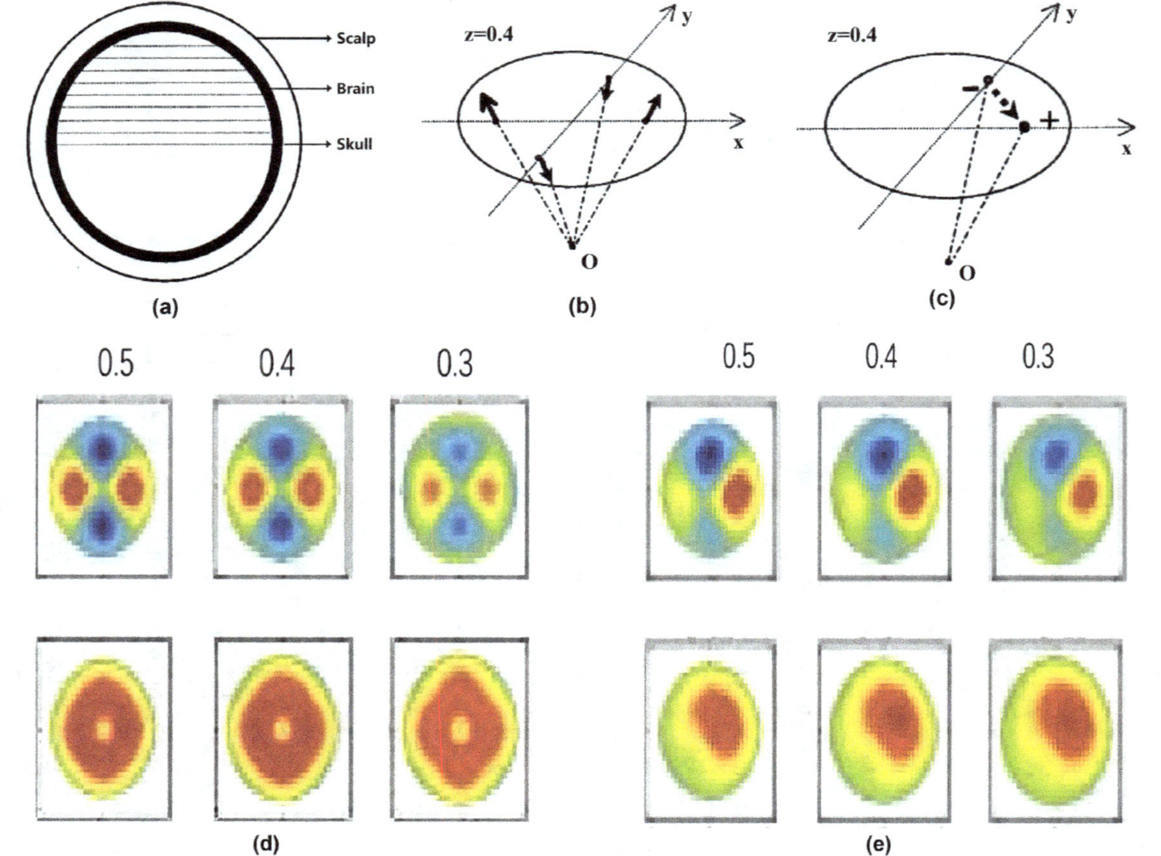

FIGURE 12.1 Illustration of the simulated source configurations and LORETA results. (a) The typical concentric three-sphere volume conductor model, the forward theory is given in Chapter 4 (Rush and Driscoll, 1969); (b) four radial dipoles; (c) one pair of point current source/sink; (d) CLORETA (up) and DLORETA (down) of the dipole source model (b); (e) CLORETA (up) and DLORETA (down) of the point source model (c). Where numbers 0.5, 0.4, and 0.3 are *z*-axis values of the section. (Modified from He et al. (2002).)

the original dipole LORETA (DLORETA). The results clearly show the overperformance of the CLORETA than DLORETA (the dipole strength) (Yao and He, 1998; He et al., 2002).

12.2.2.3 CFOCUSS

Different from the LORETA weight strategy, the weighted minimum norm solution is iteratively constructed with the previous solution involved in the update weight in FOCUSS (Gorodnitsky and Rao, 1997). The weight matrix W_k in the kth iteration is constructed by the prior iteration solution X_{k-1} as

$$W_k = \left(\text{diag}\left(X_{k-1}\right)\right) \tag{12.44}$$

With this weight, the inverse problem can be iteratively updated to reach a sparse and focal solution. In detail, let's have the linear problem as

$$\Phi = GX = GWW^{+}X = Dq, \quad \text{with} \quad X = Wq, \ D = GW \tag{12.45}$$

Where q is an auxiliary variable. Then, FOCUSS is implemented with the following three steps:

$$\begin{aligned} &1) \qquad W_k = \left(\text{diag}\left(X_{k-1}\right)\right) \\ &2) \qquad q_k = \left(GW_k\right)^{+} \Phi \\ &3) \qquad X_k = W_k q_k \end{aligned} \tag{12.46}$$

The initial source distribution X_0 is usually provided by (*C*/*D*) LORETA at the beginning of the iteration procedure. Then the FOCUSS procedure, the three steps in equation (12.46), is executed repeatedly; when the iteration number is above a predefined maximum iteration number or the solution difference between the neighboring iterations is smaller than the termination tolerance error, the iteration is to be terminated and a sparse and energy localized solution will be achieved.

Similar to that of CLORETA, when the leadfield matrix G is calculated with the charge source model, we call it charge FOCUSS (CFOCUSS), and the calculation burden of the inverse problem can be lowered than that of DFOCUSS.

12.2.2.4 CMOSS

In the kth iteration of CFOCUSS, the objective function can be stated as

$$\min_{X_k} \left\| W_k^{+} X_k \right\|^2 = \min_{X_k} \left\| q_k \right\|^2 = \min_{X_k} \sum_{i=1, w_{kii} \neq 0}^{N} \left(\frac{x_{ki}}{w_{kii}} \right)^2 \tag{12.47}$$

where w_{kii} is the ith diagonal element of W_k, and x_{ki} is the ith element of X_k. The above equation shows that a large weight can lower the contribution of the sources at the corresponding spatial position and therefore avails for the estimation of a larger x_{ki} in the sense of the minimum norm solution. As the weights of CFOCUSS are constructed directly with the sources estimated in the last iteration, the CFOCUSS iteration procedure will enhance the sources at those spatial source positions where sources are estimated with large powers in the previous iteration, and simultaneously, the sources on those positions with small strength will be iteratively degraded. Apparently, if the initial estimation is of spatial bias, it is not easy for FOCUSS to modify the bias effectively and may result in a biased source distribution (Liu and Gao, 2004); to relieve this problem, we have the following CMOSS.

For each iteration of the CFOCUSS procedure, it is quite possible to localize the sources to other positions near the actual source position; thus, we can construct a weight not just from the solution of the point itself but also from the information of the neighbor points to modify the possible bias during iterations. By considering the sources in neighbors, we defined a new diagonal weighting matrix W_N with its ith diagonal element w_{Nii} as

$$w_{Nii} = \begin{cases} \sqrt{\max\left(|x_j|, x_j \in \Omega_i\right) \times x_i}, & x_i > 0 \\ -\sqrt{\max\left(|x_j|, x_j \in \Omega_i\right) \times |x_i|}, & x_i \leq 0 \end{cases}, \quad 1 \leq i \leq N \qquad (12.48)$$

where Ω_i is defined as a 26-neighbor domain of ith point in the solution space. With the new weighting matrix W_N, the weighted form of the inverse problem is the same as that in equation (12.41). In the matrix W_N, if the current source at the neighbor center is the strongest one in the neighbor domain, the weight is the current source strength similar to that of the CFOCUSS weight; however, when the strongest source is not the one located at the neighbor center, the weight is enlarged compared with that in CFOCUSS. The main difference between these two weight strategies is in the case when the strongest source is not located at the neighbor center. For this case, the CFOCUSS strategy will impose a weaker weight on this spatial position, and the source at this position will be degraded in the consecutive CFOCUSS iterations; thus, if this position is actually the source position, CFOCUSS may localize it on a near position with certain bias or even lose it. Based on equations (12.47) and (12.48)_defined new construction strategy, even the current solution x at the true source position is weaker than the neighbors, the new weight gave to it by equation (12.8) still is the largest one in the pre-defined neighboring region, and such an enhanced weight can provide the chance to correct the source bias induced in the former iterations.

During the first several iterations, the estimated sources are distributed on many solution space positions, and accordingly, there may have been several sources in a neighbor domain; thus, the correction for source bias is obvious. With the iteration on, the source distribution becomes sparse with many null entries existing in the solution space; therefore, the function of the weights constructed in these iterations is very similar to those in CFOCUSS. Besides, the regularization techniques introduced in above can be adopted to deal with the effect of noise contamination in signal (Gorodnitsky and Rao, 1997; Tikhonov and Arsenin, 1977). In summary, the CMOSS (Xu et al., 2008) can be realized with the following iteration procedure:

1. *Preparation*: Generate the 26-neighbor information of solution space based on the segmented discrete solution space and calculate the charge leadfield matrix G.
2. *Initialization*: Set $k=1$, iteration termination error ε, and the maximum iteration number $T_{\max}$; initialize source distribution $X_0 = X_{k-1}$ with CLORETA solution.
3. Update the diagonal weight matrix W_{Nk} according to the formula (12.48).
4. Estimate the value of the auxiliary variable q_k: $q_k = \left(AW_{Nk}\right)^{+} \Phi$.
5. Update source distribution: $X_k = W_k q_k$.
6. Judge termination condition. Comparing the difference between the last and the last source distribution, if $\|X_k - X_{k-1}\| \le \varepsilon$ or $k \ge T_{\max}$, terminate the iteration, and X_k is the final source distribution else $k = k+1$, and jump to step 3) and go on.

12.2.2.5 Numeric Comparison

A three-shell realistic head model is used for EEG source localization, whose conductivities for cortex, skull, and scalp are $1.0\,\Omega^{-1}\mathrm{m}^{-1}$, $1/80\,\Omega^{-1}\mathrm{m}^{-1}$, and $1.0\,\Omega^{-1}\mathrm{m}^{-1}$, respectively (Rush and Driscoll, 1969). The solution space is restricted to cortical gray matter, hippocampus, and other possible source activity areas, consisting of 910 cubic mesh voxels with 10 mm inter-distance. A standard 128-electrode system was registered on this head model. The leadfield matrix A is calculated with the charge model by the boundary element method (BEM) (Fuchs and Drenckhahn, 1998) for the 128 electrode system and the leadfield matrix with dimension of 128×910. The origin of the coordinate system is defined as the midpoint between the left and right pre-auriculars, and the directed line from the origin through the nasion defines the +*X*-axis, and the +*Y*-axis is the directed line from the origin through the left pre-auricular. Finally, the +*Z*-axis is the line from the origin toward the top of the head (through electrode Cz). The maximum iteration number $T_{\max}$ for both FOCUSS and CMOSS is 50, and the tolerance error ε is *1.0E-6* for both of them.

In this simulation (Xu et al., 2008), three charge sources with strengths of 6.0, 4.0, and 3.9 were placed at three isolated positions (7.0, −21.0, 71.7) (mm), (−13.0, −11.0, 51.7) (mm), and (67.0, 19.0, 11.7) (mm), among which the first and the third were two superficial sources and the second one was a deep source. CLORETA, CFOCUSS, and CMOSS were used to localize the three sources, and the results are shown in Figure 12.2.

Figure 12.2 shows that, with different weight assumptions, the performances of the corresponding CLORETA, CFOCUSS, and CMOSS are different. It can be expected that if the Gaussian weight (Lei et al., 2009), the self-coherence enhancement algorithm (SCEA) (Yao and He, 2001), etc. are adopted, the solutions will be different too.

12.3 SPARSE SOURCE IMAGING

As mentioned above, the solution to the EEG inverse problem is a core issue in EEG studies. Considering its serious illness, to obtain a solution in line with the actual physiological characteristics, researchers often have to apply constraints or

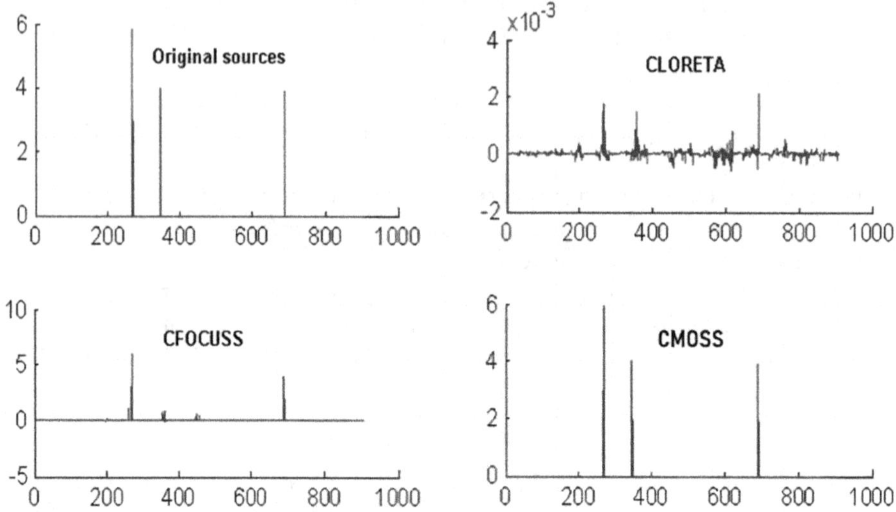

FIGURE 12.2 Localization for configuration with three sources (Xu et al., 2008).

assumptions to the inverse problem. These include limited source locations, such as a few dipoles in non-white matter regions, or cortical equivalent sources located only on the cortical surface as described previously (chapters 6 and 7). The other assumptions may be that the source is smooth and distributed, as adopted by CLORETA, or based on the energy saving principle of the brain in working, the sources should be sparsely distributed, as adopted by FOCUSS and CMOSS. This section introduces other sparse solutions except for FOCUSS and CMOSS. Therefore, we first introduce the multiple penalized least squares (MPLS) model (Valdes-Sosa et al., 2006) and some of its special cases (Vega-Hernandez et al., 2008).

12.3.1 Multiple Penalized Least Squares Model

The preceding equation (12.23) can be called a regularization approach, and it may also be interpreted as a penalized least squares regression, in which the penalty term (prior) is a quadratic function of coefficients. This is known in the linear regression domain as ridge regression with different operator W. Such quadratic function estimate usually offers smooth solution. Recent advances in penalized least squares regression have brought attention to the use of non-convex penalty functions, which can produce sparser or more concentrated solutions (Fan and Li, 2001). Here, we take the MPLS model to introduce a general formulation of the inverse problem of the EEG (Valdes-Sosa et al., 2006; Vega-Hernandez et al., 2008).

A general penalized linear regression can be written in the following form with a penalty term represented generically by $P(X)$:

$$\hat{X} = \arg\min\left\{\|\Phi - GX\|_2^2 + \lambda P(X)\right\} \tag{12.49}$$

When different priors/penalizations are introduced, equation (12.49) may be generalized to

$$\hat{X} = \arg\min\left\{\|\Phi - GX\|_2^2 + \sum \lambda_i P_i(X)\right\} \tag{12.50}$$

Equation (12.50) describes the MPLS model (Valdes-Sosa et al., 2006). Apparently, many of the above approaches can be shown to be special cases of MPLS, such as MN ($\hat{X} = \arg\min\left\{\|\Phi - GX\|_2^2 + \|IX\|_2^2\right\}$, with I as the identity matrix) and LORETA ($\hat{X} = \arg\min\left\{\|\Phi - GX\|_2^2 + \|LX\|_2^2\right\}$, with L as a discrete version of the 3D Laplacian operator). However, the use of non-convex penalty functions for $P(X)$ may derive several new types of inverse solutions, such as the series of the least absolute shrinkage selection operator (LASSO) (Tibshirani, 1996). The original LASSO penalty is the l_1-norm of the unknowns; the estimator takes the form $\hat{X} = \arg\min\left\{\|\Phi - GX\|_2^2 + \lambda\|X\|_1\right\}$. This penalty forces the solution to be sparse to provide concentrated sources. The LASSO fusion takes the form $\hat{X} = \arg\min\left\{\|\Phi - GX\|_2^2 + \lambda\|DX\|_1\right\}$, where D is a 3D gradient operator (Land and Friedman, 1996). If the LASSO is fused with $\hat{X} = \arg\min\left\{\|\Phi - GX\|_2^2 + \lambda_1\|X\|_1 + \lambda_2\|LX\|_1\right\}$, where L may be the Laplacian operator in LORETA, then it offers blurred solutions with intermediate properties between LORETA and MN. The LASSO fusion also may take the Elastic Net with $\hat{X} = \arg\min\left\{\|\Phi - GX\|_2^2 + \lambda_1\|W_1X\|_1 + \lambda_2\|W_2X\|_2^2\right\}$, which is based on the combined use of a quadratic penalty and an l_1-norm term. Apparently, with different $W_{1,2}$, a little different solutions may be obtained (Vega-Hernandez et al., 2008; Zou and Hastie, 2005).

The MPLS model can be identified with the use of modified Newton-Raphson (NR) algorithm, such as minorization-maximization (MM) algorithm (Valdes-Sosa et al., 2006). An important issue is the determination of the regularization parameter. In the case of an inverse solution with multiple penalty terms, one must decide a few optimal regularization parameters simultaneously. To avoid an exhaustive search in high-dimensional space, one may rewrite the parameters as $\lambda_1 = \lambda f_1$ and $\lambda_2 = \lambda f_2$ for 2D, and then, select only a few pairs of proportions f_1 and f_2 such that $f_1 + f_2 = 1$ and estimate an optimal value of λ by generalizing cross-validation function. In general, it is still a challenge to optimize the MPLS, and that's why one term model is still the most popular model. For more related details, please refer to Vega-Hernandez et al. (2008) and for the regularization parameters, refer to Duan et al. (2016).

12.3.2 l_p Norm Iterative Sparse Solution

l_p norm is a commonly used index to measure signal sparsity in sparse decomposition. Compared with the minimum norm solution, the l_p norm solution is more sparse and energy concentrated. If the l_p ($p \le 1$) norm is taken as the constraint condition, the solution of the inverse problem is transformed into the following optimization problem:

$$\arg\min\|\Phi - LX\|_2 + \lambda\|X\|_p \tag{12.51}$$

Where λ is a regularization parameter, which can be determined by some regularization parameter determination methods, such as the L-Curve, etc. The most direct and effective sparsity measure of the signal is l_0 norm, namely on the number of non-zero component of statistical signal directly, but this l_0 case of equation (12.51) is discontinuous and needs special treatment (see Section 12.3), and a lot of effective optimization algorithm based on objective function gradient cannot be used to solve the l_0 problem, so in practice, l_1 is commonly used in inverse problem, and this is the standard LASSO algorithm described above. Here, we introduce a combination of the traditional FOCUSS algorithm and the l_p norm sparse solution: l_p norm iterative sparse solution (LPISS) (Xu et al., 2007).

As can be seen from the basic iteration process of CFOCUSS algorithm, equation (12.46) introduced in Section 12.2, the second step will estimate q_k by taking the inverse of GW_k. However, GW_k may be seriously ill-conditioned that it can only be obtained by some special regularization methods. If the ill-conditioned matrix is not regularized, the FOCUSS method is very susceptible to the influence of source position and noise. Therefore, an important aspect of improving FOCUSS algorithm is to improve the performance of leadfield matrix inverse. LPISS uses sparse l_p $(p<1)$ norm constraint. The sparse estimation of the intermediate auxiliary variable q_k is directly solved by the following equation:

$$\arg\min \|\Phi - GW_k q_k\|_2 + \lambda \|q_k\|_p \tag{12.52}$$

Where λ is the regularization parameter and W_k is the weighted matrix updated according to the way in FOCUSS algorithm. In this way, through sparse estimation of q_k, in the process of obtaining the sparse solution of X_k, the noise can be suppressed by the objective function of l_p $(p<1)$ norm constraint. The choice of λ is related to the level of noise; generally speaking, the stronger the noise, the greater the λ; the best λ value can be determined by some regularization selection algorithm. When the noise standard deviation is σ and the dimension of the source solution space is N, $\lambda = \sigma\sqrt{2\log N}$ can be selected to effectively remove the interference of noise. In particular, when $\lambda = 0$, LPISS is the basic FOCUSS algorithm; when weighted matrix $W_k = I$, LPISS is reduced to a basic l_p $(p<1)$ norm constraint problem. For the case of $p = 1$, the simulation effect is shown in Figure 12.3, and more details are referred to Xu et al. (2007).

12.3.3 Solution Space Sparse Coding Optimization Method

12.3.3.1 Compressing the Solution Space

For EEG inverse problem, $\Phi = GX$, where Φ is the scalp EEG recordings of $M \times 1$, M is the number of scalp electrodes. G is the leadfield matrix with the size of $M \times N$, where N is the dimension size of the solution space Θ. X is the source vector to be estimated. Usually, $N \gg M$; thus, there are infinite number of solutions satisfying the underdetermined system. For practical EEG problems, a sparse solution is reasonable and desired. Based on the sparsity of the solution, the solution space Θ can be adequately compressed by removing the redundant sub-solution space, which has no

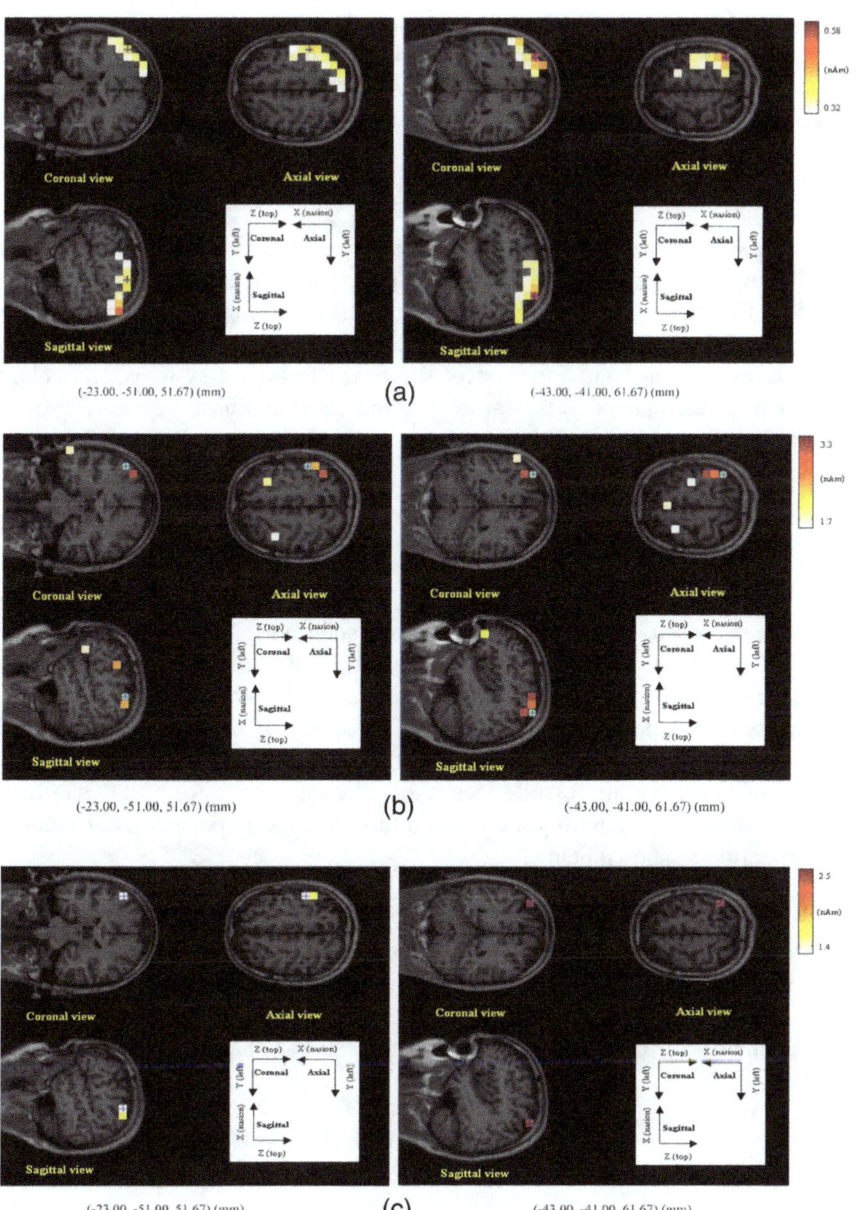

FIGURE 12.3 Localization results of LORETA, FOCUSS, and LPISS on MRI slices for two isolated sources under the noise of NSR = 0.15. (a) LORETA-localized sources; (b) FOCUSS-localized sources; (c) LPISS-localized sources. The colorful rectangle area is the estimated source location; the blue cross line within the colorful rectangle area indicates the overlapping area of the simulated source and the estimated source; the blue cross line within the green circle indicates those simulated source locations that are not overlapped with the positions of the estimated sources. (Modified from Xu et al. (2007).)

contribution to the observed signal. Let Θ_s represent the compressed solution space with dimension of K, and X_s represents the solution in space Θ_s, and Θ_s is supposed to include the whole useful spatial information of the neural electrical sources. The compressed form of formula $\Phi = GX$ in Θ_s can be denoted as

$$\Phi = G_s X_s \tag{12.53}$$

Based on the sparse assumption of the solution for an EEG inverse problem, it is reasonable to have a compressed space Θ_s with its dimension K smaller than M, and the ideal case is that K, the number of columns in G_s, is equal to the number of sources. When K is smaller than M, the original underdetermined system is transformed to an overdetermined one, and for most of the overdetermined systems, the column rank of the leadfield matrix G_s is near to full, and the solutions are more certain and stable (Michel et al., 2004), so in the compressed solution space Θ_s, the sources can be easily estimated as

$$X_s = (G_s)^+ \Phi \tag{12.54}$$

where $(G_s)^+$ denotes the inverse operator of G_s. Finally, the estimated X_s in the compressed space Θ_s is projected back to the original solution space Θ to get the possible source positions in the brain for physiological explanation. For a real EEG inverse problem, the prior information about the actual positions of the activated neural electrical sources is very poor, and if the compressed space Θ_s, a sub-space of the intact solution space Θ, is not chosen suitably, the sources may not be correctly estimated. The choice of a proper compressed space is the vital step of a successful EEG inverse approach, and that is the main point tackled in this work by particle swarm coding and optimization as shown below.

12.3.3.2 Coding and Compressing the Solution Space with Particles

Particle swarm optimization (PSO) has been proven to be powerful in searching for global optima (Wachowiak et al., 2004). In this work, a particle swarm is adopted to search for an optimal compressed space Θ_s.

For the underdetermined system $\Phi = GX$, the solution space is encoded by a particle as,

$$\begin{gathered} G = \left(G(1), G(2), \ldots, G(N)\right) \\ \Downarrow \\ E_i = \left(e_i(1), e_i(2), \ldots, e_i(N)\right) \end{gathered} \tag{12.55}$$

Where $G(n)$ ($n = 1, \ldots, N$) is the nth column of the lead field G. E_i with the same dimension as that of the original solution space $(E_i \in \Theta)$ is a particle code-chain in a particle swarm. The element $e_i(n) \in [-1, 1]$ $(1 \le n \le N)$ represents the activation probability of the source at the nth position in the original solution space Θ.

For a given source activation threshold ε, if $\|e_i(n)\| \geq \varepsilon$, the nth position is assumed to have the source. By scanning each element in the code-chain E_i from the left to the right, all the positions with $\|e_i(n)\| \geq \varepsilon$ are selected to form the compressed space Θ_s, and the corresponding columns in leadfield matrix $G \in \Theta$ are taken out to form the sub-leadfield matrix $G_s \in \Theta_s$. For convenience to describe the algorithm, the procedure of judging the possible source positions and forming the sub-space is represented by an operator $\text{Com}(.,.,.)$ as

$$G_s = \text{Com}\left(G, E_i, \varepsilon\right) \tag{12.56}$$

Based on the compressed space Θ_s and the corresponding leadfield matrix G_s, a solution X_s is readily obtained by the above equation (12.54). Apparently, the dimension K of the sub-space Θ_s is not a constant, but a dynamic variable along with the evolution of the particle state.

To get a well-compressed sub-space, Q particles, a particle swarm, are adopted to encode the solution space. Let $\{E_1, E_2, \ldots, E_Q\}$ denote the particle swarm with each $E_i\,(1 \leq i \leq Q)$ representing a coding of the solution space. And for each coding chain E_i and its corresponding solution X_s, the l_0 norm-constrained fitness function, $f(X_s) = \|\Phi - G_s X_s\|_2 + \lambda \|X_s\|_0$, is adopted to evaluate the quality of the compressed sub-space Θ_s.

12.3.3.3 Adaptive Threshold Function to Reduce the Noise Effect in Θ_s

As the leadfield matrix G_s is almost in a normal state with nearly full rank in the column, the inverse calculation of the leadfield matrix is relatively stable and the effect of noise is suppressed with only some weak fake sources remaining. In this work, to tackle those relatively weak fake sources induced by the noise in the estimated source X_s in the compressed space (Tikhonov and Arsenin, 1977; Yao and Dewald, 2005), we introduced an adaptive threshold, stimulated from the hard threshold strategy adopted in wavelet denoising (Donoho and Johnstone, 1994). We assume that the threshold is adaptively varied with the current maximum source power. Let J be the maximum absolute strength in X_s; the adaptive noise effect reduction function is defined as

$$R\left(x(k)\right) = \begin{cases} x(k), & |x(k)| > \gamma J \\ 0, & |x(k)| \leq \gamma J \end{cases} \tag{12.57}$$

Where $x(k)$ is the kth element in X_s, γ is the noise reduction factor. In this work, $\gamma = 0.1$.

12.3.3.4 Procedure of 3SCO Algorithm

Suppose the dimension of the original solution space Θ is N and the swarm is with Q particles; set the source activation threshold ε, the maximum generation number $H_{\max}$, the sparsity punishment factor λ, and the noise reduction factor γ. The detailed values of these variables are listed in the paper (Xu et al., 2010). The procedure of the solution space sparse coding optimization (3SCO) for EEG inverse problem can be summarized as follows:

Step 1: Initialize the variables: Initialize the particle swarm $\{E_1, E_2, \ldots, E_Q\}$ by initializing each particle E_i with N random values within [0, 1]; initialize the velocity of the Q particles, $\{v_1, v_2, \ldots, v_Q\}$, by initializing each particle v_i with N random values within [0, 1] too.

Step 2: Compress the solution space represented by each particle with formula (12.56), for the ith particle E_i, $G_S = \mathrm{Com}(G, E_i, \varepsilon)$, $1 \le i \le Q$. Estimate the sources in the sub-space as $X_s = (G_s)^+ \Phi$, and find the maximum power of sources in X_s, and then eliminate the noise effect by formula (12.57). Finally, X_s is mapped to the original source space, and the fitness value $f(i)$ corresponding to the ith particle is evaluated.

Step 3: Update the best position of each particle: Compare the fitness value $f(i)$ of the ith particle (code chain E_i) with the ever-achieved best fitness value by this particle code chain P_i; if $f(i)$ is better, P_i will be replaced by the current code chain E_i, else P_i will remain Where $1 \le i \le Q$.

Step 4: Update P_g, the best position ever achieved by all the particles: Compare the renewed best fitness value of the ith particle at position P_i with the global optimal fitness value at P_g; if the fitness value at P_i is better, P_g will be replaced by P_i, else P_g will be remained.

Step 5: Update the velocity and position of each particle as

$$\begin{cases} v_i(n) = wv_i(n) + c_1 r_1 \left(p_i(n) - e_i(n)\right) + c_2 r_2 \left(p_g(n) - e_i(n)\right) \\ e_i(n) = e_i(n) + v_i(n) \end{cases} \tag{12.58}$$

Where $1 \le n \le N$, $1 \le i \le Q$. $e_i(n)$ and $v_i(n)$ are the nth coding element and the nth velocity element of the ith particle, respectively; w, c_1, and c_2 are the same as those in the standard PSO. In our method, the particle velocity and the element in a code chain are constrained within the range [−1, 1] as

$$\begin{aligned} &\text{If} \quad v_i(n) > 1, \ v_i(n) = 1; \\ &\text{If} \quad v_i(n) < -1, \ v_i(n) = -1; \\ &\text{If} \quad e_i(n) > 1, \ e_i(n) = 1; \\ &\text{If} \quad e_i(n) < -1, \ e_i(n) = -1. \end{aligned} \tag{12.59}$$

Step 6: Judge the stopping criteria: If the number of generations is larger than the predefined number $H_{\max}$, the iterations will be stopped, and the solution corresponding to the global optimal position (code chain) P_g is the final solution for the underdetermined system, else return to Step 2 and go on. Figure 12.4 shows an example; for more details, please refer to Xu et al. (2010).

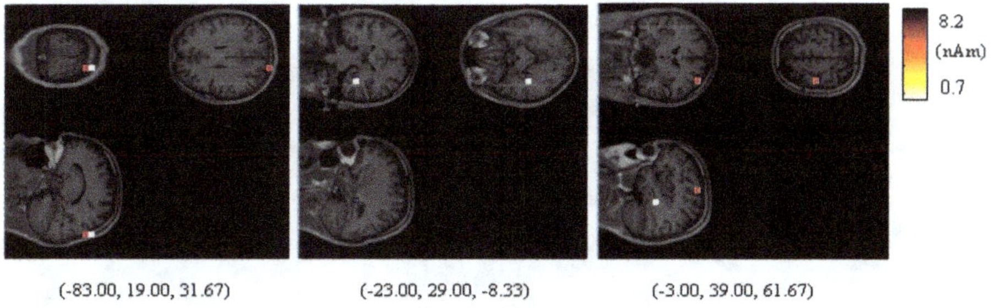

FIGURE 12.4 Localization results for three sources on the realistic head model by 3SCO. The blue cross line is the simulated source location; rectangles with different colors are the estimated source locations. (Modified from Xu et al. (2010).)

12.3.4 Least Absolute l_p ($0<p<1$) Penalized Solution

Theoretically, due to the square property of the l_2 norm, the commonly used l_2 norm loss function tends to exaggerate the outlier effect. However, the current sparse source localization methods mostly use the l_2 norm loss function; impose sparse constraints only on the source function, and ignore the influence of outliers on the EEG localization. The l_1-loss function is less sensitive to outliers than the quadratic function (Bore et al., 2019). Here, the l_1-loss function is used to measure the residual error, and the l_p ($p=0.5$) norm is used to constrain the EEG source signal. In the localization of EEG signal sources containing Gaussian noise and outliers, the proposed method shows its superiority in sparsity and robustness. The objective function is defined as follows:

$$X\_\text{LAPPS} = \min_X \left\{ \frac{1}{\eta} \|LX - \Phi\|_1 + \|X\|_p^p \right\} \tag{12.60}$$

Where the fitting error is measured in the l_1 norm space, the l_p ($0<p<1$) norm regularization is imposed onto the sources, and η is the regularization parameter. Theoretically, the first term measures the error in l_1 measurement space and thus can alleviate the effect of outliers, and the second regularization term can guarantee the attainment of the sparse EEG sources. In this work, the formulation above is solved by a modified alternating direction multiplier method (ADMM) framework (Boyd et al., 2011) with the non-convex sparsity-inducing penalty function. ADMM holds the advantage that it can split some complex objective functions into simpler sub-problems whose solutions can be easily obtained. This is a least absolute L_p ($0<p<1$) penalized solution (LAPPS) solution, and the details of the algorithm approach can be found in Bore et al. (2019).

Figure 12.5 shows a simulation result. To verify the five methods' performances on the four sources configuration shown in (a), the simulation is repeated to generate 200 simulation data for different *SNR* from -10 to 10 dB; the mean performances are calculated for each of the five approaches: the weighted minimum norm estimate

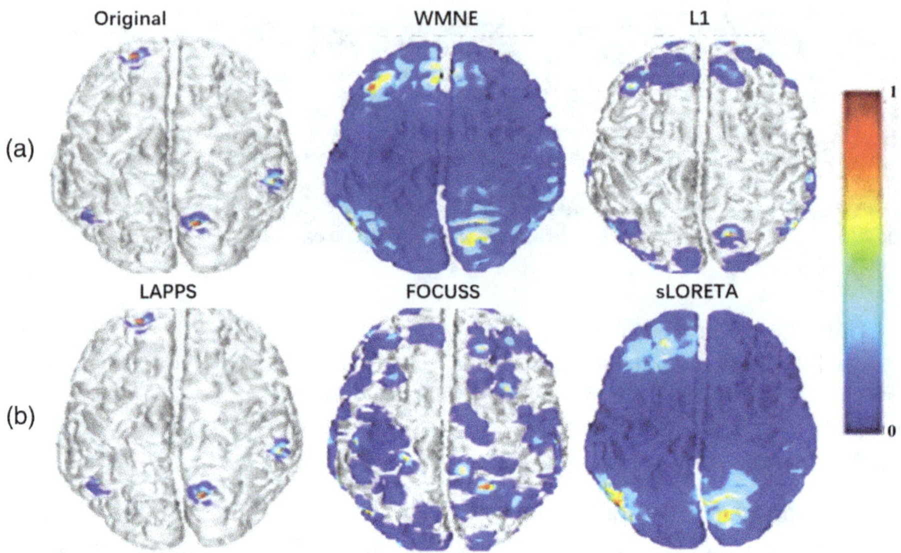

FIGURE 12.5 The inverse solutions for four sources configuration under −5 dB SNR. (a) The assumed and reconstructed Sources in the indexed solution space; (b)The assumed and reconstructed sources on the realistic head model. (Modified from Bore et al. (2018).)

(WMNE) solution, the standardized low resolution electromagnetic tomography (sLORETA), the focal underdetermined system solver (FOCUSS), LAPPS, l_1 norm solution (i.e., the least absolute shrinkage selection operator (LASSO)). The results showed that LAPPS is the best one. LAPPS attained the highest spatial resolution having the least position and energy errors among the five tested methods (Bore et al., 2018).

12.4 SELF-COHERENCE ENHANCEMENT INVERSE SOLUTION

As shown in the previous chapters, the EEG problem may be boiled down to

$$\Phi = GX \tag{12.61}$$

Φ is a vector (one-dimension matrix) of the observed potential on the scalp surface, X is a vector of the intracerebral sources distribution, and G is the leadfield matrix. Since the dimension of X is generally much larger than the dimension of Φ, numerous solutions may satisfy this equation. This section introduces a coherent weighting method to tackle this problem (Yao et al., 2002; Yao and Zhou, 2006). The method is based on the assumption that the electrical activities in the brain is usually sparse for saving energy; thus, the activity sources should be sparsely distributed in the cortex. In this paper, Tikhonov regularization method is used to regularize the algorithms. The following is an introduction to the relevant Focuss algorithm.

12.4.1 Focal Underdetermined System Solver

FOCUSS algorithm has been introduced in Section 12.3, which is the same for both dipole and charge sources. Here is a simulation example.

The assumed sources are three dipoles, which are located at (0, 0, 0.8), (−0.4, 0.4, 0.69), and (−0.4, −0.4, 0.69) respectively. They are all radial dipoles, and the intensity is 1.0, 0.6, and 0.8, respectively, and the noise intensity is 10%. The results are shown in Figure 12.6. Unless otherwise noted, the source will be the same in the following simulations in this section.

After six iterations, each intensity maximum value is already a little larger than the real intensity. If the iteration goes further, the intensity will be even bigger, and the deviation of the solution will be even bigger too. Therefore, the solution of the sixth iteration is taken as the final solution. Meanwhile, comparing the final results with the actual parameters, they are not the same, and the difference also depends on the initial solution and the noise feature which means that FOCUSS is sensitive to them. Of course, this fact also reflects that other prior information must be added to the FOCUSS algorithm in practice.

12.4.2 Self-Coherent Enhancement Algorithm (SCER)

The self-CohERence (SCER) enhancement algorithm (Yao and He, 2001) introduced here is to reduce the deviation of FOCUSS (Yao et al., 2002). In FOCUSS, the iterative process may not only correct the source position deviation in the previous solutions, but may also lose the initial correct source position along with the iteration. The SCER tries to hold the initial positions of sources and finds out the strengths of the sources in a non-iterative way.

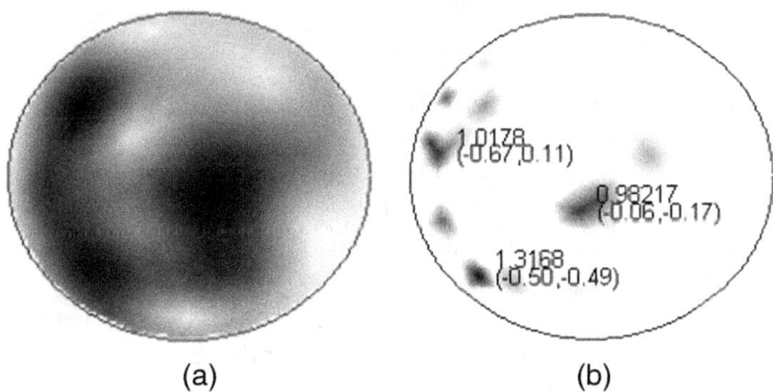

FIGURE 12.6 FOCUSS performance. Initial solution in FOCUSS: minimum norm solution (a); Final solution after six iterations (b), the pair of numbers inside the parentheses in *b* indicates the (x, y) coordinates of the point with local maximum strength, and the number above the parentheses indicates its strength.

The self-coherent weighted solution is based on an initial solution, which leads directly to a non-iterative solution by the following equation:

$$X = \alpha\left\{\boldsymbol{x}_1^T f\left(\|\boldsymbol{x}_1\|\right), \boldsymbol{x}_2^T f\left(\|\boldsymbol{x}_2\|\right), \boldsymbol{x}_3^T f\left(\|\boldsymbol{x}_3\|\right), \ldots, \boldsymbol{x}_M^T f\left(\|\boldsymbol{x}_M\|\right)\right\}^T = \alpha X'$$
$$X' = \left\{\boldsymbol{x}_1^T f\left(\|\boldsymbol{x}_1\|\right), \boldsymbol{x}_2^T f\left(\|\boldsymbol{x}_2\|\right), \boldsymbol{x}_3^T f\left(\|\boldsymbol{x}_3\|\right), \ldots, \boldsymbol{x}_M^T f\left(\|\boldsymbol{x}_M\|\right)\right\}^T \tag{12.62}$$

Here, M is the dimension of solution space, α is an undetermined constant, $X_0 = \left(\boldsymbol{x}_1^T, \boldsymbol{x}_2^T, \ldots, \boldsymbol{x}_M^T\right)^T$ is the initial solution, and everything else is determined by the initial solution and the function f, $\boldsymbol{x}_{i,}$ including the three components in the Cartesian coordinate system The initial solution may be from sLORETA or eLORETA which may give a good approximation of the spatial positions of the sources. Suppose that the initial solution is unique too (e.g., the minimum norm solution or the weighted minimum norm solution), and the function f is a deterministic function, then, Then, if α is uniquely determined, the above solution is a uniquely determinable solution of this underdetermined system.

As is known to all, the correlation between two signals in signal processing is an important means to measure the similarity degree of these two signals, and signal detection is one of its applications. Its main application is to detect a severely weakened signal by assigning a threshold value to the correlation coefficient between the severely weakened signal and the original signal. For example, in radar detection, the correlation coefficient between the signal returned from the target and the signal emitted is used to determine the presence of the target. If the two signals are the same signal, the correlation is autocorrelation. Here, we attempt to obtain real information on brain activity from the weak initial solution of the equivalent distribution source. Assuming that the real electrical activity is related to the local maximum of the initial solution, the weak brain activity can be detected by amplifying its intensity by introducing the concept of autocorrelation. Thus, the function f in equation (12.62) can be defined as

$$f\left(\|\boldsymbol{x}_i\|\right) = \|\boldsymbol{x}_i\|^K / \max\left\{\|\boldsymbol{x}_1\|^K, \ldots, \|\boldsymbol{x}_M\|^K\right\}, \quad i = 1, \ldots, M; \ K = 0, 1, 2, \ldots \tag{12.63}$$

Clearly, the function f is a function of the solution space whose value is positively related to the intensity of electrical activity represented by the value of the initial solution at each point. It can often be thought of as a higher-order autocorrelation function of the initial solution.

It is easy to see that the larger the K is, the greater the difference between the local maximum and the background value, and the higher the resolution. Thus, parameter K can be called the control parameter of resolution. When $K=0$ and $f(*)=1$, equation (12.62) is just the initial solution.

Substitute equations (12.62) and (12.63) into equation (12.61), and shape it into a least square problem as

$$h(\alpha, K) = \|\Phi - \mathbf{G}X\|^2 = \|\Phi - \alpha\mathbf{G}X'\|^2 \to \min \tag{12.64}$$

Where the parameter K can be determined by the normalized blur index (NBI) defined in Chapter 11. Set $\frac{\partial h(\alpha,K)}{\partial \alpha} = 0$, we get

$$\alpha = \frac{\Phi^T \Phi'}{\Phi'^T \Phi'}, \quad \Phi' = \mathbf{G}X'$$

Suppose $\Phi = \alpha\Phi' + N$ and $N^T\Phi' = 0$ (the potential Φ' generated by the new solution on the scalp is independent of the noise N). Then,

$$\alpha = \frac{\Phi^T (\Phi - N)/\alpha}{\Phi'^T \Phi'} = \frac{\Phi^T \Phi - (\alpha\Phi' + N)^T N}{\alpha\Phi'^T \Phi'} \approx \frac{\Phi^T \Phi - N^T N}{\alpha\Phi'^T \Phi'}$$

Thus,

$$\alpha \approx \sqrt{\frac{\Phi^T \Phi - N^T N}{\Phi'^T \Phi'}} = \sqrt{\frac{\Phi^T \Phi - E_0}{\Phi'^T \Phi'}}, \quad \Phi' = \mathbf{G}X', \quad E_0 = N^T N \tag{12.65}$$

Here, E_0 is the noise energy, which is known in the simulation experiment. In practical applications, such as in the evoked potential experiment, the data before stimulation can be used as the background noise, or the record of one single stimulation can be used as the noise (Srebro, 1996), so the noise energy can be estimated.

In general, the higher the K, the higher the resolution. However, the actual sources of neural activity are to some extent smooth. Thus, the source distribution reconstructed by a method should be with the proper smoothness degree, rather than being too smooth or too concentrated. The NBI, equation (11.12), introduced in Chapter 11 can be used to measure the smooth level. For a given distribution of the dipoles in simulation, we know the true positions of the dipoles. For the inverted source distribution, select the local maximum value of the inversion result, which can be found by observation or algorithm. By comparing the NBI of the inverse solution with the NBI of the assumed sources under different values of K, a better value of K can be obtained. In the simulated experiment, the real NBI is calculated directly from the known source distribution. In practice, it must be determined by other physiological conditions.

In summary, the algorithm consists of two parts: one is to test the parameter K of a high-order autocorrelation function, to find a K which makes the final solution have a roughly correct *NBI* similar to that of the actual sources; the second is to calculate the parameter α, which makes the source strengths in the final solution roughly close to the actual sources. In this algorithm, the centers of the distributed source are assumed to be located at the local maximum of the smooth initial solution. It can be seen from equation (12.65) that the maximum of the initial solution will continue to remain the maximum in the new solution. Thus, the maximum position deviation will be the same as the initial solution.

The simulation results of the SCER algorithm are shown in Figure 12.7. It can be seen from these results that the relative strength of the local maxima in the solution obtained by SCER remains the same as that of the initial solution, while the NBI is

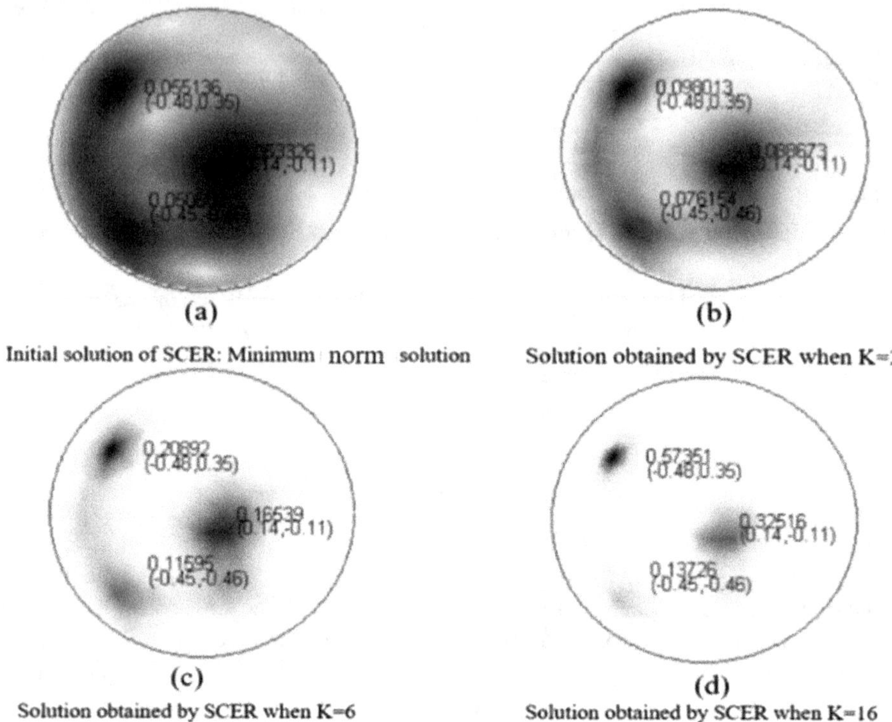

FIGURE 12.7 Performance of self-coherence enhancement algorithm. (a) Initial solution: Minimum norm solution; (b) solution obtained by SCER when $K=2$; (c) solution obtained by SCER when $K=6$; (d) solution obtained by SCER when $K=16$.

significantly reduced than that of the initial solution, and this is the value of SCER. For the details, please refer to Yao and He (2001).

12.4.3 Multi-Zone SCER

12.4.3.1 The Multi-Zone Algorithm

As we can see from Figure 12.7d, with the increase of K, sources with weaker strength may be lost. This suggests that it is difficult for SCER to reconstruct multiple sources with different intensities. It is the same K-order autocorrelation that makes the SCER result remove those points in the initial solution where the intensity is relatively small. To avoid this situation, the following scheme is proposed to segment the initial solution space to several partitions, and then implement the SCER algorithm in each partition, that is, multi-zone SCER (Yao and Zhou, 2006).

Suppose the initial solution is divided into l partitions: $(\boldsymbol{x}_{1,1},\dots,\boldsymbol{x}_{1,m_1}),(\boldsymbol{x}_{2,1},\dots,\boldsymbol{x}_{2,m_2})$ $,\dots,(\boldsymbol{x}_{l,1},\dots,\boldsymbol{x}_{l,m_l}),\ \sum_i^l m_i = M$. The partition i can be denoted as a column vector X_i with the same dimension as X, and assume that the points belonging to partition i the values in X_i takes the values in the initial solution, otherwise, the values are set to 0. Suppose l partitions are $X_1, X_2, \ldots, X_l$, then we have

$$X = \left\{X_1 f\left(\|X_1\|\right), X_2 f\left(\|X_2\|\right), X_3 f\left(\|X_3\|\right), \ldots, X_l f\left(\|X_l\|\right)\right\} \alpha = X'\alpha$$
$$X' = \left\{X_1 f\left(\|X_1\|\right), X_2 f\left(\|X_2\|\right), X_3 f\left(\|X_3\|\right), \ldots, X_l f\left(\|X_l\|\right)\right\} \quad (12.66)$$

Where α is a column vector of length l. Substitute equation (12.66) into equation (12.61), the α can be obtained by the least square method.

$$h(\alpha, K) = \|\Phi - \mathbf{G}X\|^2 = \|\Phi - \mathbf{G}X'\alpha\|^2 \rightarrow \min$$

Here, the parameter K can be determined by the NBI. And by taking $\frac{\partial h(\alpha, K)}{\partial \alpha} = 0$, we can get $\alpha = (\mathbf{G}X')^{+} \Phi$.

12.4.3.2 The Solution Space Partion

Here, Clustering (Michand, 1997) is used to divide/classify the initial solution space. As an example here, consider a 2D problem: the voxel attributes can be the x and y coordinates and intensity of the source (suppose the sources are on the cortical surface with a fixed radius; the z coordinate can be calculated from the x and y coordinates, so the z coordinate is not included). The classification may be based on partition criteria, including intraclass inertia criteria.

Intraclass inertia criterion was based on Euclidean distance; the objective function $F(P)$ was the mean square of Euclidean distance between the individual in each class and the class center (average of individual attributes); the classification is to minimize $F(P)$. The implementation of the in-class inertia criterion is relatively simple; it is adopted widely in practice, and it is used here.

12.4.3.3 The Simulation Results

The following is the result of the multi-zone SCER (the initial solution is the same as the previous SCER):

By comparing Figure 12.8b,c with Figure 12.7c,d, respectively, the multi-zone SCER almost solve the problem that the intensity deviation in the initial solution leads to an even more serious intensity deviation in the final solution of SCER, thus the multi-zone SCER may avoid the problem of SCER losing those points that should have values.

12.4.4 Multi-Zone Iterative SCER

Although the problem of intensity deviation in SCER is solved well in the multi-zone SCER, the problem of position deviation still exists. In fact, the SCER result will have the same position deviation as the original solution.

In FOCUSS, each iteration can adjust a little the position of the maximum value in the solution (corresponding to the active source). However, from the perspective of

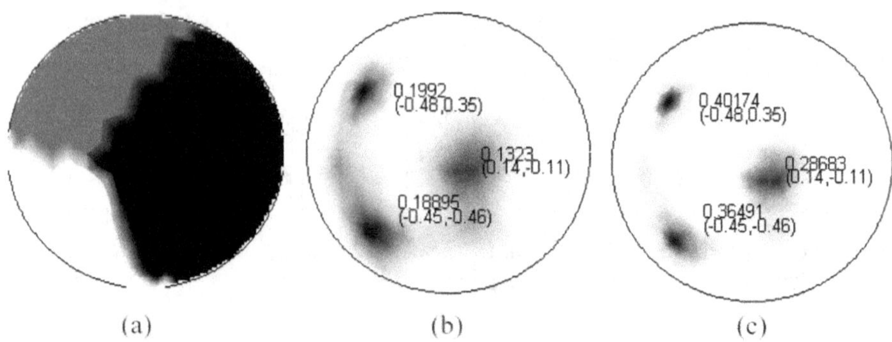

FIGURE 12.8 Performance of multi-zone SCER. (a) Partition results (three); (b) results of multi-zone OSCER when $K=6$; (c) results of multi-zone SCER when $K=16$.

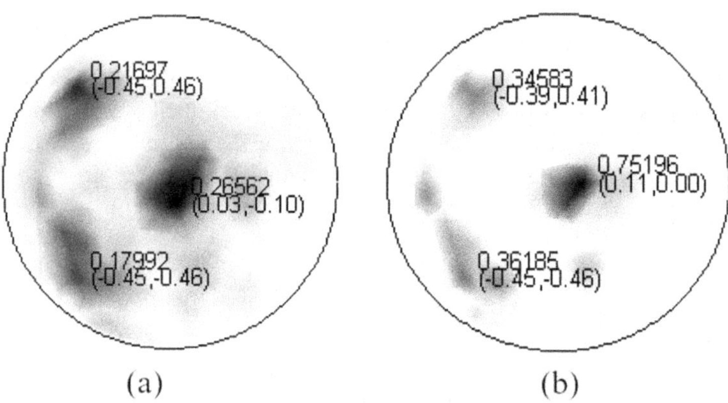

FIGURE 12.9 Performance of multi-zone iterative SCER. (a) The result of two iterations; (b) results of 12 iterations.

NBI, the decline of *NBI* in FOCUSS iteration is not controlled. This means that the convergence process of FOCUSS does not proceed smoothly, so it is possible to fall into the local extreme value and fail to reach the global optimal solution. Here, we propose to combine the multi-zone SCER with FOCUSS iteration to control the decline of *NBI* in each iteration, so that the convergence process can proceed smoothly. The weight matrix used by FOCUSS in the iteration directly takes the result of the previous step. If the multi-zone SCER is used to adjust the result X_{FOCUSS} of the previous FOCUSS step to the more reasonable X_{SCER}, and take X_{SCER} as the weight matrix for the next FOCUSS step; the final result of the fused FOCUSS-SCERwill be better (Yao and Zhou, 2006).

In this method, which combines FOCUSS and multi-zone SCER, the SCER plays a very important role: firstly, it makes the convergence process proceed smoothly; secondly, it makes the weight matrix more reasonable. Therefore, we called it the multi-zone iterative SCER. The simulation results are shown in Figure 12.9.

From the simulation results, it can be seen that the multi-zone iterative SCER does solve the problem of SCER position deviation to a certain extent. FOCUSS is a typical step-by-step iterative method based on an underdetermined minimum norm solution, while the multi-zone iterative SCER can be regarded as a mixed iteration method. Because it uses an overdetermined step to determine the weight matrix, we believe that it is this overdetermined step that ensures its better performance than FOCUSS does.

REFERENCES

Ian Goodfellow, Yoshua Bengio and Aaron Courville. 2016. Deep Learning. The MIT Press.

Baillet, S., L. Garnero. 1997. A Bayesian approach to introducing anatomo-functional priors in the EEG/MEG inverse problem. *IEEE Trans Biomed Eng* 44:374–385.

Bore, J.C., C. Yi, P. Li, et al. 2018. Sparse EEG source localization using LAPPS: Least absolute l−P (0<P<1) penalized solution. *IEEE Trans Biomed Eng*. doi: 10.1109/TBME.2018.2881092.

Bore, J.C., W.M.A. Ayedh, P. Li, et al. 2019. Sparse autoregressive modeling via the least absolute LP-norm penalized solution. *IEEE Acess* 7:2908189.

Boyd, S., N. Parikh, E. Chu, et al. 2011. Distributed optimization and statistical learning via the alternating direction method of multipliers. *Found Trends Mach Learn* 3(1):1–122.

Dong, L., C. Luo, X. Liu, et al. 2018. Neuroscience information toolbox: An open source toolbox for EEG-fMRI multimodal fusion analysis. *Front Neuroinf* 12:56.

Dong, L., Y. Zhang, R. Zhang, et al. 2015. Characterizing nonlinear relationships in functional imaging data using eigenspace maximal information canonical correlation analysis (emiCCA). *NeuroImage* 109:388–401. doi: 10.1016/j.NeuroImage2015.01.006.

Donoho, D.L., I.M. Johnstone. 1994. Ideal spatial adaptation by wavelet shrinkage. *Biometrika* 81(3):425–455.

Duan, J., C. Soussen, D. Brie, et al. 2016. Generalized LASSO with under-determined regularization matrices. *Signal Process* 127:239–246.

Fan, J., R. Li. 2001. Variable selection via nonconcave penalized likelihood and its oracle properties. *J Am Stat Assoc* 96:1348–1360.

Franzone, P.C., L. Guerri, B. Taccardi, et al. 1985. Finite element approximation of regularized solutions of the inverse potential problem of electrocardiography and applications to experimental data. *Calcolo* 22(1):91–186.

Fuchs, M., R. Drenckhahn, H. Wischmann, et al. 1998. An improved boundary element method for realistic volume-conductor medeling. *IEEE Trans Biomed Eng* 45:980–997.

Gencer, N.G., S.J. Williamson, 1998. Differential characterization of neural sources with the bimodal truncated SVD pseudo-inverse for EEG and MEG measurements. *IEEE Trans Biomed Eng* 45:827–837.

Golub, G.H., C.F. Van Loan. 1989. *Matrix Computations*, 2nd ed. Baltimore, MD: Johns Hopkins University Press.

Gorodnitsky, I.F., B.D. Rao. 1997. Sparse signal reconstruction from limited data using FOCUSS: A re-weighted minimum norm algorithm. *IEEE Trans Signal Proces* 45:600–616.

Gorodnitsky, I.F., J.S. George, B.D. Rao. 1995. Neuromagnetic source imaging with FOCUS: A recursive weighted minimum norm algorithm. *EEG Clin Neurosci* 95:231–251.

Grave de Peralta Menendez, R., O. Hauk, S. Gonzalez Andino, et al. 1998. Linear inverse solutions with optimal resolution kernels applied to electromagnetic tomography. *Hum Brain Mapp* 5:454–467.

Grave de Peralta Menendez, R., S.L. Gonzalez Andino, S. Morand, et al. 2000. Imaging the electrical activity of the brain: ELECTRA. *Hum Brain Mapp* 9:1–12

Hamalainen, M.S. 1984. Interpreting measured magnetic fields of the brain: Estimates of current distributions. Technical Report TKK-F-A559, Helsinki University of Technology.

Hamalainen, M.S., R.J. Ilmoniemi. 1994. Interpreting magnetic fields of the brain: Minimu norm estimates. *Med Biol Eng Comput* 32:35–42.

Hansen, P.C. 1992. Analysis of discrete ill-posed problems by means of the L-curve. *SIAM Rev* 34(4):561–580.

Hansen, P.C., D.P. O'Leary. 1993. The use of the L-curve in the regularization of discrete ill-posed problems. *SIAM J Sci Comput* 14(6):1487–1503.

He, B., D. Wu. 1997. A bioelectric inverse imaging technique based on surface Laplacians. *IEEE Trans Biomed Eng* 44:529–538.

He, B., D. Yao, D. Wu. 2000. *Imaging Brain Electrical Activity*. New York: Plemum Press.

He, B., D. Yao, J. Lian, et al. 2002. An equivalent current source model and Laplacian weighted minimum norm current estimates of brain electrical activity. *IEEE Trans Biomed Eng* 49(4):277–288.

Ioannides, A., J.P.R. Bolton, C.J.S. Clarke. 1990. Continuous probabilistic solutions to the biomagnetic inverse problem. *Inverse Probl* 6(4):523–542.

Jafarian, A., V. Litvak, H. Cagnan, et al. 2020. Comparing dynamic causal models of neurovascular coupling with fMRI and EEG/MEG. *NeuroImage* 216:116734.

Johnston, P.R., R.M. Gulrajani. 1997. A new method for regularization parameter determination in the inverse problem of electrocardiography. *IEEE Trans Biomed Eng* 44:19–39.

Kirsch, A. 1996. *An Introduction to the Mathematical Theory of Inverse Problems*. New York: Springer Press.

Land, S., J. Friedman. 1996. Variable fusion: A new method of adaptive signal regression. Technical Report, Stanford University.

Lawson, C.L., R.J. Hanson. 1974. *Solving Least Squares Problems*. Philadelphia, PA: Society for Industrial and Applied Mathematics.

Lei, X., P. Xu, A. Chen, et al. 2009. Gaussian source model based iterative algorithm for EEG source imaging. *Comput Biol Med* 39:978–988.

Lei, X., P. Xu, C. Luo, et al. 2011. fMRI functional networks for EEG source imaging. *Hum Brain Mapp* 32:1141–1160.

Lei, X., P.A. Valdes-Sosa, D. Yao. 2012. EEG/fMRI fusion based on independent component analysis: Integration of data-driven and model-driven methods. *J Integr Neurosci* 11(03):313–337.

Liu, H., X. Gao. 2004. A recursive algorithm for the three-dimensional imaging of brain electric activity: Shrinking LORETA-FOCUSS. *IEEE Trans Biomed Eng* 51(10):1794–1802.

Menke, W. 2012. *Geophysical Data Analysis: Discrete Inverse Theory*. London: Academic Press.

Michand, P. 1997. Clustering techniques. *Future Gener Comput Sys* 13:135–147.

Michel, C.M., M.M. Murray, G. Lantz, et al. 2004. EEG source imaging. *Clin Neurophysiol* 115:2195–2222.

Miller, K. 1970. Least squares methods for ill-posed problems with a prescribed bound. *SIAM J Math Anal* 1(1):52–74.

Morozov, V.A. 1984. *Methods for Solving Incorrectly Posed Problems*. Berlin: Springer-Verlag.

Oster, H.S., Y. Rudy. 1992. The use of temporal information in the regularization of the inverse problem of electrocardiography. *IEEE Trans Biomed Eng* 39(1):69–75.

Oster, H.S., Y. Rudy. 1997. Regional regularization of the electrocardiographic inverse problem: A model study using spherical geometry. *IEEE Trans Biomed Eng* 44:188–199.

Pascual-Marqui, R.D. 1995. Reply to comments by Hämäläinen, Ilmoniemi and Nunez. *ISBET Newsl* 6:16–28.

Pascual-Marqui, R.D., C.M. Michel, D. Lehmann. 1994. Low resolution electromagnetic tomography: A new method for localizing electrical activity in the brain. *Int J Psychophysiol* 18:49–65.

Qin, Y., N. Zhang, Y. Chen, et al. 2020. Rhythmic network modulation to thalamocortical couplings in epilepsy. *Int J Neural Syst* 30(11):2050014.

Rush, S., D.A. Driscoll. 1969. EEG electrode sensitivity: An application of reciprocity. *IEEE Trans Biomed Eng* 16:15–22.

Samuelsson, J.G., S. Khan, P. Sundaram, et al. 2019. Cortical Signal Suppression (CSS) for detection of subcortical activity using MEG and EEG. *Brain Topogr* 32:215–228.

Shim, Y.S., Z.H. Cho. 1981. SVD pseudo-inversion image reconstruction. *IEEE Trans Acoust Speech* 29:904–909.

Sidman, R.D, D.J. Vincent, D.B. Smith, et al. 1992. Experimental tests of the cortical imaging technique-applications to the response to median nerve stimulation and the localization of epileptiform discharges. *IEEE Trans Biomed Eng* 39:437–444.

Srebro, R. 1996. Iterative refinement of the minimum norm solution of the bioelectric inverse problem. *IEEE Trans Biomed Eng* 43(5):547–552.

Sullivan, B.J., B. Liu. 1984. On the use of singular value decomposition and decimation in discrete-time band-limited signal extrapolation. *IEEE Trans Acoust Speech* 32(6):1201–1212.

Tibshirani, R. 1996. Regression shrinkage and selection via the Lasso. *J R Stat Soc B* 58(1):267–288.

Tikhonov, A.N., V.Y. Arsenin. 1977. *Solutions of Ill-Posed Problems*. New York: John Wiley & Sons Press.

Twomey, S. 1963. On the numerical solution of Fredholm integral equations of the first kind by the inversion of the linear system produced by qudarature. *J ACM* 10(1):97–101.

Valdes-Sosa, P.A., J.M. Sanchez-Bornot, M. Vega-Hernandez, et al. 2006. *Granger Causality on Spatial Manifolds: Applications to Neuroimaging*. New York: John Wiley & Sons Press.

Van Veen, B.D., W. Van Drongelen, M. Yuchtman, et al. 1997. Localization of brain electrical activity via linearly constrained minimum variance spatial filtering. *IEEE Trans Biomed Eng* 44:867–880.

Vega-Hernandez, M., E. Martınez-Montes, J.M. Sanchez-Bornot, et al. 2008. Penalized least squares methods for solving the EEG inverse problem. *Stat Sinica* 18:1535–1551.

Vogel, C.R. 1996. Non-convergence of the L-curve regularization parameter section method. *Inverse Probl* 12(4):535–547.

Wachowiak, M.P., R. Smolíková, Y. Zheng, et al. 2004. An approach to multimodal biomedical image registration utilizing particle swarm optimization. *IEEE Trans Evolut Comput* 8:289–301.

Xu, P., Y. Tian, H. Chen, et al. 2007. Lp norm iterative sparse solution for EEG source localization. *IEEE Trans Biomed Eng* 54(3):400–409.

Xu, P., Y. Tian, X. Lei, et al. 2008. Equivalent charge source model based iterative maximum neighbor weight for sparse EEG source localization. *Ann Biomed Eng*. doi: 10.1007/s10439-008-9570-4.

Xu, P., Y. Tian, X. Lei, et al. 2010. Neuroelectric source imaging using 3SCO: A space coding algorithm based on particle swarm optimization and L0-norm constraint. *NeuroImage* 51:183–205.

Yao, D. 1996. The equivalent source technique and cortical imaging. *EEG Clin Neurosci* 98:478–483.

Yao, D. 2000. High-resolution EEG mappings: A spherical harmonic spectra theory and simulation results. *Clin Neurophysiol* 111:81–92.

Yao, D., B. He. 1998. The Laplacian weighted minimum estimate of three dimensional equivalent charge distribution in the brain. *Proceedings of 20th Annual International Conference on IEEE EMBS*, Hong Kong, China, vol. 4, pp. 2108–2111.

Yao, D., B. He. 2001. A self-coherence enhancement algorithm and its application to enhancing 3D source estimation from EEGs. *Ann Biomed Eng* 29(11):1019–1027.

Yao, D., L. Wang, K.D. Nielsen, et al. 2004a. Cortical mapping of EEG alpha power using a charge layer model. *Brain Topogr* 17(2):65–71.

Yao, D., S. Fu, N. Rao, et al. 2002. Computer simulation study of multi-signal classification algorithm for EEG inverse problem. *Chin J Biomed Eng* 21(1):53–58.

Yao, D., Y. Zhou. 2006. An EEG inverse algorithm based on solution space decomposition and iteration. *J Univ Electron Sci Technol China* 35(4): 714–716, 720.

Yao, D., Z. Yin, X. Tang, et al. 2004b. High-resolution electroencephalogram (EEG) mapping: Scalp charge layer. *Phys Med Biol* 49(22):5073–5086.

Yao, J., J.P.A. Dewald. 2005. Evaluation of different cortical source localization methods using simulated and experimental EEG data. *NeuroImage* 25(2):369–382.

Zhang, G., Y. Cui, Y. Zhang, et al. 2021. Computational exploration of dynamic mechanisms of steady state visual evoked potentials at the whole brain level. *NeuroImage* 237:118166.

Zou, H., T. Hastie. 2005. Regularization and variable selection via the elastic net. *J R Stat Soc B* 67(2):301–320.

13 Zero-Reference for Scalp EEG

The reference is a very common and fundamental problem that accompanies EEG and ECG technologies and incurs a lot of controversies in history. After years of research, the "central terminal" now is widely recognized in the field of ECG. While the EEG reference still is in a chaotic state, zero-reference (REST), average reference (AR), link-ears reference, etc., still speak their own words. This and the next chapter will systematically sort out the essence of an EEG reference and the basis of different EEG references from the perspective of mathematics and physics, hoping to provide a scientific facts-based guide for the selection of an EEG reference.

In this chapter, the first section is focused on a historical review of the reference electrode problem mainly from Geselowitz (1998), which covers both ECG reference and EEG reference. The second section introduces the zero-reference (Yao, 2001).

13.1 HISTORICAL DEBATES ON EEG REFERENCE

13.1.1 Introduction

In practice, evoked potential (EP) and spontaneous EEG are components that reflect the activity of different neural sources. Each component can be described by characteristics such as polarity, scalp region, frequency spectrum, latency, and voltage amplitude. In order to obtain objective values of these properties, an inactive voltage point as the reference electrode is the most desirable reference, because only the potential difference between an active electrode and a reference electrode can actually be measured, and setting a reference electrode in the scalp records is unavoidable (Geselowitz, 1998). For head and non-head reference electrodes, such as earlobe, neck, and average reference, etc., each has a specific influence on recording, and it has been suggested to use different reference electrodes for different potentials (Desmedt et al., 1990), but this ignores the dilemma of comparing results of different references, in the later stage.

For Electrocardiogram (ECG), Wilson et al. (1934) published a classific paper titled "Electrocardiogram with Single Electrode Potential Changes", which proposed the "central terminal reference", i.e., a point where the electrode connects the right arm, left arm, and right leg through three equal resistances. They assumed that "the central terminal is zero potential throughout the cardiac cycle", then the unipolar ECG can be obtained by using the central terminal as a reference. The central terminal, later named the Wilson central terminal (WCT), was traditionally used to record six precardiac leads and three limb leads in a standard 12-lead ECG.

The quasi-static field generated by ECG and EEG source activities is the gradient of the potential scalar field and is not changed by any addition or subtraction of a

DOI: 10.1201/9781032639260-13

constant to the potential, resulting always in an arbitrary constant of the potential not necessarily determined and only the potential difference, which is independent of the constant, is recorded. In practice, we can choose a convenient point as the reference, even if its potential is not zero but an unknown constant, because what we care for is the potential difference and its fluctuation. As charge is the fundamental element of any electric phenomenon such as current, dipole, quadruple, etc., for a charge in free space, we can set the potential at infinity as zero, and it is actually a well-known and accepted physical definition. However, the infinity point with zero potential is not available in the potential observation of a finite-volume conductor like the human body. Meanwhile, no point on the surface of the human body can be regarded as zero potential. In practice, we either ignore its influence (assuming that reference electrode potential fluctuation is much smaller than the active electrode potential fluctuation), by assuming a point as the reference and assigning it a value of zero, or to find a way to correct it in the post-processing, that is, make a re-reference by a computer such as REST, AR, etc., and these are the main topic of this and the next chapter.

13.1.2 Einthoven Triangle and Wilson Center in ECG

Wilson's view of WCT as the zero potential can be illustrated by the Einthoven triangle. As the father of the electrocardiogram, Einthoven has not only developed the galvanometer that made the electrocardiogram a diagnostic tool in practice but also proposed a theoretical framework for understanding the relationship between cardiac source and the recording of potential on the skin surface. The bioelectrical activity in the heart or brain generates current in the volume conductor, as well as the corresponding potential on the skin surface. As shown in Chapter 3, the bioelectricity problem can be taken as a quasi-static problem in which the human tissue follows Ohm's law, that is, it is resistive, and the impedance is independent of the current intensity and fluctuation frequency. Therefore, the volume conductor can be regarded as a linear medium, so that the potential generated by all the electric activities in the heart can be regarded as the superposition of the potential generated by each of the electrical sources in the heart. Einthoven pointed out that, in a first-order approximation, the heart acts like a dipole in a fixed position but with varying intensity over time during a cardiac cycle, and that the potential on the skin can be expressed as the sum of the potential generated by the three coordinate components of this dipole.

In practice, a pair of electrodes on the body surface is called a lead; the corresponding lead voltage can be expressed as

$$\Phi_i(t) = \vec{H}(t) \cdot \vec{C}_i \tag{13.1}$$

Here $\vec{H}$ is the dipole moment or the heart vector, and $\vec{C}$ is the leadfield vector, which represents the relationship between the lead voltage and the dipole. The lead vector depends on the electrode position, volume conductor, and dipole position.

The usual ECG leads are limb leads I, II, and III, which were first proposed by Einthoven. Let the potentials of the left arm, right arm, and left leg relative to any reference point be Φ_R, Φ_L and Φ_F. Then, the limb leads can be expressed as

$$
\begin{aligned}
e_{\mathrm{I}} &\equiv \Phi_L - \Phi_R \\
e_{\Pi} &\equiv \Phi_F - \Phi_R \\
e_{\mathrm{III}} &\equiv \Phi_F - \Phi_L
\end{aligned}
\tag{13.2}
$$

Let's define a rectangular coordinate system on the human body, setting the *X*-axis horizontally from right to left, the *Y*-axis vertically down, and the *Z*-axis from front to back. Einthoven pointed out that the vectors of the three limb leads form an equilateral triangle. This Einthoven triangle is located in the frontal plane, namely the *X–Y* plane, in which the lead vector for lead I is horizontal, from right to left. This formulation provides a solution to the forward problem with the known heart vector to the limb leads. It also allows the inverse problem, i.e., the determination of the heart vector from leads I, II, and III. However, only two components of the heart vector can be obtained from the limb lead. The Z-axis component *Hz* cannot be determined. Besides, only two of the three equations of equation (13.1) are independent, and the three equations of equation (13.2) together satisfy (13.2a).

$$
e_{\mathrm{I}} + e_{\mathrm{III}} = e_{\Pi} \tag{13.2a}
$$

It is called the Einthoven's law.

Let's revisit Wilson's argument for the zero potential at WCT. Let the central terminal potential be Φ_T,

$$
\Phi_T = \frac{\Phi_R + \Phi_F + \Phi_L}{3} \tag{13.3}
$$

Then, with (13.2) and (13.3), we have

$$
\begin{aligned}
\Phi_R - \Phi_T &= -\frac{e_{\mathrm{I}} + e_{\Pi}}{3} \\
\Phi_L - \Phi_T &= \frac{e_{\mathrm{I}} - e_{\mathrm{III}}}{3} \\
\Phi_F - \Phi_T &= \frac{e_{\Pi} + e_{\mathrm{III}}}{3}
\end{aligned}
\tag{13.4}
$$

and if $\Phi_T = 0$, Wilson obtained (Wilson et al., 1931)

$$
\begin{aligned}
\Phi_R &= -\frac{e_{\mathrm{I}} + e_{\Pi}}{3} \\
\Phi_L &= \frac{e_{\mathrm{I}} - e_{\mathrm{III}}}{3} \\
\Phi_F &= \frac{e_{\Pi} + e_{\mathrm{III}}}{3}
\end{aligned}
\tag{13.5}
$$

Describing it differently, if we specifically define Φ_R, Φ_L and Φ_F as equation (13.5), then the average of these three potentials, namely Φ_T in equation (13.3) is zero, thus it is regarded as a zero -reference. To further confirm this hypothesis. Wilson et al. built a flume model to test his theory, and the results matched the theoretical predictions very well.

Another demonstration of why WCT is a zero potential point appeared in an appendix of (Craib and Canfield, 1927), which gave a potential expression on the surface of a sphere with radius R for a dipole at the center with a dipole moment $\vec{p}$

$$\Phi = \frac{3\vec{p}\cdot\vec{R}}{4\pi\sigma R^3} + K \tag{13.6}$$

where $\vec{R}$ is the radial vector of the observation point. The derivation may look at that of equation (4.21). Here, Einthoven's theory is equivalent to the concept that the right arm, left arm, and left leg are three points on the sphere surface, and they are the three vertexes of an equilateral triangle in a vertical circle of the sphere, then $\vec{R}_R + \vec{R}_L + \vec{R}_F = 0$, and then $\Phi_R + \Phi_L + \Phi_F = 3K$. Let $K=0$, i.e., the sum of Φ_R, Φ_L, and Φ_F is 0, then the WCT potential is 0.

The electrocardiogram began as a recording of potential differences, or leads, at the limbs. For practical diagnostic reasons, it was agreed that a limited number of leads should be recorded at selected skin locations relative to a common reference point. The advantage of WCT is that it provides an easily available average potential on the surface of the human body as a reference. Historically, the right arm reference competed with WCT for a time (Wolferth and Wood, 1932), but WCT was ultimately the winner. The key reason in my opinion should be the above sphere model evidence, equation (13.6). Though the body is not a sphere, it provides us with a theoretical reassurance. The reference point may shape the recorded waveform, and the diagnostic criteria obviously depend on the reference point. People often rely on experience to determine whether an ECG/EEG reference is superior or not. Subjective selection of reference electrodes is more common in EEG research, and REST introduced in this chapter provides the chance to standardize the selection.

13.1.3 Surface Potential Topography

The distribution of body surface potential as a function of time, contains important information about the electrical activity in the body. A body surface potential diagram based on an electrode array can be plotted as a series of equipotential lines at each moment of a cardiac cycle. The choice of reference potential will affect the magnitude but not the shape of the isoline. Since the number of contours that can be drawn is limited, the appearance of the isolines will depend on the adopted reference and the number and spacing of contours.

In theory, the inverse problem, solved by using human body surface (chest or scalp) potentials to determine the sources inside, is not affected by the selection of reference points. However, in practice, the solution of the inverse problem is affected by the noise contained in the potential recording. The selection of reference points may affect the SNR of some specific electrode positions, and thus indirectly the solution of the inverse problem. For example, Gencer et al. (1996) took spatial resolution as the criterion and concluded that the optimal position of the reference electrode depended on the cortical region we were interested in. In addition, the generation of a topographic map needs the help of a potential interpolation algorithm. Obviously, the interpolation potential depends on the number and location of electrodes and will also be affected by the noise level especially when the number of electrodes is limited. Thus, the appearance of the topographic map will be affected in principle by the algorithm, and thus indirectly by the selection of reference potential. Of course, these phenomenal effects are at the algorithm level. At the physical essence level, all spatial analysis is not affected by the selection of reference electrodes.

13.1.4 Arm, Link-Ears, Neck Ring, and Other Unusual Reference

The scalp surface potential recording is often considered a unipolar record with respect to a reference potential, which is assumed to be a point of constant potential during the electrical activity of the brain. This assumption implies that the distance between the reference electrode and these sources in the brain is much greater than the distribution region of the EEG source. However, the usual distributed electrical activities in the brain often span tens of centimeters, so the electrodes on the surface of the head or even at the earlobe cannot be far enough away from the sources. Katznelson pointed out (1981) that the current from the brain, through the neck to other parts of the body, is very small because the cervical pathway is highly resistive. Therefore, the point on the neck almost has the same potential as the point on the arm. However, the potential of the arm or the neck is influenced by the electrical activity of the heart (in part because the electrical activity of the heart is much stronger than that in the brain), so it is also inappropriate to use the arm or neck as a reference point for EEG recording.

It is important to remove ECG activity from the EEG record. Katznelson (1981) showed that surface current generated by the heart may circulate through the head, causing spurious potential changes at EEG electrodes. In the early EEG literatures, physically linked ears were used as a common reference point for EEG recording. The low resistance path between the two ears short-circuits the heart's electrical current, preventing it from flowing through the head. Therefore, linking ears as a reference has the effect of reducing ECG interference. Unfortunately, this practice of forcing the same potential in both ears also will greatly distort the distribution of potential generated by the EEG sources on the surface of the head. Katznelson pointed out that the best way is to wear a conductive ring around the neck, which acts as a barrier to short-circuit interfering currents from other parts of the body. However, this operation will also change the potential distribution of the brain, so it has not been applied in practice. The other references tested before included electrodes located on the angle of the jaw, on the chin, the tip of the nose. These efforts have not been successful, and the major problem is the ECG and electromyographic (EMG) effects on all channels. Besides, there is another noncephalic reference which is a combination of two electrodes, one electrode is over the subject's C7 vertebra on the back, and the other is over the suprasternal notch, they are connected through a balancing potentiometer, and the balance point with minimum ECG is taken as the reference (Duffy et al., 1989). However, such a reference is somewhat complex for routine EEG use, as rebalancing may become necessary when electrode impedance changes or the position of the head vs. the heart is altered. Especially, as ECG and EMG are much larger than the EEG, any small residual may notably destroy the EEG recordings. And even if ECG is totally balanced out, the potential at the reference electrode still may not be zero due to the time-varying EEG signal and EMG, etc., arrived there by conduction.

13.1.5 Average Reference

Since no appropriate EEG reference point can be found on the body surface, the average potential (AR) of all channels has been widely used as a reference in the EEG analysis. In fact, this reference is obtained by subtracting moment by moment the average value of the observed potentials of all electrodes. The main advantage of AR is that the dependence on the actual potential of a specific reference point in the record is avoided, that is, the reference potential no longer varies with the actual change at a specific reference point, but, instead, the potential at a given electrode will depend on the potential at all other electrodes. If we only study the distribution of the scalp surface potential at a given moment, the effect of the average reference is only to change the level of potential. But when we need to look at brain activity over time and space, we will have a time-varying problem with the average reference, making the interpretation of EEG time series complicated. Why AR is adopted so widely? Perhaps, it is due to the illusion that the bias is shared and diluted by all electrodes, with no principal culprit to be criticized. And due to this phenomenon, some papers even stated that AR is "reference free", apparently, it's a false pretense and psychological consolation.

Actually, in the history of heart and brain electrical research in the past 100 years, the debate about the zero potential point has never stopped. In 1986, a short article was published in the American Journal of Physics, entitled "Where is the equipotential value of V=0?" (Baski and Bartlett, 1986). This paper considered the case of a dipole. In the years since then, a lot of EEG literatures appeared again in the debate about the zero potential point.

In EEG, one contention concerns the potential distributions. Lehmann (1988) pointed out that "the change of the reference potential will add or subtract a constant at all electrode positions, just like raising or lowering the water level of a lake without changing the surface". Tomberg et al. (1990) believed that "an improper selection of reference points will cause graphic distortion and pseudo potential fields". The irony is that the inappropriate reference that has been fiercely criticized by Tomberg and others is the average potential mentioned above. While Bayley (1957, 1959) believed that the average potential is the true zero potential in an ECG.

Desmedt pointed out (1990) that "ghost potential" will appear when different reference electrodes are used. By careful analysis, it is not difficult to find that the appearance of this "pseudo potential" is related to the following conditions: First, if the contours of the topographic map are sparse, some topographic features such as local extreme values may exist in one picture but not in the other (contour maps are rarely used nowadays), the other situation is when people pay special attention to the amplitude and sign (positive or negative) of the potential, even if the two patterns have exactly the same equipotential line (the isoline passes through the same point on the scalp), the interpretation may be quite different.

Desmedt et al. (1990) questioned the rationality of Lehmann's analogy of the zero potential point to the description of the earth's topography with reference to sea level. He thought that this was simply two different things. Geselowitz (1998) made a proper compromise that Desmedt et al. (1990) correctly pointed out that the reference potential of EEG changes over time, and at each moment Lehmann's analogy is also correct.

13.1.6 Discussions

The human body is a finite insulating volume conductor. It is impossible to find a real zero or constant potential point on the human body, so any effort to find a neutral point is meaningless. For the spatial characteristics of ECG/EEG, the selection of a specific reference point only adds a constant with no physical meaning, and will not in any way change the potential distribution, nor the relationship between the source and potential. However, the standardization of reference will help more direct comparisons of data in different laboratories. For example, when diagnosing with a finite number of leads, we must use the same reference point. In an ECG, the accepted reference is WCT. In a multichannel EEG, the average reference is the most commonly used one in recent years, the linked ears are the second, and the new REST zero-reference (Yao, 2001) got more and more acceptance, with the possibility of being the standard reference in the future.

Obviously, the most important issue of reference selection is its effect on the temporal aspect, such as the waveform, power spectra, time-frequency analysis, and dynamics and network analysis. And, along with the development of functional magnetic resonance imaging (fMRI), the value of the spatial analysis of EEG may be partly replaced, but the high temporal resolution of EEG is still continuously rolling the other imaging modalities, thus a good reference to guarantee a reliable temporal analysis becomes even more important.

13.2 REFERENCE ELECTRODE STANDARDIZATION TECHNOLOGY

13.2.1 Introduction

Since neural electrical activity is a spatiotemporal process, the influence of the active reference electrode exists in both space and time. The spatial effect of the reference electrode can be summarized as follows: the reference electrode does not affect the use of noiseless scalp potentials to solve the inverse problem, that is, the location of the source of neural activity is not dependent on the reference electrode (Pascual-Marqui and Lehmann, 1993). A change in the reference electrode will add or subtract a constant at all distribution positions, as in raising or lowering the horizontal level, but does not change the surface shape (Geselowitz, 1998). It can be seen that the influence of reference on the spatial properties of an EEG is the epidermis, not fundamental.

The influence of the reference electrode in the time domain is attributed to the electrical activity of the reference electrode position. If a point or average on the surface of the human body is active, its potential varies with time, and when it is used as a reference, this process of change in the time domain, which is unknown, is introduced into all the recording electrodes. It can be seen that the active reference electrode will affect the time-domain dynamic analysis and spectrum analysis of the EEG. In order to solve this problem, people naturally expect a neutral point for the reference electrode. Since the infinity point is far away in space from the source of neural activity, and will have no influence on EEG records, its potential can be assumed to be zero and can be taken as an ideal reference electrode point. In fact, infinite is always the reference point in electromagnetics.

In the literatures, the neutral reference electrode or zero potential reference electrode is always a controversial topic. We should say that the content of this section is irrelevant to these controversies, as it is not in our interest to find a neutral reference point or zero potential on the scalp, which we fully understand is an insoluble problem, in fact, there is no point on the surface of the scalp that can be considered a zero potential point (Geselowitz, 1998).

The aim of this section is to introduce a new method, called the Reference Electrode Standardization Technique (REST), which approximately transforms a record with a reference point or average potential on the scalp to a record with a reference at infinity in space. The physical basis is that the potential before and after the transformation is generated by the same actual neural sources in the brain or their equivalent sources, thus they can be related by the equivalent neural source. Apparently, before this REST reference was put forward, people took the reference problem as a time-domain problem from the perspective of reference affecting the EEG time-domain analysis. It is the first time to present the reference problem as a spatial problem as it is due to the volume conduction of the current generated by neural electrical sources, so it needs to be solved with the help of the head model with brain electrical sources or the equivalent sources. The deduced REST scheme is naturally a completely different and subversive new scheme. Here we will first introduce the relevant principle, and then use examples to illustrate its effectiveness.

13.2.2 EEG Reference Electrode Standardization Technology (REST) and Zero Reference

13.2.2.1 Scalp EEG Recording Model

The scalp potential can be expressed as

$$\Phi = GX \tag{13.7}$$

Here, the matrix Φ of size $L \times K$ represents K scalp sampling potential records at L electrodes, and the matrix X of size $M \times K$ represents K samples of M neural sources in the head model. The $L \times M$ matrix G is a leadfield matrix determined by the head model, the source model, and the electrode configuration. In general, the derivation of the leadfield matrix implicitly assumes that the potential generated by the dipole is zero at infinity, as it is implicitly assumed in the electric theory of the brain as shown in Chapters 3–5. Thus, the potential in equation (13.7) takes the infinite point as a reference. In practice, only the potential difference can be measured, so a physical point must be placed on the body surface including the scalp surface as a reference electrode. Suppose an electrode close to the earlobe is used as the reference electrode, and its potential with respect to the infinite point is

$$\mathbf{v}_e = g_e X \tag{13.8}$$

Here the row vector g_e of size $1 \times M$ consists of the row corresponding to the reference electrode in the leadfield matrix G. The row vector $\mathbf{v}_e$ of size $1 \times K$ is composed of the row corresponding to the reference electrode in the matrix V. According to equations (13.7) and (13.8), we can obtain the EEG recording model with scalp electrical reference

$$\Phi_e = \Phi - \mathbf{t}\mathbf{v}_e = GX - \mathbf{t}\mathbf{g}_e X = (G - \mathbf{t}\mathbf{g}_e) X = G_e X \tag{13.9}$$

Here $\mathbf{t}$ is a column vector of size $L \times 1$, and each of its entries is unit 1. G_e is the leadfield matrix. Similarly, we can select the average scalp potential as a reference electrode. The average value of all the electrodes is

$$\mathbf{v}_a = \text{Average}(\Phi) = \frac{1}{L}\mathbf{t}'\Phi = \frac{1}{L}\mathbf{t}'GX \tag{13.10}$$

Where $\mathbf{v}_a$ is the row vector of size $1 \times K$, and "Average" represents the average of all electrode records moment by moment. $\mathbf{t}'$ is $\mathbf{t}$ transpose. Finally, we obtained a scalp EEG recording model with an average electrode as the reference point

$$\Phi_a = \Phi - \mathbf{t}\mathbf{v}_a = GX - \frac{1}{L}\mathbf{t}\mathbf{t}'GX = \left(G - \frac{1}{L}\mathbf{t}\mathbf{t}'G\right)X = G_a X \tag{13.11}$$

Equations (13.7), (13.9), and (13.11) provide EEG recording models with infinity, scalp electrode, and average potential as reference electrodes, respectively. In fact, the average electrode potential is not obtained from the unknown potential Φ, but from the actual recorded potential Φ_e as it is easy to prove

$$\Phi_a = \Phi_e - \mathbf{t}\,\text{average}(\Phi_e) \tag{13.12}$$

Meanwhile, we have

$$\Phi_e = \Phi_a - \mathbf{t}\mathbf{v}_{ae} \tag{13.13}$$

The row vector $\mathbf{v}_{ae}$ is composed of the row corresponding to the reference electrode in the matrix Φ_a. Equations (13.12) and (13.13) indicate that Φ_a and Φ_e can be derived from each other, and therefore they provide the same physiological information.

13.2.2.2 Equivalent Source-Based Reconstruction Algorithm

Based on the equivalent source theory and technique (Chapter 6), the source X of equations (13.7), (13.9), and (13.11) are the true sources or their equivalent sources.

Since the true sources are unknown, we use their equivalent distributed dipole sources here. Actually, it is very difficult to find the true sources (Chapters 10–12), and fortunately, we do not need the true sources as their equivalent sources can help us to reconstruct the potential with zero-reference. From equations (13.7) and (13.11), we can get

$$\Phi = GX \approx \Phi' = G\left(G_a^+\Phi_a\right) = \left(GG_a^+\right)\Phi_a = R_a\Phi_a \tag{13.14}$$

Here Φ' is the Φ obtained from an approximately reconstructed equivalent source $X \approx G_a^+\Phi_a$, and the matrix Ra is the standardization matrix of the average reference electrode records. The symbol "+" denotes the generalized inverse, and in this work "+" is accomplished by singular value decomposition (SVD). Certainly, it can be realized by other regularization algorithms (Chapter 12 and Appendix E). Similarly, for earlobe reference electrode recording, we have

$$\Phi = GX \approx \Phi' = G\left(G_e^+\Phi_e\right) = \left(GG_e^+\right)\Phi_e = R_e\Phi_e \tag{13.15}$$

R_e is the matrix of the earlobe reference electrode. In addition, because $\Phi = \Phi_a + \mathbf{t}\,\text{average}(\Phi) = \Phi_a + \mathbf{t}v_a$, $\Phi' = \Phi'_a + \mathbf{t}\,\text{average}(\Phi') = \Phi'_a + \mathbf{t}v'_a$ And given v_a, we can also choose the standardized result to be

$$\Phi' = \Phi_a + \mathbf{t}v'_a \tag{13.16}$$

Obviously, the error between Φ_a and Φ'_a is eliminated after equation (13.16) is selected as the final result, and the remaining error is only the difference between the average (Φ) and the reconstructed average (Φ'). R_a and R_e in equations (13.14) and (13.15) are determined by four factors: the volumetric conductor model, the equivalent source model, the electrode distribution, and the calculation of the generalized inverse solution. They are independent of the actual neural sources within the equivalent source layer, so they can be applied to both simulated records with known sources and actual records with unknown sources.

In the primary work, the head model is the usual concentric three-layer sphere model, as shown in Figure 13.1, $L=128$ electrodes are evenly distributed on the surface of the spherical cap with the lowest latitude of 100°. Later, we used some different head models and electrode distributions to study the influence of the head model and electrode number. The discrete equivalent dipole layer is composed of a spherical cap located at the radius $r=0.869$ and a cross-section located at $z=-0.076$ (the dotted line in Figure 13.1). The 2,600 radial dipoles are uniformly distributed on the spherical cap and 400 radial dipoles are uniformly distributed on the cross-section, so the total number of equivalent sources is $m=2{,}600+400=3{,}000$.

Because of the limitation of the electrode distribution on the upper spherical cap surface, it is difficult to reconstruct the closed dipole layer well. Therefore, the equivalence between the dipole layer and the neural source within the layer is approximate, and the degree of the equivalence is related to the position and orientation of the dipole. So REST effectiveness is also related to the location and

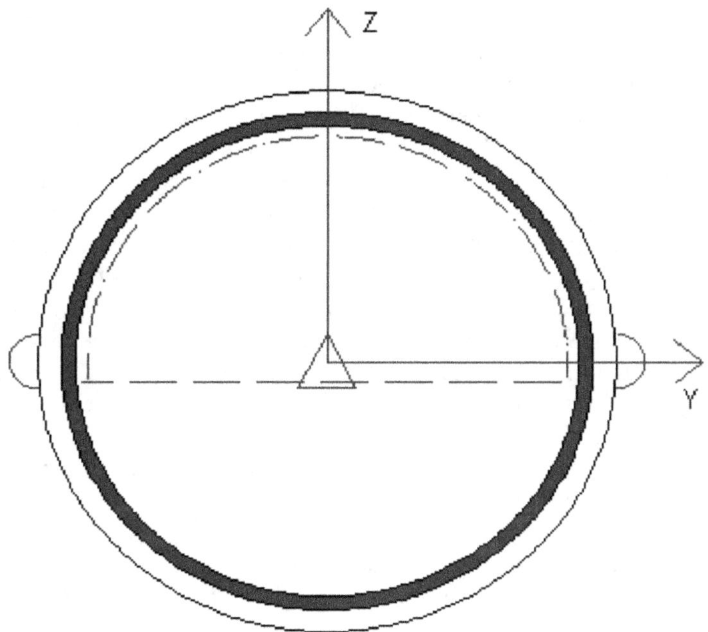

FIGURE 13.1 Sphere head volume conductor model. The triangle represents the nose, the center of the sphere is defined as the origin of coordinates, the axis from the origin to the left ear is defined as the positive *Y*-axis and the axis from the origin to the root of the nose is defined as the positive *X*-axis. The positive direction of the *Z* axis is the axis perpendicular to these two axes and pointing overhead from the origin. The radii of the three concentric spheres are 0.87 (inside the skull), 0.92 (outside the skull), and 1.0 (head), with electrical conductivity of 1.0 (brain and scalp) and 0.0125 (skull), respectively. The dotted line indicates the location of the closed equivalent layer, which consists of a spherical cap of radius 0.869 and a cross-section at $z = -0.076$ (Yao, 2001).

orientation of the real dipole source, which can be evaluated by exhaustive simulation studies as shown below.

The generalized inverse solutions of equations (13.14) and (13.15) vary with different singular value truncations, which helps to suppress the effect of noise in scalp records on the inverse solution. However, while truncation reduces the influence of measurement noise, it also has the problem of serious loss of high-frequency component information. For REST, our goal is to provide an objective tool for approximating the standardization of the reference electrode, so that any subjective singular value truncation should be avoided. Certainly, if the noise level is really a problem, then truncation or regularization will be inevitable, and REST may be called regularized REST (rREST) as to be introduced in Chapter 14. According to Appendix E, SVD generalized inverse, minimum solution length, and maximum likelihood estimate are equivalent in reaching the final solution. For more details on rREST, please refer to Section 14.2 and Appendix G.

The condition number (CN) is defined as the ratio of the singular value at the cut-off point to the first singular value. For the leadfield matrix Ga with 128 electrodes, we find that CN smoothly changes from $CN(1)=1.0$ to $CN(127)=8.4\times10^{-3}$, but after 127, there is a sharp decline, $CN(128)=4.5284\times10^{-6}$, so the cut-off point is chosen on 127. For Ge, CN smoothly changes from the initial $CN(1)=1.0$ to the final $CN(127)=3.9\times10^{-3}$, so all 127 singular values of Ge are retained in the inverse solution. In fact, the number 127 is $L-1$ which is the largest independent observation number among the 128 electrodes when the earlobe electrode or average potential is used as a reference. Because the potential of the earlobe electrode is assumed to be zero when the earlobe electrode is used as a reference, only 127 electrodes remained to be recorded. When the average potential is used as a reference, a zero average potential means that the record of any one of the L electrodes is a linear combination of the records of the other $L-1$ electrodes, so that the number of independent observations among them is only $L-1$. When other electrode configuration is tested below, for the same reason, the truncation point of the relevant Ga is objectively selected at the corresponding $L-1$.

In summary, the REST calculation steps are as follows

Given electrode distribution and scalp record Φ_a;

Suppose a head model, such as a concentric three-layer sphere model, as shown in Figure 13.1;

Suppose an equivalent source model, such as the discrete dipole layer model shown by the dotted line in Figure 13.1;

On the basis of known electrode distribution, head model, and equivalent source model, the forward theory is used to calculate the leadfield matrix G of equation (13.7);

From the leadfield matrix G to the Ga of equation (13.11);

Using SVD to calculate the generalized inverse of leadfield matrix Ga;

Calculate $\Phi' = R_a\Phi_a$ with equation (13.14), then calculate $\Phi'_a = \text{Average}(\Phi')$;

According to the known record Φ_a and the obtained record Φ'_a, the final reconstructed record Φ' is calculated by equation (13.16).

For other reference electrodes, such as the earlobe reference electrode, the calculation procedure is similar.

13.2.2.3 Simulation Evaluation (*I*): Effectiveness

13.2.2.3.1 Description of Simulation Method

The time-domain process of the dipole source is assumed to be a decaying Gaussian function

$$H(t_i) = \exp\left(-\left(2\pi f \frac{t_i - t_0}{\gamma}\right)^2\right)\cos\left(2\pi f(t_i - t_0) + \alpha\right), \quad i = 1,\ldots,K \tag{13.17}$$

$t_i = i \times dt$, $K=256$, and $dt = 0.004$ (seconds) = 4 ms. The reason we chose this function is that it looks like an evoked potential, and the reconstruction algorithm itself is independent of the selection of the process in the time domain. The values of the parameters t_0, f, γ, and α in equation (13.17) will be given in the following example.

With the function H and the previous model equations (13.7), (13.9), and (13.11), we can obtain the space-time records $\mathbf{\Phi}$, $\mathbf{\Phi}_a$ and $\mathbf{\Phi}_e$ of size $L \times K$.

Relative Error (RE) is used to measure the effectiveness of REST and is defined as

$$\mathrm{RE} = \|\mathbf{\Phi} - \mathbf{\Phi}_*\|_F \Big/ \|\mathbf{\Phi}\|_F \tag{13.18}$$

Here $\mathbf{\Phi}$ is the space-time record measured with a reference at the infinite point, and $\mathbf{\Phi}_*$ is one of the records of $\mathbf{\Phi}_a$, $\mathbf{\Phi}_e$, and recovered $\mathbf{\Phi}'$. The Frobenious norm of the matrix $\|*\|_F$ is defined as $\|\mathbf{\Phi}\|_F = \left(\sum_i \sum_{j=1}^{k} v_{ij}^2\right)^{1/2}$. In the following calculation, $\mathbf{\Phi}$ and $\mathbf{\Phi}_*$ usually correspond to all L channels ($i = 1, \ldots L$), but in some places, they are also used as single-channel signals to obtain the single-channel-based RE.

Since the transformation from $\mathbf{\Phi}_a$ or $\mathbf{\Phi}_e$ to $\mathbf{\Phi}'$ in equations (13.14)–(13.16) is linear, we only need to examine the individual properties of the single dipole potential and pure noise, while the properties corresponding to various dipole combinations or dipole field-noise combinations can be deduced from the linearity of the transformation. In order to understand the effectiveness of REST for different dipole positions and different dipole orientations, we exhaustively simulated each unit of the discrete cubic grid in the brain region space. Each grid has three unit dipoles (P_x, P_y, P_z) orienting pointing in the (x, y, z) directions of the three-dimensional Cartesian coordinates. The entire discrete cubic net consists of 1,269 units located in the brain region of radius $r \leq 0.86$ and $z \geq 0.0$ (radius 0.87 within the skull, and 0.869 between 0.86 and 0.87 for the dipole layer), with a grid spacing of 0.105. The total number of dipoles tested is 1,269 (voxels) * 3 (dipoles per voxel) = 3,807. For each dipole, we derive its $\mathbf{\Phi}$, $\mathbf{\Phi}_a$, and $\mathbf{\Phi}_e$ from the equations (13.7), (13.10), (13.12), and (13.17). Then its $\mathbf{\Phi}'$ is obtained from the equations (13.14) or (13.15) and (13.16), Let $\mathbf{\Phi}_* = \mathbf{\Phi}_a$, $\mathbf{\Phi}_* = \mathbf{\Phi}_e$ and $\mathbf{\Phi}_* = \mathbf{\Phi}'$ in equation (13.18) respectively to obtain *RE*(average), *RE*(earlobe), and *RE* (standard). In short, for each cell grid, we obtain nine *RE* belonging to three dipoles (P_x, P_y, *and Pz*) and three reference electrodes (average, earlobe, and standard).

13.2.2.3.2 The Simulation Results

13.2.2.3.2.1 Comparison of Earlobe Reference and Average Reference

In this simulation, the parameter of equation (13.17) is $t_0 = 35 * dt$, $f = 10$ Hz, $\gamma = 5$, $\alpha = \pi / 2$. Assume that the earlobe reference electrode is located on the scalp (X, Y, Z) = (0.049413, 0.964661, 0.258819), that is, the left earlobe. Figure 13.2 (top row) shows a section of the *RE* (earlobe). Obviously, *RE* (earlobe) depends not only on the direction of the dipole (the difference between the three subgraphs in the upper row), the position of the dipole (the different values in each subgraph), but also on the position of the reference electrode (the *RE* of each subgraph is obviously maximum near the reference electrode). The smaller the distance between the source and the left earlobe is, the greater the value of $\mathbf{v}_e$ (equation 13.8) is and the greater the distortion caused by subtracting $\mathbf{v}_e$ (equation 13.9) is. For comparison, the lower row of Figure 13.2 shows the volume distribution of *RE* (average). By comparing *RE* (earlobe) and *RE* (average), it can be seen that the maximum of *RE* (earlobe) is 358.23%

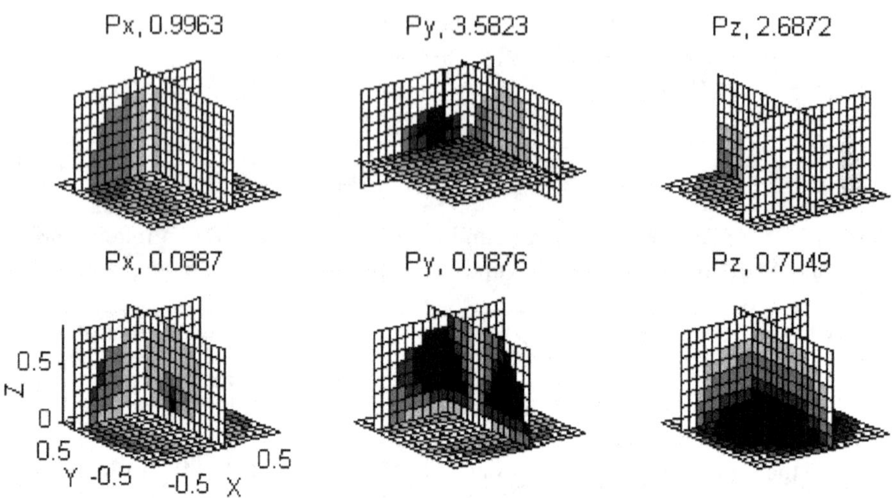

FIGURE 13.2 The slice of relative error (RE) volume distribution. The upper row represents the RE (earlobe) between the potential with the infinite point as reference and the potential with the left earlobe as reference. The next row shows the RE (average) between the potential with the infinite as a reference and the potential with the average as a reference. The white color represents zero, and the blackest dot represents the maximum value shown in each subgraph. P_x, P_y, and P_z represent dipoles pointing to the three Cartesian axes: x, y, and z, respectively (Yao, 2001).

of the corresponding P_y dipole, much larger than the maximum of *RE* (average), 70.49% of the corresponding P_z dipole. Furthermore, in the case of the average reference, *RE* (average) depends on the virtual average reference electrode and does not depend on a specific scalp point. Therefore, if the volume conductor and electrode distribution have rotational symmetry, as in the case of our simulation study here, *RE* (average) also has rotational symmetry as shown in the lower right of Figure 13.2. In fact, we find that the maximum *RE* (average) 8.87% of the P_x dipole is almost the same as that of the P_y dipole 8.76%. However, the maximum *RE* (earlobe) 99.63% of the P_x dipole is quite different (Figure 13.2 upper row) from that 358.23% of the P_y dipole. These differences between the average reference and the earlobe reference or scalp reference point are the basic reasons why the average reference has been widely used in EEG/ERP practice in recent years.

13.2.2.3.2.2 The RE Distribution and Effect of Electrode Number Due to the equivalence of equations (13.12) and (13.13), the standardized results obtained from the earlobe reference are almost identical to those obtained from the average reference. So from now on, we will only present the results for the average reference. Furthermore, since the *RE* depends on a virtual point in infinite space rather than a specific point on the scalp, and the head model used here and the electrode distribution are rotationally symmetric, the volume distribution of the *RE* is theoretically rotationally symmetric. Based on rotational symmetry, we only need to look at the RE (average) and *RE* (standard) of a quarter-Cartesian coordinate plane.

For example, the $x > 0$ and $z > 0$ part of the *XOZ* coordinate plane only, certainly, because of the finite number of electrodes, *RE* (average) and *RE* (standard) would not be strictly rotational symmetry. However, from our simulation results, the approximation is good. To simplify the representation, we assume that they are symmetric, and then only give the results on the quarter-Cartesian coordinate plane.

Figure 13.3a–c shows the distribution of *RE* in a quarter-Cartesian coordinate plane. The upper row of each subgraph is *RE* (average), and the lower row is *RE* (standard). *RE* (average) is caused by the fact that the mean of the scalp potential equation (13.10) is not equal to zero. The deviation of this average from zero is different for dipoles in different positions and directions. Figure 13.3a–c shows that the *RE* (average) of a P_z dipole is generally much larger than that of a P_x or P_y dipole. The maximum *RE* (average) of a P_z dipole is ~70%, and the minimum *RE* (average) of a P_z dipole is ~40%. The maximum *RE* (average) of P_x and P_y dipoles is only about 8% and 1%, respectively, and the minimum is almost zero. The reason is that the spherical cap electrodes are mainly distributed in the positive or negative direction of the vertical dipole (P_z), so the mean potential is generally very different from zero. The positive and negative polarity fields of the dipoles (P_x, P_y) on the *XOZ* coordinate plane, especially the tangent dipoles (P_y), can be measured by the spherical cap electrode distribution, so the total average will be approximately zero.

The bottom row of Figure 13.3a–c is the REST processing results of potentials with the average reference. The difference between the two rows of subgraphs reveals the effectiveness of REST. Based on these figures, we get: (1) if *RE* (average) is large, such as in the case of P_z dipole, then as shown in the third column of Figure 13.3a, the effectiveness of REST is quite obvious; (2) If the *RE* (average) is small, and the maximum *RE* (average) of the P_y dipole shown in the second column of Figure 13.3a is only 0.15%, then the application of REST has little effect on the error, and may even increase the error to about 1% due to the approximation of the algorithm; (3) *RE* (average) and *RE* (standard) have different dependence on source location. For *RE* (average), in the case of P_x and P_y dipoles, the larger *RE* (average) is located in the superficial brain region, on the corresponding *RE* (standard) map.

The largest *RE* (standard) is located in deep brain regions. This means that REST is particularly effective for important cortical areas; (4) The difference between Figure 13.3a–c indicates that the denser electrode array is more favorable to REST to reduce *RE* (average). Specifically, the denser electrode array is much better than the sparse electrode array in reducing *RE* (average) on P_x and P_y dipoles. However, since only half of the field of the P_z dipole can be recorded by the electrodes on the spherical cap, no matter how dense the electrode array is, there is not much difference between dense and sparse arrays in reducing the *RE* (average) of the P_z dipole. The only way to reduce the *RE* (average) of the P_z dipole is to use a global electrode array. Our tests show that with a global array of 128 electrodes, the maximum *RE* (average) of the P_z dipole is only about 1%, which is almost equal to that of P_x or P_y. However, the global distribution of electrodes is impractical for clinical use.

13.2.2.3.2.3 Effects of Volume Conductor Model The impact of the volumetric conductor model is evaluated by assuming that Φ_a is calculated from the above concentric three-layer sphere model, but the matrix R_a in equation (13.14) is formed with

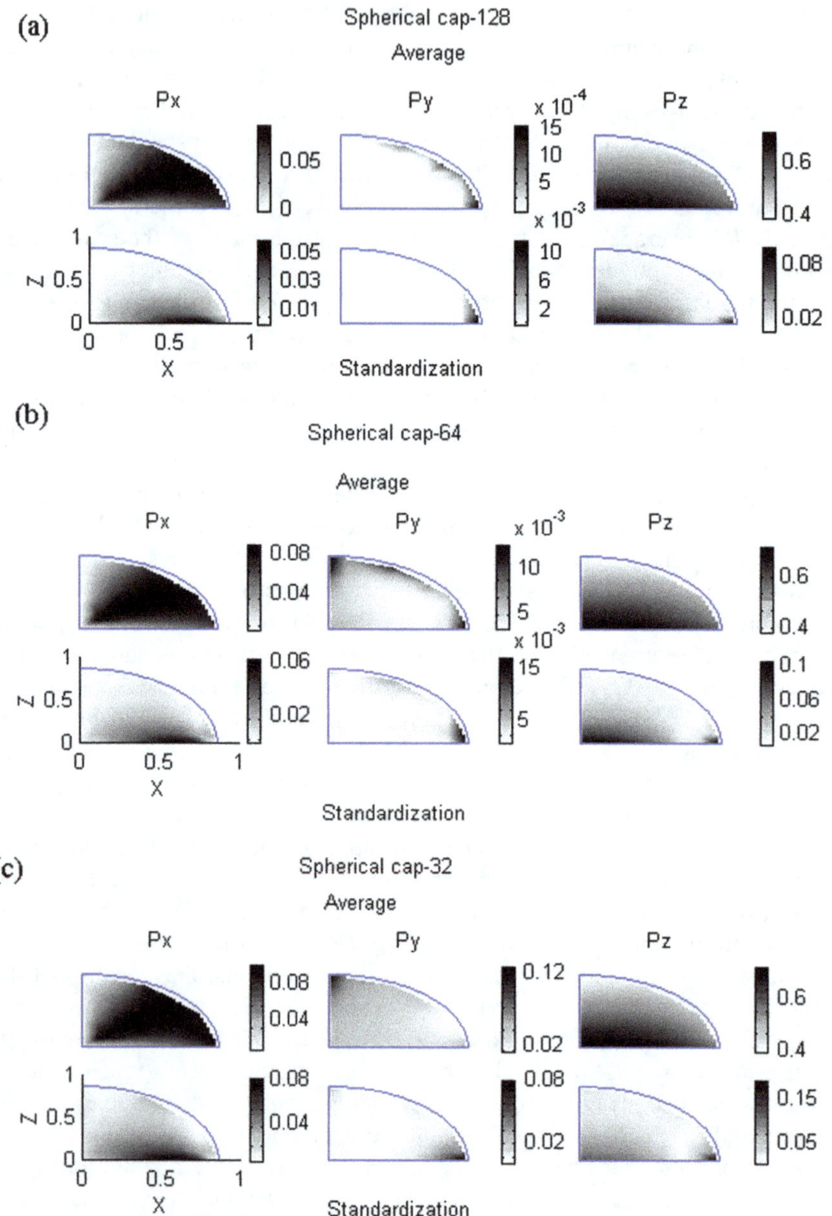

FIGURE 13.3 The effect of the number of electrodes. Subgraphs a,b and c are correspondings to the three cases with 128, 64 and 32 electrodes, respectively. The top row of each subgraph (a or b or c) represents RE (average), and the bottom row represents RE (standard). This is the XOZ plane with positive X and Z of the Cartesian coordinates. The white dot corresponds to the smallest value and the blackest dot corresponds to the largest value. Dipoles along the x- (P_x), y- (P_y), and z- (P_z) axes in the corresponding directions. Relative to the XOZ plane, each P_y is a tangential dipole, while P_x and P_z are the two coordinate components of a radial dipole. The electrodes are evenly distributed on the spherical cap, and the number of electrodes is after the title "Spherical Cap –" in each subgraph (Yao, 2001).

another model to simulate the situation where we do not know the true head model. In our study, we used two models to obtain R_a, one was a concentric three-layer sphere model with some modifications with a radius of 0.87 (brain surface), 0.95 (skull surface), and 1.0 (scalp surface), and conductivity of 1.0 (brain and scalp) and 0.2 (skull). The other is a single sphere with a radius of 1.0 and a conductivity of 1.0. The forward theory of a single homogeneous sphere is described in Section 4.2 above. Figure 13.4 shows the corresponding results, in which the number of electrodes used for calculation is 128. Other parameters are the same as in Figure 13.3. As can be seen from the figure, the *RE* (average) in the shallow area is larger than the *RE* (standard), indicating that REST plays a positive role, but the effect is less than that in Figure 13.3. For example, the *RE* (average) of P_z dipole in the shallow cortex is about 40%, while the corresponding standardized *RE* (standard) is about 15% for a single sphere model (Figure 13.4b), 8% for the modified three-layers concentric sphere model (Figure 13.4a), and 2% for the accurate concentric three-layer sphere

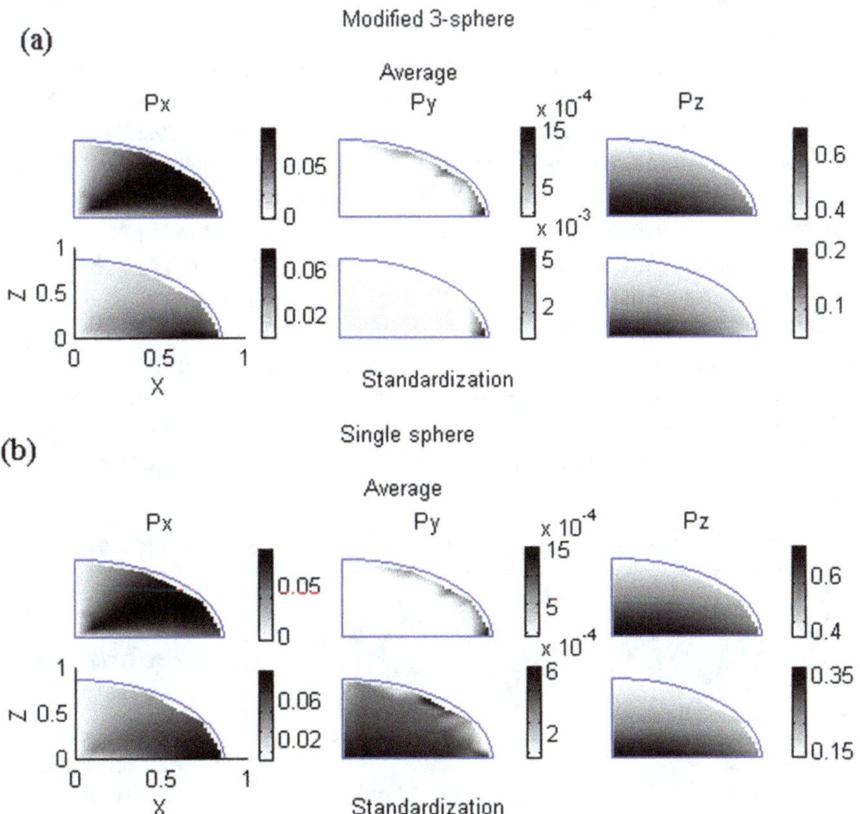

FIGURE 13.4 Influence of head volume conductor model. The number of electrodes is 128. The meanings of other parameters are the same as those in Figure 13.3. (a) The situation after modifying the parameters of the skull layer in the three-layer concentric spherical head model. The modified radii are 0.87, 0.95, and 1.0, and the conductivity is 1.0 (brain and scalp) and 0.2 (skull). (b) Homogeneous spherical model after the skull layer was removed (Yao, 2001).

model (Figure 13.3a). On one hand, the results showed that a detailed head model can obtain a better standardization of the reference electrode than an approximate model does, while on the other hand, it also indicated that REST still played a positive role in the simplified head model, i.e., the homogeneous model, for example, *RE* (standard) of P_z dipole reduced to less than half of *RE* (average).

13.2.2.3.2.4 Influence of Noise As both equations (13.14) and (13.15) are linear, and in order to examine the influence of noise separately, only noise is adopted as input in this experiment study, i.e., Gaussian white noise generated by MATLAB to simulate the noise of the environment and the instrument. According to equations (13.14) and (13.16), we get $N' = N + t\,\text{average}(N')$. The standard deviation and mean of the original input noise (corresponding to 128 (channel)×256 (sampling)=32,768 points) are 0.993382 and 0.00, respectively, while the standard deviation and mean of the output N' are 1.044484 and 0.006930, respectively. The ratio of 1.051 between the standard deviations and the difference of 0.006930 between the means implies that the re-reference leads to little change in the total amount of noise. However, the difference between N and N', namely $\text{Average}(N')$ varies over time. For all 32,768 points, the RE calculated by $\Phi = N$ and $\Phi_* = N'$ in equation (13.18) is 32.4%. If the calculation is carried out one by one channel at a time, then in this example the largest RE occurring on channel 48 is 35.73% and the minimum is 29.25% on channel 37. The left and middle figures in Figure 13.5 respectively represent the N (original noise) and N' (post-transformation noise) of 128 channels at a certain moment. At this moment, the difference between N and N' of each channel is 0.1845 $(\text{Average}(N'))$. However, as shown in the right curve of Figure 13.5, the $\text{Average}(N')$ changes over time and looks like a random sequence with a standard deviation and a mean of 0.3338 and −0.0021, respectively. These results suggest that when there is some noise in the data, re-reference may introduce an instantaneous constant at each channel, which positive or negative depends on the instantaneous noise distribution on the scalp.

Ideally, if $\text{Average}(N)$ is zero, REST should not introduce a non-zero $\text{Average}(N')$. However, REST assumes that the source of the scalp surface record is inside the

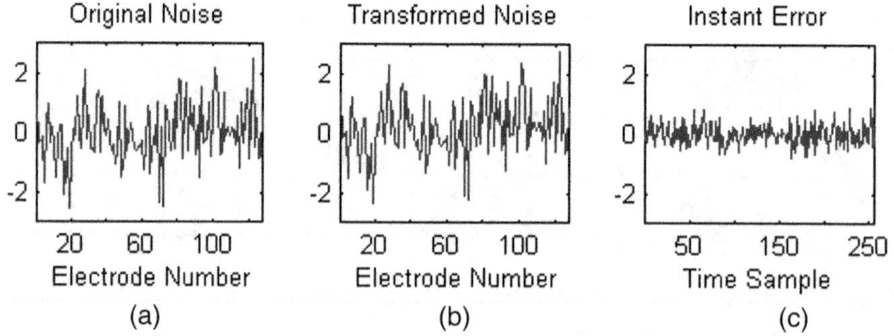

FIGURE 13.5 The effect of noise. The panel (a) (original noise) is the noise of 128 electrodes at a moment, and the panel (b) (transformation noise) is the output of the original noise after standardized operation. The panel (c) is the 256 normalized "correction" constants of the 256 instantaneous noises used in our study, with the first point equal to 0.1845, the difference between the middle figures on the left (Yao, 2001).

equivalent dipole layer, and the pure noise record of the scalp surface cannot be completely generated by some source within the closed equivalent layer, and therefore in principle, they cannot be normalized by REST. In this experiment, the pure noise was used as Φ_a in equation (13.14), so it was assumed to be a potential generated by sources within the closed equivalent layer, and the standardization resulting a " Average(N')", whose value was determined by the instantaneous noise distribution on the scalp surface. Since the noise distribution on the scalp usually changes randomly over time, the "Average(N')" is also the same, as shown in the right subgraph of Figure 13.5.

13.2.2.4 Simulation Evaluation (II): Value for EEG Temporal Analysis

13.2.2.4.1 Dynamic Effect of Artifacts Induced by Average Reference

Since the true potential of the scalp reference electrode point or the average reference electrode varies with the EEG temporal fluctuation, the time and frequency domain analysis of the EEG will be related to the selection of the reference electrode. Here, we focus on the dynamic process of the artifacts caused by average reference and the situation after REST correction.

Suppose there are three dipoles in the volume conductor, with coordinates at (−0.42, −0.21, 0.525) (r=0.7044), (−0.21, 0.42, 0.630) (r=0.7857), and (−0.315, −0.105, 0.735) (r=0.8065) respectively, each of them is a radial dipole, with a unit strength for the first two and a half strength for the third one. The time-domain process is defined by equation (13.17), where the parameters are $t_0 = 35 * dt$, $f = 10$ Hz, $\gamma = 5$, $\alpha = \pi/2$ for dipole 1, and $t_0 = 40 * dt$, $f = 11$ Hz, $\gamma = 4$, $\alpha = \pi / 2$ for dipole 2, and $t_0 = 80 * dt$, $f = 8$ Hz, $\gamma = 6$, $\alpha = 0$ for dipole 3.

Figure 13.6a is the records of nine electrodes, selected from channels 1 to 128 with an interval of 16, including channel 41 which has the largest difference before and after standardization. In this example, the total *RE* (average) is 35.5%, while the total *RE* (standard) is only 0.6%. If using equation (13.18) to calculate the *RE* of each channel, we can obtain the RE distribution based on a single channel, as shown in Figure 13.6b. It can be seen from this figure that the maximum single-channel *RE* (average) and *RE* (standard) are located in channel 41, because this channel has the minimum signal energy, while the minimum single-channel *RE* (average) and *RE* (standard) are located in channel 15 because this channel has the maximum signal energy. The reason is that the numerator in the equation for calculating the single-channel *RE* is the reference electrode error or the incomplete standard error, which is the same for each channel, and the only difference is the denominator of the formula for calculating the *RE*, which is the signal energy of each channel. Therefore, the greater the channel signal energy is, the smaller the *RE* is. As can be seen from Figure 13.6a, the waveform recorded by the electrode with the average reference is quite different from that recorded by the electrode with the infinite reference, and the standardized record is almost the same as that recorded by the electrode with the infinite reference. Figure 13.6b shows that the single-channel *RE* decreases from a range of 3.91% (minimum) −690.61% (maximum) to 0.24% (minimum) −11.76% (maximum) respectively. Clearly, REST attains great success because the assumed sources here are in the shallow areas where REST works well (see Figures 13.2 and 13.3).

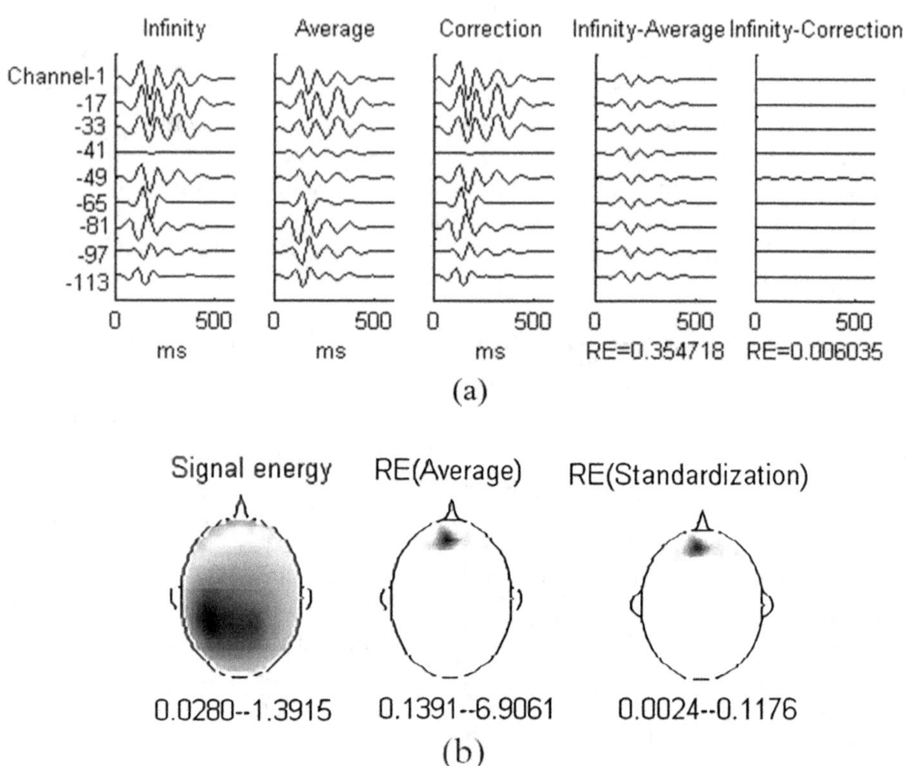

FIGURE 13.6 Simulated scalp records and standardized results. (a) Time-domain process analysis, the horizontal axis is in milliseconds, and the channel number is listed on the left side of the image in column 1. From left to right, the first is the infinite reference, followed by the average potential reference and the standardized record. The last two are related difference graphs. The total RE between the infinite reference and the standardized potential was 0.6035%. The total RE induced by the average reference was 35.4718%. (b) Channel-based signal energy and relative error distribution. Below each subgraph are the maximum (blackest dot) and minimum (whitest dot). From left to right, the first is the distribution of L2 mode signal energy on the scalp surface, with the maximum value of 1.3915 located in channel −15 and its Cartesian coordinates (−0.355361, 0.494277, 0.793353) and the minimum value of 0.0280 located in channel-41 and its Cartesian coordinates (0.762777, 0.070682, 0.642788). The second figure is the distribution of single-channel RE (average), with a maximum value of 690.61% and a minimum value of 13.91% in channel −41. The third figure is the relative error distribution after standardization, with a maximum value of 11.79% and a minimum value of 0.24% in channel −15 (Yao, 2001).

Figure 13.7 shows the normalized intensity of the Fourier spectrum for nine channels. It is clear that the difference between the ideal potential with the infinite reference and the actual potential with the average reference is obvious. This kind of spectrum difference in EEG rhythm analysis may lead to the misinterpretation of the dynamic process in the time domain. The results in Figure 13.7 also show that REST does a good job of reconstructing the actual spectrum, suggesting that EEG spectrum analysis can be more objective and realistic by applying REST.

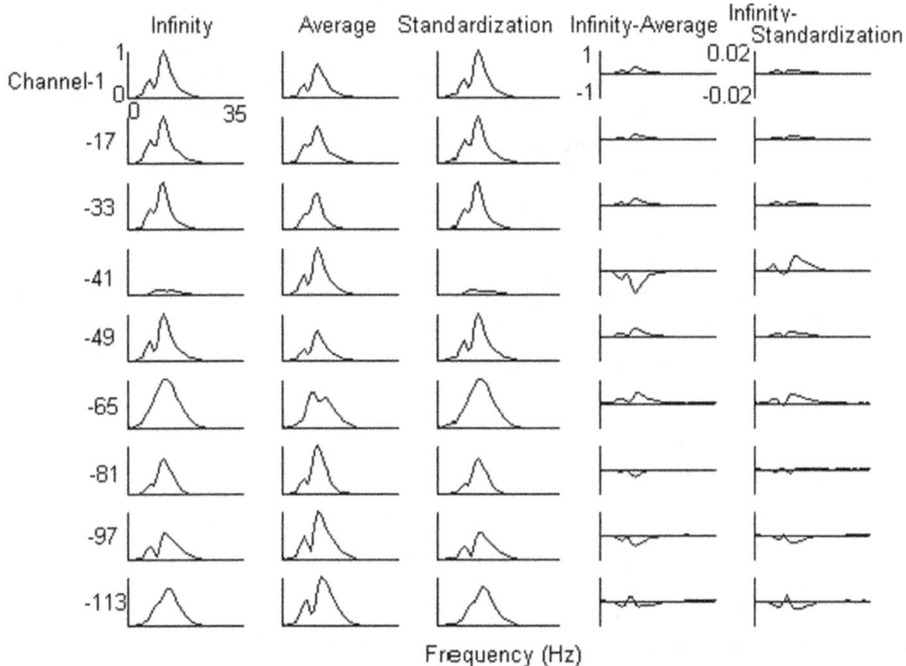

FIGURE 13.7 The normalized intensity of the Fourier spectrum on nine scalp recordings as in Figure 13.6. The first three sets of curves are normalized by the maximum value in each set respectively. The horizontal axis represents the frequency and displays a range from 0 to 35 Hz. The five groups of curves are one-to-one corresponding to the five groups of curves in Figure 13.6 from left to right. The first three-column graph is shown in the range [0, 1] because they are always positive, and the fourth column is shown in [−1, 1]. The fifth column is shown in the range [−0.02, 0.02] because their values are small (Yao, 2001).

13.2.2.4.2 The Actual Example

Figures 13.8 and 13.9 show an actual example. The number of electrodes in the experiment is 120, and their specific locations are composed of the projection of the average electrode positions of ten subjects on the best-fitting spherical surface, which is normalized into a concentric three-layer sphere model. The potential is also a set of average VEP data of ten subjects. Due to the obvious artifact on channel 86, only the remaining 119 channels are actually used. Visual stimulation is a checkerboard pattern that flashes in the right visual field. The recording period was from −200 ms before the stimulus to 1,200 ms after the stimulus. Based on the electrodes here (119 channels), the equivalent dipole layer model in Figure 13.1, and the concentric three-layer sphere model, R_a is recalculated for the model.

Figure 13.8 shows 9 out of 119 channels. Average is the actual record. Standardization is the result of REST. Standardization-Average is the component of reconstruction. Using the equation (13.18), let, $\Phi = Standardization$ and $\Phi_a = Average$, then the calculated RE is 58.31%. The largest RE for a single channel is 135.63% in channel 56 with coordinates of $(x, y, z) = (0.326, -0.861, 0.390)$.

Figure 13.9 shows the spectrum of nine channels before and after standardization. Figures 13.8 and 13.9 both show that REST introduces some large values into certain channels. As can be seen from Figure 13.8, some change can be found not only in the "potential value", but also in the polarity of the peak and the latency period. Since these parameters are basic parameters for interpreting the VEP data, standardization may have an important impact on the application of VEP.

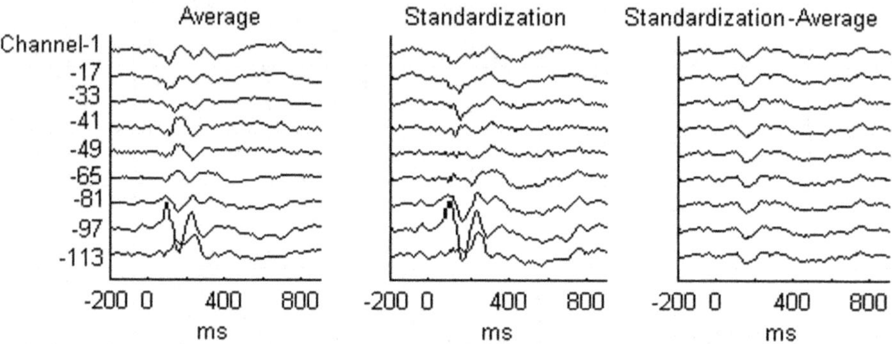

FIGURE 13.8 Actual VEP records. The unit of the horizontal axis is milliseconds, the average is the actual record, and the standardization is the result after REST (Yao, 2001).

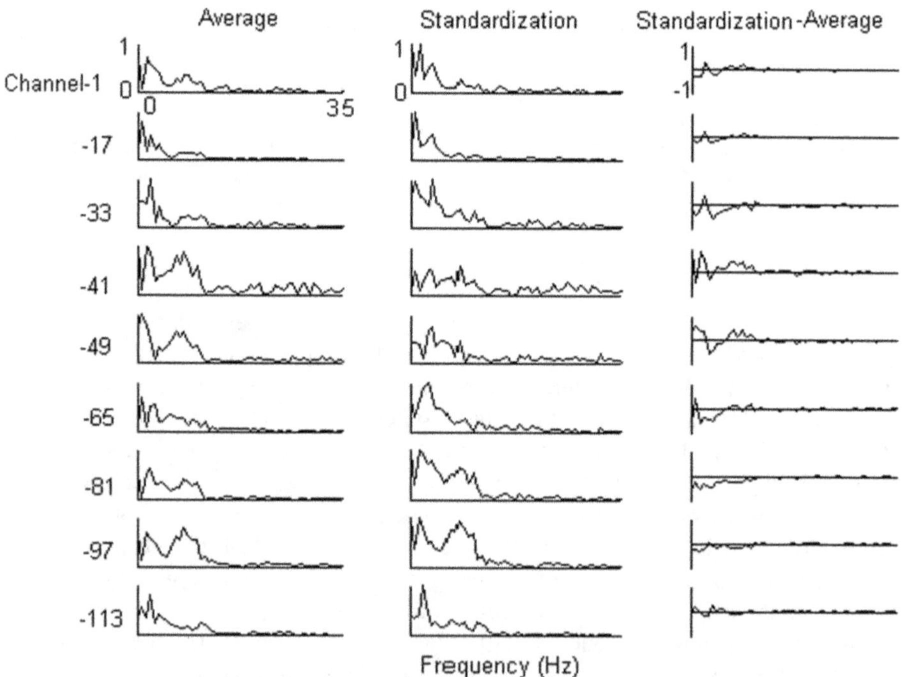

FIGURE 13.9 The normalized spectrum of nine scalp records in Figure 13.8. The horizontal axis represents the frequency, with a range from 0 to 35 Hz. The three sets of curves are normalized by the maximum value in each set respectively. The range is for the first two columns, and for the third column (Yao, 2001).

13.2.2.5 Summary

The above work is based on the fact that (1) the choice of a specific reference electrode does not change the theoretical connection between the source and scalp potential; (2) Since the real sources within the brain and their equivalent sources generate the same scalp potential, the potential with the infinity reference can be reconstructed from the equivalent source, which can be reconstructed from the recordings with a scalp surface point as reference or that with the average reference. The approximation of equivalent source reconstruction is determined by the approximation of the head model and the fact that the finite electrode array cannot provide comprehensive information. The simulation results show that the degree of approximation varies with the position of the source and the orientation of the dipole moment. Optimistically, the simulations show that REST works best in the superficial cortex, the most important area for various higher brain functions.

13.2.2.6 Discussions

13.2.2.6.1 Reference Problem in Sense of Algebra

Suppose there are n electrodes, we have one as reference such as x_n, then we actually will have $n-1$ recordings and $n-1$ equations, $y_1 = x_1 - x_n, \ldots, y_{n-1} = x_{n-1} - x_n$. However, what we want is the whole n recording $x_1, \ldots, x_n$. It means we have n unknown but we only have $n-1$ equation. Therefore it is an undetermined problem, we need supplementary information. Case 1, if we know the potential at electrode n, for example, we assumed that it is zero, then we get $x_i = y_i$, for $i = 1, \ldots, n-1$, and $x_n = 0$. This is what the various online references adopted, they assumed somewhere the potential is a constant, such as nose, chin, etc., but it's not true. Case 2, if we know that the sum of all the channels is zero, then we have another equation $\text{sum}(x_1, x_2, \ldots, x_n) = 0$ or $\text{Average}(x_1, \ldots, x_n) = 0$ thus totally we have n equations and n unknowns, the equation set is solvable, we can get $x_1, \ldots, x_n$. This is the case when the average reference is assumed. However, $\text{the sum} = 0$ is just an assumption that is not true in general. Case 3, if $\text{the sum}(\text{two mastoids}) = 0$, then we have the nth equation too, and the total equation set is n that the problem is solvable too. This is the assumption of the linked-mastoids/ears (*LM*) reference. Apparently, such an assumption is totally a beautiful desire, they are not zero in general. Case 4, the scalp recordings are generated by electrical sources or their equivalent sources in the brain, then we have the equation (13.10) to get the on-line reference channel x_n or $\text{Average}(x_1, \ldots, x_n) = v_a \neq 0$, and so we actually have n equations for the n variables x_{1-n}. This is the strategy in REST. Apparently, each re-reference strategy tries to find an additional equation to supplement the "lost channel" based on a special assumption. However, among the all efforts, only REST's assumption is not a hypothesis but a physical fact, and that's why REST is the best in theory and should be best in practice. There are two crucial points that REST originally understood and realized in solving the reference problem. First, it corrects the wrong idea that the reference problem is a problem in the time domain, and returns it to the physical fact that it is a problem in the space domain, due to volume conduction; Second, when facing the nonunique inverse problem in spatial domain, the equivalent source principle is timely introduced, which solves the dilemma of the EEG inverse problem. These two points are indispensable for the success of REST.

13.2.2.6.2 Postscript

The main purpose of REST is to reconstruct the true potential of the reference electrode in the EEG record. In another study, we evaluated REST for a realistic head, the effectiveness of REST is quite similar to the concentric three-sphere model (Zhai and Yao, 2004). These results led us to rethink the question of whether the true spectral characteristics of different brain states with zero-reference are consistent with what we have seen in previous recordings with non-zero reference electrodes and whether the EEG network obtained with other references is similar to the truth.

In 2007, Marzetti et al. conducted a very important work on large-scale interactions in the brain from EEG signals. In this work, the artifactual self-interaction by volume conduction is eliminated by using the imaginary part of the complex coherency as a measure of interaction, and the REST is used for the approximate standardization of the reference of scalp EEG recordings to a point at infinity that, being far from all possible neural sources, acts like a neutral virtual reference. After careful comparative studies of simulated and real EEG data, they argued that "the reference choice has, therefore, substantial effects on the analysis and the interpretation of EEG data". "The influence of an active reference on the coherency maps can be significantly reduced by the offline transformation of the EEG data obtained with the standardization procedure". "Moreover, since the REST transformation is a standardized procedure, a spectral and coherency mapping based on such data facilitates the comparison of results obtained from different EEG laboratories or stored with respect to different references in databases collected over time".(Marzetti et al., 2007).

Non-zero reference is a common component mixing in all channels, thus definitely changing the EEG correlation, phase synchrony, and coherence (Hu et al., 2007). In 2010, we conducted another important work on the effect of various references on the EEG network (Qin et al., 2010). A total of 100 randomly positioned source configurations, each consisting of two dipoles with coherent waveforms, were adopted for simulating EEG networks, meanwhile, a dense (129-channel), eyes-closed EEG was recorded from 15 subjects. Simulated data with infinity as reference and the real data were re-referenced to reconstructed infinity (REST), average (AR), linked mastoids (LM), and left mastoid (L) references. For simulated data, the effects of different references on coherence and network were investigated. For real data, the spectral properties of seven conventional EEG frequency bands were first analyzed, and then a default mode network (DMN) was constructed based on the coherence. The simulation showed that REST can exactly recover the true EEG network configuration. For real EEG data, significant differences among references were found for the power spectra, coherence, and DMN configuration. Compared with REST, the long-distance connectivity between anterior and posterior areas was strengthened by AR, and the connectivity over posterior areas was destroyed when LM and L were employed. Moreover, all comparisons demonstrated frequency-dependent reference effects. In summary, a non-neutral reference influences the power spectra, coherence, as well as network analysis. REST demonstrates its validity in data referencing, and meanwhile, AR is much closer to REST than the other references in terms of spectra and coherence. However, the DMN alters a great deal with AR. Due to this work, Professors Jürgen Kayser and Craig E. Tenke take REST as a "Rosetta Stone for scalp

EEG" (Kayser et al., 2010) in that it translates the mysterious recordings (blurred by the unknown reference signals) to the physically determined signals generated by the neural electrical activities. Meanwhile, Professor PL Nunez confirmed that REST is a good idea on the right track, though he expect some more evaluations(Nunez, 2010). Since then, REST got more and more attention and more and more critical evaluations, and the results all confirmed its unique value in recovering the true EEG potential, such as Liu et al. (2015) stated that "both AR and REST have relatively low reconstruction errors compared to LM, and that REST is less sensitive than AR and LM to artifacts mixed in the EEG data. For both AR and REST, high electrode density yields low re-referencing reconstruction errors. A realistic head model is critical for REST, leading to a more accurate estimate of a neutral reference compared to spherical head models. With a low-density montage, REST shows a more reliable reconstruction than AR either with a realistic or a three-layer spherical head model" (Liu et al., 2015).

In a simulation study, you may occasionally meet a case that AR is better than REST. Even so, we do not think that's a surprise. In fact, AR is a sum of all recordings, it is possible that the potential distribution over the scalp surface is antisymmetric by chance or positive and negative can be offset, such as a dipole oriented in cross-section and located at the origin of a sphere, the scalp surface potential sum will be quite close to zero, then the reconstruction error in REST will be an additional noise. However, for practice, who knew such an occasional neural source configuration appeared, thus we have to follow the general performance as shown by the simulations illustrated in this and the next chapter where REST is usually much better than AR, and that's enough for us to choose REST. Certainly, the head model in REST should not be too far away from the actual head shape, such as a homogeneous single sphere violates the core advantage of the REST algorithm by utilizing the head model to supplement "deficiency" of the other algorithms including AR and LM. Based on the reported simulation, an individual realistic head model is the best and the three concentric spheres head model is acceptable.

REFERENCES

Baski , A.A., A.A. Bartlett. 1986. Where is the V=0 equipotential. *Am J Phys* 54(9):854–854.

Bayley, R.H. 1957. Exploratory lead systems and zero potentials. *Ann NY Acad Sci* 65(6):1110–1126.

Bayley, R.H. 1959. Unipolar potential measurements in the electric field produced by an arbitrary dipole in a circular homogeneous lamina. *Circulation* 7:537.

Craib, W.H., R. Canfield. 1927. A study of the electrical field a surrounding active heart muscle. *Heart* 14:71–109.

Desmedt, J.E., V. Chalklin, C. Tomberg. 1990. Emulation of somatosensory evoked potential (SEP) components with the 3-shell head model and the problem of 'ghost potential fields' when using an average reference in brain mapping. *EEG Clin Neurophysiol* 77:243–258.

Duffy, F.H., V.G. Iyer, W.W. Surwillo. 1989. *Clinical Electroencephalography and Topographic Brain Mapping: Technology and Practice*. Berlin: Springer Science & Business Media.

Gencer, N.G., S.J. Williamson, A. Gueziec, et al. 1996. Optimal reference electrode selection for electric source imaging Electroenceph. *Clin Neurophysiol* 99:163–173.

Geselowitz, D.B. 1998. The zero of potential. *IEEE Eng Med Bio* 1:128–132.

Hu, S., M. Stead, G.A. Worrell. 2007. Automatic identification and removal of scalp reference signal for intracranial EEGs based on independent component analysis. *IEEE Trans Biomed Eng* 54:1560–1572.

Katznelson, R.D. 1981. Normal modes of the brain: Neuroanatomical basis and a physiological theoretical model. In P.L. Nunez (eds.), *Electric Fields of the Brain: The Neurophysics of EEG.* New York: Oxford University Press.

Kayser, J., C.E. Tenke, D. Malaspina, et al. 2010. Neuronal generator patterns of olfactory event-related brain potentials in schizophrenia. *Psychophysiology* 47(6):1075–1086.

Lehmann, D. 1988. The view of an EEG-EP mapper. *Brain Topogr* 1:77–78.

Liu, Q., J.H. Balsters, M. Baechinger, et al. 2015. Estimating a neutral reference for electroencephalographic recordings: The importance of using a high-density montage and a realistic head model. *J Neural Eng* 12(5):056012. doi: 10.1088/1741-2560/12/5/056012.

Marzetti, L., G. Nolte, M.G. Perrucci, et al. 2007. The use of standardized infinity reference in EEG coherency studies. *NeuroImage* 36:48–63. doi: 10.1016/j.NeuroImage2007.02.034.

Nunez, P.L. 2010. REST: A good idea but not the gold standard. *Clin Neurophysiol* 121:2177–2180. doi: 10.1016/j.clinph.2010.04.029.

Pascual-Marqui, R.D., D. Lehmann. 1993. Topographical maps, sources localization inference, and the reference electrode: Comments on a paper by Desmedt et al Electroenceph. *Clin Neurophysiol* 88:532–533.

Qin, Y., P. Xu, D. Yao. 2010. A comparative study of different references for EEG default mode network: The use of the infinity reference. *Clin Neurophysiol* 121:1981–1991. doi: 10.1016/j.clinph.2010.03.056.

Tomberg, C., P. Noël, I. Ozaki, et al. 1990. Inadequacy of the average reference for the topographic mapping of focal enhancements of brain potentials. *EEG Clin Neuro* 77:243–258.

Wilson, F.N., A.G. Macleod, P.S. Barker. 1931. The potential variations produced by the heart beat at the apices of Einthoven's triangle. *Am Heart J* 7:207–211.

Wilson, F.N., F.D. Johnston, A.G. Macleod, et al. 1934. Electrocardiograms that represent the potential variations of a single electrode. *Am Heart J* 9(4):447–458.

Wolferth, C.C., F.C. Wood. 1932. The electrocardiographic diagnosis of coronary occlusion by the use of chest leads. *Am J Med Sci* 183:30.

Yao, D. 2001. A method to standardize a reference of scalp EEG recordings to a point at infinity. *Physiol Meas* 22(4):693–711.

Zhai, Y., D. Yao. 2004. A radial-basis function based surface laplacian estimate for a realistic head model. *Brain Topogr* 17(1):55–62.

14 EEG Reference Selection

14.1 BACKGROUND

Which reference should we choose? This problem has been a debated issue since the first report of the human EEG in 1929 (Berger, 1929; Gloor, 1969). Berger had paied attention to this question at the very beginning. Instructively, from 1929 to 1938, in his primary 14 papers, Berger tested various types of electrodes, recording sites, and both bipolar and unipolar references (La Vaque, 1999). Thus, it can be seen that unipolar and bipolar recordings were almost simultaneously introduced from the discovery of human EEG.

At an early stage, there are some misleading examples of poor references. A case in point is the ever-popular linked-mastoids/ears (LM) reference, which was used in the 1930s by Gibbs and Lennox to study grand mal and psychomotor (partial complex) seizures (Gibbs et al., 1936). This work spurred international interest in the role of EEG in clinical epilepsy and firmly linked the term "psychomotor epilepsy" to a specific EEG pattern. However, the authors failed to accurately localize the origin of psychomotor seizures as the LM reference distorted the field map (Faux et al., 1990; Feindel et al., 2009; Stone and Hughes, 2013). Subsequently, it was shown that LM seriously biases EEG power (Niedermeyer, 1987) and coherence spectra (Fein et al., 1988), confounding the interpretation of results (Shaw, 1984; Travis, 1994). Furthermore, due to near the neck, this reference tends to pick up electromyography (EMG) and electrocardiogram (ECG) artifacts (Luck, 2014). It illustrates that an intuitively appealing reference may be fraught with difficulties. The problem is not limited to the LM. Many other body sites have been explored, such as the angle of the jaw, the chin, the tip of the nose, the neck, etc. These attempts were similarly problematic due to the contamination from EMG and ECG and the difficulty of interpreting the field maps.

In the 1940s, pursuing a better reference was inspired by ECG technology. Notably, the Wilson ECG common terminal reference sought a zero-potential reference by combining leads from three limbs (Section 13.1). This suggested the average reference (AR), first connecting all EEG electrodes through high resistances in order and then taking the common junction as a reference. In 1950, the first clinical use of AR was reported (Goldman, 1950), and it stated that "if the EEG sources consist of a large number of randomly placed and randomly oriented dipoles, a rather constant zero average will be obtained over the surface of the scalp. Experience with the average monopolar reference electrode shows that this is usually approached in practice" (Offner, 1950). The AR is currently one of the most widely adopted references. Also, it is now implemented by offline re-referencing instead of the original online recording setup. The assumption of 'sum to zero' behind AR was partly buttressed by the demonstration that the surface potential integral of a dipole in a layered spherical surface is zero (Bertrand et al., 1985). Hereafter, this theoretical result has been

DOI: 10.1201/9781032639260-14

thought to be true for the actual human head, as "it is important to note that the dipolar nature of ERP components means that every component is actually positive over some parts of the head and negative over other parts, summing to zero over the entirety of the head" (Luck, 2014). Given this belief, the AR has been advocated as the best reference option, as Nunez stated, "when used with large numbers of electrodes…, it often performs reasonably well" (Nunez and Srinivasan, 2006), and "AR errors are due to (1) limited electrode density and (2) incomplete electrode coverage (sampling only the upper part of the head). If these errors were fully eliminated (only possible in detached heads), the AR would provide the desired gold standard, that is, the nominal reference with respect to infinity" (Nunez, 2010). Alas, this sweet dream has recently been shown to be just a dream. The surface integral of EEG potential may not be zero for a homogeneous and isotropic realistic head geometry shape that deviates from a sphere (Yao, 2017) (Appendix F). Thus, AR as an approximation to zero potential is subject to many conditions and is not universally valid.

The most recent reference is REST, which was proposed to approximately reconstruct EEG potential with an infinity reference (Yao, 2001). REST utilizes the fact that EEG recordings are generated by neural electric sources but attenuated and mixed by the volume conductor. These electric sources are independent of any reference. Thus, the neural electric sources may be taken as the bridge to transform one reference recording to another, including to transfer a nonzero reference recording to a recording with zero reference. Besides, REST has been extensively evaluated with various simulations (Zhai and Yao, 2004a; Marzetti et al., 2007; Qin et al., 2017, 2010; Liu et al., 2015; Chella et al., 2017; Huang et al., 2018). Recently, the advantages of REST have been underscored by the demonstration that both REST and AR are particular cases of a unified reference estimator under the Bayesian statistical framework (Hu et al., 2018b) (Appendix G); the difference is that the prior requirement for REST is that the neural sources are independent identical distributions (IIDs), whereas that of AR is that the multichannel EEG recordings are iid, and such a uncorrelation assumption is not practical as the EEG recordings are generated by the same sources, and the volume conduction further enlarges the similarity among channels, thus they are definitely correlated. Such an essential difference will certainly have consequences for the relative performance of the two references. Before proceeding, we must emphasize that even if a zero reference is obtained, the mixing effect of volume conduction is still present in the signal. However, as we shall elaborate in the following, using a correct reference will solidify the basis for subsequent de-mixing and temporal analysis.

An approach that sidesteps the unipolar reference is to avoid studying electric potential but rather calculate the current source density (CSD) (Hjorth, 1975), an estimate of current flow through the scalp surface (Yao, 2002a) (Chapter 8). CSD is independent of the reference and helps localize brain activity close to the electrodes. However, as a Laplacian spatial filter, CSD is more sensitive to noise with broadband spatial spectra than to physiological sources, and it probes local shallow neural activities at the expense of neglecting widely distributed deep source activities. It should be noted that, in some literature, CSD was called "Laplacian reference", which is obviously a misunderstanding of reference as reference is a special issue for voltage and CSD (Laplacian) is not voltage at all.

Additionally, the bipolar recording widely used in clinic practice is not a unipolar reference. Bipolar recordings are the potential differences of two nearby electrodes; they are proportional to the local current through these two electrodes.

In summary, "the reference is an absolutely fundamental aspect of EEG/ERP recordings. If you don't fully understand referencing, you won't understand the signal that you are recording" (Luck, 2014). In practice, everyone agrees that the lack of consensus in reference choice causes considerable confusion. This chapter is to review recent progress in this domain (Yao, 2017; Hu et al., 2018a,b, 2019; Nunez and Srinivasan, 2006; Nunez, 2010). This chapter is modified from Yao et al., 2019. It can also be seen as a deepening of the previous Chapter 13. In general, "if you really want to get an estimate of the absolute voltage, REST seems like the best current approach" (Luck, 2022). Meanwhile, we will provide the available options for different scenarios.

14.2 EEG REFERENCE PROBLEM

14.2.1 No Constant Point on the Scalp Surface

Assuming a head with several compartments, each with homogeneous and isotropic conductivity, Poisson's equation is valid for EEG and ERP potentials (Gulrajani, 1998):

$$\nabla^2\Phi(\vec{r}) = -s(\vec{r}) \tag{14.1}$$

where Φ is the potential recorded by the electrode placed on the scalp and s is an equivalent current source density (Plonsey and Heppner, 1967; Malmivuo and Plonsey, 1995; Yao and He, 1998), located anywhere in the brain that generates the potential Φ. Combining (14.1) with the boundary conditions over the scalp surface and internal interfaces of compartments, the scalp EEG/ERP may be linked to the neural current source inside the brain, laying the basis for the EEG/ERP forward and inverse problems (Yao and He, 2003). Based on equation (14.1) and the head model with boundary conditions, the physiological potential Φ with the infinity zero reference related to the neural electric sources S is

$$\Phi = G_{\infty}S \tag{14.2}$$

Here, G_{∞} is known as the leadfield matrix, expresses the forward model theoretically with the infinity zero reference.

For a dipolar source, the scalp potential is positive over one part of the head and negative over the other, and the potentials at the boundary between the positive and the negative are zero. In Figure 14.1, the simulated EEG was generated by two dynamical sources inside a realistic head model. The head model was built using FEM with the SimBio pipeline (Vorwerk et al., 2018), and the conductivities of gray matter, skull, and scalp were set as 0.33, 0.01, and 0.43, respectively. The electrode set is the EGI GSN-HydroCel-129.sfp. The source space is totally consisted of 3,471 * 3 dipoles, oriented to the x-, y-, and z-axes. Two dipoles in the right and left hemispheres with different time series but orientations fixed to the x-axis were

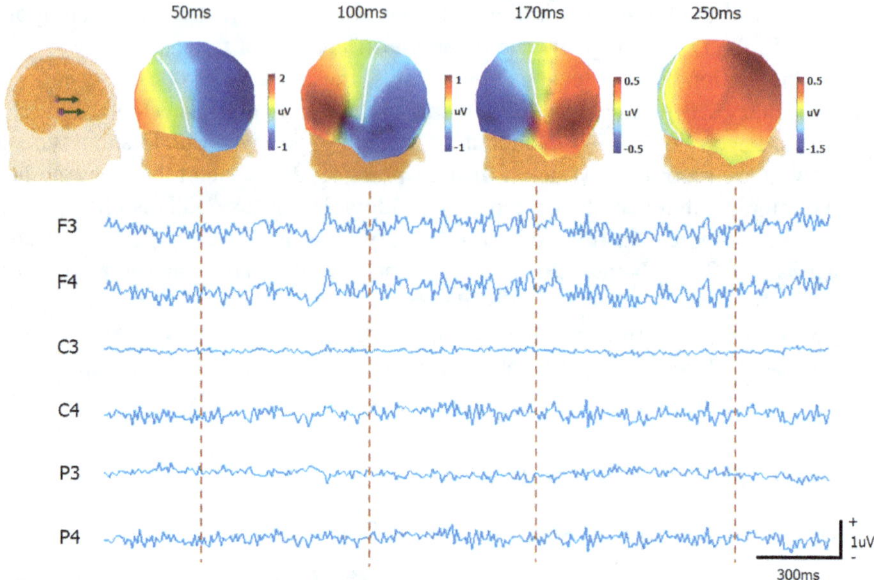

FIGURE 14.1 Simulated data shows the inconsistency of the zero potential lines at different temporal moments. The heads show the position of two dipoles and the topographic maps at four time samples, where the white curve over the scalp is the zero-potential curve, which is dynamical momently. The traces show the EEG temporal processes over six electrodes (Yao et al., 2019).

taken as the active neural sources to produce the scalp EEG without considering the recording noise. The topographic maps of EEG/ERP at different instants were displayed. However, due to the time-varying nature of the neural electric sources, this zero-boundary curve is never static, and one cannot practically use it as a reference; that is, there is no time-invariant 'zero' potential point on the head.

In the cognitive ERP domain, this physical phenomenon is termed the "no-Switzerland principle" (Luck, 2014): there is no electrically neutral site on the head or body surface. It is underscored that any online reference deviates from the infinity zero reference. Therefore, any waveform recorded with an online unipolar reference is the potential difference between one active electrode and a non-zero reference electrode, which is actually another active electrode, and both the two electrodes record the potential generated by the same neural source activities. In other words, it is the neural current distribution in the whole brain that produces EEG activities at all electrodes, including the reference electrode, as shown in equation (14.2).

14.2.2 General Form of the Reference Problem

In practice, one can never observe $\Phi \in R^{N_c \times 1}$, what one observes instead is the referenced data $\Phi_r \in R^{N_c \times 1}$. That is the linear transformation via pre-multiplying the reference operator $T_r \in R^{N_c \times N_c}$ with the clean physiological potentials Φ adding the sensor noise ε. The general form of the reference problem is modeled as

$$\Phi_r = T_r \Phi + \varepsilon_r \tag{14.3}$$

The EEG reference problem in equation (14.3) is apparently an underdetermined linear regression problem, where T_r is a non-stochastic matrix of observations, Φ are potentials with the infinity reference, supposed to be a deterministic, fixed but unknown vector, $\varepsilon_r = T_r \varepsilon$, ε are non-observable random sensor noise disturbances.

Without loss of generality, Φ_r and ε_r are considered to have a multivariate normal distribution. If the sensor noise has an IID across channels, the covariance of the sensor noise in the referenced data will be $\Sigma_{\varepsilon_r \varepsilon_r} = \sigma^2 T_r T_r^T$ because referencing effect is taken on the noise as well during recording (Pascual-Marqui et al., 1994; Hu et al., 2018b).

In this study, T_r of unipolar references is the overwhelming body of the EEG reference issue. Furthermore, T_r can also be the first derivative in the bipolar recordings, which is proportional to the local current density between two adjacent electrodes, and the second differential operator in the scalp Laplacian, a possible approximation to the current source density. The latter two are different physical quantities from EEG potentials.

In (Hu et al., 2018b), it was demonstrated that any estimator of the potentials referenced to infinity is the maximum a posteriori (MAP) estimator

$$\hat{\Phi}_r = \Sigma_{\Phi\Phi} T_r^T \left(T_r \Sigma_{\Phi\Phi} T_r^T + \sigma^2 T_r T_r^T \right)^+ \Phi_r \tag{14.4}$$

where + denotes the Moore-Penrose pseudoinverse. The difference between $\hat{\Phi}_r$ estimators is the prior covariance $\Sigma_{\Phi\Phi}$ assumed for the potentials at infinity. This result is surprising since it brings many techniques till now considered totally different under a unique Bayesian statistical framework. If the potentials Φ with infinity reference are priori IID over all electrodes, the estimator will correspond to the AR, whatever the unipolar reference is. By contrast, if the potentials Φ with an infinity reference are generated by the IID neural sources, the REST estimator is derived. Physically, IID over all electrodes for AR is unreasonable for practical EEG as the scalp signals are generated by the common sources underlying the scalp, and IID neural sources for REST are possible as all the actual sources together (a source distribution) can be decomposed into a few independent source distributions mathematically (Hu et al., 2019). Thus, the assumption of REST is much more reasonable than that of AR.

14.2.3 Indeterminacy Principle of Scalp EEG

The origin of the EEG reference problem is volume conduction, which leads to an "indeterminacy principle of scalp EEG". Looking at equation (14.2), on the one hand, it is only by volume conduction that we can observe the scalp potentials; on the other hand, volume conduction obstructs precise knowledge of source positions and their time series. From the work by Helmholtz in the 19th century, it is well known that the EEG inverse problem does not admit a unique solution (Chapter 10). In fact, any surface potential distribution can be equivalently generated by a closed source layer containing all the true sources, or alternatively, by a multipole series expansion at the coordinate origin (Yao, 2000; Yao and He, 2003). This equivalence makes it generally very difficult to estimate the precise source activities.

It is clear that any solution to the EEG reference problem must be based on volume conduction, thus transforming it into a physically spatial problem (Yao, 2001; Hu et al., 2018b). Obviously, this fundamental principle has been ignored in previous efforts, that considered the reference problem a purely mixing issue over time. For example, there are some recent efforts that make the unrealistic assumption that the reference signal is statistically independent from (or orthogonal to) the true EEG signals. This leads to mathematically tractable solutions based on various blind source separation methods, e.g., temporal independent or principal component analysis. Unfortunately, the indeterminacy principle indicates that the reference signal is generated by the same source as all other active electrodes. Apparently, the temporal evolution of the reference signal should be quite similar to that of nearby channels, breaking the assumptions of temporal independence. Thus, the reference problem is not because of any temporal process but rather due to spatial volume conduction. An upshot of this is to tackle the reference problem by considering volume conduction.

14.3 UNIPOLAR REFERENCES

Unipolar reference is regarded if all electrodes are referenced to a unique physical reference or a unique virtual reference (Figure 14.2). The physical reference is usually an electrode (e.g., Cz, Fz, Oz, and FCz) placed over the scalp or the body surface during online recording. The virtual reference is the linearly combined signal of the recordings from all or part of the electrodes during offline processing after the EEG data acquisition. Typical examples of virtual references are REST, AR, and LM.

The reference operator in equation (14.3) for unipolar references (Hu et al., 2018a) is commonly as

$$T_r = I_{N_c} - 1 f_r^T \tag{14.5}$$

where I is a vector of ones and $f_r \in R^{N_c \times 1}$ consists of the linear combination weights of all the electrodes. The brief communication (Hu et al., 2019) demonstrated that the unipolar references have the 'no memory' property and the 'rank deficient by 1' property, namely, T_r are all full rank deficient by 1 for all the unipolar references (Appendix G).

14.3.1 Recording Reference

It is prevalent that the EEG is recorded with respect to a single physical reference electrode, such as the Cz, left or right earlobe, chin, etc. For these online recording references, the reference operator equation (14.5) is

$$T_{RR} = I_{N_c} - 1 f_{RR}^T, \quad f_{RR} = [0,\ldots,1,\ldots,0]^T \tag{14.6}$$

which is a vector of zeros except for a unique entry being 1 at the corresponding index of the reference electrode.

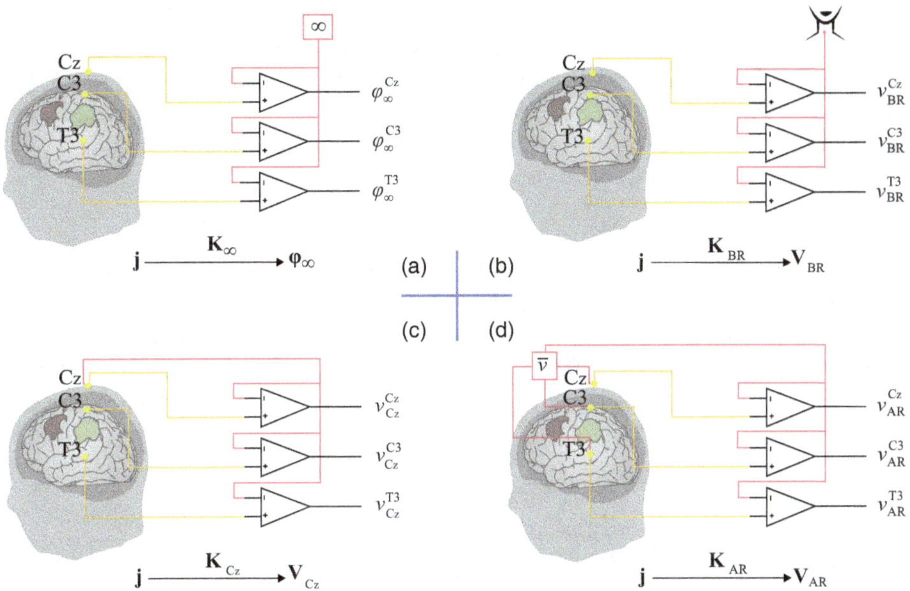

FIGURE 14.2 Schematic representation of unipolar references. (a) Infinity/zero reference; (b) recording reference electrode at neck; (c) recording reference electrode Cz; and (d) average reference. (Reproduced from Hu et al. (2018b).)

Early on this single site was chosen with the guess that it would be less or not active compared to the other sites. A seemingly promising approach is to select the reference as far as possible from electrodes presumed to reflect the neural activity of interest. For instance, to study the state of the left temporal lobe, the right ear may be taken as the reference. The main problem is that all channels record contributions from both the active electrodes as well as the reference site. The no-Switzerland principle – 'no point on the scalp or the body surface with the neutral potential', makes it a fantasy to separate the EEG activity with an infinity reference from that with a body reference.

However, as one can now easily re-reference digital EEG offline and given the 'no memory' property of unipolar references (Appendix G), the importance of selecting a proper recording reference has decreased. In current practice, Cz is widely adopted as the online recording reference since it is easy to secure the electrode contact, avoiding additional artifacts being injected. Offline digital processing is a feasible way to correct the recording reference by reconstructing a neutral reference. There are several typical attempts, such as REST, AR, and LM, as discussed in the following.

14.3.2 Linked Mastoids

LM assumes that the average of the potentials recorded over two mastoids (ears) is close to zero or neutral, where the reference operator in equation (14.5) is

$$T_{LM} = I_{N_c} - 1 f_{LM}^T, \quad f_{LM} = [0,\ldots,0.5,\ldots,0.5,\ldots,0]^T \tag{14.7}$$

which is a vector of zeros except for two entries being 0.5 at the corresponding indexes of two mastoids (ears). The studies using LM are usually exploring the data recorded from electrodes at the middle line of the scalp, such as F3, Fz, F4, C3, Cz, C4, P3, Pz, P4, O1, Oz, O2, etc. However, this subjective choice, not based on any physical principle, resulted in ambiguities, for example, in studies of N170.

14.3.3 Average Reference

One attempt to estimate the potentials Φ with infinity reference in equation (14.3) is by means of the AR. It is justified that for a perfectly layered spherical head with neural currents inside, the integral of the potentials over the scalp surface is zero (Bertrand et al., 1985; Yao, 2017). Thus, the average potential over all electrodes might tend to zero, and then AR would be suitable as the reference signal. In our formulation (14.5), it is

$$T_{AR} = I_{N_c} - 1 f_{AR}^T, \quad f_{AR} = 1/N_c \tag{14.8}$$

where f_{AR} is a vector of $1/N_c$ for all the electrodes.

Given the 'rank deficient by 1' property, the T_r is always singular. The estimation of Φ in equation (14.3) is thus a generalized linear inverse problem. The minimum norm solution is the special case of equation (14.4) with the prior $\Sigma_{\Phi\Phi} = \alpha^2 I_{N_c}$ and the assumption σ^2 tending to zero (Hu et al., 2018b)· Thus, the solution finally simplifies to

$$\hat{\Phi}_r = T_r^{+}\Phi_r = T_r^{+}T_r\Phi = T_{AR}\Phi = \Phi_{AR} \tag{14.9}$$

Here, we adopted the orthogonal projector centering property $T_r^{+}T_r = T_{AR}$ (Hu et al., 2018b, 2019).

This means that the minimum norm solution of equation (14.3) with any T_r is same as applying the AR to the potentials. This fact shows that AR has a unique position in the unipolar reference family, that is, starting from any unipolar reference, under the assumption that the scalp surface potentials are IID, the ceiling of the approximate zero reference potential estimate is the AR potential.

14.3.4 REST Zero Reference

As presented in the last chapter, REST recognized the fact that EEG potential recordings are ultimately generated by the same sources S in equation (14.2), whatever reference is used. Therefore, the following version of equation (14.3) is valid

$$\Phi_r = G_r S + \varepsilon_r \tag{14.10}$$

where $G_r = T_r G_\infty$ is the modified forward model with the same reference as in the EEG data. With the covariance of the equivalent source over time as Σ_{ss}, similar to (14.4), the solution of equation (14.10) may be expressed as

$$\hat{\Phi}_{\text{rREST}} = G_\infty \Sigma_{ss} G_r^T \left(G_r \Sigma_{ss} G_r^T + \sigma^2 T_r T_r^T\right)^{+} \Phi_r \tag{14.11}$$

This is the regularized version of REST (rREST). If assuming the equivalent source covariance is $\Sigma_{ss} = \alpha^2 I_N$, and σ^2 tends to 0, which is the case of noise-free data, it turns into the REST transforming (Hu et al., 2018b)

$$\hat{\Phi}_{\text{REST}} = G_\infty \left(G_r^+ \Phi_r \right) = \left(G_\infty G_r^+ \right) \Phi_r = R_r \Phi_r \tag{14.12}$$

where $R_r = G_\infty G_r^+$ is the reference standardization matrix depending on T_r and the equivalent source is approximately estimated as $\hat{S} = G_r^+ \Phi_r$ (Yao, 2001).

The REST operator is defined as

$$T_{\text{REST}} = G_\infty \left(T_r G_\infty \right)^+ T_r \tag{14.13}$$

It is demonstrated in Appendix G (Hu et al., 2019) that the REST operator is a unipolar reference and admits the ‘no memory’ property with

$$T_{\text{REST}} = I_{N_c} - 1 f_{\text{REST}}^T, \quad f_{\text{REST}} = G_\infty^{+T} G_\infty^{+} 1 \Big/ \left[1^T G_\infty^{+T} G_\infty^{+} 1 \right] \tag{14.14}$$

When one assumes that EEG data are generated by brain sources, REST is, in theory, the optimal for estimating the potentials at infinity. With additional channels in forward calculation, the EEG potentials at the missing channels or bad channels can be recovered by the interpolation function of REST (Dong et al., 2021).

The goal of REST is not to find the actual sources, which we do not need to disentangle. One may take a closed distributed dipole layer with all actual sources inside the layer as the equivalent sources (Yao, 2000; Yao and He, 2003). Then it is a linear equation from the scalp data Φ_r to the strengths of the equivalent sources with fixed positions. Since the number of sources on the equivalent layer is usually much larger than the independent recordings, equation (14.10) is an undetermined system. Thus, the pseudoinverse of G_r can be adopted to get the minimum norm solution of s. Equations (14.10) and (14.12) also show us that the sources s just play a role of bridge from s to Φ and that is the chance for REST from any unipolar reference recordings to Φ at infinity (Yao, 2001). This argument can be phrased in the Bayesian framework: the source distribution only enters the estimation as a prior. REST is the maximum a posterior estimator for which the equivalent source approach offers the protection against the interference of the actual source mislocalization.

14.4 COMPARISON OF UNIPOLAR REFERENCES

14.4.1 Reconstruction of Reference Signal

Figure 14.3 is the diagram of the unipolar references, such as AR, LM, Cz, and REST. The simulation scheme is the same as that in Figure 14.1. The referenced EEG and the reference signal traces (1–300 ms) are displayed. The signals of AR and LM are obtained from the average of potentials over all channels and two channels of mastoids, respectively. The head model in REST is built by FEM, but the equivalent source space is consisted of 27,921 * 3 discrete dipoles with x, y, and z directions. The signal of REST is the difference between the forward recordings with the infinity reference in Figure 14.1 and the REST reconstructed recordings. The results showed

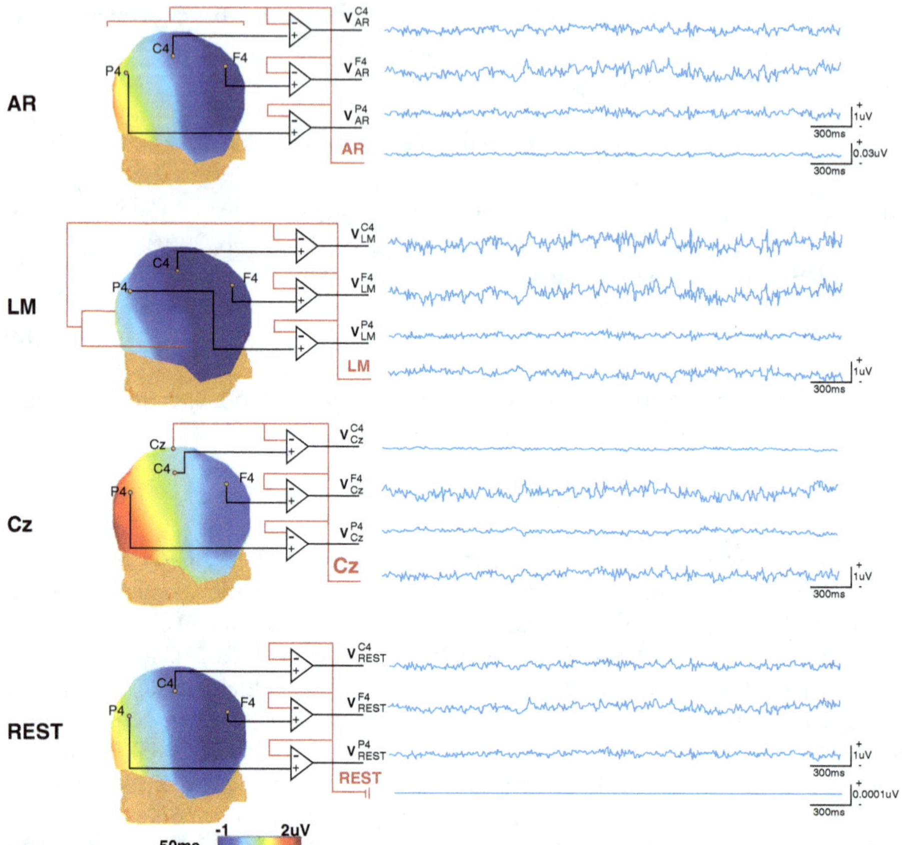

FIGURE 14.3 Simulated data illustrates the reference signals. With the same simulation procedure as in Figure 14.1, only two dipoles were used to simulate the EEG potentials, and the sensor noise was not considered. The re-referenced scalp potentials and the topographies at 50 ms were displayed. The reference signals are displayed at the last trace of each panel. (From Yao et al. (2019).)

the signals of LM and Cz are evident, and that of AR is small but nonnegligible, while REST almost recovers the actual zero potential. Recently, the AR signal in Figure 14.3 was named the "yielded average oscillation" (YAO) by REST and found as an electrophysiological signature of the resting-state fMRI global signal (Huang et al., 2018). This fact also means that the YAO signal (the lost AR signal) does have physiological significance.

14.4.2 Sensitivity to Errors in Head Model

The theoretical advantage of REST is its use of the volume conductive model. In practice, the volume conduction model depends on the following factors: (1) the co-registration of electrodes with the scalp surface built from a structural MRI T1 image; (2) the neural sources model; (3) the head geometry model; and (4) the conductivities of head tissues. One can only approximate that the living human head consists of complex biological tissues and structures. No matter how fine the numerical head model is, it will still deviate from the truth in the sense of anatomy and physiology. In addition, the neural sources can only be modeled with the assumption on the number and position of the actual sources. Furthermore, the electrode location deviation will introduce an error to the volume conduction model as well. The preciseness of the volume conduction model is a common issue not only for REST but also for the source inverse solution. For AR and REST, one may conjecture that the additional 'forward' step to the 'inverse' step, that is, the estimation to the equivalent source in REST, makes REST probably more sensitive to the accuracy of the head model than AR.

To mitigate this concern, some primary evaluations have been presented in the last chapter, and what (Hu et al., 2018a) tested here is taking a very fine volume conduction model with infinity reference in generating EEG potentials, then reconstructing the potentials using REST, where the volume conduction model should be an alternative or perturbed by injecting errors to that in the simulation (Hu et al., 2018a). Investigated five alternatives shown in Figure 14.4a and perturbated volume

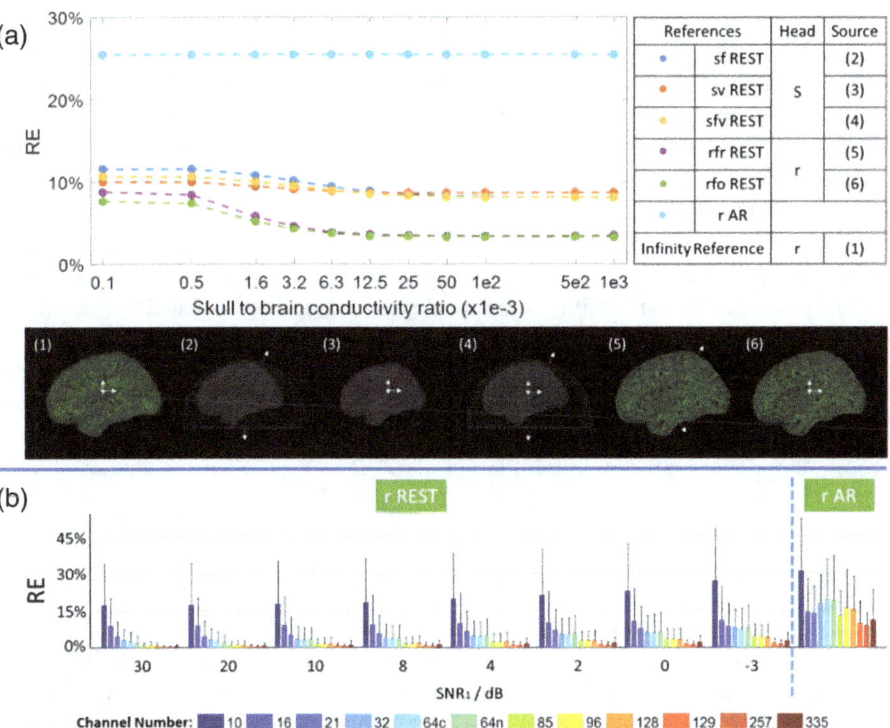

FIGURE 14.4 Simulated data tests the robustness of REST to errors in head models. (a) RE – the relative error; 1e2, 5e2, 1e3, and 1e-3 mean 100, 500, 1,000, and 0.001, respectively. (b) The numbers in the *x*-axis are the SNR in dB. Gray bars -- the standard deviation of the REs over different sources. More detailed explanations are presented in the text. (Reproduced from Hu et al. (2018a).)

conduction models by injecting the gaussian noise at different levels shown in Figure 14.4b. Specifically, the very fine volume conduction model in generating the EEG potentials is displayed in Figure 14.4a (1), and the one used in REST is illustrated in Figure 14.4a (2)–(6) with different source numbers and orientations. Using the prefixes to indicate the head shape and source configuration, the five alternatives are: 'sf' – homogenous spherical head and cortical surface dipoles with radial orientation, 'sv' – homogenous spherical head and brain volume dipoles with orthogonal direction, 'sfv' – homogenous spherical head and cortical surface dipoles together with volume dipoles, 'rfr' – realistic head and cortical surface dipoles with perpendicular (R) orientation, and 'rfo' – realistic head and cortical surface dipoles with orthogonal directions. Figure 14.4b shows perturbating the volume conduction model used in generating the EEG potentials with the gaussian noise at different levels than the one used in REST. The results demonstrate that REST is robust to reach a less potential error than the AR.

14.4.3 Sensitivity to Neural Source Position

To test the sensitivity of unipolar references to the source position, simulations were newly conducted with each source position repeatedly, as shown in Figures 13.2 and 13.3. Using the same volume conduction model in Figure 14.1, the scalp potentials were generated by each of the 3,471*3 dipolar sources individually. For REST, we use the same head model built by FEM, but the equivalent sources were 27,921*3 dipoles with x, y, and z directions. Relative error (RE) for each dipole between the forward noisy free scalp potential with an infinity reference and the potential with a different reference is calculated and displayed at its location. The display of the RE distribution of the potentials generated by each source with different references is shown in Figure 5 in Yao et al. (2019). It is shown that REST is always of the closest in contrast to the forward infinity reference potential, and AR is usually much better than LM and Cz. For AR, LM, and Cz references, the errors depend on the dipole location and orientation. In addition to the FEM-based forward model, simulated results of potential errors before and after referencing were also investigated by using the spherical head model and the boundary element method (BEM)-based realistic head model (Zhai and Yao, 2004a; Hu et al., 2018b). And all these studies showed similar results.

14.4.4 Sensitivity to Signal Noise Ratio and Head Model

There are two factors that might potentially affect AR or REST estimates of the potential at infinity. The first is the EEG signal to the sensor noise ratio (SNR), which will always affect the performance of inverse methods. Actual EEG measurement would not only record the physiological potentials but also unavoidably introduce the sensor noise. The second factor is the sensitivity of AR and REST to the underlying assumptions that lead to the estimator. As discussed before, AR conceptually depends on the measurement of potentials over a spherical head as well as the assumption of IID recordings. REST, in turn, utilizes the assumption of IID sources and a specific volume conduction model. The regularized versions of AR and REST, termed rAR and rREST (Hu et al., 2018b), therefore allow us to access these factors.

We carried out a simulation based on 89 individual realistic lead fields obtained from 89 subjects in the Cuban Human Mapping Project database (Hernandez-Gonzalez et al., 2011). Two patches of 150 dipolar sources with a four-order autoregressive model are used to produce the source time series; the individual lead fields are used to generate the potentials for each subject; and the different SNRs were set as 20, 8, 4, and 2 dB. For rAR and rREST, generalized cross validation (GCV) is used to select the denoising parameter (Chung et al., 2014). A usual direct measure to evaluate references is the RE of referenced potentials against the simulated potentials with reference at infinity. Four alternative head models were explored, and for the details of these four head models, please refer to (Hu et al., 2018b). The results are shown in Figure 14.5.

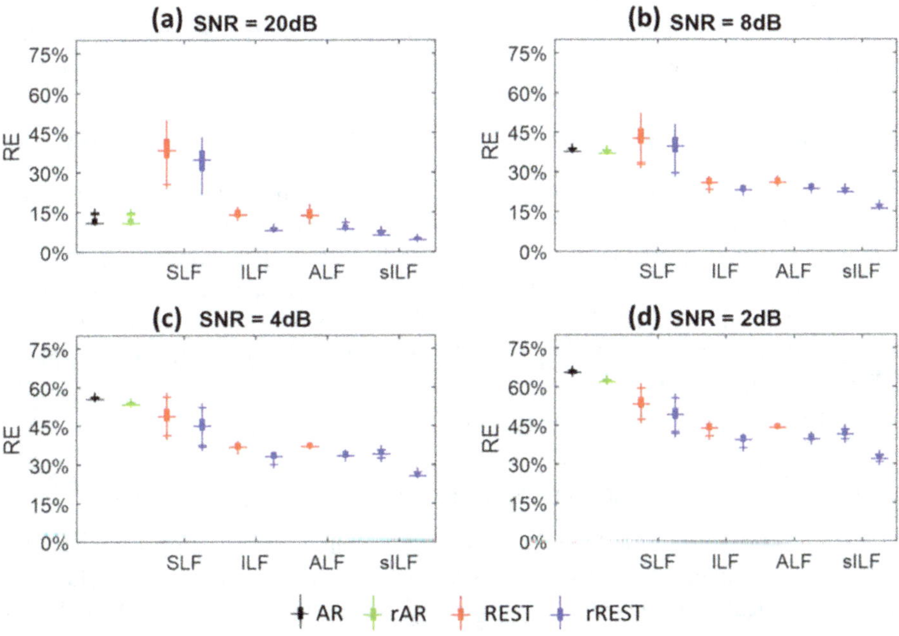

FIGURE 14.5 Simulated data tests the robustness of AR and REST on the head model and sensor noise. SLF: three-layer concentric spherical head model; ILF: normalized individual head model; ALF: averaged head model over 89 subjects; sILF: individual head model with sparse sources; rAR: AR with denoising; and rREST: REST with denoising parameter selection and a finer head model than SLF. The panels (a)–(d) are with different noisy levels. (Reproduced from Hu et al. (2018b).)

The adopted leadfields are from concentric three spheres head model (SLF), individual realistic head model (ILF), average realistic head model (ALF), and sparse individual ead model (sILF) for which the voxels not used were switched off in the simulations. In high SNRs, the matching of the head model used for REST and the one used for simulation becomes the main factor affecting the relative error. Therefore, REST with SLF is worse than AR with SNR = 20 and 8 dB.

With the SNR decreased, the impacts of noise overwhelmed those of the matching of head models for REST. For SNRs = 4 and 2 dB, any of the REST models performs better than the AR. It is the denoising technique that greatly reduces the relative error of REST, although using the SLF.

The rREST is more robust than the rAR with the regularization technique in terms of denoising.

The rREST models achieves less relative error than REST; for sophisticated studies, better accuracy is achieved with the most accurate head and source models by rREST.

Unless the real EEG recording has an extremely high SNR, REST with SLF can also be used without the expense of building the realistic head models for which a structural MRI is definitely needed.

In general, the averaged lead field (ALF) over a population and the denoising technique of GCV should be used in the rREST practice.

14.5 IMPACT OF UNIPOLAR REFERENCE ON REAL DATA ANALYSIS

Improper unipolar reference may introduce an unknown nonneutral value momently to all active channels, as shown in Figure 14.3, thus we need information criteria for reference model selection (Hu et al., 2018b) and evaluate the effects of reference on waveform-related parameters, such as amplitude, latency, spectra, and their derived measures like coherence (Marzetti et al., 2007), network (Qin et al., 2010; Chella et al., 2016; Huang et al., 2017), bi-spectra (Chella et al., 2017), and statistical test (Tian and Yao, 2013). Here are examples related to the information criterion, spectra, amplitude, and latency of ERP shown below.

14.5.1 Evaluation of References by Statistical Information Criteria

In Section 14.4, we reported the results of a statistical comparison of different references using simulated EEG data. However, it is much more important to evaluate the actual statistical adequacy of the different REST models against real data. This is carried out in Hu et al. (2018b) by employing several statistical information criteria that balance the goodness of fit and model complexity (rAR, rREST-SLF, rREST-ILF, and rREST-ALF). Involved are the three information criteria for the model selection: GCV, Akaike information criteria (AIC), and Bayesian information criteria (BIC). The results of this evaluation for the 89 subjects of the Cuban Human Brain Mapping database (Hernandez-Gonzalez et al., 2011) are shown in Figure 14.6, where GCV, AIC, BIC and residual sum square error(RSS), degree of freedom (DF) and regularization parameter(LMD) are all defined according to parameters in the general ridge regression form solution of equation (14.10) (Hu etal., 2018). In statistical model

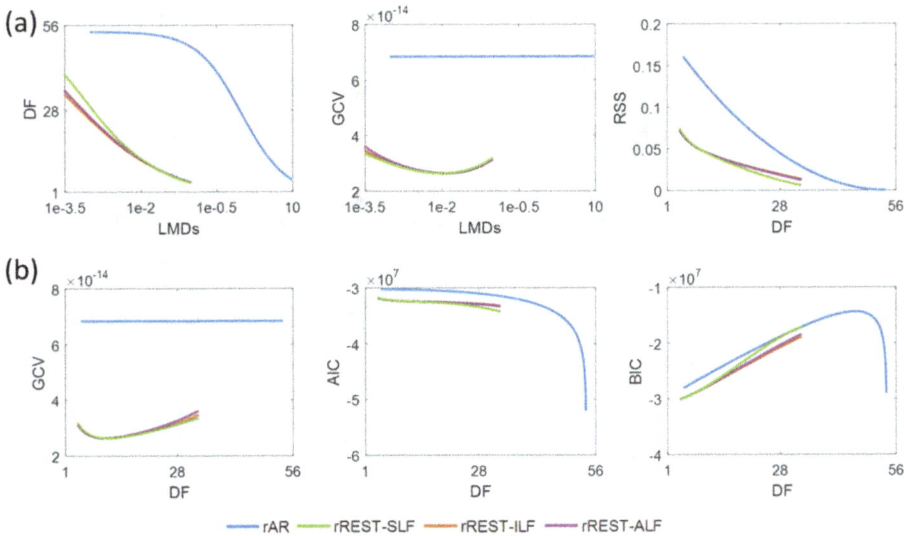

FIGURE 14.6 Recorded data derives the evaluation of references by the statistical information criteria. (a) DF and GCV against LMD, and RSS varying with DF; (b) model selection criteria (GCV, AIC, and BIC) against DF. DF: degree of freedom; GCV: generalized cross validation; RSS: the potentials residual sum of square; AIC: Akaike information criterion; BIC: Bayesian information criterion; and LMDs: the regularization denoising parameters. SLF: spherical head model; ILF: individual head models; ALF: averaged head models over 89 subjects; rAR: denoising average reference; and rREST: regularized reference electrode standardization technique. (This figure is reproduced from Hu et al. (2018b), and for the details of these parameters in Figure 14.6, please also refer to Hu et al. (2018b).)

selection, the model with smaller information criteria is preferred (Robert, 2007; Konishi and Kitagawa, 2008). As is evident from the curves, any of the REST models achieves much smaller values for the statistical criterion than the AR, except for BIC of rREST with SLF when the DF is around 28. And for all practical purposes, the performances of all the rREST with realistic head models (ILF, ALF, and sILF) are equivalent. The last plot in panel B shows that when DF is around 28, the BIC of rREST with SLF coincides with the BIC curve of rAR, reinforcing our simulated result in Section 14.4 that unless there is an extremely high SNR, one can use the simplest SLF.

14.5.2 EEG Power Spectra

For the spontaneous EEG with a nonneutral reference signal mixed in other channels, the scalp power spectra map might be altered systematically. Figure 14.7 shows the results of alpha-1 (7.5–9.5 Hz) and alpha-2 (10.0–12.5 Hz) of 11 subjects in EEG with their eyes open. These results confirm that different references result in systematic changes in the distribution of EEG spectra power. It is therefore necessary to adopt a common prevalent reference and reduce the effect of such systematic shifts, guaranteeing the interpretation to be consistent from different Labs.

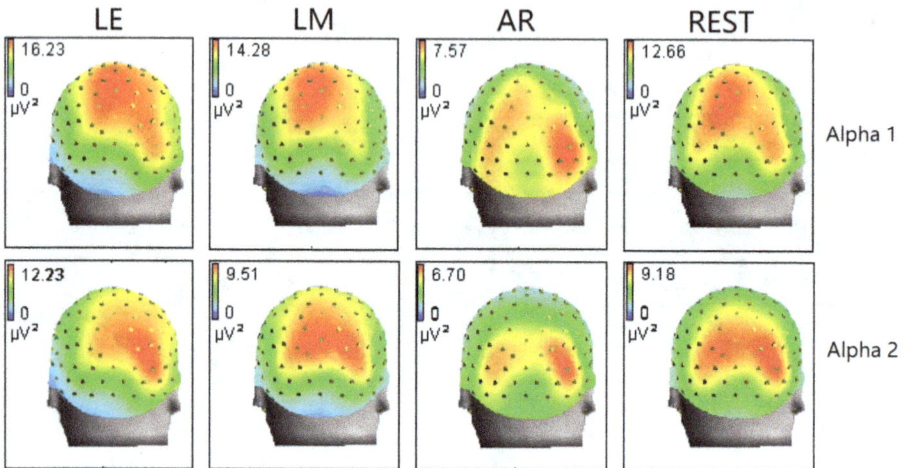

FIGURE 14.7 The power spectra maps of EEG with different references. From left to right, the references are left ear (LE), LM, AR, and REST. Compared to REST, the power spectra maps show shifts to the right positions with LE, to frontal and superficial positions with LM, and to a deeper position with AR (Yao et al., 2005).

14.5.3 ERP Amplitude

Waving the sea level will instantly change the height of a mountain contrast to the sea-level but not alter the mountain shape. Analogously, a nonzero reference will change the amplitude of the ERP component but not alter the topographic distribution for a moment. Could the amplitude change induce a different interpretation?

In psychological study, subtracting one ERP from another is a common strategy to get a different response in two stimulus cases. As the nonzero reference values of the two cases may not be the same, the difference of ERP will depend on the reference adopted. Here is an example using ERPs in an audiovisual (AV) stimulus (Tian and Yao, 2013). Three references, AR, LM, and REST, were comparatively investigated via ERPs and statistical parametric scalp mappings (SPSM), which is the scalp distribution of the significant statistical difference between two conditions (Tian and Yao, 2013). Specifically, for the N1 (170–190 ms), the SPSM results showed an anterior distribution for LM, a posterior distribution for REST, and both anterior and posterior distributions for AR (Figure 14.8). In Tian and Yao (2013), the result of REST is consistent with that by LORETA (Pascual-Marqui et al., 1994). Such a distinct difference might mislead the interpretation of the underlying mechanism, and actual amplitude information would be the starting point for any following studies.

14.5.4 ERP Latency

Referenced ERP is obtained by subtracting the reference signal from the active electrodes. If the reference signal is nonzero, the subtraction would distort the amplitude;

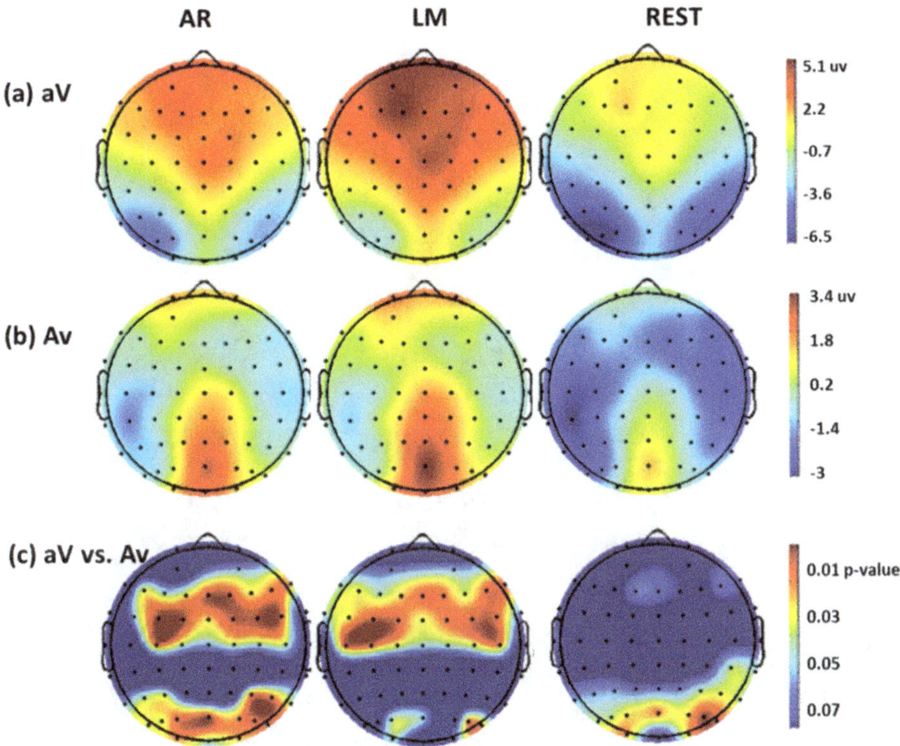

FIGURE 14.8 The potential topographies and SPSM of N1 peaks (at 170–190 ms). (a) Potential topographies of attending V in an audiovisual stimulus (aV) at 170–190 ms after the stimulus onset; (b) potential topographies of attending A in an audiovisual stimulus (Av); and (c) SPSM (aV vs. Av) at 170–190 ms (Tian and Yao, 2013).

if the reference signal has a different phase compared with the other active channels, it would affect the latency as well.

N170 is a negative ERP component that appeared about 170 ms elicited by the human face. The influence of the references on N170 was investigated using the scalp time-varying network method (Li et al., 2016). As the mastoids may be problematic for the N170 and other components that are largest at lateral posterior electrode sites (Luck, 2014). Two references, AR and REST, were comparatively investigated via the time-varying network processing of N170. Both AR and REST-based networks show transfer functions from the right P8 channel to the left. However, REST-based results are more robust and earlier than AR-based results (Figure 14.9). This phenomenon is further confirmed by a simulation study (Tian et al., 2017). This means that reference is an important issue in the precise investigation of the spatial-temporal dynamics of ERP, and REST-based zero-reference would be the first step for the following explanation of various ERPs.

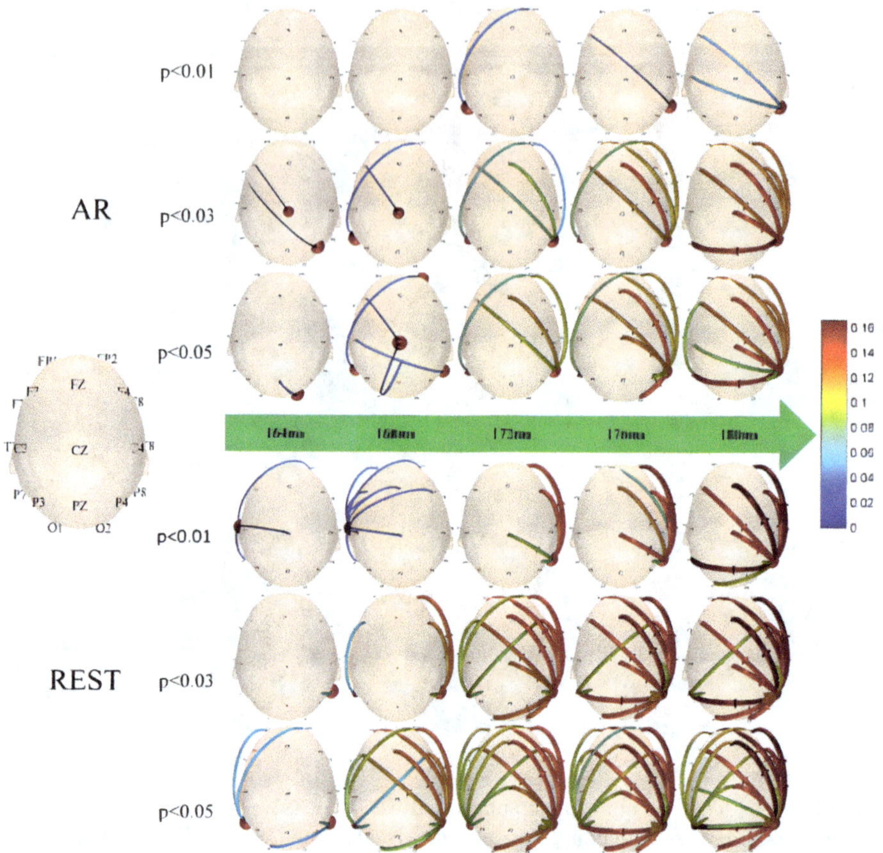

FIGURE 14.9 Recorded data derives the time-varying networks of N170. Hubs and connection mode change over time to near 170 ms under the three different significance levels (Tian et al., 2017).

14.6 NON-UNIPOLAR REFERENCES

14.6.1 Bipolar Recordings

At the very beginning of EEG, Berger had only two electrodes for recordings. So, he located the two electrodes within a part for the patients with partial skulls missing and "front to back" mostly for the healthy subjects (Berger, 1929; Stone and Hughes, 2013). This has evolved as nowadays unipolar recording. Differently, the bipolar recordings are to estimate the potential differences between two adjacent electrodes. Any two electrodes may be subtracted to obtain one channel of bipolar recording. Currently, bipolar recordings are still widely used in clinical evaluation for epilepsy, where each electrode is typically referenced to an adjacent electrode. The bipolar montage may be in the longitudinal/anteroposterior direction or the transverse/coronal direction (Niedermeyer and Da Silva, 2005), as illustrated in Figure 14.10.

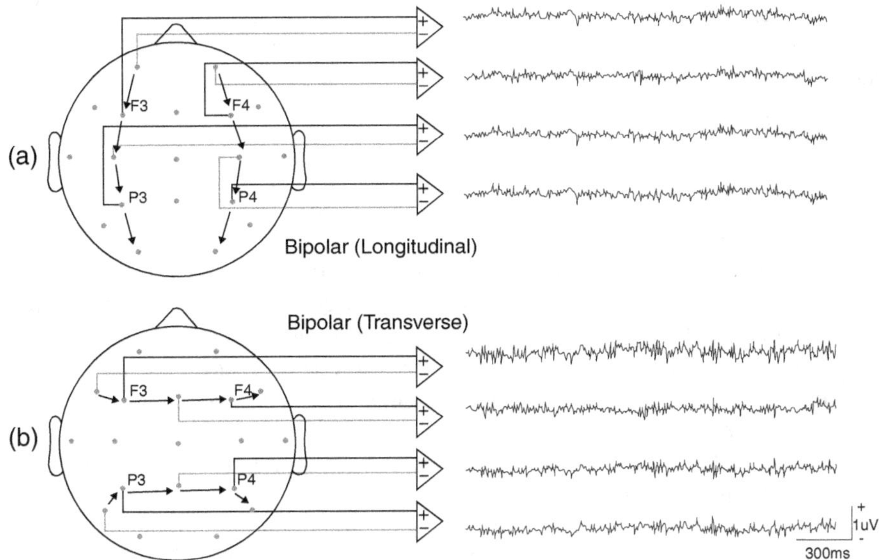

FIGURE 14.10 Bipolar recordings. With the same simulation procedure as in Figure 14.1, it displays the bipolar longitudinal (a) and transverse (b) recordings, respectively. The waveforms and polarities may be different at a moment for different montages. The waveforms here are distinct from the unipolar recordings in Figure 14.3 (Yao et al., 2019).

In cognitive and affective neuroscience experiments, bipolar recordings are often used to measure the electrooculogram (EOG), which is the electrical potential caused by blinks and eye movements (Luck, 2014). Mathematically, bipolar recording is a neighbor derivative, thus it is proportional to the local current density shown in the following:

$$\begin{aligned}\Phi_b &\approx \Phi(n+1)-\Phi(n)=\left(\Phi(n+1)-c\right)-\left(\Phi(n)-c\right),\\ &=\Phi_r(n+1)-\Phi_r(n)\propto\frac{\partial\Phi}{\partial d}\end{aligned} \tag{14.15}$$

$$\vec{j}\cdot\vec{e}_d-\sigma\vec{E}\cdot\vec{e}_d=-\upsilon\nabla\Phi\cdot\vec{e}_d=-\sigma\frac{\partial\Phi}{\partial d}\propto\Phi_b \tag{14.16}$$

where vector $\vec{j}$ is the current density, vector $\vec{E}$ is the electric field, σ is the conductivity of the scalp layer, $\vec{e}_d$ is a unit vector from electrode n to $n+1$, d is a unit distance scalar from electrode n to $n+1$, $\Phi(n)$ is the potential at electrode n, and Φ_b is the bipolar recording between electrode n and $n+1$. According to equation (14.15), the difference between the two points detected by bipolar recording is actually an approximation of the first-order derivative of the potential. Φ_b is invariant to a constant c or say that it is reference-free. According to the theory of electric field, it is a metric related to tangential current density over the scalp surface, not a potential at all, as illustrated by equation (14.16). Obviously, it depends on the montage. It is more sensitive to noise than to EEG signals and less sensitive to signals from deep neural sources because the derivative-like operation acts as a high pass filter. Bipolar recordings are mainly used in clinics to "enhance" focal activity (Niedermeyer and Da Silva, 2005).

14.6.2 Scalp Laplacian

Hjorth (1975) proposed the use of a mathematical procedure for the estimation of brain generators of scalp EEG potentials. The procedure tried to estimate the orthogonal current through the skull entering (source) or exiting (sink) the scalp at each electrode site (Chapter 8), so the result was originally called "orthogonal source derivation". Scalp/Surface Laplacian (SL) is a discretization of the planar Laplacian operator, i.e., the difference between the potential at each electrode and the averaged potential of its nearest four neighbors.

In practice, SL can be estimated by a simple subtraction of a channel $\Phi^{(n,m)}$ from its four neighbors

$$\begin{aligned} \mathrm{SL} = i_{xy} &= \frac{\partial^2 \Phi^{(n,m)}}{\partial x^2} + \frac{\partial^2 \Phi^{(n,m)}}{\partial y^2} = \frac{\partial^2 \Phi_r^{(n,m)}}{\partial x^2} + \frac{\partial^2 \Phi_r^{(n,m)}}{\partial y^2} \\ &\propto \left(\Phi_r^{(n+1,m)} + \Phi_r^{(n-1,m)} + \Phi_r^{(n,m+1)} + \Phi_r^{(n,m-1)} - 4\Phi_r^{(n,m)} \right) \end{aligned} \tag{14.17}$$

where $\Phi^{(n,m)}$ is the potential at a two-dimensional grid point (n, m). Apparently, SL is reference-free. Alternatively, we may also first fit the scalp discrete data to a continuous function, such as a spherical harmonic function for a spherical surface (Pascual-Marqui et al., 1988; Perrin et al., 1989), a spline function for a realistic head model (Babiloni et al., 2001), a radial-basis function (Yao, 2002a; Zhai and Yao, 2004b), and then conduct second-order analytical derivatives of the function.

The physical meaning of SL depends on the head model. Equation (14.17) implicitly assumes the scalp as a plane, and then, combining with equation (14.1), SL is an estimate of the current source density (CSD) (Hjorth, 1975; Yao, 2002a). However, if the scalp layer of the human head is modeled as a cubic element, SL will be an estimate of local current density/flux (CD) through the skull into the scalp (Yao, 2002a; Nunez and Srinivasan, 2006). If the scalp layer is a more realistic spherical shell model, SL and local CD are related by a complex and nonlinear function of spatial frequency. But for practically low spatial frequencies, they are approximately linearly related, so one may consider SL as an approximate CD in practice (Yao, 2002a). These indicate that the physical meaning of SL, CSD (Tenke and Kayser, 2005), and CD (Giard et al., 2014) is undetermined but dependent on the head model assumed.

Anyway, SL/CSD/CD is not potential in nature, free of the potential unipolar reference puzzle. As a different metric of the neural activities, approximately the normal current (CD) passes through the skull into the scalp layer or the local CSD of a scalp point. SL may be used to illustrate local activities and is called a high-resolution spatial imaging method (Yao, 2000; Yao et al., 2019). However, as a second-order derivative of the potential in equation (14.17), it is highly sensitive to the noise with wide spectra and low sensitive to the deep sources. Actually, either direct measurement (Besio et al., 2006) or numeric calculation of SL is still a problem in debate. Depending on the head model shape, noise level, and electrode density, various methods are developed, e.g., local numeric derivatives (Hjorth, 1975), spherical harmonic expansion (Pascual-Marqui et al., 1988), a global spherical spline approach (Perrin et al., 1989), and a moderate-scale radial-basis function approach (Yao, 2002b). Due to the pros and cons of SL, someone recommended examining both the potential waveforms and the current density waveforms together (Luck, 2014).

14.7 REFERENCE SELECTION IN PRACTICE

Due to the different physics, each reference would be used under suitable and valuable situations. Here, a comprehensive recommendation is provided. A non-trivial and important advantage of unipolar references is the 'no memory' property, which means one can always safely re-reference the EEG/ERP recordings with different unipolar reference techniques without worrying about whether re-referencing multiple times will accumulatively introduce artifacts (Hu et al., 2018b). Before applying a different unipolar reference, it is better to check if the data at present is with a unipolar reference. Transforming from a non-unipolar reference to a unipolar reference needs a special algorithm (Dong et al., 2023), and it is no problem to transform the data within the family of unipolar references. In addition, when Laplacian is infeasible and bipolar is unacceptable, unipolar reference is a proper choice.

Table 14.1 provides with a summary of the prevalent unipolar references and the frequently noted factors: electrode setup (density and coverage) and head model (shape and volume conduction). Apparently, online recording reference and offline LM are independent to the recording montage, and they are totally determined by the signals at the reference electrodes. Their main problem is the fact that the potential at the reference electrode is not constant, as it is also generated by the dynamic sources inside the brain. For REST and AR, they are hoping to recover the infinity reference. The accuracy depends on the assumptions behind them and the montage available, which means the channel information available for the calculation affects their performances. Their assumptions are based on the volume conduction model, therefore affecting the feasibility of the method.

TABLE 14.1
Factors Impacting the Unipolar Reference Signal

	Unipolar Reference	Electrode Density	Electrode Coverage	Head Model (Shape, Inner Conductivity)
Online	Cz, Pz, etc.			
Offline	LM			
	REST	√	√	√
	AR	√	√	√

14.7.1 Recording Reference

EEG reference mainly adopted online reference before the digital EEG era. One needs to pre-choose the reference point, such as the nose, chin, ear, etc., relying on the guessed site, which is relatively inactive. For example, to explore the neural mechanism of visual cognition, some researchers may assume the activities around the ears are weak, then an ear (mastoid) is taken as a reference, and usually they only analyze the channels on the middle line, such as Pz, Oz, and Fz as they are a little far away from the ears. In the current digital EEG era, if the available channel number is limited (<10) or the coverage is partial to a local region, such as in a wearable EEG device, the online recording reference would be a compromising choice, especially when the offline re-reference is infeasible. In general, the online reference site should be far away from the active electrode, and it should be in a position where the electrode is in contact with the skin all the time.

14.7.2 Linked Mastoids

As noted in Luck (2014), "the whole reference issue is a bit of a pain, but one nice thing is that you can easily change the reference offline after the data has been recorded. And you can do this many times to see what your data looks like with different references (which I highly recommend you do)". Offline unipolar re-references are the main options in current EEG studies. Among the three typical offline unipolar references, *LM* was the earliest one (Gibbs et al., 1935). It was believed to be better than a nose reference (Faux et al., 1990). However, LM was later criticized due to failing to localize the origin of the psychomotor seizure since the reference electrode linked to the ears distorted temporal activity (Feindel et al., 2009). In cognitive neuroscience studies, *LM* is still one of the most widely used references. But the papers using *LM* mainly study the channels at the middle line of the scalp as people are aware of the distortions near the two ears. Online LM recording reference is not recommended now, as physically linking the wires from these two electrodes creates a zero-resistance electrical bridge as a short-circuit between the two hemispheres, which may distort the distribution of voltage over the scalp and reduce hemispheric differences (Nunez and Srinivasan, 2006; Luck, 2014). Now, when should we use LM offline? The following notes should be considered: (1) the activities near the two ears are believed to be weak or possibly cancelled by each other, and the channels of interest are mainly the middle line electrodes; (2) the recording channels are limited ($\leq$10), making it difficult to implement the REST, AR, or Laplacian.

14.7.3 Average Reference

Offline LM is not to approach zero potential but just because of the guess that the averaged potentials over the two ears are close to zero; that is a subjective empirical assumption without theoretical proof. In contrast, inspired from the Wilson common terminal reference in ECG, AR was reported (Goldman, 1950; Offner, 1950) and there was a theoretical proof confirming the surface potential integral of a layered spherical sphere being zero (Bertrand et al., 1985; Yao, 2017). It was thus widely used in both EEG and ERP.

However, the integral may not be zero when a homogeneous and isotropic head is non-spherical (Yao, 2017), and no one knows the situation for a non-spherical, inhomogeneous, and anisotropic head. As shown by Table 14.1, the accuracy of AR depends on: whether a whole surface observation is feasible? Whether the electrode density is high enough to approximate the theoretical integral (Nunez, 2010)? And whether the head is a homogeneous and isotropic spherical conductor (Bertrand et al., 1985; Yao, 2017)? If all the answers are yes, it would be a golden standard (Nunez, 2010).

However, the measurement cannot be a whole scalp surface, the actual available surface is mainly the upper semi-scalp surface; the head shape is not spherical, homogeneous, and isotropic but usually much more complex; and the electrode arrays are usually not very dense (Hu et al., 2018a). Comparatively, our recent work showed that the performance of AR has no close relation to the electrode density, which is different from the usual understanding to AR based on its zero integral assumption, or say, coverage is a more important factor than the electrode density (Hu et al., 2018a). Therefore, AR cannot be a golden standard but an approximation.

So, when should we take AR into practice? Usually, if the montage has good coverage, such as being equal to or wider than a scalp surface as the EGI system has enough density, such as >128 channels (Hu et al., 2018a), and the subject head is approximately close to a sphere, then AR may be an acceptable approximation. In general, we do not recommend using AR if REST is available. Besides, in the current digital EEG era, online AR is not recommended for the same reason noted in the above LM reference. The additional limitation of AR is that one has to be sure the EEG data at hand is with unipolar references before applying AR (Hu et al., 2018b). In general, if AR is adopted, the whole electrodes involved in the average should be reported so that the reader can follow and repeat the results in their later studies (Luck, 2014).

14.7.4 REST Zero Reference

As confirmed, with the physical fact that all physiological scalp signals at both active electrodes and the reference electrode are generated by the same brain sources, REST (Yao, 2001) performs much better in recovering the actual potential on the scalp surface with the approximated infinity reference. In general, the accuracy of REST depends on the equivalence of the reconstructed equivalent sources and the unknown actual sources in generating the scalp potential, it can be applied to any complex head model. However, the lead fields in equation (14.12) involve the three factors in Table 14.1. Thus, the accuracy of REST may be improved with wider coverage, denser observation, and a more realistic head model. Till now, based on the publications from many groups in the world, REST would be the best choice for such a case: the electrode montage with a nice coverage that is at least the upper hemi-head surface, necessary electrode density (≥16), acceptable approximate head model (the concentric three-sphere head model or MRI image-based realistic head model). Generally, REST would be the best for most cognitive studies and clinic EEG problems, which were repeatedly confirmed by a series of simulation studies (Zhai and Yao, 2004a; Marzetti et al., 2007; Qin et al., 2010; Liu et al., 2015; Chella et al., 2016). Its rationality in processing various real data was also proven step by step (Yao et al., 2005; Bonfiglio et al., 2013; Tian and Yao, 2013; Xu et al., 2014; Kugiumtzis and Kimiskidis, 2015; Chella et al., 2016; Mumtaz and Malik, 2018; Pidal-Miranda et al., 2019; Folstein and Monfared, 2019; Henke and Meyer, 2021; Tian et al., 2021; Peng et al., 2021; Wang et al., 2021; Li et al., 2021; Koutlis et al., 2021; Song et al., 2021; Halder et al., 2021; Semprini et al., 2021; Agger et al., 2022; Peng et al., 2022; He et al., 2022; Sun et al., 2022; Perez et al., 2022).

Two prominent advantages of REST are that (1) it adapts to the EEG data with unipolar, bipolar recordings, and even Laplacian transformed, whereas the strict prerequisite in using AR is that the EEG data at hand needs to be with unipolar reference (Hu et al., 2018b), which means that only REST can be adopted to reunificate of various reference electrodes to a specific reference electrode after data acquisition; and (2) with the additional channels in forward calculation, the EEG potentials at the missing channels rejected as bad channels can be recovered with the interpolation function of REST (Dong et al., 2021). Besides, one may worry about the possible limitations of REST: (1) sensor noise problem and (2) the inaccurate head model. To address these two problems, we have conducted systematic simulation studies as shown in Figure 13.5 and introduced generalized cross-validation as the criterion to select the denoising parameter in the use of REST noted as rREST (Hu et al., 2018b), and found that the perturbations to the head model do not affect the robustness of REST (Hu et al., 2018a,b).

14.7.5 Non-Unipolar References

Bipolar reference recordings are not the way to get the actual potential, but the first order derivative of potentials shows the local surface potential variance of underlying neural activities. So, if one is not interested in the actual potential, or one has insufficient (<10) channels to apply REST, and the main concern is local activities instead of the whole scene, then bipolar reference recordings may be acceptable, such as in a neurological clinic where interictal epileptic spikes or local abnormal electric current are interested (Reilly, 2005). However, for cognitive and psychological studies, such a reference montage is never used. The distinct advantage of bipolar reference recording is that it is free of the influence from electrode number and density.

As one of the unknown reviewers of our paper ever noted:

> Bipolar recordings are generally of very limited value when electrode separation is large. However, with small separations bipolar recordings provide estimates of the tangential electric field halfway between the electrodes. This approach has been used effectively to estimate the propagation speed of traveling waves of electric field across the scalp for both resting EEG and evoked potentials.

Laplacian montage is often recommended due to its reference-free nature and relatively higher spatial resolution. However, Laplacian is not a physical measure but the second-order derivative of the scalp potential (Lai and Yao, 2009)· Dense electrode arrays (>64) and high SNR are necessary to get a valuable estimation of Laplacian over the whole scalp surface. In addition, Laplacian is more sensitive to shallow local sources than to distributed deep sources, and its estimates at the boundary channels are usually unreliable (Yao, 2002a; Zhai and Yao, 2004b). Thus, if the interested activities are located deeply or are distributed and the concerned channels are close to the boundary channels, cautions should be taken when using Laplacian. Otherwise, it might be an accredited choice to get a reference-free signal. Certainly, if online direct measurement of Laplacian by tripolar electrode approach is easily realized in the future, Laplacian may be used specifically for some points interested on the scalp surface even in wearable EEG system (Besio et al., 2004, 2006). As one of the unknown reviewers of our paper ever noted:

> Laplacian should not generally replace the reference potential, rather it provides estimates of smaller scale source regions, thereby yielding additional and complementary information. The issue of noise depends very much on application. For example, the resting state alpha band consists of multiple source regions of different sizes and locations. A large local Laplacian, even if its magnitude is somewhat inaccurate, can indicate the presence of local sources within a much larger synchronous region.

Finally, in 2014, Lepage et al. proposed a statistical procedure and claimed to mitigate the reference effect in their examples. Unfortunately, the effort is based on a zero assumption on the sum of the scalp recordings; however, this is not the truth in practice (Chapter 13 and Appendix F). As we have shown in these two chapters, the EEG reference problem is a determined physical problem that the volume conduction causes both active and reference electrodes to detect the signal from the source simultaneously. It is not a random signal processing problem, though there is some noise that is additionally mixed. In other words, the lost reference signal itself in practice recording is not a random signal but a determined physical signal. If we give up this physical background and forcibly treat it as a statistical signal processing problem, it is similar to "losing watermelon and picking sesame", of a Chinese proverb. The other statistical method, such as independent component analysis (Hu et al., 2007), also fell into a similar trap because no matter how effective the illustrated examples are, it cannot change the fact that it does not conform to the physical fact that the reference electrode signal and other electrode signals cannot be independent to each other as all of them are from the same sources with mixing of the volume conduction, so the method cannot be a universally applicable method.

14.8 SUMMARY

Many studies have shown that nonzero reference has distinct effects on waveforms and related parameters, such as information criteria, amplitude, latency, power, phase, and further derived parameters, e.g., coherence, correlation, network, symmetry, covariance, and statistic test. A neutral unipolar reference is fundamentally important to comparison among different labs and the data collected and stored with different references over time.

The reference problem is a special issue for the potential difference over reference electrode and the active electrode. As the observed multichannel recording is 'rank deficient by 1' (Appendix G), namely, the lost signal of the reference electrode cannot be recovered from itself, thus the information content of the offline LM or AR-based recordings is the same as that of the online unipolar reference recordings such as Cz,

Pz, etc. Differently, REST tackles the problem by realizing that volume conduction is the nature of the reference problem, it utilizes volume conduction just like adding an equation so that total information is close to find the lost channel. All the known simulation studies, with golden standard data as the ground truth, confirmed that REST is the best way to approach the ideal unipolar infinity reference, and it was recommended by the IFCN Guidelines (Babiloni et al., 2019), the "Best Practices in Data Analysis and Sharing in Neuroimaging using MEEG" of OHBM (https://cobidasmeeg.wordpress.com/), and the textbook on event-related potential data analysis (Luck, 2022). Now, REST is acknowledged as the best reference (Delorme. 2023), and a REST toolbox has been integrated into EEGLAB, Fieldtrip, WeBrain, BEAPP, and HAPPILEE.

Meanwhile, non-unipolar references such as bipolar reference recordings and Laplacian may be alternatives for clinic practice and shallow source-focused situations, respectively, since they are not potential but the derivatives of the potential, thus far away from reference problems.

REFERENCES

Agger, M.P., M.S. Carstensen, M.A. Henney, et al. 2022. Novel invisible spectral flicker induces 40 Hz neural entrainment with similar spatial distribution as 40 Hz stroboscopic light. *J Alzheimers Dis* 88(1):335–344.

Arnaud Delorme. 2023. What is the best EEG reference? https://www.youtube.com/watch?v=ioIETUX4G4k

Babiloni, C., R.J. Barry, E. Başar, et al. 2019. International Federation of Clinical Neurophysiology (IFCN) guidelines for topographic and frequency analysis of resting state electroencephalographic rhythms. *Clin Neurophysiol* 129:e208.

Babiloni, F., C. Babiloni, L. Fattorini, et al. 2001. Spatial enhancement of EEG data by surface Laplacian estimation: The use of magnetic resonance imaging-based head models. *Clin Neurophysiol* 112:724–727. doi: 10.1016/S1388-2457(01)00494-1.

Berger, H. 1929. Uber das Elektrenkephalogramm des Menschen. *Arch Psychiatr Nervenkr* 87(1):527–570.

Bertrand, O., F. Perrin, J. Pernier. 1985. A theoretical justification of the average reference in topographic evoked-potential studies. *EEG Clin Neurophysiol* 62:462–464.

Besio, G., K. Koka, R. Aakula, et al. 2006. Tri-polar concentric ring electrode development for laplacian electroencephalography. *IEEE Trans Biomed Eng* 53(5):926–933.

Besio, W., K. Koka, R. Patwardhan. 2004. Computer simulation and tank experimental verification of concentric ring electrodes. *The 26th Annual International Conferece on IEEE EMBS*, vol. 1, pp. 2243–2246. doi: 10.1109/IEMBS.2004.1403653.

Bonfiglio, L., U. Olcese, B. Rossi, et al. 2013. Cortical source of blink-related delta oscillations and their correlation with levels of consciousness. *Hum Brain Mapp* 34:2178–2189. doi: 10.1002/hbm.22056.

Chella, F., A. D'Andrea, A. Basti, et al. 2017. Non-linear analysis of scalp EEG by using bispectra: The effect of the reference choice. *Front Neurosci* 11:1–15. doi: 10.3389/fnins.2017.00262.

Chella, F., V. Pizzella, F. Zappasodi, et al. 2016. Impact of the reference choice on scalp EEG connectivity estimation. *J Neural Eng* 13(3):1–21. doi: 10.1088/1741-2560/13/3/036016.

Chung, J., M.I. Español, T. Nguyen. 2014. Optimal regularization parameters for generalform tikhonov regularization. arXiv preprint arXiv.1407.1911,121.

Dong, L., L. Zhao, Y. Zhang, et al. 2021. Reference Electrode Standardization Interpolation Technique (RESIT): A novel interpolation method for scalp EEG. *Brain Topogr* 34:403–414.

Dong, L., Y. Lai, M. Duan, et al. 2023. Rereferencing of clinical EEGs with nonunipolar mastoid reference to infinity reference by REST. *Clin Neurophysiol* 151:1–9.

Faux, S.F., M.E. Shenton, R.W. McCarley, et al. 1990. Preservation of P300 event-related potential topographic asymmetries in schizophrenia with use of either linked-ear or nose reference sites. *EEG Clin Neurophysiol* 75:378–391.

Fein, G., J. Raz, F.F. Brown, et al. 1988. Common reference coherence data are confounded by power and phase effects. *EEG Clin Neurophysiol* 69:581–584.

Feindel, W., R. Leblanc, A.N. De Almeida. 2009. Epilepsy surgery: Historical highlights 1909-2009. *Epilepsia* 50:131–151.

Folstein, J.R., S.S. Monfared. 2019. Extended categorization of conjunction object stimuli decreases the latency of attentional feature selection and recruits orthography-linked ERPs. *Cortex* 120:49–65.

Giard, M.H., J. Besle, P.E. Aguera, et al. 2014. Scalp current density mapping in the analysis of mismatch negativity paradigms. *Brain Topogr* 27:428–437. doi: 10.1007/s10548-013-0324-8.

Gibbs, F.A., H. Davis, W.G. Lennox. 1935. The electro-encephalogram in epilepsy and in conditions of impaired consciousness. *Arch Neuro Psychiatr* 34:1133–1148.

Gibbs, F.A., W.G. Lennox, E.L. Gibbs. 1936. The electroencephalogram in diagnosis and in localization of epileptic seizures. *Arch Neuro Psychiatr* 36(6):1225–1235.

Gloor, P. 1969. Hans Berger on electroencephalography. *Am J Eeg Technol* 9:1–8.

Goldman, D. 1950. The clinical use of the "average" reference electrode in monopolar recording. *EEG Clin Neurophysiol* 2:209–212.

Gulrajani, R. 1998. *Bioelectricity and Biomagnetism*. New York: John Wiley & Sons Press.

Halder, S., B.E. Juel, A.S. Nilsen, et al. 2021. Changes in measures of consciousness during anaesthesia of one hemisphere (Wada test). *NeuroImage* 226:117566.

He, Z., N. Zhuang, G. Bao, et al. 2022. Cross-day EEG-based emotion recognition using transfer component analysis. *Electronics* 11(4):651. doi: 10.3390/electronics11040651.

Henke, L., L. Meyer. 2021. Endogenous oscillations time-constrain linguistic segmentation: cycling the garden path. *Cereb Cortex* 31(9):4289–4299.

Hernandez-Gonzalez, G., M.L. Bringas-Vega, L. Galán-Garcia, et al. 2011. Multimodal quantitative neuroimaging databases and methods: The Cuban human brain mapping project. *Clin EEG Neurosci* 42:149–159. doi: 10.1177/155005941104200303.

Hjorth, B. 1975. An on-line transformation of EEG scalp potentials into orthogonal source derivations. *EEG Clin Neurophysiol* 39:526–530.

Hu, S., D. Yao, M.L. Bringas-Vega, et al. 2019. The statistical estimation of unipolar EEG references: Derivations and properties. *Brain Topogr* 32:696–703.

Hu, S., D. Yao, P.A. Valdes-Sosa. 2018b. Unified Bayesian estimator of EEG reference at infinity: rREST (Regularized Reference Electrode Standardization Technique). *Front Neurosci* 3(12):297. doi: 10.3389/fnins.2018.00297.

Hu, S., M. Stead, G.A. Worrell. 2007. Automatic identification and removal of scalp reference signal for intracranial EEGs based on independent component analysis. *IEEE Trans Biomed Eng* 54:1560–1572.

Hu, S., Y. Lai, P.A. Valdes-Sosa, et al. 2018a. How do reference montage and electrodes setup affect the measured scalp EEG potentials? *J Neural Eng* 15(2):026013. doi: 10.1088/1741-2552/aaa13f.

Huang, X., Z. Long, X. Lei. 2018. Electrophysiological signatures of the resting-state fMRI global signal: A simultaneous EEG-fMRI study. *J Neurosci Meth* 311:351–359.

Huang, Y., J. Zhang, Y. Cui, et al. 2017. How different EEG references influence sensor level functional connectivity graphs. *Front Neurosci* 11:1–12. doi: 10.3389/fnins.2017.00368.

Konishi, S., G. Kitagawa. 2008. *Information Criteria and Statistical Modeling*. New York: Springer Science & Business Media.

Koutlis, C., V.K. Kimiskidis, D. Kugiumtzis. 2021. Comparison of causality network estimation in the sensor and source space: Simulation and application on EEG. *Front Netw Physiol* 1:706487. doi: 10.3389/fnetp.2021.706487.

Kugiumtzis, D., V.K. Kimiskidis. 2015. Direct causal networks for the study of transcranial magnetic stimulation effects on focal epileptiform discharges. *Int J Neural Syst* 25(5):1550006. doi: 10.1142/S0129065715500069.

LaVaque, T.J. 1999. The history of EEG Hans Berger. *J Neurother* 3(1):1–9.

Lai, Y., D. Yao. 2009. Local laplacian estimate on spherical scalp surface. *Chin J Electron* 18(4):681–685.

Lepage, K.Q., M.A. Kramer, C.J. Chu. 2014. A statistically robust EEG re-referencing procedure to mitigate. *J Neurosci Meth* 235:101–116. doi: 10.1016/j.jneumeth.2014.05.008.

Li, F., B. Chen, H. Li, et al. 2016. The time-varying networks in P300: A task-evoked EEG study. *IEEE Trans Neur Sys Reh* 24:725–733. doi: 10.1109/Tnsre.2016.2523678.

Li, Z., X. Wang, W. Shen, et al. 2021. Objective recognition of tinnitus location using electroencephalography connectivity features. *Front Neurosci* 15:784721. doi: 10.3389/fnins.2021.784721.

Liu, Q., J.H. Balsters, M. Baechinger, et al. 2015. Estimating a neutral reference for electroencephalographic recordings: The importance of using a high-density montage and a realistic head model. *J Neural Eng* 12(5):056012. doi: 10.1088/1741-2560/12/5/056012.

Luck, S. 2022. *Applied Event-Related Potential Data Analysis*. Davis, CA: LibreTexts.

Luck, S.J. 2014. *An Introduction to the Event-Related Potential Technique*. Boston, MA: MIT Press.

Malmivuo, J., R. Plonsey. 1995. *Bioelectromagnetism: Principles and Applications of Bioelectric and Biomagnetic Fields*. New York: Oxford University Press.

Marzetti, L., G. Nolte, M.G. Perrucci, et al. 2007. The use of standardized infinity reference in EEG coherency studies. *NeuroImage* 36:48–63. doi: 10.1016/j.NeuroImage2007.02.034.

Mumtaz, W., A.S. Malik. 2018. A comparative study of different EEG reference choices for diagnosing unipolar depression. *Brain Topogr* 31:875–885. doi: 10.1007/s10548-018-0651-x.

Niedermeyer, E. 1987. The normal EEG of the waking adult. In E. Niedermeyer, F. H. Lopes da Silva (eds.), *Electroencephalography: Basic Principles, Clinical Applications, and Related Fields*, pp. 97–118. Baltimore, MD: Lippincott Williams & Wilkins.

Niedermeyer, E., F. Da Silva. 2005. *Electroencephalography: Basic Principles, Clinical Applications and Related Fields*, 5th edn. Philadelphia, PA: Lippincott Williams & Wilkins.

Nunez, P.L. 2010. REST: A good idea but not the gold standard. *Clin Neurophysiol* 121:2177–2180. doi: 10.1016/j.clinph.2010.04.029.

Nunez, P.L., R. Srinivasan. 2006. *Electric Fields of the Brain: The Neurophysics of EEG*. New York: Oxford University Press.

Offner, F.F. 1950. The EEG as potential mapping: The value of the average monopolar reference. *EEG Clin Neurophysiol* 2:213.

Pascual-Marqui, R.D., C.M. Michel, D. Lehmann. 1994. Low resolution electromagnetic tomography: A new method for localizing electrical activity in the brain. *Int J Psychophysiol* 18:49–65.

Pascual-Marqui, R.D., S.L. Gonzalez-andino, P.A. Valdes-sosa. 1988. Current source density estimation and interpolation based on the spherical harmonic Fourier expansion. *Int J Neurosci* 43:237–249. doi: 10.3109/00207458808986175.

Peng, M., Z. Xu, H. Huang. 2021. How does information overload affect consumers' online decision process? An event-related potentials study. *Front Neurosci* 15:695852.

Peng, M., Z. Xu, H. Huang. 2022. To each their own: The impact of regulatory focus on consumers' response to online information load. *Front Neurosci* 16:757316. doi: 10.3389/fnins.2022.757316.

Perez, A., M.H. Davis, R.A.A. Ince, et al. 2022. Timing of brain entrainment to the speech envelope during speaking, listening and self-listening. *Cognition* 224:105051.

Perrin, F., J. Pernier, O. Bertrand, et al. 1989. Spherical splines for scalp potential and current density mapping. *EEG Clin Neurophysiol* 72:184–187.

Pidal-Miranda, M., A.J. González-Villar, M.T. Carrillo-de-la-Peña. 2019. Pain expressions and inhibitory control in patients with fibromyalgia: Behavioral and neural correlates. *Front Behav Neurosci* 12:323. doi: 10.3389/fnbeh.2018.00323.

Plonsey, R., D.B. Heppner. 1967. Considerations of quasi-stationarity in electrophysiological systems. *Bull Math Biophys* 29:657–664.

Qin, Y., P. Xu, D. Yao. 2010. A comparative study of different references for EEG default mode network: The use of the infinity reference. *Clin Neurophysiol* 121:1981–1991. doi: 10.1016/j.clinph.2010.03.056.

Qin, Y., X. Xin, H. Zhu, et al. 2017. A comparative study on the dynamic EEG center of mass with different references. *Front Neurosci* 11:509. doi: 10.3389/Fnins.2017.00509.

Reilly, E.L. 2005. EEG recording and operation of the apparatus. In E. Niedermeyer, F. H. Lopes da Silva (eds.), *Electroencephalography: Basic Principles, Clinical Applications, and Related Fields*, 5th ed. Baltimore, MD: Lippincott Williams & Wilkins, 198–225.

Robert, C.P. 2007. *The Bayesian Choice: From Decision-Theoretic Foundations to Computational Implementation*. New York: Springer Press.

Semprini, M., G. Bonassi, F. Barban, et al. 2021. Modulation of neural oscillations during working memory update, maintenance, and readout: An hdEEG study. *Hum Brain Mapp* 42:1153–1166.

Shaw, J.C. 1984. Correlation and coherence analysis of the EEG: A selective tutorial review. *Int J Psychophysiol* 1:255–266. doi: 10.1016/0167-8760(84)90045-X.

Song, P., D. Cao, S. Li, et al. 2021. Effects of hyperventilation with face mask on brain network in patients with epilepsy. *Epilepsy Res* 176:106741.

Stone, J.L., J.R. Hughes. 2013. Early history of electroencephalography and establishment of the American Clinical Neurophysiology Society. *J Clin Neurophysiol* 30:28–44. doi: 10.1097/WNP.0b013e31827edb2d.

Sun, Y., Y. Xu, J. Lv, et al. 2022. Self- and situation-focused reappraisal are not homogeneous: Evidence from behavioral and brain networks. *Neuropsychologia* 173:108282.

Tenke, C.E., J. Kayser. 2005. Reference-free quantification of EEG spectra: Combining current source density (CSD) and frequency principal components analysis (fPCA). *Clin Neurophysiol* 116:2826–2846. doi: 10.1016/j.clinph.2005.08.007.

Tian, F., Y. Zhang, Y. Li. 2021. From 2D to VR film: A research on the load of different cutting rates based on EEG data processing. *Information* 12(3):130. doi: 10.3390/info12030130.

Tian, Y., D. Yao. 2013. Why do we need to use a zero reference? Reference influences on the ERPs of audiovisual effects. *Psychophysiology* 50(12):1282–1290.

Tian, Y., W. Xu, H. Zhang, et al. 2017. The scalp time-varying networks of N170: Reference, latency and information flow. *Front Neurosci* 2018:250. doi: 10.3389/fnins.2018.00250.

Travis, F. 1994. A second linked-reference issue: Possible biasing of power and coherence spectra. *Int J Neurosci* 75:111–117.

Vorwerk, J., R. Oostenveld, M.C. Piastra, et al. 2018. The field trip SimBio pipeline for EEG forward solutions. *Biomed Eng OnLine* 17:37.

Wang, X., H. Wu, J. Huang, et al. 2021. Reward mechanism of depressive episodes in bipolar disorder: Enhanced theta power in feedback-related negativity. *J Affect Disorders* 292:217–222. doi: 10.3389/fnins.2021.695852.

Xu, P., X.C. Xiong, Q. Xue, et al. 2014. Recognizing mild cognitive impairment based on network connectivity analysis of resting EEG with zero reference. *Physiol Meas* 35(7):1279–1298.

Yao, D. 2000. High-resolution EEG mappings: A spherical harmonic spectra theory and simulation results. *Clin Neurophysiol* 111:81–92.

Yao, D. 2001. A method to standardize a reference of scalp EEG recordings to a point at infinity. *Physiol Meas* 22(4):693–711.

Yao, D. 2002a. The theoretical relation of scalp Laplacian and scalp current density of a spherical shell head model. *Phys Med Biol* 47(12):2179–2185.

Yao, D. 2002b. High-resolution EEG mapping: A radial-basis function based approach to the scalp Laplacian estimate. *Clin Neurophysiol* 113(6):956–967.

Yao, D. 2017. Is the surface potential integral of a dipole in a volume conductor always zero? A cloud over the average reference of EEG and ERP. *Brain Topogr* 30:161–171. doi: 10.1007/s10548-016-0543-x.

Yao, D., B. He. 1998. The Laplacian weighted minimum estimate of three dimensional equivalent charge distribution in the brain. *Proceedings of the 20th Annual International Conference on IEEE EMBS*, Hong Kong, China, vol. 4, pp. 2108–2111.

Yao, D., B. He. 2003. Equivalent physical models and formulation of equivalent source layer in high-resolution EEG imaging. *Phys Med Biol* 48(21):3475–3483.

Yao, D., L. Wang, R. Oostenveld, et al. 2005. A comparative study of different references for EEG spectral mapping: The issue of the neutral reference and the use of the infnity reference. *Physiol Meas* 26(3):173–184.

Yao, D., Y. Qin, S. Hu, et al. 2019. Which reference should we use for EEG and ERP practice? *Brain Topogr* 32:530–549.

Zhai, Y., D. Yao. 2004a. A study on the reference electrode standardization technique for a realistic head model. *Comput Meth Prog Bio* 76:229–238.

Zhai, Y., D. Yao. 2004b. A radial-basis function based surface Laplacian estimate for a realistic head model. *Brain Topogr* 17:55–62. doi: 10.1023/B Brat.0000047337.25591.32.

15 EEG Cloud Platform – WeBrain

The field of brain science research is at the beginning of a new era that combines the burgeoning capabilities of the IT revolution with powerful emerging data acquisition technologies. Neuroscience communities, as well as brain projects promoted by national initiatives (e.g., the Human Connectome Project (HCP) (Glasser et al., 2016), the European Union Human Brain Project (HBP) (Amunts et al., 2016), and Chinese Brain Project (Poo et al., 2016), etc.), produce more complex and massive multimodal (e.g., electroencephalography (EEG), magnetoencephalography (MEG), magnetic resonance imaging (MRI), etc.) brain information datasets, which may imply the arrival of "big data" age in neurosciences (Sejnowski et al., 2014; Van Horn and Toga, 2014). By addressing these large-scale multi-modal datasets, an increasing number of neuroscience researches are being transformed through the application of cloud computing resources including computational power, memory, and storage capabilities to process these datasets (Bouchard et al., 2016). Obviously, this era in neurosciences is experiencing a substantial shift from isolated single efforts with limited data, tools, and computing resources to more efforts with larger datasets, more comprehensive pipelines, and cyberinfrastructures. With the recommendation of open science (Koch and Jones, 2016), this transition further generates "cloud neuroscience," realized by neuroimaging cloud platforms comprising categories of data, infrastructure, apps, algorithms, and education, to accelerate the brain science research (Vogelstein et al., 2016). In this part, to our knowledge, we will introduce the current cloud platforms and pipeline tools in the neuroscience field first, and then, introduce the EEG cloud platform, WeBrain (https://webrain.uestc.edu.cn/) in detail.

15.1 CURRENT NEUROIMAGING CLOUD PLATFORMS

This part will introduce the main neuroimaging cloud platforms worldwide. Table 15.1 lists main cloud platforms in the neuroscience fields, and Table 15.2 lists the main EEG pipeline tools to processing large-scale EEG data.

15.1.1 WeBrain

WeBrain (Dong et al., 2021) is a web-based computing platform that enables large-scale EEG and EEG-fMRI multi-modal data storing, exploring, and analyzing using cloud high-performance computing (HPC) facilities. WeBrain connects researchers of different fields to EEG and multimodal tools and processing power required to handle the large datasets that have become the norm in the field. It also aims to construct an International Virtual Community of Brainformatics (IVCB) to set the scene for more ambitious multinational initiatives and co-operations in the

DOI: 10.1201/9781032639260-15

TABLE 15.1
Current Neuroimaging Platforms

Platforms	Brief Instruction	Home Page
WeBrain	A web-based brainformatics platform of computational ecosystem for EEG big data analysis	https://webrain.uestc.edu.cn/
XNAT	An informatics platform for managing, exploring, and sharing neuroimaging data	https://www.xnat.org/
CBRAIN	A web-based, distributed computing platform for collaborative neuroimaging research	https://cbrain.ca/
BrainLiner	A neuroinformatics platform for sharing time-aligned brain-behavior data	http://brainliner.jp/
NSG	A portal to facilitate access and use of cloud computing resources	https://www.nsgportal.org/
HBP-JP	A new European scientific research joint platform	https://www.humanbrainproject.eu/en/
volBrain	An open-access platform for MRI volumetric brain analysis	https://volbrain.net//
Brain-CODE	A neuroinformatics platform for management, federation, sharing, and analysis of multidimensional neuroscience data	https://www.braincode.ca/
NiftyNet	A deep-learning platform for medical imaging	https://niftynet.io/
OpenNeuro	An open resource for sharing of neuroscience data	https://openneuro.org/
NimiBrainCloud	A brain simulation platform	http://nimibrain.cuc.edu.cn

TABLE 15.2
EEG Tools/Methods and File Formats Currently Supported by the WeBrain Platform

	Supported Tools/File Formats
EEG tools/methods	REST
	Bad block marking
	ICA running
	Quality assessment
	Preprocessing of continuous EEG raw data
	Power spectrum analysis
	Event-related potential analysis
	EEG network calculation
	Network topology analysis
	Generation of real head conduction model (BEM)
	EEG source imaging
	Time-frequency analysis
	Nonlinear analysis
	Microstate analysis

(*Continued*)

TABLE 15.2 (*Continued*)
EEG Tools/Methods and File Formats Currently Supported by the WeBrain Platform

	Supported Tools/File Formats
EEG file formats (manufacturer/file format)	ASCII/Float file (*.txt)
	MATLAB (*.mat/*.dat)
	EEGLAB ({*.set, *.fdt} for or *.fdt)
	Curry 6/7 ({*.dat, *.dap, *.rs3})
	Curry 8/9 ({*.cdt, *.cdt.ceo, *.cdt.dpa})
	Brain Products/Brain Vision ({*.vhdr, *.vmrk, *.dat} or {*.vhdr, *.vmrk, *.eeg})
	NeuroScan (*.cnt or *.EEG)
	Biosemi/European Data Format (*.bdf or *.edf)
	BIOSIG (*.edf, *.edf+, *.gdf or *.bdf)
	EGI MFF file (.mff)
	Neuracle EEG Recorder ({data.bdf, evt.bdf})
	ANT (*.cnt)

brain research. It does at the same time reduce the technical expertise required to use these resources. It provides an easy-to-use for novice users (even no computer programming skills) and flexibility for experienced researchers. It is not necessary to install any software or system for users, and all that is needed is a modern web browser of any kind. A range of resources including neuroimaging analysis tools are available as well as documents related to WeBrain. The WeBrain was developed by the University of Electronic Science and Technology of China (UESTC).

15.1.2 XNAT

XNAT (the Extensible Neuroimaging Archive Toolkit) (Marcus et al., 2007) is an open-source neuroimaging informatics software platform originally developed by the Neuroinformatics Research Group at Washington University and currently located at Harvard University. It was designed to facilitate performing imaging-based research. XNAT enables convenient ways to upload, organize, share, view, and download user's clinical neuroimaging data such as MRI images. It supports quality control procedures and secure ways for user to access, store, management, and share clinical data. Meanwhile, XNAT has good extensibility to support a wide range of imaging-based projects including a powerful pipeline engine allowing for running complex data processing workflows.

15.1.3 CBRAIN

CBRAIN (the Canadian Brain Imaging Research Platform) (Sherif et al., 2014) is a web-based research platform developed by the ACElab, McConnell Brain Imaging Centre, Montreal Neurological Institute, and McGill University. It allows users to perform computationally intensive analyses on data by connecting them to HPC facilities across Canada and around the world. CBRAIN offers access to the tools and processing power required to handle the large neuroimaging datasets that have become the norm in the field. Its web interface reduces the technical expertise required to large-scale data processing. For users, no computer programming skills are required, no software is necessary to be installed, and all that is needed is a modern web browser of any kind. A number of neuroimaging tools are also available on the platform as well as 2D and 3D real-time visualization to view the brain imaging data.

15.1.4 BrainLiner

BrainLiner (Takemiya et al., 2016) is a web platform developed by Department of Neuroinformatics, ATR Computational Neuroscience Laboratories, Kyoto University, for sharing time-aligned and brain-behavior data. It has used a common data file format named HDF5 to simplify data processing and analyses. Meanwhile, the BrainLiner platform allows users to explore, search, upload, and download for data from the web platform as well as provides a number of tools for data sharing and data-driven modeling. A WebGHL-based data explorer was also developed to visualize neurophysiological data from the web browser in a highly detailed way.

15.1.5 NSG

NSG (the Neuroscience Gateway) is a portal to facilitate access and use of high-performance computing (HPC), high throughput computing (HTC), open science grid (OSG) resources, and academic cloud computing resources for simulation and data processing. NSG also provides access to a number of neuroscience tools, libraries, and pipelines including EEGLAB, Freesurfer, NEURON, MATLAB, Python, PyTorch, TensorFlow, etc. The NSG was developed by University of California San Diego, Yale University, and University College London together.

15.1.6 HBP-JP

The HBP-JP (Human Brain Project Joint Platform) (Amunts et al., 2016) is a new European scientific research joint platform containing six platforms including Neuroinformatics, Brain Simulation, High-Performance Analytics and Supercomputing, Medical Informatics, Neuromorphic Computing, and Neurorobotics, which are connected through the "Collaboratory" (COLLAB) interface. These platforms are inspired by and developed in co-design with neuroscientists and IT engineers from Human Brain Project, which is a multinational European brain research initiative funded by the European Commission Directorate General for Communications Networks, Content, and Technology (DG CONNECT).

15.1.7 volBrain

volBrain (Manjón and Coupé, 2016) platform is an online open MRI brain volumetric system developed by Universitat Politècnica deValència. It was designed to help neuroscience researchers to obtain automatically volumetric brain information from MRI data without the need of learning complex software packages or having expensive computational infrastructures. The platform works in a fully automatic manner and provides volumetric brain analysis without any human interaction in few minutes. Several pipelines dealing with different brain areas and diseases from MRI data were deployed on the volBrain platform.

15.1.8 Brain-CODE

Brain-CODE (Vaccarino et al., 2018) is an extensible and neuroinformatics platform designed to support collection, storage, analysis, and sharing of multidimensional neuroscience data (e.g., clinical, neuroimaging, and molecular) collected from patients with a variety of brain disorders such as cerebral palsy, epilepsy, neurodevelopmental disorders, neurodegeneration, depression, etc. It was therefore designed to help researchers understand common underlying causes of brain dysfunction and develop novel approaches to improve patient care. The Brain-CODE platform was developed by Ontario Brain Institute, Toronto, Canada.

15.1.9 NiftyNet

NiftyNet (Gibson et al., 2018) is a TensorFlow-based open-source convolutional neural networks (CNNs) platform developed by the University College London for medical image analysis and image-guided therapy. Its modular structure was designed for sharing networks and pretrained models. Using the platform, users can quickly start with established pretrained networks, adapt existing networks to imaging data, and build new solutions to image analysis problems.

15.1.10 OpenNeuro

OpenNeuro (Markiewicz et al., 2021) is a free and open platform for validating and sharing various neuroimaging data including MRI, PET, MEG, EEG, and iEEG data. It was supported by the Brain Research through Advancing Innovative Neurotechnologies (BRAIN) Initiative, National Institutes of Health, USA. The OpenNeuro platform currently offers 911 public datasets including more than 35,000 participants, comprising multiple species, measurement modalities, and a broad range of phenotypes.

15.1.11 NiMiBrainCloud

NiMiBrainCloud (NMBC) is a brain simulation platform developed by the Communication University of China. It provides a high-speed brain simulation platform of large-scale brain neural network with the advantages of high precision, fast computing, good parallelism, and strong interactivity for researchers. Several demos including visual perception, multimodal integration, and simulation of Parkinson's disease-related brain regions were provided on the platform. Note that the current Chinese version of the NiMiBrainCloud is available only.

In addition, to handle large-scale EEG data, a number of EEG tools and pipelines are expected and have been developed. For example, while most traditional

MATLAB-based EEG tools, such as EEGLAB (Delorme and Makeig, 2004) and FieldTrip (Oostenveld et al., 2011), were designed to detail and batch process small-scale EEG data for offline personal computer users, many MATLAB-based pipeline tools, including PREP (Bigdely-Shamlo et al., 2015), CTAP (Cowley et al., 2017), HAPPE (Gabard-Durnam et al., 2018) and Automagic (Pedroni et al., 2019), or standards, such as EEG Study Schema (ESS, http://www.eegstudy.org/), have been developed to pre-process large-scale EEG raw data and assess the quality of preprocessed EEG data offline. These tools often require familiarity with programming skills, the understanding of different tools with different underlying design philosophies, and knowledge of configuration parameters and inputs relative to both EEG and toolboxes.

15.2 WEBRAIN PLATFORM

WeBrain is a web-based computing platform that enables large-scale EEG and EEG-fMRI multimodal data storing, exploring, and analyzing using cloud high-performance computing (HPC) facilities across UESTC. It connects researchers of different fields to EEG and EEG-fMRI multimodal tools (especially developed by UESTC) and processing power required to handle the datasets. On the WeBrain platform, users can manage their EEG and multimodal data, use basic/specific EEG tools and methods that have become the norm in the EEG field, use the power of cloud high-performance computing on the data, and benefit from a set of worldwide brain researchers in the community of brainformatics, IVCB. Noting that the WeBrain is a companion piece of the mature IT infrastructure CBRAIN (Sherif et al., 2014) and LORIS (Das et al., 2012) developed by Prof. Alan Evans at MNI Group of McGill University. The WeBrain is also sponsored by the CCC-Axis (Joint Brain Research of Canada–China–Cuba, initiated by Profs. Pedro A. Valdes-Sosa, Alan Evans, and Dezhong Yao) and the later Global Brain Consortium (GBC)(https://globalbrainconsortium.org/). In the future, the WeBrain team will keep on cooperating with members of CCC-Axis and GBC to further improve the WeBrain to realize interconnections between platforms including the CBRAIN and LORIS. More EEG and EEG-fMRI fusion tools will be integrated into the WeBrain. And more computing nodes/clusters will be added into the WeBrain to increase the computing and analytics capabilities of WeBrain. The integration of these tools and computing resources will further benefit the EEG and EEG-fMRI studies with big data (Figure 15.1).

15.2.1 WeBrain Computational Ecosystem

To efficiently integrate IT infrastructures and researchers in different fields including administrators/IT engineers, brain information tool developers and neuroscientists, the WeBrain platform was designed as a sustainable brainformatics platform for a computational ecosystem. The architecture of WeBrain is composed of three main layers. (1) The user layer represents the service consumption requested from users through web browsers. (2) The platform layer hosts the WeBrain service that is responsible for all requests from the user layer. The platform layer realizes user, data, and tool management and computing resource scheduling. (3) The cloud resource layer represents the wide cloud resources including tool repositories, data storage,

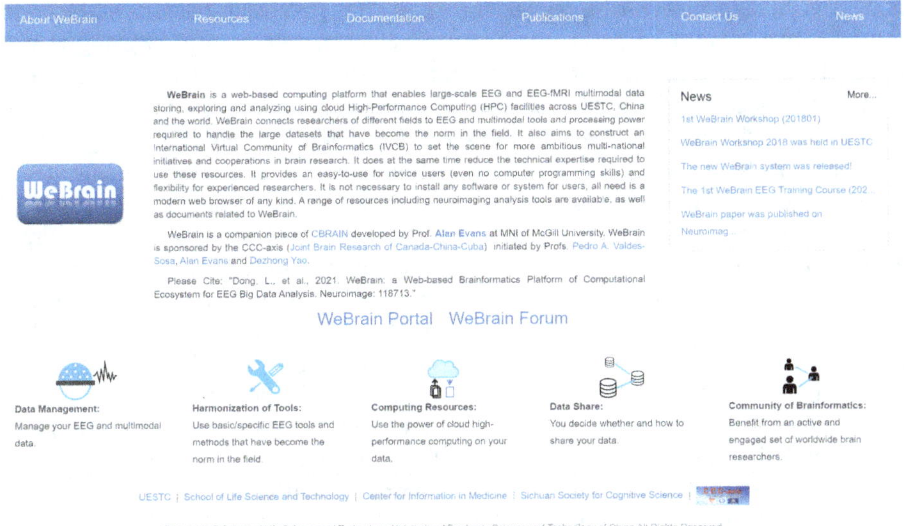

FIGURE 15.1 Home page of the WeBrain platform (https://webrain.uestc.edu.cn/).

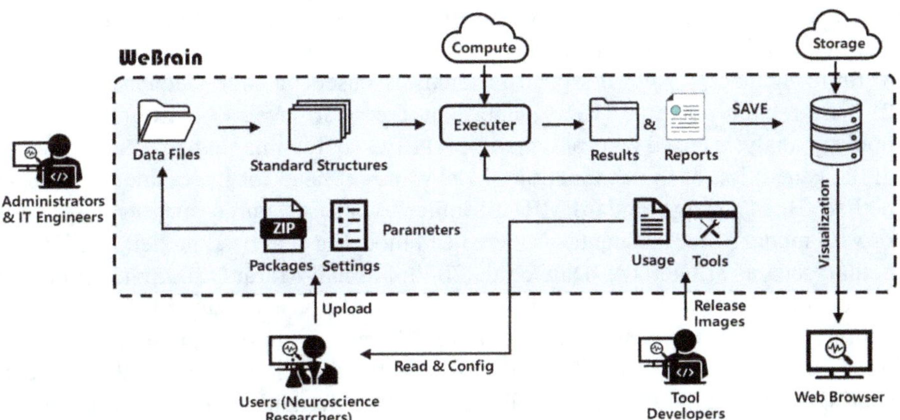

FIGURE 15.2 Computational ecosystem in the WeBrain platform. According to the different demands and experiences of researchers, three roles including IT engineers/administrators, tool developers, and neuroscience researchers are categorized.

and computing nodes. Meanwhile, in the WeBrain ecosystem (Figure 15.2), three roles including (1) administrators and IT engineers, (2) tool developers, and (3) neuroscience researchers are categorized according to the different demands and experiences of researchers. These roles are associated in the WeBrain ecosystem and could work independently to some degree. The role of administrators and IT engineers is responsible for the daily management and maintenance of the WeBrain system. And, most of the daily operations including registration and privilege management, data files management, tool upgrade and management, monitoring computing tasks, and homepage management are all realized on the WeBrain website. The role of the brain information method/tool developers is responsible for releasing tools using a flexible and conceptual application programming interface (API) of standardized units (container images) defined by WeBrain. The tool is a standardized unit (container image) that includes everything needed to run the application (e.g., operating system, environments, and codes) for development and shipment using the container technique; therefore, tools developed by researchers can be easily integrated and called in WeBrain, no matter what programming languages and environments are used. The role of neuroscience researchers is to be regular registered users benefited from those tools and cloud computing resources that are easy to use on the WeBrain. The basic workflow of WeBrain for users contains (1) organizing and uploading local data files to WeBrain, (2) creating and running a computing task, and (3) checking and downloading results to a personal computer. In addition, these roles further help to form a WeBrain virtual community of brainformatics across worldwide researchers.

Currently, a number of basic/specific EEG tools are integrated in the WeBrain platform, since they are commonly used or computationally expensive and generally complex to use for the novice user (Table 15.1). For EEG preprocessing, WeBrain includes (1) an EEG re-referencing tool for the reference electrode standardization technique (REST) developed by our group (Dong et al., 2017; Yao, 2001; Yao et al., 2005), (2) a stable quality assessment (QA) tool for large-scale continuous EEG raw data, and (3) standardized and specific EEG preprocessing pipelines of large-scale continuous EEG raw data to remove various artifacts. For further EEG analysis, WeBrain includes (1) power spectrum analysis based on time-frequency analysis; (2) event-related potential (ERP) analysis at the scalp level; (3) EEG network and topology analysis based on graph theory; (4) EEG source imaging analysis including (a) the generation of the conduction model of a real head for the boundary element method (BEM) using standard MRI T1 images, (b) EEG source imaging based on a forward model (three-concentric sphere head model or real head model) and inverse method such as sLORETA (Dale et al., 2000; Pascual-Marqui, 2002); (5) time-frequency analysis to reveal time-varying spectrum of nonstationary EEG signals; (6) nonlinear analysis of calculating Lempel-Ziv Complexity indices to reveal the capacity of information in the EEG signal fragments; and (7) microstate analysis of calculating microstate indices to reveal a sparse characterization of the spatiotemporal features of large-scale brain network activity.

15.2.2 Usage Example of WeBrain

Typical data workflow on WeBrain for registered users is (1) organizing and uploading local data files to WeBrain; (2) creating and running a computing task; and (3) checking and downloading results to a personal computer. Here, as an example, the method and usage of the source imaging tool (WB_EEG_sourceimage) is introduced in detail.

15.2.2.1 The Source Imaging Tool

WB_EEG_sourceimage is a tool to estimate source signals of scalp EEG/ERP data based on a forward model and inverse method (e.g., sLORETA). Source imaging estimation consists of (Figure 15.3):

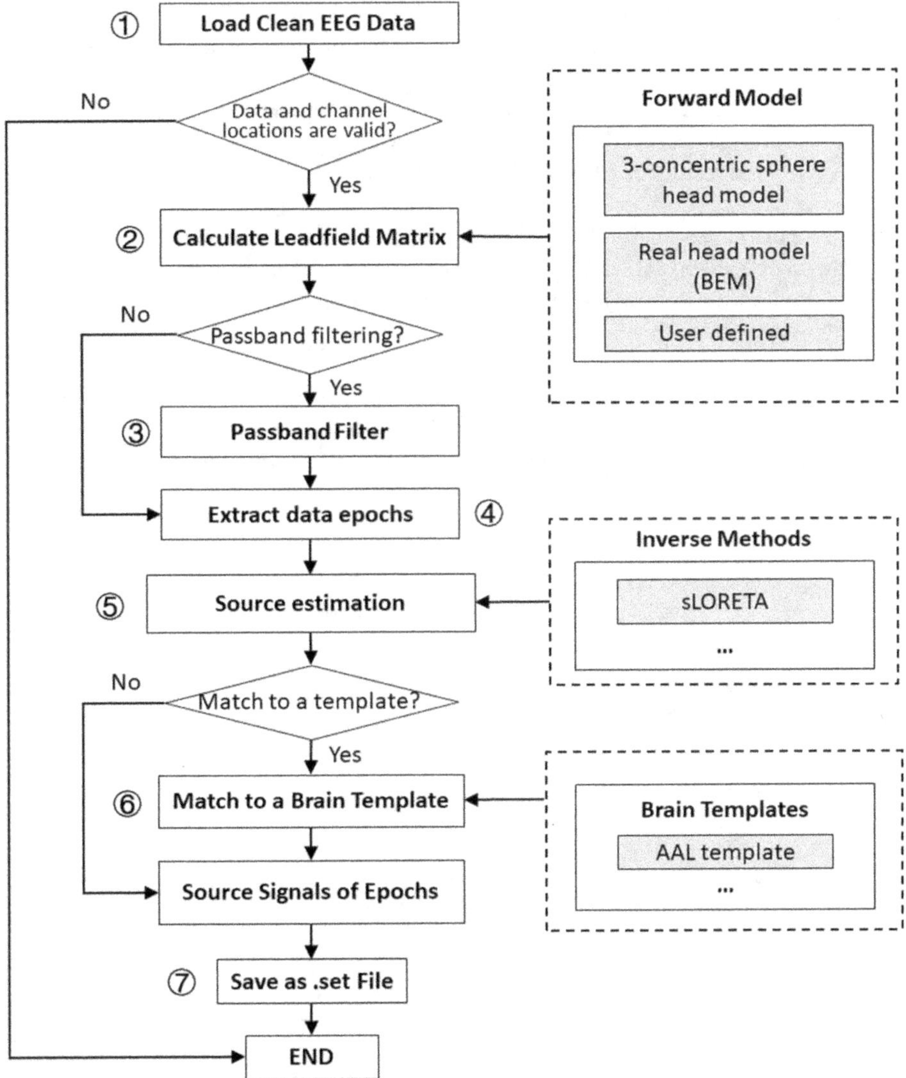

FIGURE 15.3 The pipeline of EEG source imaging.

1. Loading EEG data and check the items including data, channel locations, and sampling rate.
2. Calculating the leadfield matrix by solving forward problem based on selected head model and channel locations. Or, obtaining a user defined leadfield matrix.
3. If needed, passband filtering the EEG data. Default is no filtering.
4. Specific event data can be extracted according to the input "eventlabel." If the input "eventlabel" is empty, all data will be used. And, specific EEG signals will be divided into small epochs. If applicable, EEG segments in bad block (label 9999, marked by WB_EEG_Mark) will be rejected automatically, and NOT used to estimate sources.

5. EEG data of each epoch (default is 5-s epoch) is subjected to estimate sources to obtain the EEG signals in the source space using an inverse method such as "sLORETA."
6. If needed, matching source signals to a brain template (e.g., AAL template) to obtain the averaged source signals of brain regions.
7. Saving the results of EEG source signals and parameters as a.set file.

Here, as an example, default parameters are inputted on the web interface and used to estimate source signals at trial level. After calculation is finished, the result files will be generated on the detail page of "My Project" (Figure 15.4). The details of source imaging tool parameters are as follows:

epochLenth: Length of small epochs to calculate EEG source signals. Unit is second. Default is 5s. If "epochLenth" is negative, it means that if possible, data before event labels (eventlabel) will be used (no overlapped).

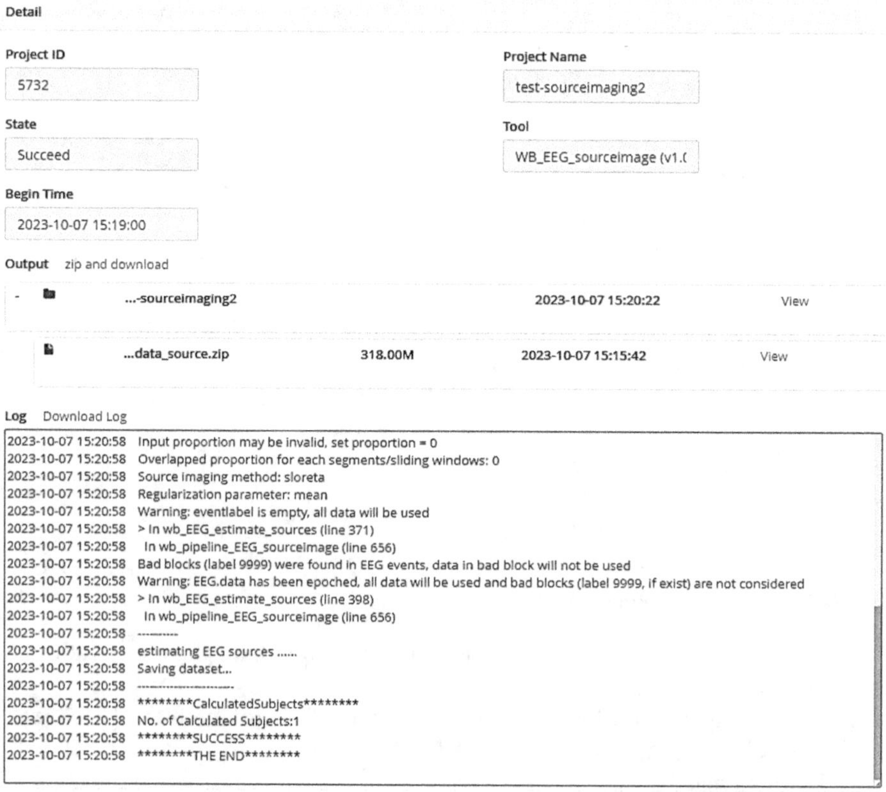

Detail

Project ID 5732

Project Name test-sourceimaging2

State Succeed

Tool WB_EEG_sourceimage (v1.(

Begin Time 2023-10-07 15:19:00

Output zip and download

-	...-sourceimaging2		2023-10-07 15:20:22	View
	...data_source.zip	318.00M	2023-10-07 15:15:42	View

Log Download Log

```
2023-10-07 15:20:58 Input proportion may be invalid, set proportion = 0
2023-10-07 15:20:58 Overlapped proportion for each segments/sliding windows: 0
2023-10-07 15:20:58 Source imaging method: sloreta
2023-10-07 15:20:58 Regularization parameter: mean
2023-10-07 15:20:58 Warning: eventlabel is empty, all data will be used
2023-10-07 15:20:58 > In wb_EEG_estimate_sources (line 371)
2023-10-07 15:20:58   In wb_pipeline_EEG_sourceimage (line 656)
2023-10-07 15:20:58 Bad blocks (label 9999) were found in EEG events, data in bad block will not be used
2023-10-07 15:20:58 Warning: EEG.data has been epoched, all data will be used and bad blocks (label 9999, if exist) are not considered
2023-10-07 15:20:58 > In wb_EEG_estimate_sources (line 398)
2023-10-07 15:20:58   In wb_pipeline_EEG_sourceimage (line 656)
2023-10-07 15:20:58 ----------
2023-10-07 15:20:58 estimating EEG sources ......
2023-10-07 15:20:58 Saving dataset...
2023-10-07 15:20:58 ----------------------
2023-10-07 15:20:58 ********CalculatedSubjects********
2023-10-07 15:20:58 No. of Calculated Subjects:1
2023-10-07 15:20:58 ********SUCCESS********
2023-10-07 15:20:58 ********THE END********
```

FIGURE 15.4 The detail page of source imaging results on the WeBrain. Results and logs will be shown on this page. A P300 ERP sample data from a healthy subject were used to demonstrate the usage of source imaging tool.

eventlabel: Event label which means specific event data. Default is empty. If it is empty, all data will be used. If "eventlabel" is not found, NO data will be segmented and calculated. If structure event ("eventlabel") does not include duration, the duration will be equal to "epochLenth." You can also input multiple event labels at once, split by comma (e.g., "S22, S23"); and all these event data will be calculated.

seleChanns: a string number with indices of the selected channels (e.g., "[1:4,7:30]" or "all"). Default is "all."

passband: Passband of filtering. Default is no filtering (i.e., "[]").

ForwardMethod: Forward method used to calculate the leadfield matrix.

ForwardMethod = 0: use a user defined leadfield matrix. While ForwardMethod = 0, the leadfield file must be inputted.

ForwardMethod = 1 (default): if EEG contains correct channel locations, it will automatically calculate leadfield matrix based on three-concentric sphere head model.

ForwardMethod = 2: if EEG contains correct channel locations, it will automatically calculate leadfield matrix based on real head modal (BEM modal) using FieldTrip.

proportion: overlapped percentage for each segments/sliding windows. It should be [0,1). Default is 0 (no overlapped).

SourceMethod: Inverse method used to calculate EEG source signals. Default is "sloreta."

"sloreta": EEG source imaging using sLORETA method (based on Dale et al. standardization). The sLORETA is a tomographic method for electric neuronal activity, where localization inference is based on images of standardized current density. The method is denoted as standardized low-resolution brain electromagnetic tomography (sLORETA). Noting that the Dale et al. standardization is used in the standardized estimation of sLORETA in the WeBrain.

alpha: regularization parameter of inverse method.

alpha >= 0: use user-defined value.

alpha = "mean": use averaged alpha with Tikh regularization (default);

alpha = "timevarying": use time-varying alpha with Tikh regularization. The time cost of "time-varying" alpha is extremely high for EEG time courses. It is used for the situation of less topographies.

elecDirecFlag: It is valid while calculating leadfield. It is optional.

elecDirecFlag = 0: *XYZ* coordinates is the electrode array with their Cartesian x (the left ear is defined as $-x$ axis), y (the nasion is the $+y$ axis), z coordinates in three columns.

elecDirecFlag = 1: *XYZ* coordinates is the electrode array with their Cartesian x (the nasion is the $+x$ axis), y (the left ear is the $+y$ axis), z coordinates in three columns. Default is 1.

matchflag: Matching the source signals to a brain template (e.g., AAL template) to obtain the averaged source signals of brain regions.

matchflag = [] (default): no matching.

matchflag = "aal": matching to AAL template and obtaining averaged source signals of brain regions.

erpflag: Estimating the source signals for averaged ERP waves? It is useful for the segmented ERP epoch data (i.e., the dimension of EEG.data is channels × time points × epochs).

erpflag = 0 (default): estimating the source signals of each epoch.

erpflag = 1: averaging trials and then estimating the source signals of averaged ERP waves. It is valid, while the data are segmented.

leadfieldfile: a user-defined lead field file (optional). It is valid, while the ForwardMethod = 0 only. The file could be a MATLAB.mat file, containing a matrix named "leadfield" (channels × sources/dipoles, e.g., 60 channels × 6,144 sources/dipoles) or a MATLAB structure containing leadfield of x-, y-, z-orientations (e.g., leadfield.X with dimension channels × dipoles (x-orientation); leadfield.Y with dimension channels × dipoles (y-orientation); leadfield.Z with dimension channels × dipoles (z-orientation)) which is calculated by using the forward theory, based on the electrode montage, head model, and equivalent source model. It can also be the output of ft_prepare_leadfield.m (e.g., lf.leadfield, dipoles contain x-, y-, z-orientations, 60 channels × 6,144*3 dipoles) based on real head modal (BEM modal) using FieldTrip.

Gridresolution: The grid resolution of dipoles (sources) inside the brain (optional). It is valid, while ForwardMethod = 2. If it is empty or ≤0, the default dipoles are vertices which are little smaller than brain, and the orientations of dipoles are their normal vector directions, i.e., the normals of the brain mesh. If it is >0 (unit is mm), the dipoles are distributed on regular 3D grid inside the brain mesh. The orientations of dipoles are X-, Y-, and Z-orientations, i.e., there are X-, Y-, and Z-oriented dipoles. Default is empty.

srate: Sampling rate of EEG data (optional). Default is obtained from EEG data ("[]").

The output of source imaging tool is an EEGLAB.set file which contains a structure of EEG including results of EEG source imaging, as well as parameters used to estimate source signals (Figure 15.5). The details of the results are as follows:

EEG.data: EEG sources with dimensions no. of dipoles × no. of time points

EEG.parameter.eventlabel: An event label which means good quality data

EEG.parameter.selechanns: An array with selected channels

EEG.parameter.epochLenth: Length of small epochs, and unit is time point

EEG.parameter.srate: Sampling rate of EEG data

EEG.parameter.sourcemethod: Inverse method used to estimate EEG source signals

EEG.parameter.proportion: Overlapped percentage for each segments/sliding windows

EEG.parameter.chanlocs: EEG channel locations on the scalp

EEG.parameter.ref: Original EEG reference

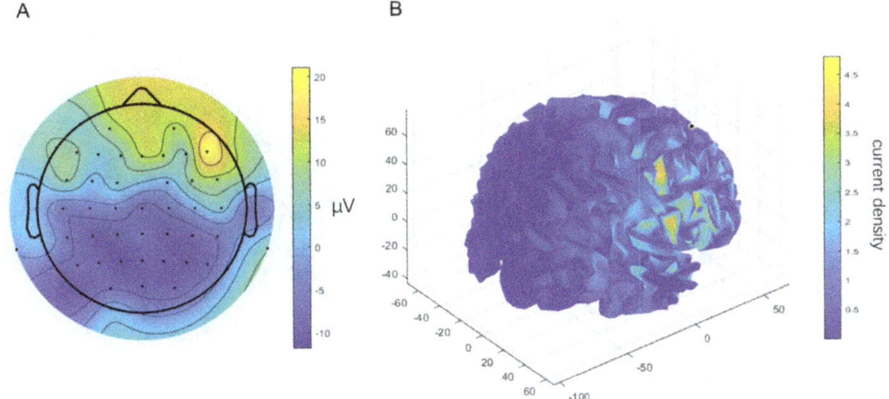

FIGURE 15.5 Example results of EEG sources at a time point of one trial. (a) EEG signals at a time point on the scalp. (b) Estimated current density of EEG sources at the same time point.

EEG.parameter.leadfield: Leadfield matrix
EEG.parameter.alpha: Regularization parameter
EEG.parameter.elecDirecFlag: Flag of channel directions
EEG.parameter.chaninfo: Channel information
EEG.parameter.erpflag: Flag of average ERP
EEG.parameter.headmodel: Default head model, and for details, see below
EEG.parameter.ForwardMethod: Forward method used in the tool
EEG.parameter.passband: Passband of filtering
EEG.parameter.gridresolution: The grid resolution of dipoles (sources) inside the brain
EEG.parameter.elec_aligned: Aligned channel locations of electrodes
EEG.parameter.BrainRegionInd: Brain region indices according to a template such as AAL
EEG.parameter.template: The template used to obtain the averaged source signals of the brain regions
EEG.parameter.matchflag: Flag of matching source signals to a brain template

15.2.2.2 Important Issues in Source Imaging Tool

15.2.2.2.1 Three-Concentric Sphere Head Model

Default head models are based on three-concentric spheres, in which the radii of spheres are normalized by the largest sphere. i.e., [0.87, 0.92, 1], corresponding to the brain, skull, and scalp. Conductivities of three-concentric spheres are normalized by the largest resistivity, and the default is [1, 0.0125, 1] (Figure 15.6). A high-density canonical cortical mesh was used to define the dipoles. These meshes were obtained by warping a template mesh to the T1-weighted structural anatomy of an individual subject, as described in Mattout et al. (2007). This warping is the inverse of the transformation derived for the spatial normalization of the subject's structural MRI image. The template mesh was generated by Fieldtrip (http://

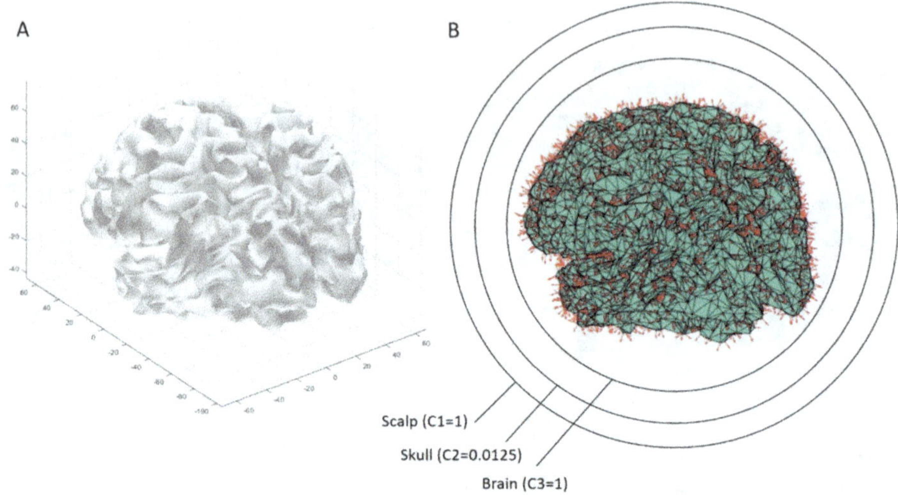

FIGURE 15.6 The default three-concentric sphere head model in the WeBrain. (a) The default geometrically triangular grid in the WeBrain, which is based on the standard brain (MNI space) and contains 6,144 dipoles. (b) Three-concentric spheres and their conductivities. Nodes and their normal vector directions are showed.

fieldtrip.fcdonders.nl/download.php) and was extracted from a structural MRI of a neurotypical male. The wrapping procedure provided a high-density mesh with 33,001 vertexes, which was uniformly distributed on the gray–white matter interface. The mesh was further downsampled to 6,144 vertexes (SPM MNI space) to reduce the computational load (Lei et al., 2012). Finally, the leadfield matrix was calculated analytically based on spherical harmonic spectra theory as shown in Chapter 4 (Yao, 2000; Yao et al., 2004).

15.2.2.2.2 Real Head Model Based on Boundary Element Method

The details about BEM (Section 5.4) and real head model used in the WeBrain are as follows:

A standard head model using "dipoli" method based on boundary element method (BEM) was used in the WeBrain (Figure 15.7). The head model contains a standard BEM volume conduction model of the head that can be used for EEG forward and inverse computations. The geometry is based on the "colin27" template that is described further down. The BEM model is expressed in MNI coordinates in mm. For more details, see: http://www.fieldtriptoolbox.org/template/headmodel/. More details about the real head model and BEM can also be seen in the WeBrain tool instruction 3.12 WB_EEG_CalcLeadfield_standardBEM. The "colin27" anatomical MRI and its relation to the TT and MNI template atlas are described in detail in http://imaging.mrc-cbu.cam.ac.uk/imaging/MniTalairach.

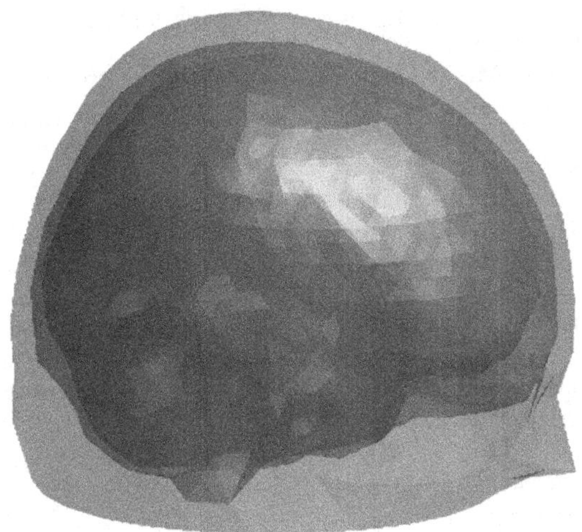

FIGURE 15.7 A standard BEM volume conduction model of the real head based on the "colin27" anatomical MRI. The conductivities of tissues are brain = 0.33, skull = 0.0041, and scalp = 0.33, and the number of vertices of tissues are brain = 1,500 points, skull = 1,000 points, and scalp = 500 points.

REFERENCES

Amunts, K., C. Ebell, J. Muller, et al. 2016. The human brain project: Creating a European research infrastructure to decode the human brain. *Neuron* 92(3):574–581.

Bigdely -Shamlo, N., T. Mullen, C. Kothe, et al. 2015. The PREP pipeline: Standardized preprocessing for large-scale EEG analysis. *Front Neuroinf* 9:16.

Bouchard, K.E., J.B. Aimone, M. Chun, et al. 2016. High-performance computing in neuroscience for data-driven discovery, integration, and dissemination. *Neuron* 92(3):628–631.

Cowley, B.U., J. Korpela, J. Torniainen. 2017. Computational testing for automated preprocessing: A MATLAB toolbox to enable large scale electroencephalography data processing. *PeerJ Comput Sci* 3:e108.

Dale, A.M., A.K. Liu, B.R. Fischl, et al. 2000. Dynamic statistical parametric mapping: Combining fMRI and MEG for high-resolution imaging of cortical activity. *Neuron* 26(1):55–67.

Das, S., A.P. Zijdenbos, J. Harlap, et al. 2012. LORIS: A web-based data management system for multi-center studies. *Front Neuroinf* 5:37.

Delorme, A., S. Makeig. 2004. EEGLAB: An open source toolbox for analysis of single-trial EEG dynamics including independent component analysis. *J Neurosci Methods* 134(1):9–21.

Dong, L., F. Li, Q. Liu, et al. 2017. MATLAB toolboxes for reference electrode standardization technique (REST) of scalp EEG. *Front Neurosci* 11:601.

Dong, L., J. Li, Q. Zou, et al. 2021. WeBrain: A web-based brainformatics platform of computational ecosystem for EEG big data analysis. *NeuroImage* 245:118713.

Gabard-Durnam, L.J., A.S. Mendez Leal, C.L. Wilkinson, et al. 2018. The Harvard Automated Processing Pipeline for Electroencephalography (HAPPE): Standardized processing software for developmental and high-artifact data. *Front Neurosci* 12:97.

Gibson, E., W. Li, C. Sudre, et al. 2018. NiftyNet: A deep-learning platform for medical imaging. *Comput Methods Programs Biomed* 158:113–122.

Glasser, M.F., S.M. Smith, D.S. Marcus, et al. 2016. The human connectome project's neuroimaging approach. *Nat Neurosci* 19(9):1175–1187.

Koch, C., A. Jones. 2016. Big science, team science, and open science for neuroscience. *Neuron* 92(3):612–616.

Lei, X., J. Hu, D. Yao. 2012. Incorporating FMRI functional networks in EEG source imaging: A Bayesian model comparison approach. *Brain Topogr* 25:27–38.

Manjón, J.V., P. Coupé. 2016. volBrain: An online MRI brain volumetry system. *Front Neuroinf* 10:30.

Marcus, D.S., T.R. Olsen, M. Ramaratnam, et al. 2007. The extensible neuroimaging archive toolkit: An informatics platform for managing, exploring, and sharing neuroimaging data. *Neuroinformatics* 5:11–33.

Markiewicz, C.J., K.J. Gorgolewski, F. Feingold, et al. 2021. The OpenNeuro resource for sharing of neuroscience data. *Elife* 10:e71774.

Mattout, J., R.N. Henson, K.J. Friston. 2007. Canonical source reconstruction for MEG. *Comput Intel Neurosci* 2007:67613.

Oostenveld, R., P. Fries, E. Maris, et al. 2011. FieldTrip: Open source software for advanced analysis of MEG, EEG, and invasive electrophysiological data. *Comput Intel Neurosci* 2011:1–9.

Pascual-Marqui, R.D. 2002. Standardized low-resolution brain electromagnetic tomography (sLORETA): Technical details. *Methods Find Exp Clin Pharmacol* 24(Suppl D):5–12.

Pedroni, A., A. Bahreini, N. Langer. 2019. Automagic: Standardized preprocessing of big EEG data. *NeuroImage* 200:460–473.

Poo, M.M., J.L. Du, N.Y. Ip, et al. 2016. China brain project: Basic neuroscience, brain diseases, and brain-inspired computing. *Neuron* 92(3):591–596.

Sejnowski, T.J., P.S. Churchland, J.A. Movshon. 2014. Putting big data to good use in neuroscience. *Nat Neurosci* 17(11):1440–1441.

Sherif, T., P. Rioux, M.E. Rousseau, et al. 2014. CBRAIN: A web-based, distributed computing platform for collaborative neuroimaging research. *Front Neuroinf* 8:54.

Takemiya, M., K. Majima, M. Tsukamoto, et al. 2016. BrainLiner: A neuroinformatics platform for sharing time-aligned brain-behavior data. *Front Neuroinf* 10:3.

Vaccarino, A.L., M. Dharsee, S. Strother, et al. 2018. Brain-CODE: A secure neuroinformatics platform for management, federation, sharing and analysis of multi-dimensional neuroscience data. *Front Neuroinf* 12:28.

VanHorn, J.D., A.W. Toga. 2014. Human neuroimaging as a "Big Data" science. *Brain Imaging Behav* 8:323–331.

Vogelstein, J.T., B. Mensh, M. Häusser, et al. 2016. To the cloud! A grassroots proposal to accelerate brain science discovery. *Neuron* 92(3):622–627.

Yao, D. 2000. High-resolution EEG mappings: A spherical harmonic spectra theory and simulation results. *Clin Neurophys* 111(1):81–92.

Yao, D. 2001. A method to standardize a reference of scalp EEG recordings to a point at infinity. *Physiol Meas* 22(4):693.

Yao, D., L. Wang, L., R. Oostenveld, et al. 2005. A comparative study of different references for EEG spectral mapping: The issue of the neutral reference and the use of the infinity reference. *Physiol Meas* 26(3):173.

Yao, D., Z. Yin, X. Tang, et al. 2004. High-resolution electroencephalogram (EEG) mapping: Scalp charge layer. *Phys Med Biol* 49(22):5073.

Appendix A
δ *Functions and Legendre Functions*

The δ and Legendre functions are very basic mathematic tools for understanding the brain's electric phenomena and developing the algorithms for EEG forward and inverse problems. Here, the following materials are extracted from the literature (Arfken et al., 2005; Fu, 1993; Zhang, 2003).

A.1 POINT SOURCE, δ FUNCTION, AND PRINCIPLE OF LINEAR SUPERPOSITION

According to the electromagnetic theory introduced in Chapter 2, the electric potential $\Phi(\vec{r})$ satisfies Poisson's equation:

$$\nabla^2\Phi(\vec{r}) = -\frac{1}{\epsilon}\rho(\vec{r}) \tag{A.1}$$

where $\rho(\vec{r})$ is the charge density at $\vec{r}$ and ϵ is the dielectric constant (the dual equation due to current source density in a conductor is (2.86)). This is a non-homogeneous linear partial differential equation of an unknown field function $\Phi(\vec{r})$. The non-homogeneous term $\frac{1}{\epsilon}\rho(\vec{r})$ is the source function, which is usually given, and $\Phi(\vec{r})$ is known as the (potential) field generated by the source $\frac{1}{\epsilon}\rho(\vec{r})$. According to the linearity of equation (A.1), if $\Phi_1(\vec{r})$ is the field produced by the source $\frac{1}{\epsilon}\rho_1(\vec{r})$ and satisfies the equation $\nabla^2\Phi_1(\vec{r}) = -\frac{1}{\epsilon}\rho_1(\vec{r})$, and $\Phi_2(\vec{r})$ is the field produced by the source $\frac{1}{\epsilon}\rho_2(\vec{r})$ and satisfies the equation $\nabla^2\Phi_2(\vec{r}) = -\frac{1}{\epsilon}\rho_2(\vec{r})$, then the combination $\frac{1}{\epsilon}\rho(\vec{r}) = c_1\frac{1}{\epsilon}\rho_1(\vec{r}) + c_2\frac{1}{\epsilon}\rho_2(\vec{r})$ will produce a combinational field $\Phi(\vec{r}) = c_1\Phi_1(\vec{r}) + c_2\Phi_2(\vec{r})$.

This property suggests that the fields generated by all sources can be found as long as the field generated from each point source is known and these fields are superposed together. This is the principle of linear superposition determined by linear equations. Thus, the field problem of sources combination is decomposed into the problem of finding the field generated from each point source in the space. In general, the solution of Poisson's equation (A.1) comes down to the solution of Poisson's equation for a point source. $\delta(\vec{r} - \vec{r}')$ is used to represent the location $\vec{r}$-dependent distribution of the point source density at $\vec{r}'$, while $G(\vec{r}-\vec{r}')$ is used to represent the field generated from it. Their relationship satisfies Poisson's equation:

$$\nabla^2 G(\vec{r}-\vec{r}') = -\delta(\vec{r}-\vec{r}') \tag{A.2}$$

where the rectangular coordinate form of the Laplacian operator is $\nabla^2 = \frac{\partial^2}{\partial x^2} + \frac{\partial^2}{\partial y^2} + \frac{\partial^2}{\partial z^2}$ and the derivation is calculated relative to the coordinates x, y, and z of the point $\vec{r}$. As the integral of the density of a variable in a volume represents the quantity in the volume, the point source density function $\delta(\vec{r}-\vec{r}')$, with a total amount of 1, has an integral property as follows:

$$\int_V \delta(\vec{r}-\vec{r}')dV = \begin{cases} 1, & \text{if } \vec{r}' \in V \\ 0, & \text{or else} \end{cases} \tag{A.3}$$

It implies that there is a quantity 1 source in the infinitesimal volume element including point $\vec{r}'$, while there is quantity 0 source in the volume free of $\vec{r}'$. That is the implication of the point source at $\vec{r}'$. Equation (A.3) defines a function $\delta(\vec{r}-\vec{r}')$, called a 3D δ function, which can be decomposed into the product of three unary functions: $\delta(\vec{r}-\vec{r}') = \delta(x-x')\delta(y-y')\delta(z-z')$, where x', y', and z' are the rectangular coordinates of the point $\vec{r}'$. The three factors on the right are apparently of the same function, except that the independent variables are different. They share the same properties. For example,

$$\int_a^b \delta(x-x')dx = \begin{cases} 1, & \text{if } x' \in (a,b) \\ 0, & \text{or else} \end{cases} \tag{A.4}$$

This is the definition of the one-dimensional $\delta(x-x')$ function. Obviously, the δ function and 3D δ function are different from the ordinary function. It does not contain the function value corresponding to each independent variable value but is defined according to the integral value of the independent variable in each region. This type of function is called a generalized function. An ordinary function can also be defined by the integral value in each region and is thus a generalized function. But a generalized function is not necessarily an ordinary function.

Applying the δ function, Poisson's equation can be rewritten as below:

$$\nabla^2 \Phi(\vec{r}-\vec{r}') = -\frac{q}{\epsilon}\delta(\vec{r}-\vec{r}') \tag{A.5}$$

It can be proved that the point charge q at $\vec{r}'$ in the unbounded space expressed by this formula has an electric potential at $\vec{r}$ as follows:

$$\Phi(\vec{r}-\vec{r}') = \frac{q}{4\pi\epsilon|\vec{r}-\vec{r}'|} \tag{A.6}$$

In summary, here the point source, δ function, is a small volume source dwindling into a point, and the integral of its density over the volume is of the finite value; in order to express the density distribution of a unit point source, Dirac proposed using the δ function to describe it. Obviously, the δ function is a symbolic function defined to describe a "physical phenomenon or model." It is also called Dirac δ function. The field generated from a point source is called the Green function. Traditionally, a unit point source is adopted in the Green function.

A.2 PROPERTIES OF δ FUNCTION

A.2.1 Integral Property

$\delta(x)$ is a generalized function defined by integral. It is assumed to be 0 at $x \neq 0$, and to approach infinity under the condition of $x \to 0$,

$$\int_a^b \delta(x)\,dx = 1, \quad \text{if} \quad 0 \in (a,b) \tag{A.7}$$

A.2.2 Value Selection

The δ function can be used to express the charge density with the electric quantity of q at $\vec{r}'$ as $\rho(\vec{r}) = q\delta(\vec{r} - \vec{r}')$. So, the total charge quantity in any volume V containing the point $\vec{r}'$ is $\int_V \rho(\vec{r})dV = \int_V q\delta(\vec{r} - \vec{r}')dV = q$. This equation can be mathematically understood as the integral of the product of a constant function q, and $\delta(\vec{r} - \vec{r}')$ in volume V containing $\vec{r}'$ is equal to the value of the constant function q at $\vec{r}'$. In other words, $\delta(\vec{r} - \vec{r}')$ acts as a value selection (sampling) for the function multiplied by it. Because $\delta(\vec{r} - \vec{r}') = 0$ at points outside $\vec{r}'$, the integral would not be affected, regardless of the value of the function is at other points, i.e., for any continuous function $f(\vec{r})$, there is

$$\int_V f(\vec{r})\delta(\vec{r} - \vec{r}')dV = \begin{cases} f(\vec{r}'), & \text{if } \vec{r}' \in V \\ 0, & \text{if } \vec{r}' \notin V \end{cases} \tag{A.8}$$

A.2.3 Expansion of δ Function

The δ function was originally proposed by Dirac in quantum mechanics. Because of its usefulness in mathematical physics, many mathematicians later studied it, endowing it with a strict mathematical meaning, obtaining many useful properties, and forming a mathematical branch known as "Generalized Function Theory." As proved mathematically, the δ function can be regarded as the generalized limit of some classical functions. So, it can be processed in the same way as that of classical functions. For example, the δ function can be decomposed into corresponding factor functions in different coordinate systems.

$$\delta(\vec{r}-\vec{r}')=\delta(x-x')\delta(y-y')\delta(z-z')=\frac{1}{\rho}\delta(\rho-\rho')\delta(\varphi-\varphi')\delta(z-z')$$
$$=\frac{1}{r^2\sin\theta}\delta(r-r')\delta(\theta-\theta')\delta(\varphi-\varphi') \tag{A.9}$$

A.2.4 Symmetry

$$\delta(x)=\delta(-x) \tag{A.10}$$

A.2.5 Scaling (or Similarity)

$$\delta(ax)=\frac{1}{|a|}\delta(x) \tag{A.11}$$

A.2.6 Differential of Position Vector

$$\nabla\cdot\frac{\vec{r}}{r^3}=-\nabla^2\left(\frac{1}{r}\right)=4\pi\delta(\vec{r}) \tag{A.12}$$

A.2.7 Approximation of δ Function

Some functions can be considered approximations of the Dirac function. Though none of them is the Dirac function itself, they can be treated as the Dirac function in some mathematical calculations. For example,

$$\delta(x)=\lim_{a\to 0^+}\frac{1}{a\sqrt{\pi}}e^{-x^2/a^2} \tag{A.13}$$

A.3 Legendre Polynomial and Spherical Harmonic Functions

A.3.1 Definition

$$P_l(\xi)=\frac{1}{2^l l!}\frac{d^l}{d\xi^l}\left(\xi^2-1\right)^l=\frac{1}{2^l}\sum_{v=0}^{\left[\frac{l}{2}\right]}(-1)^v\frac{(2l-2v)!}{v!(l-v)!(l-2v)!}\xi^{l-2v} \tag{A.14}$$

For this Legendre polynomial, $\left[\frac{l}{2}\right]$ is the maximum integer not greater than $\frac{l}{2}$.

$$P_l^m(\xi)=\left(1-\xi^2\right)^{\frac{m}{2}}\frac{d^m}{d\xi^m}P_l(\xi)=\frac{1}{2^l l!}\left(1-\xi^2\right)^{\frac{m}{2}}\frac{d^{l+m}}{d\xi^{l+m}}\left(\xi^2-1\right)^l \tag{A.15}$$

This is the associated Legendre polynomial, m is an integer and $0 \le m \ll l$. It is a polynomial of ξ when m is an even number.

$$Y_{lm}(\theta,\varphi) = \pm\sqrt{\frac{2l+1}{4\pi}\frac{(l-|m|)!}{(l+|m|)!}}P_l^{|m|}(\cos\theta)e^{im\varphi} \tag{A.16}$$

This is the spherical harmonic function, m is an integer and $-l \le m \le l$. The sign before the radical sign is positive when $m \le 0$, and positive or negative for even or odd m, respectively, when $m > 0$. The spherical harmonics satisfy the identity related to its complex conjugation function.

$$Y_{lm}^*(\theta,\varphi) = (-1)^m Y_{l-m}(\theta,\varphi) \tag{A.17}$$

The spherical harmonic functions form a complete orthonormal set of functions in the sense of the Fourier series.

A.3.2 Differential Equation

It can be directly verified by the definition (A.14) that the Legendre polynomial satisfies the Legendre equation.

$$\frac{d}{d\xi}(1-\xi^2)\frac{dP_l(\xi)}{d\xi} + l(l+1)P_l(\xi) = 0 \tag{A.18}$$

Similarly, by differentiating this formula m times with respect to ξ and multiplying it by $(1-\xi^2)^{\frac{m}{2}}$, it can be found that the associated Legendre polynomial (A.15) satisfies the associated Legendre equation.

$$\frac{d}{d\xi}(1-\xi^2)\frac{dP_l^m(\xi)}{d\xi} + \left[l(l+1) - \frac{m^2}{1-\xi^2}\right]P_l^m(\xi) = 0 \tag{A.19}$$

The angular part of the Laplacian operator is represented by the spherical coordinates,

$$\nabla_{\theta\varphi}^2 = \frac{1}{\sin\theta}\frac{\partial}{\partial\xi}\sin\theta\frac{\partial}{\partial\theta} + \frac{1}{\sin^2\theta}\frac{\partial^2}{\partial\varphi^2}$$

and it can be changed by transformation $\xi = \cos\theta$ as

$$\nabla_{\theta\varphi}^2 = \frac{\partial}{\partial\xi}(1-\xi^2)\frac{\partial}{\partial\xi} + \frac{1}{1-\xi^2}\frac{\partial^2}{\partial\varphi^2}$$

Apply it to the spherical harmonic function (A.16), and use the associated Legendre equation (A.19) to obtain the differential equation that the spherical harmonic function satisfies:

$$\nabla_{\theta\varphi}^2 Y_{lm}(\theta,\varphi) + l(l+1)Y_{lm}(\theta,\varphi) = 0 \tag{A.20}$$

A.3.3 Orthogonal Relationship and Normalization

Legendre polynomials are orthogonal:

$$\int_{-1}^{1} P_l^m(\xi) P_{l'}^m(\xi) d\xi = \frac{(l+m)!}{(l-m)!} \frac{2}{2l+1} \delta_{ll'} \tag{A.21}$$

According to equation (A.21) and $\int_0^{2\pi} \left(e^{im\varphi}\right)^* e^{im'\varphi} d\varphi = 2\pi\delta_{mm'}$, it can be got

$$\begin{aligned}
\int Y_{lm}^*(\theta,\varphi) Y_{l'm'}(\theta,\varphi) d\Omega &= \int_0^{\pi} \sin\theta d\theta \int_0^{2\pi} d\varphi Y_{lm}^*(\theta,\varphi) Y_{l'm'}(\theta,\varphi) \\
&= \pm\sqrt{\frac{2l+1}{4\pi}\frac{(l-|m|)!}{(l+|m|)!}}\sqrt{\frac{2l'+1}{4\pi}\frac{(l'-|m'|)!}{(l'+|m'|)!}} \cdot \int_{-1}^{1} P_l^{|m|}(\xi) P_{l'}^{|m'|}(\xi) d\xi \\
&\int_0^{2\pi} \left(e^{im\varphi}\right)^* e^{im'\varphi} d\varphi \\
&= \sqrt{\frac{2l+1}{2}\frac{(l-|m|)!}{(l+|m|)!}}\sqrt{\frac{2l'+1}{2}\frac{(l'-|m|)!}{(l'+|m|)!}} \cdot \int_{-1}^{1} P_l^{|m|}(\xi) P_{l'}^{|m'|}(\xi) d\xi \delta_{mm'} \\
&= \delta_{ll'}\delta_{mm'}
\end{aligned} \tag{A.22}$$

where $d\Omega = \sin\theta d\theta d\varphi$ is a solid angle element. The integral variable has a transformation of $\xi = \cos\theta$ at the second equal sign. Equation (A.22) represents the orthogonality normalization of spherical harmonic functions. The radical sign operation on the right end of the definition of spherical harmonic function (A.16) is to make it unity. This step of multiplying by an appropriate factor to normalize a function is called normalization, and the multiplied factor is called the normalization factor.

A.3.4 Generating Function

$$G(\xi,t) \equiv \left(1 - 2\xi t + t^2\right)^{-\frac{1}{2}} = \sum_{l=0}^{\infty} P_l(\xi) t^l \tag{A.23}$$

where $-1 \le \xi \le 1$, $|t| < 1$. The wonderful thing about generating function is that it is often a very simple function but can indeed derive a complex sequence of functions.

A.3.5 Plus Theorem

If $\hat{\theta}$ is the angle between the directions of (θ,φ) and (θ',φ'), then

$$P_l\left(\cos\hat{\theta}\right)=\frac{4\pi}{2l+1}\sum_{m=-l}^{l}Y_{lm}(\theta,\varphi)Y_{lm}^{*}(\theta',\varphi') \tag{A.24}$$

REFERENCES

Arfken, G.B., H.J. Weber, F.E. Harris. 2005. *Mathematical Methods for Physicists*, 6th edn, section 1.1. London: Academic Press.

Fu, G. 1993. *Green's Function Method in Electromagnetic Field (In Chinese)*. Beijing: Higher Education Press.

Zhang, Q. 2003. *Classical Field Theory (In Chinese)*. Beijing: Science Press.

Appendix B
Analytical Potential Solutions for Regular Head Model

B.1 POTENTIAL FOR THE CONCENTRIC FOUR-LAYER SPHERICAL HEAD MODEL

B.1.1 Derivation of the Four-Layer Model

The four-layer concentric sphere model is taken as an example here (Yao et al., 2003a,b). The four layers include the brain, cerebrospinal fluid (CSF), skull, and scalp. The four-layer concentric sphere is considered to be an improvement of the three-layer concentric sphere owing to the introduction of CSF. Some studies show that the four-layer concentric sphere model remains an effective approximate model even when compared with the finite element model of a realistic head model. It can provide reasonable head–surface potential estimation, with an error of 10%–20%. Below is the derivation process of the potential formula for the four-layer concentric sphere model. The model is shown in Figure B.1, and the parameters can be specifically selected from the following data or other new experimental data.

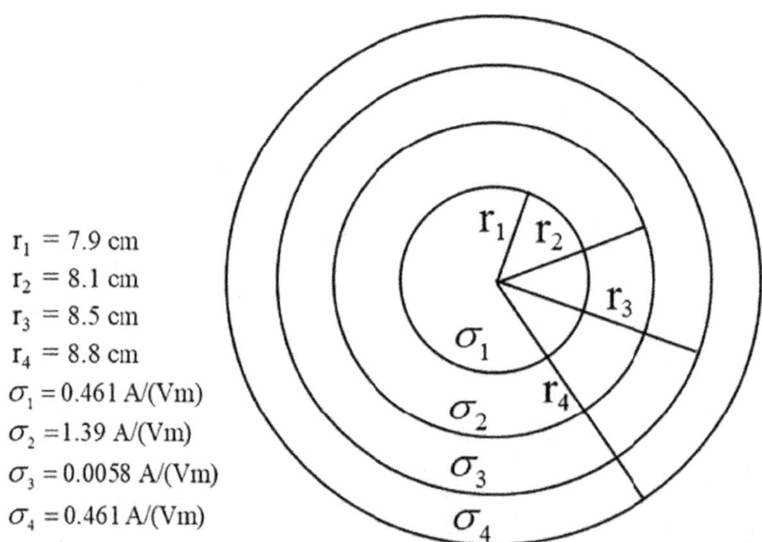

FIGURE B.1 Concentric four-layer sphere model (Yao, 2003a,b).

For the situation in Figure B.1, based on the preparation in Section 4.5, suppose the potentials of each layer from the inside to the outside are represented by Φ_1, Φ_2, Φ_3, and Φ_4, respectively, then they can be written as

$$\Phi_1 = \sum_{l,m} \left[A_l^m \cdot r^l + K_l^m \cdot \frac{1}{r^{l+1}} \right] \cdot S_l^m \tag{B.1}$$

$$\Phi_2 = \sum_{l,m} \left[C_l^m \cdot r^l + D_l^m \cdot \frac{1}{r^{l+1}} \right] \cdot S_l^m \tag{B.2}$$

$$\Phi_3 = \sum_{l,m} \left[E_l^m \cdot r^l + F_l^m \cdot \frac{1}{r^{l+1}} \right] \cdot S_l^m \tag{B.3}$$

$$\Phi_4 = \sum_{l,m} \left[G_l^m \cdot r^l + H_l^m \cdot \frac{1}{r^{l+1}} \right] \cdot S_l^m \tag{B.4}$$

The corresponding boundary conditions are

$$\Phi_1 = \Phi_2 \Big|_{r=r_1} \tag{B.5}$$

$$\sigma_1 \frac{\partial \Phi_1}{\partial r} = \sigma_2 \frac{\partial \Phi_2}{\partial r} \Bigg|_{r=r_1} \tag{B.6}$$

$$\Phi_2 = \Phi_3 \Big|_{r=r_2} \tag{B.7}$$

$$\sigma_2 \frac{\partial \Phi_2}{\partial r} = \sigma_3 \frac{\partial \Phi_3}{\partial r} \Bigg|_{r=r_2} \tag{B.8}$$

$$\Phi_3 = \Phi_4 \Big|_{r=r_3} \tag{B.9}$$

$$\sigma_3 \frac{\partial \Phi_3}{\partial r} = \sigma_4 \frac{\partial \Phi_4}{\partial r} \Bigg|_{r=r_3} \tag{B.10}$$

$$\frac{\partial \Phi_4}{\partial r} \Bigg|_{r=r_4} = 0 \tag{B.11}$$

Under the known K_l^m and S_l^m, we adopted the boundary conditions to derive the formulas of the coefficients A_l^m, C_l^m, D_l^m, E_l^m, F_l^m, G_l^m, and H_l^m in the formulas (B.1)–(B.4) (refer to Section 4.5.1).

From the boundary conditions (B.11), we can get

$$G_l^m = H_l^m \cdot \frac{l+1}{l} \cdot \frac{1}{r_4^{2l+1}} \tag{B.12}$$

According to the boundary conditions (B.9), use formula (B.12) to obtain

$$E_l^m \cdot r_3^l + F_l^m \cdot \frac{1}{r_3^{l+1}} = H_l^m \cdot \left(r_3^l \cdot \frac{l+1}{l} \cdot \frac{1}{r_4^{2l+1}} + \frac{1}{r_3^{l+1}} \right) \tag{B.13}$$

According to the boundary conditions (B.10), use formula (B.12) to get

$$\sigma_3 \left[E_l^m \cdot l \cdot r_3^{l-1} - F_l^m \cdot (l+1) \cdot \frac{1}{r_3^{l+2}} \right] = \sigma_4 \cdot H_l^m \cdot (l+1) \cdot \left[\frac{1}{r_4^{2l+1}} \cdot r_3^{l-1} - \frac{1}{r_3^{l+2}} \right] \tag{B.14}$$

Use formula (B.13) to get

$$H_l^m = \frac{E_l^m \cdot r_3^l + F_l^m \cdot \frac{1}{r_3^{l+1}}}{r_3^l \cdot \frac{l+1}{l} \cdot \frac{1}{r_4^{2l+1}} + \frac{1}{r_3^{l+1}}} = \frac{E_l^m \cdot r_3^{2l+1} + F_l^m}{\left(\frac{r_3}{r_4} \right)^{2l+1} \cdot \frac{l+1}{l} + 1} \tag{B.15}$$

Use formula (B.14) to get

$$H_l^m = \frac{\sigma_3 \cdot \left[E_l^m \cdot l \cdot r_3^{l-1} - F_l^m \cdot (l+1) \cdot \frac{1}{r_3^{l+2}} \right]}{\sigma_4 \cdot (l+1) \cdot \left(\frac{1}{r_4^{2l+1}} \cdot r_3^{l-1} - \frac{1}{r_3^{l+2}} \right)} = \frac{\sigma_3}{\sigma_4} \cdot \frac{E_l^m \cdot \frac{l}{l+1} \cdot r_3^{2l+1} - F_l^m}{\left(\frac{r_3}{r_4} \right)^{2l+1} - 1} \tag{B.16}$$

Use formulas (B.15) and (B.16) to get

$$\frac{E_l^m \cdot r_3^{2l+1} + F_l^m}{\left(\frac{r_3}{r_4} \right)^{2l+1} \cdot \frac{l+1}{l} + 1} = \frac{\sigma_3}{\sigma_4} \cdot \frac{E_l^m \cdot \frac{l}{l+1} \cdot r_3^{2l+1} - F_l^m}{\left(\frac{r_3}{r_4} \right)^{2l+1} - 1} \tag{B.17}$$

Simplify and organize the above formula to get

$$\begin{aligned} &E_l^m \cdot \left[\left(\frac{r_3}{r_4} \right)^{2l+1} \cdot \left(1 - \frac{\sigma_3}{\sigma_4} \right) - \left(1 + \frac{l}{l+1} \cdot \frac{\sigma_3}{\sigma_4} \right) \right] \\ &= F_l^m \cdot \left[\frac{1}{r_3^{2l+1}} \cdot \left(1 - \frac{\sigma_3}{\sigma_4} \right) - \frac{1}{r_4^{2l+1}} \cdot \left(1 + \frac{l+1}{l} \cdot \frac{\sigma_3}{\sigma_4} \right) \right] \end{aligned} \tag{B.18}$$

The above formula can be written as

$$p \cdot E_l^m = q \cdot F_l^m \tag{B.19}$$

where

$$p = \left(\frac{r_3}{r_4}\right)^{2l+1} \cdot \left(1 - \frac{\sigma_3}{\sigma_4}\right) - \left(1 + \frac{l}{l+1} \cdot \frac{\sigma_3}{\sigma_4}\right) \tag{B.20}$$

$$q = \frac{1}{r_3^{2l+1}} \cdot \left(1 - \frac{\sigma_3}{\sigma_4}\right) - \frac{1}{r_4^{2l+1}} \cdot \left(1 + \frac{l+1}{l} \cdot \frac{\sigma_3}{\sigma_4}\right) \tag{B.21}$$

According to the boundary conditions (B.7), use formula (B.20) to obtain

$$C_l^m \cdot r_2^l + D_l^m \cdot \frac{1}{r_2^{l+1}} = E_l^m \cdot r_2^l + F_l^m \cdot \frac{1}{r_2^{l+1}} = E_l^m \cdot r_2^l + \frac{p}{q} \cdot E_l^m \cdot \frac{1}{r_2^{l+1}} = E_l^m \cdot \left(r_2^l + \frac{p}{q} \cdot \frac{1}{r_2^{l+1}}\right) \tag{B.22}$$

According to the boundary conditions (B.8), use the formula (B.19) to obtain

$$\begin{aligned} \sigma_2 \cdot \left[C_l^m \cdot l \cdot r_2^{l-1} - D_l^m \cdot (l+1) \cdot \frac{1}{r_2^{l+2}}\right] &= \sigma_3 \cdot \left[E_l^m \cdot l \cdot r_2^{l-1} - F_l^m \cdot (l+1) \cdot \frac{1}{r_2^{l+2}}\right] \\ &= \sigma_3 \cdot \left[E_l^m \cdot l \cdot r_2^{l-1} - \frac{p}{q} \cdot E_l^m \cdot (l+1) \cdot \frac{1}{r_2^{l+2}}\right] \\ &= \sigma_3 \cdot E_l^m \left[l \cdot r_2^{l-1} - \frac{p}{q} \cdot (l+1) \cdot \frac{1}{r_2^{l+2}}\right] \end{aligned} \tag{B.23}$$

Use formula (B.22) to get

$$E_l^m = \frac{C_l^m \cdot r_2^l + D_l^m \cdot \frac{1}{r_2^{l+1}}}{r_2^l + \frac{p}{q} \cdot \frac{1}{r_2^{l+1}}} = \frac{C_l^m \cdot r_2^{2l+1} + D_l^m}{r_2^{2l+1} + \frac{p}{q}} \tag{B.24}$$

Use formula (B.23) to get

$$E_l^m = \frac{\sigma_2 \cdot \left[C_l^m \cdot l \cdot r_2^{l-1} - D_l^m \cdot (l+1) \cdot \frac{1}{r_2^{l+2}}\right]}{\sigma_3 \cdot \left[l \cdot r_2^{l-1} - \frac{p}{q} \cdot (l+1) \cdot \frac{1}{r_2^{l+2}}\right]} = \frac{\sigma_2 \cdot \left[C_l^m \cdot \frac{l}{l+1} \cdot r_2^{2l+1} - D_l^m\right]}{\sigma_3 \cdot \left[\frac{l}{l+1} \cdot r_2^{2l+1} - \frac{p}{q}\right]} \tag{B.25}$$

Use formulas (B.24) and (B.25) to get

$$\frac{C_l^m \cdot r_2^{2l+1} + D_l^m}{r_2^{2l+1} + \frac{p}{q}} = \frac{\sigma_2 \cdot \left[C_l^m \cdot \frac{l}{l+1} \cdot r_2^{2l+1} - D_l^m \right]}{\sigma_3 \cdot \left[\frac{l}{l+1} \cdot r_2^{2l+1} - \frac{p}{q} \right]} \tag{B.26}$$

Organize formula (B.26) to get

$$\begin{aligned} & C_l^m \cdot \left[\frac{l}{l+1} \cdot r_2^{2l+1} \cdot \left(r_2^{2l+1} + \frac{p}{q} \right) \cdot \frac{\sigma_2}{\sigma_3} - r_2^{2l+1} \cdot \left(\frac{l}{l+1} \cdot r_2^{2l+1} - \frac{p}{q} \right) \right] \\ & = D_l^m \cdot \left[\left(\frac{l}{l+1} \cdot r_2^{2l+1} - \frac{p}{q} \right) + \left(r_2^{2l+1} + \frac{p}{q} \right) \cdot \frac{\sigma_2}{\sigma_3} \right] \end{aligned} \tag{B.27}$$

In the above formula, let

$$\alpha = r_2^{2l+1} \cdot \left[\frac{l}{l+1} \cdot r_2^{2l+1} \cdot \left(\frac{\sigma_2}{\sigma_3} - 1 \right) + \frac{p}{q} \cdot \left(\frac{l}{l+1} \cdot \frac{\sigma_2}{\sigma_3} + 1 \right) \right] \tag{B.28}$$

$$\beta = r_2^{2l+1} \cdot \left(\frac{l}{l+1} + \frac{\sigma_2}{\sigma_3} \right) + \frac{p}{q} \cdot \left(-1 + \frac{\sigma_2}{\sigma_3} \right) \tag{B.29}$$

Then, formula (B.27) can be written as

$$\alpha \cdot C_l^m = \beta \cdot D_l^m \tag{B.30}$$

According to the boundary conditions (B.5), use the formula (B.30) to obtain

$$A_l^m \cdot r_1^l + K_l^m \cdot \frac{1}{r_1^{l+1}} = C_l^m \cdot r_1^l + D_l^m \cdot \frac{1}{r_1^{l+1}} = C_l^m \cdot r_1^l + \frac{\alpha}{\beta} C_l^m \cdot \frac{1}{r_1^{l+1}} = C_l^m \cdot \left[r_1^l + \frac{\alpha}{\beta} \cdot \frac{1}{r_1^{l+1}} \right] \tag{B.31}$$

According to the boundary conditions (B.6), use the formula (B.30) to obtain

$$\begin{aligned} \sigma_1 \cdot \left[A_l^m \cdot l \cdot r_1^{l-1} - K_l^m \cdot (l+1) \cdot \frac{1}{r_1^{l+2}} \right] &= \sigma_2 \cdot \left[C_l^m \cdot l \cdot r_1^{l-1} - D_l^m \cdot (l+1) \cdot \frac{1}{r_1^{l+2}} \right] \\ &= \sigma_2 \cdot \left[C_l^m \cdot l \cdot r_1^{l-1} - \frac{\alpha}{\beta} \cdot C_l^m \cdot (l+1) \cdot \frac{1}{r_1^{l+2}} \right] \\ &= \sigma_2 \cdot C_l^m \cdot \left[l \cdot r_1^{l-1} - \frac{\alpha}{\beta} \cdot (l+1) \cdot \frac{1}{r_1^{l+2}} \right] \end{aligned} \tag{B.32}$$

Use formula (B.31) to get

$$A_l^m = \frac{1}{r_1^l} \cdot \left[C_l^m \cdot \left(r_1^l + \frac{\alpha}{\beta} \cdot \frac{1}{r_1^{l+1}} \right) - K_l^m \cdot \frac{1}{r_1^{l+1}} \right] \tag{B.33}$$

Use formula (B.32) to get

$$A_l^m = \frac{1}{l \cdot r_1^{l-1}} \cdot \left\{ \frac{\sigma_2}{\sigma_1} \cdot C_l^m \cdot \left[l \cdot r_1^{l-1} - \frac{\alpha}{\beta} \cdot (l+1) \cdot \frac{1}{r_1^{l+2}} \right] + K_l^m \cdot (l+1) \cdot \frac{1}{r_1^{l+2}} \right\} \tag{B.34}$$

Use formula (B.33) and (B.34) to get

$$\frac{1}{r_1^l} \cdot \left[C_l^m \cdot \left(r_1^l + \frac{\alpha}{\beta} \cdot \frac{1}{r_1^{l+1}} \right) - K_l^m \cdot \frac{1}{r_1^{l+1}} \right]$$

$$= \frac{1}{l \cdot r_1^{l-1}} \cdot \left\{ \frac{\sigma_2}{\sigma_1} \cdot C_l^m \cdot \left[l \cdot r_1^{l-1} - \frac{\alpha}{\beta} \cdot (l+1) \cdot \frac{1}{r_1^{l+2}} \right] + K_l^m \cdot (l+1) \cdot \frac{1}{r_1^{l+2}} \right\}$$

$$C_l^m = \frac{K_l^m \cdot (2l+1)}{l \cdot r_1^{2l+1} \cdot \left(1 - \frac{\sigma_2}{\sigma_1} \right) + \frac{\alpha}{\beta} \cdot \left[l + \frac{\sigma_2}{\sigma_1} \cdot (l+1) \right]} \tag{B.35}$$

Substitute (B.35) into (B.33) to get

$$A_l^m = K_l^m \cdot \left[\frac{(2l+1) \cdot \left(1 + \frac{\alpha}{\beta} \cdot \frac{1}{r_1^{2l+1}} \right)}{l \cdot r_1^{2l+1} \cdot \left(1 - \frac{\sigma_2}{\sigma_1} \right) + \frac{\alpha}{\beta} \cdot \left[l + \frac{\sigma_2}{\sigma_1} \cdot (l+1) \right]} - \frac{1}{r_1^{2l+1}} \right] \tag{B.36}$$

Use formulas (B.22) and (B.30) to get

$$E_l^m = \frac{C_l^m \cdot r_2^l + \frac{\alpha}{\beta} \cdot C_l^m \cdot \frac{1}{r_2^{l+1}}}{r_2^l + \frac{p}{q} \cdot \frac{1}{r_2^{l+1}}} = \frac{C_l^m \cdot \left(r_2^l + \frac{\alpha}{\beta} \cdot \frac{1}{r_2^{l+1}} \right)}{r_2^l + \frac{p}{q} \cdot \frac{1}{r_2^{l+1}}} \tag{B.37}$$

Use formula (B.23) to get

$$H_l^m = \frac{E_l^m \cdot r_3^l + F_l^m \frac{1}{r_3^{l+1}}}{r_3^l \cdot \frac{l+1}{l} \cdot \frac{1}{r_4^{2l+1}} + \frac{1}{r_3^{l+1}}}$$

According to formula (B.19), the above formula can be written as

$$H_l^m = \frac{E_l^m \cdot \left(r_3^l + \frac{p}{q} \cdot \frac{1}{r_3^{l+1}} \right)}{r_3^l \cdot \frac{l+1}{l} \cdot \frac{1}{r_4^{2l+1}} + \frac{1}{r_3^{l+1}}} \tag{B.38}$$

Substitute (B.38) into (B.12) to get

$$G_l^m = H_l^m \cdot \frac{l+1}{l} \cdot \frac{1}{r_4^{2l+1}} = \frac{r_3^l + \frac{p}{q} \cdot \frac{1}{r_3^{l+1}}}{r_3^l + \frac{l}{l+1} \cdot \frac{r_4^{2l+1}}{r_3^{l+1}}} \cdot E_l^m$$

Substitute (B.37) into the above formula to get

$$G_l^m = \frac{r_3^l + \frac{p}{q} \cdot \frac{1}{r_3^{l+1}}}{r_3^l + \frac{l}{l+1} \cdot \frac{r_4^{2l+1}}{r_3^{l+1}}} \cdot \frac{r_2^l + \frac{\alpha}{\beta} \cdot \frac{1}{r_2^{l+1}}}{r_2^l + \frac{p}{q} \cdot \frac{1}{r_2^{l+1}}} \cdot C_l^m$$

Substitute (B.35) into the above formula to get

$$G_l^m = \frac{r_3^l + \frac{p}{q} \cdot \frac{1}{r_3^{l+1}}}{r_3^l + \frac{l}{l+1} \cdot \frac{r_4^{2l+1}}{r_3^{l+1}}} \cdot \frac{r_2^l + \frac{\alpha}{\beta} \cdot \frac{1}{r_2^{l+1}}}{r_2^l + \frac{p}{q} \cdot \frac{1}{r_2^{l+1}}} \cdot \frac{(2l+1)}{l \cdot r_1^{2l+1} \cdot \left(1 - \frac{\sigma_2}{\sigma_1}\right) + \frac{\alpha}{\beta} \cdot \left[l + \frac{\sigma_2}{\sigma_1} \cdot (l+1) \right]} \cdot K_l^m \tag{B.39}$$

Take $K_l^m \equiv 1$ corresponding to equations (4.67)–(4.68), then all the parameters A_l^m, C_l^m, D_l^m, E_l^m, F_l^m, G_l^m, and H_l^m are irrelevant to m, so they can be replaced by A_l, C_l, D_l, E_l, F_l, G_l, and H_l, respectively.

Use formula (B.1) to get

$$\Phi_1 = \sum_{l,m} \left[A_l \cdot r^{2l+1} + 1 \right] \cdot \frac{S_l^m}{r^{l+1}} = \sum_{l,m} W_l(1) \cdot \frac{S_l^m}{r^{l+1}} \tag{B.40}$$

where

$$W_l(1) = A_l \cdot r^{2l+1} + 1 = \left[\frac{(2l+1) \cdot \left(1 + \frac{\alpha}{\beta} \cdot \frac{1}{r_1^{2l+1}}\right)}{l \cdot r_1^{2l+1} \cdot \left(1 - \frac{\sigma_2}{\sigma_1}\right) + \frac{\alpha}{\beta} \left[l + \frac{\sigma_2}{\sigma_1} \cdot (l+1) \right]} - \frac{1}{r_1^{2l+1}} \right] \cdot r^{2l+1} + 1 \tag{B.41}$$

Use formula (B.2) to get

$$\Phi_2 = \sum_{l,m} \left[C_l \cdot r^{2l+1} + D_l \right] \cdot \frac{S_l^m}{r^{l+1}} = \sum_{l,m} W_l(2) \cdot \frac{S_l^m}{r^{l+1}} \tag{B.42}$$

where

$$W_l(2) = C_l \cdot r^{2l+1} + D_l \tag{B.43}$$

Use formula (B.30) to get

$$D_l = \frac{\alpha}{\beta} \cdot C_l = \lambda \cdot C_l \tag{B.44}$$

where

$$\lambda = \frac{\alpha}{\beta} \tag{B.45}$$

Substitute formula (B.44) into (B.43) to get

$$W_l(2) = C_l \cdot \left(r^{2l+1} + \lambda \right) \tag{B.46}$$

where

$$C_l = \frac{(2l+1)}{l \cdot r_1^{2l+1} \cdot \left(1 - \frac{\sigma_2}{\sigma_1} \right) + \frac{\alpha}{\beta} \cdot \left[l + \frac{\sigma_2}{\sigma_1} \cdot (l+1) \right]} \tag{B.47}$$

Use formula (B.3) to get

$$\Phi_3 = \sum_{l,m} \left[E_l \cdot r^{2l+1} + F_l \right] \cdot \frac{S_l^m}{r^{l+1}} = \sum_{l,m} W_l(3) \cdot \frac{S_l^m}{r^{l+1}} \tag{B.48}$$

where

$$W_l(3) = E_l \cdot r^{2l+1} + F_l \tag{B.49}$$

Use formula (B.34) to get

$$F_l = \frac{p}{q} \cdot E_l = \xi \cdot E_l \tag{B.50}$$

where

$$\xi = \frac{p}{q} \tag{B.51}$$

Substitute formula (B.50) into (B.49) to get

$$W_l(3) = E_l \cdot \left(r^{2l+1} + \xi\right) \tag{B.52}$$

where E_l is obtained by formulas (B.37) and (B.35)

$$E_l = \frac{(2l+1)}{l \cdot r_1^{2l+1} \cdot \left(1 - \frac{\sigma_2}{\sigma_1}\right) + \frac{\alpha}{\beta} \cdot \left[l + \frac{\sigma_2}{\sigma_1} \cdot (l+1)\right]} \cdot \frac{r_2^l + \frac{\alpha}{\beta} \cdot \frac{1}{r_2^{l+1}}}{r_2^l + \frac{p}{q} \cdot \frac{1}{r_2^{l+1}}} \tag{B.53}$$

Use formula (B.4) to get

$$\Phi_4 = \sum_{l,m} \left[G_l \cdot r^{2l+1} + H_l\right] \cdot \frac{S_l^m}{r^{l+1}} = \sum_{l,m} W_l(4) \cdot \frac{S_l^m}{r^{l+1}} \tag{B.54}$$

where

$$W_l(4) = G_l \cdot r^{2l+1} + H_l \tag{B.55}$$

Use formula (B.12) to get

$$H_l = \frac{1}{l+1} \cdot r_4^{2l+1} \cdot G_l = \chi \cdot G_i \tag{B.56}$$

where

$$\chi = \frac{l}{l+1} \cdot r_4^{2l+1} \tag{B.57}$$

Substitute formula (B.57) into (B.55) to get

$$W_l(4) = G_l \cdot \left(r^{2l+1} + \chi\right) \tag{B.58}$$

where G_l is obtained by formulas (B.39) by removing K_m^l

In summary, it is known

$$S_l^m = K_l^m \left(u_l^m \cdot \cos\theta + v_l^m \cdot \sin\theta\right) \cdot P_l^m(\cos\theta) \tag{B.59}$$

Substitute formula (B.59) into formulas (B.41), (B.43), (B.48), and (B.54) to get

$$\Phi_i = \sum_{l,m} \frac{W_l(i)}{r^{l+1}} \cdot S_l^m = \sum_{l,m} \frac{P_l^m(\cos\theta)}{r^{l+1}} \cdot \left(U_l^m \cdot \cos\theta + V_l^m \cdot \sin\theta\right), \tag{B.60}$$

$$i = 1,2,3,4$$

where

$$U_l^m(i) = W_l(i)\cdot u_l^m K_l^m \tag{B.61}$$

$$V_l^m(i) = W_l(i)\cdot v_l^m K_l^m \tag{B.62}$$

For dipoles, refer to equations (4.48), (4.55) and (4.56) in Chapter 4 to obtain

$$\begin{aligned} u_l^m &= l\cdot P_r\cdot P_l^m(\cos\theta)\cdot\cos(m\varphi_0) - \frac{m\cdot P_\varphi}{\sin\theta_0}\cdot P_l^m(\cos\theta)\cdot\sin(m\varphi_0) \\ &-\frac{P_\theta}{2}\cdot\left[(l-m+1)\cdot(l+m)\cdot P_l^{m-1}(\cos\theta_0) - P_l^{m+1}(\cos\theta_0)\right]\cdot\cos(m\varphi_0) \end{aligned} \tag{B.63}$$

$$\begin{aligned} v_l^m &= l\cdot P_r\cdot P_l^m(\cos\theta)\cdot\sin(m\varphi_0) + \frac{m\cdot P_\varphi}{\sin\theta_0} P_l^m(\cos\theta)\cdot\cos(m\varphi_0) \\ &-\frac{P_\theta}{2}\cdot\left[(l-m+1)\cdot(l+m)\cdot P_l^{m-1}(\cos\theta_0) - P_l^{m+1}(\cos\theta_0)\right]\cdot\sin(m\varphi_0) \end{aligned} \tag{B.64}$$

$W_l(i)$, $i = 1,2,3,4$, are calculated by formulas (B.41), (B.46), (B.52), and (B.58), respectively. Refer to equations (4.57)–(4.60) or (4.62)–(4.66) in Chapter 4, the corresponding general expression can be written.

B.2 THE FORMULA OF THE CONCENTRIC THREE-LAYER SPHERE MODEL SIMPLIFIED FROM THE FORMULA OF THE CONCENTRIC FOUR-LAYER SPHERE MODEL

Hypothesis 1: $r_3 = r_4 = r$, then from equations (B.20) and (B.21), we get

$$\frac{p}{q} = \frac{l}{l+1} r^{2l+1} \tag{B.65}$$

If taking $\sigma_3 = \sigma_4 = \sigma$, formula (B.65) can also be obtained from equations (B.20) and (B.21). Substitute (B.65) into the formula of G_l in equation (B.58) of the four-layer sphere model, where the filter coefficient becomes

$$G_l = \frac{r_2^l + \frac{\alpha}{\beta}\cdot\frac{1}{r_2^{l+1}}}{r_2^l + \frac{p}{q}\cdot\frac{1}{r_2^{l+1}}}\cdot\frac{(2l+1)}{l\cdot r_1^{2l+1}\cdot\left(1-\frac{\sigma_2}{\sigma_1}\right) + \frac{\alpha}{\beta}\cdot\left[l+\frac{\sigma_2}{\sigma_1}\cdot(l+1)\right]} \tag{B.66}$$

At this moment, (B.66) and the third layer filter coefficient (B.53) are all the same, and the coefficients ξ in equation (B.52) and χ in equation (B.58) are also the same. It can be seen that at this moment, the potential calculation in the sense of four layers has completely degenerated into that in the sense of three layers. Substitute (B.65) into (B.28) and (B.29) to get

$$\begin{aligned}\alpha &= r_2^{2l+1}\cdot\left[\frac{l}{l+1}\cdot r_2^{2l+1}\cdot\left(\frac{\sigma_2}{\sigma_3}-1\right)+\frac{l}{l+1}\cdot r^{2l+1}\cdot\left(\frac{l}{l+1}\cdot\frac{\sigma_2}{\sigma_3}+1\right)\right]\\ &= -r_2^{2l+1}\cdot r^{2l+1}\cdot\frac{l}{l+1}\cdot\left[f^{2l+1}\cdot(1-s)-\left(\frac{l}{l+1}\cdot s+1\right)\right]\\ &= -r_2^{2l+1}\cdot r^{2l+1}\cdot\frac{l}{l+1}\cdot\alpha_3\end{aligned} \tag{B.67}$$

$$\begin{aligned}\beta &= r_2^{2l+1}\cdot\left(\frac{l}{l+1}+s\right)+r_2^{2l+1}f^{-2l-1}\cdot\frac{l}{l+1}\cdot(-1+s)\\ &= -r_2^{2l+1}\cdot\frac{l}{l+1}\cdot\left[-\left(1+s\frac{l+1}{l}\right)+f^{-2l-1}\cdot(1-s)\right]\\ &= -r_2^{2l+1}\cdot r^{2l+1}\cdot\frac{l}{l+1}\cdot\beta_3\end{aligned} \tag{B.68}$$

where $f=\frac{r_2^{2l+1}}{r^{2l+1}}$, $s=\frac{\sigma_2}{\sigma_3}$, $\alpha_3=\left[f^{2l+1}\cdot(1-s)-\left(\frac{l}{l+1}\cdot s+1\right)\right]$, and $\beta_3=r^{-2l-1}\cdot\left[f^{-2l-1}\cdot(1-s)-\left(1+s\frac{l+1}{l}\right)\right]$.

Obviously,

$$\gamma=\frac{\alpha}{\beta}=\frac{\alpha_3}{\beta_3} \tag{B.69}$$

Substitute (B.69) into (B.66) to get

$$G_l^m=\frac{(2l+1)}{l\cdot r_1^{2l+1}\cdot(1-s_1)+\gamma\cdot[l+s_1\cdot(l+1)]}\cdot\frac{(l+1)\left(r_2^{2l+1}+\gamma\right)}{(l+1)r_2^{2l+1}+l\cdot r^{2l+1}} \tag{B.70}$$

where $s_1=\frac{\sigma_2}{\sigma_1}$. Equation (B.70) is now the same as the one for the concentric three-layer model in Section 4 (In Yao, 2000b,2001 and He et al., 2002, the factor r^{-2l-1} in β_3 was missed).

This result shows that the formula and software for the four-layer sphere model can be simultaneously applied to the calculation of the 1~3-layer model with specific parameter settings, which can be achieved by amalgamating the radii into one or eliminating the difference in conductivities.

REFERENCES

He, B., D. Yao, J. Lian, et al. 2002. An equivalent current source model and laplacian weighted minimum norm current estimates of brain electrical activity. *IEEE Trans Biomed Eng* 49(4):277–288.

Yao, D. 2000b. High-resolution EEG mappings: A spherical harmonic spectra theory and simulation results. *Clin Neurophysiol* 111:81–92.

Yao, D. 2001. A method to standardize a reference of scalp EEG recordings to a point at infinity. *Physiol Meas* 22(4):693–711.

Yao, D. 2003a. *Electric Theory and Method of Brain Function Prosprcting*. Beijing: Science Press.

Yao, D. 2003b. High-resolution EEG mapping: An equivalent charge-layer approach. *Phys Med Biol* 48(13):1997–2011.

Appendix C
Green Function, Green Integral, and Reciprocity Theorem

Green function, Green integral, and reciprocity are the fundamental basis of the boundary element method (Chapter 5) and the equivalent source technique (Chapter 6), and we believe they are of even more value in various new method developments in the future. Here, the contents are extracted from textbooks (Fu, 1993; Zhang, 2003; Jackson, 1962; Arfken et al., 2005; Smythe, 1968).

C.1 SOLUTION OF POISSON'S EQUATION OF UNIT POINT SOURCE

The following is Poisson's equation for the unit point source:

$$\nabla^2 G(\vec{r}-\vec{r}') = -\delta(\vec{r}-\vec{r}') \tag{C.1}$$

which has a solution as

$$G(\vec{r}-\vec{r}') = \frac{1}{4\pi|\vec{r}-\vec{r}'|} \tag{C.2}$$

Based on the properties of the δ function (Appendix A), it can be proved that (C.2) indeed meets with equation (C.1).

Here, the field function of a point source with unit strength is called Green function of the field equation. In fact, Green function for solving the electromagnetic field problem is of universal significance. Since the field equation is linear, any field source distribution can be decomposed into a collection of point sources; thus, the field generated by any field source distribution under given boundary conditions is equal to a linear superposition of the field generated by all these point sources under the same boundary conditions. Therefore, after obtaining the field of a point source (i.e., Green function) under the given boundary conditions, it can be used to solve the field generated by any field source distribution under the same boundary conditions.

C.1.1 Time-Domain Green Function in Unbounded Space

In the infinite space with a linear, homogeneous, and isotropic medium, the scalar wave function, generated by any instantaneous space charge distribution $s(\vec{r},t) = q$ at the observation point *r*, should satisfy the non-homogeneous scalar wave equation.

$$\nabla^2\Psi(\vec{r},t)-\frac{1}{\upsilon^2}\frac{\partial^2\Psi(\vec{r},t)}{\partial^2 t}=-q(\vec{r},t) \tag{C.3}$$

where υ is the wave speed in the medium. The scalar wave function $\Psi(\vec{r},t)$ can represent either the scalar potential or a rectangular coordinate component of a vector field. The specific solution of formula (C.3) has been given in electromagnetics as follows:

$$\Psi(\vec{r},t)=\frac{1}{4\pi}\int_V \frac{q\left(\vec{r}',t-\frac{1}{\upsilon}|\vec{r}-\vec{r}'|\right)}{|\vec{r}-\vec{r}'|}dV' \tag{C.4}$$

Equation (C.4) shows that $\Psi(r,t)$ at the time t is generated by the source at an earlier time $t-R/\upsilon$, with a lagging time of $R/\upsilon\left(R=|\vec{r}-\vec{r}'|\right)$. So it is called the lag potential, being usually expressed as $[q]=q(\vec{r}',t-R/\upsilon)$. Mathematically, the equation may have an advanced potential, but it is not physically reasonable. So it should not be considered in general.

If a charge is located at point $\vec{r}_0'$, and its temporal characteristic behaves as $f_0(t)$, that is, $q(\vec{r}',t)=\delta(\vec{r}-\vec{r}_0')f_0(t)$. Let $\Psi(\vec{r},t)$ in this particular situation be $G_0(\vec{r},\vec{r}';t)$, then from equation (C.4), we can get

$$G_0(\vec{r},\vec{r}_0';t)=\frac{1}{4\pi}\frac{1}{|\vec{r}-\vec{r}_0'|}f_0\left(t-\frac{1}{\upsilon}|\vec{r}-\vec{r}_0'|\right) \tag{C.5}$$

Generally, the changing law of $s(\vec{r}',t)=q$ over time may be different point by point. To illustrate this phenomenon, we can use $f_{\vec{r}'}(t)$ to express the time-domain change of the charge density at a point $\vec{r}'$ and write $q(\vec{r}',t)$ as $q(\vec{r}')f_{\vec{r}'}(t)$. Let $\vec{r}_0'$ in equation (C.5) to be any a point $\vec{r}'$ in volume V and substitute it into formula (C.4) to get:

$$\Psi(\vec{r},t)=\int_V q(\vec{r}')G_0(\vec{r},\vec{r}';t)dV' \tag{C.6}$$

Here, $G_0(\vec{r},\vec{r}';t)$ satisfies the following equation.

$$\nabla'^2 G_0(\vec{r},\vec{r}';t)-\frac{1}{\upsilon^2}\frac{\partial^2}{\partial t^2}G_0(\vec{r},\vec{r}';t)=-\delta(\vec{r}'-\vec{r})f_{\vec{r}'}(t) \tag{C.7}$$

In formulas (C.6) and (C.7), $G_0(r,r';t)$ is called the time-domain Green function of unbounded space. Its changing rule over time depends on the change of the source at each source point, which is obviously inconvenient to calculate. The time-domain Green function for two special cases will be given below.

1. *Electrostatic field situation*: If the field source does not change with time, i.e., $\partial/\partial t=0$, or its change is very slow, leading the time derivative to be approximately zero, then equation (C.3) is simplified to Poisson's equation,

which shows a phenomenon from source to field moment by moment with the temporal variable t being redundant, and (C.4) without the temporal term in q is the expression of the electrostatic potential (similar to equation (3.28)). In the main EEG frequency range (low frequency, long period), the EEG field can be treated as an electrostatic field with the propagation effect neglected (Section 3.2). From another angle, if the spatial scale interested is small, while the electromagnetic wave propagation velocity is very large, then the delay term in q in equation (C.4) also is very small, approximate zero, then (C.4) will also reduce to a moment by moment identity with variable t being redundant, and it behaves just like an electrostatic solution, too.

2. *Time-harmonic field*: The establishment condition for formula (C.4) is that the potential velocity is only determined by the medium parameter, irrelevant to the changing form of the source over time (non-dispersive), and the medium is lossless. So it is a solution in a specific situation. In general, we should get the Fourier transform of the field source and the field over time and solve each frequency component independently, where a source with a specific frequency corresponds to a field with the same frequency. In this case, we get the frequency-domain Green function.

   Due to the skin effect of electromagnetic waves on the surface of a conductor, there is no electromagnetic wave propagation in the brain, and the electromagnetic wave from the outside can only penetrate into a limited depth of the epidermis (Section 2.2). Compared with microwave and millimeter, etc., the near-infrared band has better penetration, and a functional near-infrared spectroscopy imaging technique has been developed accordingly. Since the EEG problem is mainly a problem under the electrostatic approximation, the following discussion will not consider the wave equation, and neglecting the propagation effect as noted in Section 3.2. This section uses the wave equation to show that the result of the electrostatic approximation can be simply obtained by degeneration from the wave equation, such as taking $\partial/\partial t = 0$ in the time-domain equation, or taking $\omega = 0$ in the frequency domain. More wave-related issues can be found in the literature on classic electrodynamics.

C.1.2 Green Function in Bounded Space

There often are boundary conditions in practical problem. Let us first consider the inner domain problem of the boundary. For example, Figure C.1 demonstrates that an unit point charge at $\vec{r}'$ will produce potential in the form of Green function $G_0(\vec{r},\vec{r}')+G_i(\vec{r},\vec{r}')$ in the volume conductor V closed by the interface S (Figure C.1a), where $G_0(\vec{r},\vec{r}')$ is the field generated by the point source charge in the infinite space as equation (C.5), $G_i(\vec{r},\vec{r}')$ is the field generated by the induced surface charge on the conductor wall. Obviously, $G_i(\vec{r},\vec{r}')$ is passive in volume V and should satisfy the equation:

$$\nabla^2 G_i(\vec{r},\vec{r}') = 0 \tag{C.8}$$

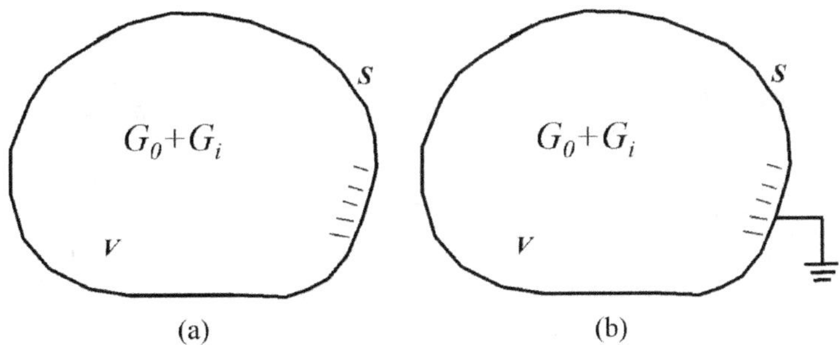

FIGURE C.1 Green function in bounded space. (a) Air or other insulating medium is outside the conductor; (b) the conductor is grounded.

Then the Green function in bounded space is

$$G(\vec{r},\vec{r}')=G_0(\vec{r},\vec{r}')+G_i(\vec{r},\vec{r}') \tag{C.9}$$

It can be proved by equations (C.7), (C.8), and (C.9) that both $G(\vec{r},\vec{r}')$ and $G_0(\vec{r},\vec{r}')$ satisfy the same equation (C.1). If the volume conductor surface is grounded as Figure C.1b, as $G(\vec{r},\vec{r}')$ is the field in volume V surrounded by the conductor wall, when the point $\vec{r}$ falls on the boundary surface S of volume V, the homogeneous boundary condition should be satisfied.

$$G(\vec{r},\vec{r}')=0,\ \vec{r}\ \text{on the}\ S \tag{C.10}$$

Here, $G_0(\vec{r},\vec{r}')$ alone is not subject to this restriction.

C.1.3 Integral Solution of Poisson's Equation and Radiation Condition of the Green Function

To use the source in volume V and the boundary value on the boundary surface S to represent the potential of a point in the same volume, it is necessary to apply the Green formula.

From the Gaussian divergence theorem, we can get:

$$\int_V \nabla'\cdot\vec{F}dV'=\oint_S \vec{F}\cdot d\vec{S}' \tag{C.11}$$

Let $\vec{F}=\Phi\nabla'\Psi$, where Φ and Ψ are arbitrary scalar functions, whose second derivative is continuous in the volume, and the first derivative is continuous on the boundary. Apply identities

$$\nabla'\cdot\vec{F}=\nabla'\cdot(\Phi\nabla'\Psi)=\Phi\nabla'^2\Psi+\nabla'\Phi\cdot\nabla'\Psi \tag{C.12}$$

to get

$$\int_V \left(\Phi\nabla'^2\Psi + \nabla'\Phi\cdot\nabla'\Psi\right)dV' = \oint_S (\Phi\nabla'\Psi)\cdot\vec{n}\,dS' = \oint_S \Phi\frac{\partial\Psi}{\partial\vec{n}'}dS' \qquad \text{(C.13)}$$

where $\vec{n}$ is the normal unit vector outward of the surface element on the boundary surface S. This formula is called Green first identity. Subtract the original formula from the identity after swapping the positions of Φ and Ψ, then we get

$$\int_V \left(\Phi\nabla'^2\Psi - \Psi\nabla'^2\Phi\right)dV' = \oint_S (\Phi\nabla'\Psi - \Psi\nabla'\Phi)\cdot d\vec{S}' = \oint_S \left(\Phi\frac{\partial\Psi}{\partial\vec{n}'} - \Psi\frac{\partial\Phi}{\partial\vec{n}'}\right)dS' \qquad \text{(C.14)}$$

This formula is called the Green second identity or Green theorem.

Now apply formula (C.14) first to the electrostatic field. Let Φ satisfy Poisson's equation

$$\nabla'^2\Phi = -q(\vec{r}') \qquad \text{(C.15)}$$

and let

$$\Psi = \frac{1}{4\pi|\vec{r} - \vec{r}'|}$$

It satisfies the equation

$$\nabla'^2\Psi = -\delta(\vec{r}' - \vec{r}) \qquad \text{(C.16)}$$

Then, it can be obtained by equation (C.14)

$$\int_V \left\{\Psi\left[-q(\vec{r}')\right] + \Phi\left[\delta(\vec{r} - \vec{r}')\right]\right\}dV' = \oint_S \left[\Psi\frac{\partial\Phi}{\partial\vec{n}'} - \Phi\frac{\partial\Psi}{\partial\vec{n}'}\right]dS'$$

Therefore,

$$\Phi(\vec{r}) = \int_V \frac{1}{4\pi|\vec{r} - \vec{r}'|}q(\vec{r}')dV' + \oint_S \frac{1}{4\pi|\vec{r} - \vec{r}'|}\left\{\frac{\partial\Phi(\vec{r}')}{\partial\vec{n}'} + \left[\frac{1}{|\vec{r} - \vec{r}'|}\Phi(\vec{r}')\right]\frac{\partial}{\partial\vec{n}'}|\vec{r} - \vec{r}'|\right\}dS'$$

or

$$\Phi(\vec{r}) = \frac{1}{4\pi}\int_V \frac{q(\vec{r}')dV'}{|\vec{r} - \vec{r}'|} + \frac{1}{4\pi}\oint\left[\frac{1}{|\vec{r} - \vec{r}'|}\frac{\partial\Phi(\vec{r}')}{\partial\vec{n}'} - \Phi(\vec{r}')\frac{\partial}{\partial\vec{n}'}\left(\frac{1}{|\vec{r} - \vec{r}'|}\right)\right]dS' \qquad \text{(C.17)}$$

Equation (C.17) is called the scalar Kirchhoff formula.

Now let us start considering the outer domain problem as shown in Figure C.2. The interface S surrounding the volume V is composed of a finite interface S_0 and an infinite sphere S_∞. As $S = S_0 + S_\infty$, S is also an infinite interface. Assuming $q(\vec{r}') = 0$ in volume V, then (C.17) is written as:

$$\Phi(\vec{r}) = \frac{1}{4\pi}\int_{S_0}\left[\frac{1}{|\vec{r}-\vec{r}'|}\frac{\partial\Phi(\vec{r}')}{\partial\vec{n}'} - \Phi(\vec{r}')\frac{\partial}{\partial\vec{n}'}\left(\frac{1}{|\vec{r}-\vec{r}'|}\right)\right]dS' + \frac{1}{4\pi}\int_{S_\infty}\left[\frac{1}{|\vec{r}-\vec{r}'|}\frac{\partial\Phi(\vec{r}')}{\partial\vec{n}'} - \Phi(r')\frac{\partial}{\partial\vec{n}'}\left(\frac{1}{|\vec{r}-\vec{r}'|}\right)\right]dS' \tag{C.18}$$

We expand S_∞ to an infinite sphere to obtain the solution of an unbounded outer space. For the surface S_∞ integral in equation (C.18), $\vec{r}'$ is from the coordinate origin 0 to a point on S_∞, and we have $\vec{r}' \to \infty$ when S_∞ being expanded to an infinite sphere, then we have $|\vec{r}-\vec{r}| \to r'\,(r' \gg r)$ and $\partial/\partial\vec{n}' \to \partial/\partial r'$. Thus

$$\int_{S_\infty}\left[\frac{1}{r'}\frac{\partial\Phi(\vec{r}')}{\partial r'} - \Phi(\vec{r}')\left(-\frac{1}{r'}\right)\frac{1}{r'}\right]dS' = \int_{S_\infty}\left[\frac{1}{r'}\frac{\partial\Phi(\vec{r}')}{\partial r'} + \Phi(\vec{r}')\frac{1}{r'^2}\right]dS'$$

When $r' \to \infty$, the integral on the surface S_∞ should tend to zero. This requires

$$\int_{S_\infty}\Phi(\vec{r}')\frac{1}{r'^2}dS' \to 0$$

and

$$\int_{S_\infty}\left[\frac{\partial\Phi(\vec{r}')}{\partial r'}\right]\frac{1}{r'}dS' \to 0$$

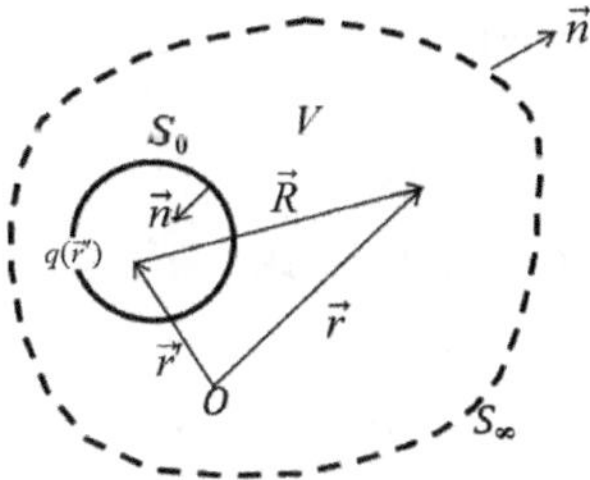

FIGURE C.2 Outer domain problem. The distance between the field point and the source point $q(\vec{r}')$ is $R = |\vec{r}-\vec{r}'|$.

As the area of S_∞ is proportional to r'^2, it requires that:

$$\left.\begin{aligned} &\lim_{r\to\infty} r\Phi(\vec{r}) \to \text{limited Value} \\ &\lim_{r\to\infty} r\left[\frac{\partial\Phi(\vec{r})}{\partial r}\right] \to 0 \end{aligned}\right\} \tag{C.19}$$

These two conditions are called radiation conditions.

In formula (C.14), if let $\Psi = G(\vec{r},\vec{r}')$, and satisfies (C.16), then we get

$$G(\vec{r},\vec{r}') = \frac{1}{4\pi|\vec{r}-\vec{r}'|} \tag{C.20}$$

Then according to the two equations (C.15) and (C.16), the formula (C.17) can be modified with the Green function as

$$\Phi(\vec{r}) = \int_V G(\vec{r}-\vec{r}')q(\vec{r}')dV' + \oint_S \left[G(\vec{r}-\vec{r}')\frac{\partial\Phi(\vec{r}')}{\partial\vec{n}'} - \Phi(\vec{r}')\frac{\partial G(\vec{r}-\vec{r}')}{\partial\vec{n}'}\right]dS' \tag{C.21}$$

Now, consider the surface integral term in formula (C.21). By applying the similar derivation and the result of the formula (C.19), the radiation condition of the Green function can be obtained as $(r \to \infty)$:

$$\left.\begin{aligned} &\lim_{r\to\infty} rG = \text{limited value} \\ &\lim_{r\to\infty} r\left(\frac{\partial G}{\partial r}\right) = 0 \end{aligned}\right\} \tag{C.22}$$

With the radiation conditions (C.19) and (C.22), the integral solution of an unbounded space V can be obtained from equation (C.21) where the S_∞ expended to infinity, and the surface S simplified to S_0. When the source q is outside of V, the first term will disappear, too. Here these radiation conditions are similar to the infinity condition shown by equation (2.43).

C.1.4 Boundary Condition of the Green Function and Green Function Method

It can be seen from the two formulas (C.17) and (C.21) that the surface integral contains two boundary value terms. If using one given boundary condition to solve another unknown boundary condition, it then becomes an integral equation. For example, in EEG forward modeling, we suppose that the current cannot flow out of the scalp surface that the normal gradient of the scalp surface potential is zero, which is called the Neumann boundary, also called as the second type of boundary condition, so as to solve the value of the scalp surface potential. In fact, Green function can be used not only to represent field or potential, but also in many cases to express integral equation of the unknown boundary value. It can be used not only on the surface of one volume when solving one boundary value under given another boundary value, but also on the continuity condition of the boundary value on the common interface of two volumes so as to solve the boundary value on the interface. Because of this, Green function is a very important tool, and the main application of Green function is to establish integral equation. But when using the scalar Kirchhoff formula to find the integral solution of the field or potential, it is required to eliminate one of the two surface integrals so as to get the integral solution according to one specific type of boundary condition. This is because the solutions are of course the same when the two types of boundary conditions given at the same time happen to be the boundary values of the specific situation, whereas there is not necessarily a solution when the two types of boundary conditions are given arbitrarily. Obviously, according to the uniqueness theorem, either of the two types of boundary conditions has a unique solution. If equation (C.17) can be processed to eliminate one surface integral term, the boundary value problem of the scalar wave equation can be uniquely solved.

In the follow derivation, we use the formula (C.21) to express the field or potential by Green function. In the general formula of expressing field or potential by Green function, if $G_i(\vec{r},\vec{r}')$ is selected appropriately, it is possible to eliminate one surface integral term, $G_i(\vec{r},\vec{r}')$ in equation (C.9) is a general solution that satisfies the homogeneous equation (C.8) (Jackson, 1975). Such an approach is often referred to as the Green function method. But when the interface is complicated, $G_i(\vec{r},\vec{r}')$ might have a very complicated form and have no analytical formula in general. Therefore, it is not easy to implement in practical operation. In Chapter 6, we developed the equivalent source approach by only modifying the coefficients while maintaining the Green function form unchanged.

Now let us specifically consider the solution of equation (C.21). In the formula, S can be an interface such as the surface of a conductor, or a closed interface fictitiously conceived in free space. If artificially stipulating that the boundary value condition $G(\vec{r},\vec{r}')=0$ (the first type of boundary condition) or $\partial G(\vec{r},\vec{r}')/\partial\vec{n}'=0$ (the second type of boundary condition) is satisfied when $\vec{r}'$ is on S, one boundary value integral can be eliminated. Therefore, the so-called Green function is the solution of equation (C.1) for $G=G_0+G_i$, and equation (C.8) for G_i under artificially prescribed boundary condition. By the way, when solving the Green function, in order to determine the solution of the equation, not only the boundary condition but also the radiation conditions should be specified for the Green function.

Based on the above discussions, we can select $G_i(\vec{r},\vec{r}')$ to let $G(\vec{r},\vec{r}')$ in equation (C.21) satisfy the two types of boundary conditions separately, and write the two solutions of $\Phi(\vec{r})$ as

$$G(\vec{r},\vec{r}')=0,\ \vec{r}'\ \text{on}\ S \tag{C.23a}$$

$$\Phi(\vec{r})=\int_V G(\vec{r},\vec{r}')q(\vec{r}')dV'-\oint_S \Phi(\vec{r}')\frac{\partial G(\vec{r},\vec{r}')}{\partial\vec{n}'}dS' \tag{C.23b}$$

and

$$\frac{\partial G(\vec{r},\vec{r}')}{\partial \vec{n}'} = 0, \quad \vec{r}' \text{ on } S, \text{ and } S \to \infty \tag{C.24a}$$

$$\Phi(\vec{r}) = \int_V G(\vec{r},\vec{r}')q(\vec{r}')dV' + \oint_S \frac{\partial \Phi(\vec{r}')}{\partial \vec{n}'} G(\vec{r},\vec{r}')dS' \tag{C.24b}$$

those formulas of equations (C.23a,b) and (C.24a,b) are very important for applying Green function to calculate scalar potential or vector field components.

If V is composed of the volumes V_1 and V_2, then in V_1 and V_2, the Green functions are $G^{(1)}(\vec{r},\vec{r}')$ and $G^{(2)}(\vec{r},\vec{r}')$ separately, where $\vec{r}$ should be in V_1 or V_2. When $\vec{r}'$ on the interface of V_1 and V_2, the boundary condition may be assumed to be a combination (Zhang, 2003):

$$\frac{\partial G(\vec{r},\vec{r}')}{\partial \vec{n}'} + aG(\vec{r},\vec{r}') = 0, \quad a \geq 0, \vec{r}' \text{ on } S \tag{C.25}$$

The above three types of boundary conditions show that in practical problems, the boundary conditions often only contain the value of the solution Φ on the surface S (the first type of boundary condition, Dirichlet boundary condition, equation (C.23a)), or the normal derivative $\frac{\partial \Phi}{\partial \vec{n}}$ of the solution on the surface S (the second type of boundary condition, Neumann boundary condition, equation (C.24a)), or a linear combination of them (equation C.25). In order to distinguish these Green functions that satisfy different boundary conditions, or different Green function methods, different subscripts are used to characterize them. Generally speaking, the subscript 0 represents an unbounded area, where only the radiation conditions are needed to be satisfied by Green function (equation C.19). The subscripts 1, 2, and 3 denote the three types of Green functions which meet the first, second, and third types of boundary conditions separately. Here, the first two of the three are (C.23a) and (C.24a). However, among them, equation (C.24a) cannot be satisfied (Zhang, 2003) in general. To understand this, we can place a point r' in equation $\nabla^2 G(\vec{r},\vec{r}') = -\delta(\vec{r}-\vec{r}')$ in V, then integrate in the entire volume V, use Gauss theorem to change the volume integral on the left into a surface integral, and swap the symbol r and r', the result is

$$\int_S \frac{\partial G(\vec{r},\vec{r}')}{\partial \vec{n}'} dS' = -1 \tag{C.26}$$

It shows that for a point $\vec{r}$ in V, $\frac{\partial G(\vec{r},\vec{r}')}{\partial \vec{n}'} dS'$ on surface S is impossible to be always equal to zero. However, we can take a step back to require $\frac{\partial G(\vec{r},\vec{r}')}{\partial \vec{n}'}$ on S to be a constant. According to equation (C.26), this constant is

$$\frac{\partial G(\vec{r}',\vec{r})}{\partial \vec{n}'} = -\frac{1}{s} \tag{C.27}$$

Substitute this boundary condition into (C.21) to get

$$\Phi(\vec{r}) = \int_V q(\vec{r}')G(\vec{r}',\vec{r})dV' + \int_S \frac{\partial\Phi(\vec{r}')}{\partial\vec{n}'}G(\vec{r}',\vec{r})dS' + \bar{\Phi}_S \tag{C.28}$$

where

$$\bar{\Phi}_S = \frac{1}{S}\int_S \Phi(\vec{r})dS \tag{C.29}$$

It is the average value of $\Phi(\vec{r})$ on the surface S, i.e., a constant. Specifically, for a dipole in a hemisphere, the surface integral of the potential does being zero as proofed in Appendix F (Yao, 2017). Since the electrostatic potential can add or subtract an arbitrary constant, it can still be expressed as (C.24b) and to be used to solve Poisson's equation under the second type of boundary condition. However, the Green function can only be got by solving Poisson's equation under the boundary condition (C.26).

For the third type of boundary condition (C.30) where both a and b are non-zero, we can solve Poisson's equation of G under the boundary condition as following:

$$aG(\vec{r}',\vec{r}) + b\frac{\partial G(\vec{r}',\vec{r})}{\partial\vec{n}'} = 0 \tag{C.30}$$

Multiply $G(\vec{r}',\vec{r})$ by $\Pi(\vec{r}') = a\Phi(\vec{r}') + b\frac{\partial\Phi(\vec{r}')}{\partial\vec{n}'}$, and $\Phi(\vec{r}')$ by equation (C.30), we get two equations, then sort out these two equations and eliminate the items related to a, we get:

$$b\left[\frac{\partial\Phi(\vec{r}')}{\partial\vec{n}'}G(r',\vec{r}') - \Phi(\vec{r}')\frac{\partial G(\vec{r}',\vec{r})}{\partial\vec{n}'}\right] = \Pi(\vec{r}')G(\vec{r}',\vec{r}) \tag{C.31}$$

Substitute (C.31) into the right end of equation (C.21), and we get

$$\Phi(\vec{r}) = \int_V q(\vec{r}')G(\vec{r}',\vec{r})dV' + \frac{1}{b}\int_s \Pi(\vec{r}')G(\vec{r}',\vec{r})dS' \tag{C.32}$$

This formula can be used to solve Poisson's equation under the third type of boundary conditions (Zhang, 2003).

The Green function only provides a method to solve the boundary value problem of Poisson's equation. Its prerequisite is to know the Green function. However, to get the Green function, it is still necessary to solve Poisson's equation under certain boundary conditions. In this approach, Poisson's equation (C.21) has a standard form and is irrelevant to the charge distribution in practical problem. The three types of boundary conditions also have standard forms, and are irrelevant to the details of the boundary conditions in the actual problem. The Green function method just replaces various specific problems in actual situations with standard forms of problems. Once the standard form of the solution is obtained, the solution under various specific conditions can be obtained by the related formulas. However, the above derivation only proves that if there is a solution for Poisson's equation that meets the first, second, or third type of boundary conditions, then it can be expressed as equations (C.23b), (C.24b), or (C.32) correspondingly. Anyway, don't forget to check whether the obtained solution meets the given boundary condition. Poisson's equation does not have a solution under some boundary condition. For example, if the total charge Q in volume V is non-zero, according to Gauss's theorem, the normal derivative $\frac{\partial \Phi}{\partial \vec{n}}$ of the electrostatic potential on the surface S cannot be always zero (equations C.26 and C.29). Then Poisson's equation will not have a solution if a zero boundary condition is rigidly included.

The Green function of Poisson's equation can always be written as

$$G(\vec{r},\vec{r}') = \frac{1}{4\pi|\vec{r}-\vec{r}'|} + f(\vec{r},\vec{r}')$$

where f satisfies the Laplace equation. Substitute this expression into (C.23b) and (C.24b) to decompose these solutions as

$$\Phi(\vec{r}) = \Phi_0(\vec{r}) + \Phi_1(\vec{r})$$

where

$$\Phi_0(\vec{r}) = \int_V \frac{q(\vec{r}')}{4\pi|\vec{r}-\vec{r}'|} dV' \tag{C.33}$$

is the solution of equation (C.1), and is also called a basic solution. $\Phi_1(\vec{r})$ is a Harmonic function as it satisfies the Laplace equation:

$$\nabla^2 \Phi_1 = 0 \tag{C.34}$$

After adding this harmonic function to the basic solution, it is still the solution of Poisson's equation. However, adjusting the harmonic function can let the solution meet the given boundary condition. It can also be said that $\Phi_0(\vec{r})$ is the electrostatic potential generated by the charge distribution in V, while $\Phi_1(\vec{r})$ is the electrostatic potential generated by the charge distribution outside V, which includes the electrostatic potential generated by the charge distribution on the interface S. Actually, such a basic solution Φ_0 plus harmonic solution Φ_1 approach has been repeatedly adopted in Sections 6.1–6.3.

C.1.5 Properties of the Green Function

When solving the boundary value problem of the Green function, its properties are often adopted. Therefore, understanding the properties of the Green function helps in its calculation.

C.1.5.1 Singularity

The Green function has a singularity at the source point, and continuity everywhere in the space outside the source point. Its first derivative has an abrupt change at the source point (Fu, 1993).

C.1.5.2 Symmetry

An important property of the Green function is the symmetry of its source and field points, that is, the scalar function at point $\vec{r}$ generated by a unit charge at point $\vec{r}'$ is equal to the scalar function at point $\vec{r}'$ generated by a unit charge at point $\vec{r}$, this phenomenon is the symmetry of Green function (Fu, 1993), and the mathematical expression is

$$G(\vec{r},\vec{r}')=G(\vec{r}',\vec{r}) \tag{C.35}$$

Obviously, this can be explained by the symmetry of the δ function itself, namely

$$\delta(\vec{r}-\vec{r}')=\delta(\vec{r}'-\vec{r}) \tag{C.36}$$

To prove (C.35), let us assume that there are Green functions $G(\vec{r}',\vec{r}_1)$ and $G(\vec{r}',\vec{r}_2)$, which are scalar functions at the field point $\vec{r}'$ generated by the different source at points $\vec{r}_1$ and $\vec{r}_2$, then in the same volume *V*, they must satisfy the following scalar equation:

$$\nabla'^2 G(\vec{r}',\vec{r}_{1,2})=-\delta(\vec{r}'-\vec{r}_{1,2}) \tag{C.37}$$

Substitute $G(\vec{r}',\vec{r}_1)$ and $G(\vec{r}',\vec{r}_2)$ into formula (C.14) to get

$$\int_V\left[G(\vec{r}',\vec{r}_2)\nabla'^2G(\vec{r}',\vec{r}_1)-G(\vec{r}',\vec{r}_1)\nabla'^2G(\vec{r}',\vec{r}_2)dV'\right]$$

$$=\oint_S\left[G(\vec{r}',\vec{r}_2)\frac{\partial G(\vec{r}',\vec{r}_1')}{\partial\vec{n}'}-G(\vec{r}',\vec{r}_1)\frac{\partial G(\vec{r}',\vec{r}_2')}{\partial\vec{n}'}\right]dS'$$

From formula (C.37), we can get

$$G(\vec{r}_1,\vec{r}_2)-G(\vec{r}_2,\vec{r}_1)=\oint_s\left[G(\vec{r}',\vec{r}_2)\frac{\partial G(\vec{r}',\vec{r}_1)}{\partial\vec{n}'}-G(\vec{r}',\vec{r}_1)\frac{\partial G(\vec{r}',\vec{r}_2)}{\partial\vec{n}'}\right]dS' \tag{C.38}$$

For Green functions $G(\vec{r},\vec{r}_{1,2})$ which meet with the first, second, and third types of boundary conditions, all the integrals on *S* are zero. So we can immediately get

$$G(\vec{r}_1,\vec{r}_2)=G(\vec{r}_2,\vec{r}_1) \tag{C.39}$$

Let $\vec{r}_1$ and $\vec{r}_2$ be $\vec{r}$ and $\vec{r}'$, respectively, (C.39) becomes equation (C.35). Of course, $\vec{r}_1$ and $\vec{r}_2$ should not be on *S* simultaneously, and the surface integral over $\vec{r}'$ of equations (C.23b) and (C.24b) can be conceived as the result when a point in *V* approaching a point on *S*.

C.1.6 Reciprocity Theorem

The reciprocity theorem was first proposed by Rayleigh (Smythe, 1968). It is also called the Green reciprocity theorem since it can be derived from the reciprocity of the Green function. Formula (C.39) can also be called the reciprocity of the Green function, and it can be extended to general scalar field (Fu, 1993).

Specifically, the reciprocity theorem describes the reciprocal relationship between two charge distributions and the two potentials generated by them in a charged system. Assuming that in a limited space V, there are n conductors with n charges $q_i\,(i=1,2,\ldots,i,\ldots,n)$ and a charging system with a charge density of ρ, and the measured potential of the ith conductor is Φ_i, and the potential distribution over the charging system is Φ. When the charges on the conductors become q_i' and the charge density is ρ', the corresponding potential become Φ_i' and Φ'. Then there is

$$\sum_{i=1}^{n} q_i\Phi_i' + \int_V \Phi'\rho\, dV' = \sum_{i=1}^{N} q_i'\Phi_i + \int_V \Phi\rho'\, dV' \tag{C.40}$$

This is the general form of the scalar reciprocity theorem. It can be regarded as a direct inference and application of the Green theorem; thus, it can also be proved by using the Green theorem.

In formula (C.14), let $\Psi = \Phi'$ to get

$$\int_V \left(\Phi\nabla'^2\Phi' - \Phi'\nabla'^2\Phi\right) dV' = \oint_S \left(\Phi\frac{\partial\Phi'}{\partial\vec{n}} - \Phi'\frac{\partial\Phi}{\partial\vec{n}}\right) dS' \tag{C.41}$$

where the closed interface S is composed of all conductor surfaces S_i and the interface S_∞ surrounding all conductors and the space charge system. Therefore, the surface integral has the following form: $\oint_S = \sum_{i=1}^{n}\oint_{S_i} + \oint_{S_\infty}$.

Let $S_\infty \to \infty$, since the charge density (charge system) is distributed in a limited area, the potential on the closed interface is zero at infinity, i.e., $\oint_{S_\infty} \to 0$. Since the conductor surface potential is a constant, the right end of the equation becomes

$$\begin{aligned}
I_{\text{right}} &= \sum_{i=1}^{n}\oint_{S_i}\left(\Phi_i\frac{\partial\Phi_i'}{\partial\vec{n}'} - \Phi_i'\frac{\partial\Phi_i}{\partial\vec{n}'}\right) dS' \\
&= \sum_{i=1}^{n}\left[\Phi_i\oint_{S_i}\left(-\frac{\partial\Phi_i'}{\partial\vec{n}}\right) dS - \Phi_i'\oint_{S_i}\left(-\frac{\partial\Phi_i}{\partial\vec{n}}\right) dS\right] \qquad \text{(C.42a)} \\
&= \sum_{i=1}^{n}\frac{1}{\varepsilon_0}\left[\Phi_i q_i' - \Phi_i' q_i\right]
\end{aligned}$$

Then, use Poisson's equation

$$\left.\begin{aligned} \nabla'^2\Phi &= -\frac{\rho}{\varepsilon_0} \\ \nabla'^2\Phi' &= -\frac{\rho'}{\varepsilon_0} \end{aligned}\right\} \tag{C.42b}$$

which changes the left end of the equation to be

$$I_{\text{left}} = \frac{1}{\varepsilon_0}\int_V (\Phi'\rho - \Phi\rho')\,dV' \tag{C.42c}$$

Formula (C.40) can be obtained from formula (C.42a,c).

Now let us consider two special cases.

When there is no charge density distribution in the entire space except for conductors, then formula (C.40) becomes

$$\sum_{i=1}^{n} q_i\Phi_i' = \sum_{i=1}^{n} q_i'\Phi_i \tag{C.43}$$

Especially, when there are only two conductors i and j, it can be given by equation (C.43) as

$$\left.\frac{q_i}{\Phi_j}\right|_{q_j=0} = \left.\frac{q_j'}{\Phi_i'}\right|_{q_i'=0}, \quad i \neq j \tag{C.44}$$

where Φ_i and Φ_j' are constants, i.e., the potentials on the conductors. Formula (C.43) is a common scalar reciprocity theorem. Equation (C.44) shows that the ratio of the charge q_i on the conductor i to the potential Φ_j induced by q_i on the uncharged conductor j is equal to the ratio of the charge q_j' on the conductor j to the potential Φ_i' induced by q_j' on the uncharged conductor i. This conclusion is always correct, regardless of the conductor's shape, size, relative position, and also whether there is a dielectric.

When there are no other conductors in the entire space except for the charge density distribution, then formula (C.40) becomes

$$\int_V \Phi'\rho\,dV' = \int_V \Phi\rho'\,dV' \tag{C.45}$$

Equations (C.43) and (C.45) are special forms of the scalar reciprocity theorem.

In fact, the scalar reciprocity theorem can also be proved directly based on the Green function, thus taking it as an application of the symmetry of the Green function. Taking (C.45) as an example, suppose that the scalar potentials Φ and Φ' satisfy Poisson's equation (C.42b) and the homogeneous boundary condition on S, and let

$G(\vec{r},\vec{r}')$ or $G(\vec{r}',\vec{r})$ satisfy the same homogeneous boundary conditions, then the surface integral of formula (C.21) is zero. So

$$\int_V \Phi' \rho \, dV = \int_V \rho(\vec{r}) \left[\int_V \rho'(\vec{r}') G(\vec{r},\vec{r}') dV' \right] dV \tag{C.46}$$

When integrating within the volume V, exchange the order of integration, and exchange $\vec{r}$ and $\vec{r}'$, notice that $G(\vec{r},\vec{r}') = G(\vec{r}',\vec{r})$, then there is

$$\int_V \rho'(\vec{r}) \left[\int_V \rho(\vec{r}') G(\vec{r},\vec{r}') dV' \right] dV = \int_V \Phi \rho' \, dV \tag{C.47}$$

Therefore, formula (C.45) is proved. For the application of the reciprocity theorem in EEG, refer to Section 10.4.

REFERENCES

Arfken, G.B., H.J. Weber, F.E. Harris. 2005. *Mathematical Methods for Physicists*, 6th edn, section 1.1. London: Academic Press.

Fu, G. 1993. *Green's Function Method in Electromagnetic Field (In Chinese)*. Beijing: Higher Education Press.

Jackson, J.D. 1962. *Classical Electrodynamics*. New York: John Wiley & Sons Press.

Jackson, J.D. 1975. *Classical Electrodynamics*, 2nd edn. New York: John Wiley & Sons Press.

Smythe, R. 1968. *Static and Dynamic Electricity*. New York: McGraw Hill Education.

Yao, D. 2017. Is the surface potential integral of a dipole in a volume conductor always zero? A cloud over the average reference of EEG and ERP. *Brain Topogr* 30:161–171. doi: 10.1007/s10548-016-0543-x.

Zhang, Q. 2003. *Classical Field Theory (In Chinese)*. Beijing: Science Press.

Appendix D
Scalp Laplacian and Skull Surface Potential

For a local planar area, suppose the z-axis with origin at the center, the value of z at the outer surface of the skull is r_{skull}, the value at the scalp surface is r_{scalp}, and the thickness $h = r_{\text{scalp}} - r_{\text{skull}}$, then we have the Taylor expansion of the potential at the skull surface as (Yao, 2002)

$$\Phi(r_{\text{skull}}) = \Phi(r_{\text{sclap}}) + (-h)\frac{\partial \Phi(z)}{\partial z}\bigg|_{z=r_{\text{sclap}}} + (h^2/2)\frac{\partial^2 \Phi(z)}{\partial z^2}\bigg|_{z=r_{\text{sclap}}} + \cdots \quad \text{(D.1)}$$

where the second term would be diminished as there is no current pass through the scalp, then

$$\begin{aligned} \Phi(r_{\text{skull}}) &= \Phi(r_{\text{sclap}}) + \frac{\partial^2 \Phi(z)}{\partial z^2}\bigg|_{z=r_{\text{sclap}}} h^2/2 + \cdots \\ &= \Phi(r_{\text{sclap}}) - L_{xy(\text{scalp})}\, h^2/2 + \cdots \end{aligned} \quad \text{(D.2)}$$

$$L_{xy(\text{scalp})} = 2\left[\Phi(r_{\text{sclap}}) - \Phi(r_{\text{skull}})\right]/h^2 = -2/h^2\left[\Phi(r_{\text{skull}}) - \Phi(r_{\text{sclap}})\right] \quad \text{(D.3)}$$

As the absolute value of $\Phi(r_{\text{skull}})$ is usually larger than $\Phi(r_{\text{scalp}})$, and with more high-frequency information, $L_{xy(\text{scalp})}$ shows more about the potential pattern on the skull surface. And due to the spatial spectra difference, the spatial distribution of the $\Phi(r_{\text{scalp}})$ is much smoother than that of the $\Phi(r_{\text{skull}})$, we may take $\Phi(r_{\text{scalp}})$ as a constant C locally, then we have $L_{xy(\text{scalp})} \sim \Phi(r_{\text{skull}}) + C$, thus we may say that the scalp Laplacian is an approximate of the skull surface potential. From a phenomenal perspective, a similar derivation can be repeated for the inner surface of the skull, then we may say that the scalp Laplacian is an approximate of the skull inner surface potential, or say the cortex surface potential, and that is the result in Nunez and Srinivasan (2006).

Besides, as the current in the scalp and/or skull layer is not a simple radial current, we also cannot simply take the difference between the two potentials as the approximation of the radial current.

REFERENCES

YaoD. The theoretic relation of scalp Laplacian and scalp current density of a spherical head model. Phys Med Biol. 2002, 47(12): 2179-85

PaulL. Nunez, Ramesh Srinivasan. Electric Fields of the Brain: The Neurophysics of EEG. Oxford University Press, 2006

Appendix E
Mathematical Theory of Linear Inverse Problems

E.1 LINEAR INVERSE PROBLEM

For EEG and MEG, we have the scalp data: $\Phi = [\phi_1, \phi_2, \phi_3, \ldots, \phi_N]^T$ from N sensors, we want to know the model parameters vector of the underlying M sources inside the brain: $s = [s_1, s_2, s_3, \ldots, s_M]^T$ from the scalp data. In general, this problem is an explicit linear problem if the source positions are fixed,

$$f(\Phi, s) = 0 = \Phi - Gs \tag{E.1}$$

Here, the static electric field or potential follows the linear superposition principle, as shown in Appendix C. This equation $Gs = \Phi$ is also the simplest and best-understood inverse problem and is the focus of discrete inverse theory, where matrix G is called the data kernel, transfer matrix, or leadfield matrix in EEG. And vector $e = (\Phi - Gs)$ is the predictive error. The following concentrated materials are mainly from a classical book (Menke, 2012); thus, for details, refer to the corresponding chapters noted in the following sections of this book. The other references may be Tarantola (2005), Lawson and Hanson (1974), Marquardt (1970), and Bertero et al. (1985, 1988).

E.2 THE LENGTH METHOD FOR LINEAR INVERSE PROBLEM

The term *norm* is used as measure of length. The following are the widely visible measures in the scientific literature:

L_1 norm: $\|e\|_1 = \left[\sum_i |e_i|^1\right]$

L_2 norm: $\|e\|_2 = \left[\sum_i |e_i|^2\right]^{1/2}$

L_n norm: $\|e\|_n = \left[\sum_i |e_i|^n\right]^{1/n}$

L_∞ norm: $\|e\|_\infty = \max|e_i|$

Apparently, a higher order norm means a larger weight on the larger element, for ∞, it only depends on the largest one.

The length measure of the solution can be defined as

$$L = \left[s - \langle s \rangle\right]^T W_m \left[s - \langle s \rangle\right] \tag{E.2}$$

where *a priori* model vector $\langle s \rangle$ and the weighting matrix W_m are introduced. Similarly, the prediction error can be defined as

$$E = e^T W_e e \tag{E.3}$$

where the matrix W_e defines the relative contribution of each individual error to the total prediction error.

If the equation $Gs = \Phi$ is completely overdetermined, then one can estimate the model parameters by minimizing E in equation (E.3) through setting zero for the derivative of the error E with respect to one of the model parameters and rewriting the result in matrix notation, and this procedure leads to the solution

$$s^{\text{est}} = \left[G^T W_e G\right]^{-1} G^T W_e \Phi \tag{E.4}$$

This is the weighted least squares solution of an overdetermined problem.

If the equation $Gs = \Phi$ is completely underdetermined, then one can estimate the model parameters by minimizing the generalized length L in equation (E.2) $\left[s - \langle s \rangle\right]^T$ subject to the constraint that $e = \Phi - Gs = 0$ by the method of Lagrange multipliers, and the result is

$$s^{\text{est}} = \langle s \rangle + W_m^{-1} G^T \left[G W_m^{-1} G^T\right]^{-1} \left[\Phi - G\langle s \rangle\right] \tag{E.5}$$

which is called weighted minimum length solution of a completely underdetermined problem.

If the equation $Gs = \Phi$ is slightly underdetermined, we can have the object function as $O(s) = E + \varepsilon^2 L$, L is the norm of the solution (Franklin, 1970), and the result is

$$s^{\text{est}} = \left[G^T W_e G + \varepsilon^2 W_m\right]^{-1} \left[G^T W_e \Phi + \varepsilon^2 W_m \langle s \rangle\right] \tag{E.6}$$

This is called a weighted damped least squares. The parameter ε, acting as a regularization parameter, may be chosen by trial and error to yield a reasonable solution as shown in section 12.1.5.

Apparently, by choosing $\langle s \rangle$, W_m, and W_e, we can have the solution applied to various situations. Linking to Section 12.1, these three terms correspond to source priors, weights in solution space, and measurement space, respectively. For more details of this section, refer to Chapter 3 in Menke (2012) or Section 5.5 in Tarantola (2005).

E.3 GENERALIZED INVERSE FOR LINEAR INVERSE PROBLEM

Above equation (E.4) indicates that the estimate of model parameters s^{est} may be $s^{\text{est}} = P\Phi^{\text{obs}}$, where P is a matrix or projection independent of the observed data Φ^{obs}. As the matrix P solves, or "inverts," the inverse problem $Gs = \Phi$, it is often called the *generalized inverse* and given the symbol $P = G^{-g}$. The exact form of the generalized inverse depends on the problem at hand. In some ways, the generalized inverse is analogous to the ordinary matrix inverse. However, such an analogy is very limited, it may not be a square, and neither $G^{-g}G$ nor GG^{-g} need equal an identity matrix (Menke, 2012). For $Gs = \Phi$, we have

$$s^{\text{est}} = G^{-g}\Phi^{\text{obs}} \tag{E.7}$$

By plugging the estimate into the equation, we get the predicted data

$$\Phi^{\text{pre}} = Gs^{\text{est}} = G\left[G^{-g}\Phi^{\text{obs}}\right] = \left[GG^{-g}\right]\Phi^{\text{obs}} = N\Phi^{\text{obs}} \tag{E.8}$$

Here, the $N \times N$ square matrix $N = GG^{-g}$ is called the *data resolution matrix.* If $N = I$, then $\Phi^{\text{pre}} = \Phi^{\text{obs}}$ and the prediction error is zero. Otherwise, the prediction error is nonzero.

Similarly, plugging the expression for the observed data $Gs^{\text{true}} = \Phi^{\text{obs}}$ into the equation (E.7) gives

$$s^{\text{est}} = G^{-g}\Phi^{\text{obs}} = G^{-g}\left[Gs^{\text{true}}\right] = \left[G^{-g}G\right]s^{\text{true}} = Rs^{\text{true}} \tag{E.9}$$

Here, R is the $M \times M$ *model resolution matrix.* If $R = I$, then each model parameter is uniquely determined. If R is not an identity matrix, then the estimates of the model parameters are weighted averages of the true model parameters.

Furthermore, if the data Φ^{obs} are assumed to be uncorrelated and to have uniform variance σ^2, the unit covariance matrix of the model parameter is given by

$$\left[\text{cov}_u\, s\right] = \sigma^{-2}G^{-g}\left[\text{cov}\,\Phi\right]G^{-gT} = G^{-g}G^{-gT} \tag{E.10}$$

Even if the data are correlated, one can find a normalization of the data covariance matrix to have a *unit data covariance matrix* $\left[\text{cov}_u\,\Phi\right]$, then

$$\left[\text{cov}_u\, s\right] = G^{-g}\left[\text{cov}_u\,\Phi\right]G^{-gT} \tag{E.11}$$

Here, $\left[\text{cov}_u\, s\right]$ can be defined to characterize the degree of error amplification that occurs in the mapping from data to model parameters. Apparently, R, N, and $\left[\text{cov}_u\, s\right]$ are functions of only the data kernel, independent of the actual values of the data; they can be a useful tool in experiment design or evaluation of the generalized inverse. For example, a good G^{-g} should be of good N, R, and $\left[\text{cov}_u\, s\right]$, or a reasonable tradeoff among them.

To measure the goodness of N and R, the *Dirichlet spread functions* are introduced as

$$\text{Spread}(N) = \|N - I\|_2^2 = \sum_{i=1}^{N}\sum_{j=1}^{N}\left[N_{ij} - \delta_{ij}\right]^2$$
$$\text{Spread}(R) = \|R - I\|_2^2 = \sum_{i=1}^{N}\sum_{j=1}^{N}\left[R_{ij} - \delta_{ij}\right]^2 \tag{E.12}$$

When $R/N = I$, spread $(R/N) = 0$, otherwise, the spread is not zero. Meanwhile, the size of the unit covariance matrix can be used as a measure of the amount of error amplification mapped from data to model parameters as

$$\text{Size}\left(\left[\text{cov}_u\, s\right]\right) = \left\|\left[\text{var}_u\, s\right]^{1/2}\right\|_2^2 = \sum_{i=1}^{M}\left[\text{cov}_u\, s\right]_{ii} \tag{E.13}$$

Now, we may seek the generalized inverse G^{-g} that minimizes the weighted sum of Dirichlet measures of resolution spread and covariance size.

$$\text{Minimize}: \ \alpha_1\,\text{spread}(N) + \alpha_2\,\text{spread}(R) + \alpha_3\,\text{size}\left(\left[\text{cov}_u\, s\right]\right) \tag{E.14}$$

where α_s are arbitrary weighting factors. As N, R, and $\left[\text{cov}_u\, s\right]$ are functions of the unknown generalized inverse G^{-g}, this problem is done in exactly the same fashion as the one in length approach by performing the differentiation separately for each element of G^{-g}, and summarizing the result to matrix notation. The result is an equation for the generalized inverse:

$$\alpha_1\left[G^TG\right]G^{-g} + G^{-g}\left[\alpha_2\left[GG^T\right] + \alpha_3\left[\text{cov}_u\,\Phi\right]\right] = \left[\alpha_1 + \alpha_2\right]G^T \tag{E.15}$$

An equation of this form is called a *Sylvester equation*, and explicit solutions can be obtained for a variety of special choices of the weighting factors.

If $\alpha_1 = 1$ and $\alpha_2 = \alpha_3 = 0$, the least squares solution for overdetermined problem is obtained (with $\left[\text{cov}_u\,\Phi\right] = I$)

$$G^{-g} = \left[G^TG\right]^{-1}G^T,$$
$$N = GG^{-g} = G\left[G^TG\right]^{-1}G^T,$$
$$R = G^{-g}G = \left[G^TG\right]^{-1}G^TG = I \tag{E.15a}$$
$$\left[\text{cov}_u\, s\right] = G^{-g}G^{-gT} = \left[G^TG\right]^{-1}G^TG\left[G^TG\right]^{-1} = \left[G^TG\right]^{-1}$$

If $\alpha_1 = 0$, $\alpha_2 = 1$, and $\alpha_3 = 0$, the minimum length solution for underdetermined problem is obtained (with $[\text{cov}_u \Phi] = I$)

$$
\begin{aligned}
G^{-g} &= G^T\left[GG^T\right]^{-1}, \\
N &= GG^{-g} = GG^T\left[GG^T\right]^{-1} = I, \\
R &= G^{-g}G = G^T\left[GG^T\right]^{-1}G \\
[\text{cov}_u s] &= G^{-g}G^{-gT} = G^T\left[GG^T\right]^{-1}\left[GG^T\right]^{-1}G = G^T\left[G^TG\right]^{-2}G
\end{aligned}
\tag{E.15b}
$$

If $\alpha_1 = 1$, $\alpha_2 = 0$, α_3 equals some constant (say, ε^2), and $[\text{cov}_u \Phi] = I$, then

$$\mathbf{G}^{-g} = \left[\mathbf{G}^{\mathrm{T}}\mathbf{G} + \varepsilon^2\mathbf{I}\right]^{-1}\mathbf{G}^{\mathrm{T}} \tag{E.16}$$

It is the damped least squares inverse as the above (E.6).

If $\alpha_1 = 0$, $\alpha_2 = 1$, $\alpha_3 = \varepsilon^2$, and $[\text{cov}_u \Phi] = I$, then

$$G^{-g} = G^T\left[GG^T + \varepsilon^2 I\right]^{-1} \tag{E.17}$$

It may be termed *damped minimum length.*

For more details of this section, refers to Chapter 4 in Menke (2012).

E.4 MAXIMUM LIKELIHOOD METHOD FOR LINEAR INVERSE PROBLEM

Assume the data in $Gs = \Phi$ have a multivariate Gaussian probability density function,

$$p(\Phi) \propto \exp\left[-\frac{1}{2}(\Phi - Gs)^T[\text{cov}\,\Phi]^{-1}(\Phi - Gs)\right] \tag{E.18}$$

Here, the data covariance and data kernel are known. The maximum likelihood (ML) is used to estimate the model parameters that maximize the probability of the observed data. It can be realized by maximizing $\log p\left(\Phi^{\text{obs}}\right)$ or maximizing $p\left(\Phi^{\text{obs}}\right)$ since $\log(p)$ is a monotonic function. Apparently, such a Gaussian probability function bear to L2 norm; certainly, an exponential probability density function can be adopted, then it bears to L1 norm, and L1 norm problem can be transformed into a "linear programming" problem. For more details, refer to Chapter 4 in Taranrola (2005) or Chapter 5 in Menke (2012), here only focus on L2 issues.

If we assume that the probability density functions of the data, model parameters, and the theory are all Gaussian, so that

$$p_A(s) \propto \exp\left[-\frac{1}{2}\left(s-\langle s\rangle\right)^T [\operatorname{cov} s]_A^{-1}\left(s-\langle s\rangle\right)\right]$$

$$p_A(\Phi) \propto \exp\left[-\frac{1}{2}(\Phi-\Phi^{obs})^T [\operatorname{cov}\Phi]^{-1}\left(\Phi-\Phi^{\text{obs}}\right)\right] \tag{E.19}$$

$$p_g(s,\Phi) \propto \exp\left[-\frac{1}{2}(\Phi-Gs)^T \left[\operatorname{cov} g\right]^{-1}(\Phi-Gs)\right]$$

Here, the subscript A means "*a priori*", P_g is an envisioned probability density function centered about $g(s)=\Phi\,(Gs=\Phi)$, or say the theoretical model is represented by a Gaussian probability density function with covariance $[\operatorname{cov} g]$. The total distribution $p_T(s,\Phi)$ is the product of these three distributions, and the subscript T means the combined or total distribution. As products of Gaussian probability density functions are themselves Gaussian, so $p_T(s,\Phi)$ is Gaussian. Now ML changes to seek the maximum likelihood point of $p_T(s,\Phi)$ that defines s^{est} and Φ^{pre}.

After a detailed derivation, look at Section 5.26 in Menke (2012), we have

$$s^{\text{est}} = \langle s\rangle + G^{-g}\left(\Phi^{\text{obs}} - G\langle s\rangle\right) = G^{-g}\Phi^{\text{obs}} + [I-R]\langle s\rangle$$

With

$$G^{-g} = [\operatorname{cov} s]_A\, G^T\left\{[\operatorname{cov}\Phi]+[\operatorname{cov} g]+G[\operatorname{cov} s]_A\, G^T\right\}^{-1} \tag{E.20}$$

This generalized inverse is similar to the minimum-length inverse $G^T\left[GG^T\right]^{-1}$, the generalized inverse may also be represented as

$$G^{-g} = \left\{G^T\left([\operatorname{cov}\Phi]+[\operatorname{cov} g]\right)^{-1} G + [\operatorname{cov} s]_A^{-1}\right\}^{-1} G^T\left([\operatorname{cov}\Phi]+[\operatorname{cov} g]\right)^{-1} \tag{E.21}$$

which is similar to the least squares generalized inverse $[G^T G]^{-1} G^T$

Suppose we have uncorrelated *a priori* model parameters $\left([\operatorname{cov} s]_A = \sigma_s^2 I\right)$, data $\left([\operatorname{cov}\phi]_A = \sigma_\phi^2 I\right)$, and theory$\left([\operatorname{cov} g] = \sigma_g^2 I\right)$, we may have the following limiting cases.

1. Exact data and theory, take $\sigma_\varphi^2 = \sigma_g^2 = 0$ and a priori $\langle s\rangle = 0$, the solution (E.20) reduced to the minimum-length solution

$$s^{\text{est}} = G^T\left[GG^T\right]^{-1}\Phi^{\text{obs}} \tag{E.22}$$

Such a minimum-length form exists only when the problem is purely underdetermined.

If the *a priori* $\langle s\rangle \neq 0$, then

$$s^{\text{est}} = G^{-g}\Phi^{\text{obs}} + (I-R)\langle s\rangle = G^T\left[GG^T\right]^{-1}\Phi^{\text{obs}} + \left\{I - G^T\left[GG^T\right]^{-1} G\right\}\langle s\rangle$$

2. Exact theory and model, take $\sigma_s^2 = \sigma_g^2 = 0$ and a priori $\langle s \rangle = 0$, the solution (E.21) reduced to the least squares solution.

$$s^{\text{est}} = \left[G^T G\right]^{-1} G^T \Phi^{\text{obs}}$$

Such a least squares form exists only when the problem is purely overdetermined.

For this case, even if $\langle s \rangle \neq 0$, the solution still is the same, as $R = I$.

3. Infinitely inexact data and theory, take $\sigma_\phi^2 \to \infty$ or $\sigma_g^2 \to \infty$ (or both), then the solution becomes

$$s^{\text{est}} = \langle s \rangle \tag{E.23}$$

Since the data and theory contain no information, the solution simply recovers the *a priori* model parameters.

4. No *a priori* knowledge of the model parameters; for this case, the limit is $\sigma_s^2 \to \infty$. The solutions (E.20) and (E.21) may reduce to the following:

$$s^{\text{est}} = G^T \left[GG^T\right]^{-1} \Phi^{\text{obs}} + \left\{I - G^T \left[GG^T\right]^{-1} G\right\} \langle s \rangle = \left[G^T G\right]^{-1} G^T \Phi^{\text{obs}} \tag{E.24}$$

For more details of this section, refers to Chapter 5 of Menke (2012).

E.5 SINGULAR-VALUE DECOMPOSITION AND NATURAL GENERALIZED INVERSE

E.5.1 Vector Space

In sense of vector space, the linear equation $\Phi = Gs$ can be interpreted as a mapping of vector from model vector space $S(s)$ to the data vector space $S(\Phi)$, and its estimate solution $s^{\text{est}} = G^{-g}\Phi$ as a mapping of vector from data vector space $S(\Phi)$ to model vector space $S(s)$. By using Householder transformation, which can triangularize a matrix such as G, then the linear equation $\Phi = Gs$ can be inverted. However, if the problem is underdetermined, then the equation $Gs = \Phi$ contains information about only some of the model parameters which span a subspace $S_p(s)$, and the other part of the solution lies in a null space $S_o(s)$ which is completely "unsaw" by $Gs = \Phi$. On the other hand, if the problem is overdetermined, the product Gs may only span a subspace $S_p(\Phi)$ of the data space, and the other part is a null space $S_o(\Phi)$. For mixed-determined problems, where a look-like underdetermined problem ($N < M$) with some degree overdetermined, or a look like overdetermined problem ($N > M$) with some degree underdetermined, and $S_o(s)$ and $S_o(\Phi)$ may exist simultaneously. So, in general, we would like to have a method to identify the p and null subspace of a linear problem. Fortunately, singular value decomposition (SVD) can identify the p and null subspace of the linear problem easily.

E.5.2 Singular Value Decomposition

With SVD, any $N \times M$ matrix can be written as the product of three matrices (Penrose, 1955):

$$G = U \Lambda V^T \tag{E.25}$$

The data space $S(\Phi)$ is spanned by the matrix U which is an $N \times N$ matrix of eigenvectors:

$$U = \left[u^{(1)}, u^{(2)}, u^{(3)}, \ldots, u^{(N)} \right] \tag{E.26}$$

where the $u^{(i)}$s are the individual vectors. The vectors are orthogonal to one another and can be chosen to be of unit length so that $UU^T = U^T U = I$ (with size of $N \times N$). Similarly, the model parameter space is spanned by the matrix V which is an $M \times M$ matrix of eigenvectors:

$$V = \left[v^{(1)}, v^{(2)}, v^{(3)}, \ldots, v^{(M)} \right] \tag{E.27}$$

Here, the $v^{(i)}$s are the individual orthonormal vectors so that $VV^T = V^T V = I$ (with size of $M \times M$). The matrix $\wedge$ consisted of an $N \times M$ diagonal eigenvalue matrix whose diagonal elements called *singular values* are nonnegative. The singular values λ_i are usually arranged in order of decreasing size. As some of the singular value may be zero, Λ is naturally divided into a submatrix Λ_p of p nonzero singular values and several zero matrices as

$$\Lambda = \begin{bmatrix} \Lambda_p & 0 \\ 0 & 0 \end{bmatrix} \tag{E.28}$$

where Λ_p is a diagonal matrix with size of $p \times p$. The decomposition then becomes $U \Lambda V^T = U_p \Lambda_p V_p^T$, where U_p and V_p consist of the first p columns of U and V, respectively. The other portions are canceled by the zeros in Λ, which are just the null spaces V_0 and U_0, respectively.

The equation $\Phi = Gs = U_p \Lambda_p V_p^T s$ means that G contains no information about the part of the model parameters in the space V_0. Similarly, no matter what value $\left\{ \Lambda_p V_P{}^T s \right\}$ attains, it can have no component in the space U_0 as it is multiplied by U_p (and U_0 and U_p are orthogonal), apparently, U_p lies completely in $S_p(\Phi)$ and U_0 lies completely in $S_0(\Phi)$.

The p and null matrices are orthogonal and are normalized in the sense that $V_p^T V_P = U_p^T U_p = I$, where I is $p \times p$ in size. However, as these matrices do not in general span the complete data and model spaces, $V_p^T V_P$ and $U_p^T U_p$ are not in general identity matrices.

E.5.3 Natural Solution

The natural solution must have an s^{est} that has no component in $S_0(s)$ and a prediction error e that has no component in $S_p(\Phi)$. If we take a solution as

$$s^{\text{est}} = V_p \Lambda_p^{-1} U_p^T \Phi \tag{E.29}$$

It is easy to verify $V_0^T s^{\text{est}} = 0$ and $U_p^T e = 0$; thus, it is a natural solution.

As $V_p^T V_P = U_p^T U_p = \Lambda_p \Lambda_p^{-1} = I$, the natural solution of the inverse problem is therefore shown to be equation (E.29).

In general, even for a mixed-determined problem, we may define a *natural generalized inverse*

$$G^{-g} = V_p \Lambda_p^{-1} U_p^T \tag{E.30}$$

Equation (E.30) is so useful that it is sometimes referred to as *the* generalized inverse.

The natural generalized inverse has model resolution

$$R = G^{-g} G = \left\{ V_p \Lambda_p^{-1} U_p^T \right\} \left\{ U_p \Lambda_p V_p^T \right\} = V_p V_p^T \tag{E.31}$$

If V_p spans the complete space of model parameters, that is, if there are no zero eigenvalues and $p \geq M$, the model parameters will be perfectly resolved.

The data resolution matrix is

$$N = G G^{-g} = \left\{ U_p \Lambda_p V_p^T \right\} \left\{ V_p \Lambda_p^{-1} U_p^T \right\} = U_p U_p^T \tag{E.32}$$

If U_p spans the complete space of data and $p = N$, the data are perfectly resolved.

Finally, if the data are uncorrelated with uniform variance σ_Φ^2, the model covariance is

$$\left[\text{cov}\, s^{\text{est}} \right] = G^{-g} [\text{cov}\, \Phi] G^{-gT} = \sigma_\Phi^2 V_p \Lambda_p^{-2} V_p^T \tag{E.33}$$

The covariance of the estimated model parameters is very sensitive to the smallest nonzero eigenvalue. If one prefers a solution based on the natural inverse, it is appropriate to use the formula $s^{\text{est}} = G^{-g}\Phi + [I - R]\langle s \rangle$, where G^{-g} is the natural inverse. The covariance of this estimate is now

$$\left[\text{cov}\, s^{\text{est}} \right] = G^{-g} [\text{cov}\, \Phi] G^{-gT} + [I - R][\text{cov}\, s]_A [I - R]^T \tag{E.34}$$

which is based on the usual rule for computing covariance.

E.5.4 Truncation and Regularization

The value of the integer p for truncation must be chosen so that zero eigenvalues are excluded from the estimate; else a division-by-zero error will occur. In practice, the turning point is not very clear, one often excludes near-zero eigenvalues as well, as the resulting solution (while not quite the natural solution) is often better behaved. If one chooses p so as to include those very small singular values, the solution variance will be very large.

Instead of choosing a sharp cutoff for the singular values, it is possible to include all the singular values while damping the smaller ones by replacing the reciprocals of all the singular values by $\lambda_i/\left(\varepsilon^2+\lambda_i^2\right)$, where ε is some small number. This change has little effect on the large singular values but prevents the small ones from leading to large variances. Of course, the solution is no longer the natural solution. While its variance is improved, its model and data resolution are degraded.

E.5.5 Weighted Damped Least Squares Solution

Suppose we are searching a solution s^{est} that minimizes a weighted combination of prediction error and solution length for a mixed as

$$\text{minimize}: \quad E+L=e^TW_ee+\varepsilon^2s^TW_ss \tag{E.35}$$

As W_s and W_e usually being symmetric, have a symmetric square root, so that $W_s=W_s^{1/2}W_s^{1/2}=W_s^{1/2T}W_s^{1/2}=T_s^TT_s$ and $W_e=T_e^TT_e$, where we find two transformations $s'=T_ss$ and $e'=T_ee$, which change the problem to a damped least squares problem (Wiggins, 1972)

$$\Phi'=T_e\Phi=T_eGT_s^{-1}T_ss=G's'$$

$$s'=T_ss$$

$$G'=T_eGT_s^{-1}$$

$$E+L=e'^Te'+\varepsilon^2s'^Ts'$$

Then, we have the damped least squares solution

$$s'^{\text{est}}=\left[G'^TG'+\varepsilon^2I\right]^{-1}G^T\Phi' \tag{E.36}$$

If we have $G'=U\Lambda V^T$, then

$$
\begin{aligned}
\left[G'^T G' + \varepsilon^2 I\right]^{-1} G'^T &= \left[V\Lambda U^T U\Lambda V^T + \varepsilon^2 I\right]^{-1} V\Lambda U^T \\
&= \left[V\Lambda^2 V^T + \varepsilon^2 V I V^T\right]^{-1} V\Lambda U^T \\
&= V\left[\Lambda^2 + \varepsilon^2 I\right]^{-1} V^T V\Lambda U^T \\
&= V\left\{\left[\Lambda^2 + \varepsilon^2 I\right]^{-1} \Lambda\right\} U^T
\end{aligned}
\tag{E.37}
$$

This fact means that for a (weighted) damping least squares problem, SVD still can be an easy way to get the solution.

For more details of this section, refer to Chapter 7 of Menke (2012).

E.6 EQUIVALENCE OF THE DIFFERENT APPROACHES

The above description shows that different approaches may arrive to the same or similar results (Menke, 2012), but each one give us a special scene to understand the inverse problem.

1. Minimizing a weighted sum of the prediction error and solution length
   Minimize:

$$
e^T W_e e + \varepsilon^2 \left[s - \langle s\rangle\right]^T W_s \left[s - \langle s\rangle\right] \tag{E.38}
$$

2. Minimizing a weighted sum of the Dirichlet spreads of model resolution and data resolution and the size of the model covariance (generalized inverse)
   Minimize:

$$
\alpha_1\, \text{spread}(R) + \alpha_2 \text{spread}(N) + \alpha_3 \text{size}\left(\left[\text{cov}_u\, s\right]\right) \tag{E.39}
$$

3. maximizing the likelihood of the joint Gaussian distribution of data, *a priori* model parameters, and theory.
   Maximize:

$$
L = \log p_T(s, \Phi) \tag{E.40}
$$

4. Realizing the correspondence between the data vector space and model parameter vector space to get natural generalized inverse by singular values decomposition and transformation in vector space.

REFERENCES

Bertero, M., C. De Mol, E.R. Pike. 1985. Printed in Great Britain Linear inverse problems with discrete data.I: General formulation and singular system analysis. *Inverse Probl* 1(4):301–330.

Bertero, M., C. De Mol, E.R. Pike. 1988. Printed in Great Britain Linear inverse problems with discrete data.II: Stability and regularization. *Inverse Probl* 4(3):573–594.

Franklin, J.N. 1970. Well-posed stochastic extensions of ill-posed linear problems. *J Math Anal Appl* 31(3):682–716.

Lawson, C.L., R.J. Hanson. 1974. *Solving Least Squares Problems*. Philadelphia, PA: Society for Industrial and Applied Mathematics.

Marquardt, D.W. 1970. Generalized inverses, ridge regression, biased linear estimation and non-linear estimation. *Technometrics* 12(3):591–612.

Menke, W. 2012. *Geophysical Data Analysis: Discrete Inverse Theory*. London: Academic Press.

Penrose, R.A. 1955. A generalized inverse for matrices. *Math Proc Cambridge Philos Soc* 51(3):406–413.

Tarantola, A. 2005. *Inverse Problem Theory and Methods for Model Parameter Estimation*, Chapter 3. Philadelphia, PA: Society for Industrial and Applied Mathematics. doi: 10.1137/1.9780898717921.

Wiggins, R.A. 1972. The general linear inverse problem: Implication of surface waves and free oscillations for Earth structure. *Rev Rev Geophys* 10(1):251–285.

Appendix F
Physics of EEG Average Reference

According to previous chapters, the ideal reference potential for scalp EEG should be zero. However, as there is no point with zero potential on the head or body surface, the actual online reference electrode on the head or body surface was unable to provide an ideal zero reference. The average reference (AR) technology was developed in 1950s, based on an assumption that people believe the average of the scalp surface potential being close to zero. In this appendix, we will provide the integral study of the scalp surface potential and identify the cases under which the integral may be zero or not, thus providing a theoretical physical guidance for cautious use of the AR (Yao, 2017).

F.1 BACKGROUND

The AR (Goldman, 1950; Offner, 1950) has often been advocated as the best available reference option before, as Nunez and Srinivasan (2006) stated, "when used with large numbers of electrodes…it often performs reasonably well…". The fundamental assumption of AR is that the surface potential integral of a dipole in a volume conductor is zero; thus, the average potential of a dense and wide coverage electrode array is close to the ideal zero reference (Geselowitz, 1998). However, such a zero potential integral assumption has been theoretically proved only for a spherical surface (Bertrand et al., 1985; Yao, 2017). In this appendix, three counter-examples are given, which remind that the average may not always be zero for some specific surfaces.

According to the theory of bio-electromagnetism, biological electric source is current source density, and for any current source density distribution $s(\vec{r})$, its potential satisfies Poisson's equation:

$$\nabla^2 \Phi(\vec{r}) = -s(\vec{r}) \tag{F.1}$$

And according to the section 2.4.2, the potential produced by such source $s(\vec{r})$ can be equivalently generated by a combination of multipole sources, i.e., monopole (point current source density), dipole, quadrupole, and octapole, etc. at the origin of a coordinate system. Among them, the first monopole term (zeroth-order multipole) would be omitted due to the current conservation of a living system that requires the sum of the current source density inside a living system vanishes. The second dipole term (first-order multipole) is consisted of a positive and a negative point current source density, and the other lth-order multipoles with $l > 1$ are various complex linear combinations of dipoles. Furthermore, the linear relation between potential Φ and source $s(\vec{r})$ in equation (F.1) means that the principle of linear superposition is valid for EEG forward problem; thus, we only need to check the potential integral over the surface of a dipole in a volume conductor.

F.2 A DIPOLE IN A SPHERICAL VOLUME CONDUCTOR

For a dipole in a spherical volume conductor, it has been proved that the potential integral over the surface is zero (Bertrand et al., 1985). In fact, this conclusion is physically clear as the potential of a single dipole can be generated equivalently by a series of lth-order multipole ($l \geq 1$) at the origin (Yao, 2017, Section 2.4). And the positive and negative potential of each lth-order multipole ($l \geq 1$) always appear anti-symmetrically on the sphere surface, thus their integral must be zero. In another word, not only a dipole but also any combination of dipoles located anywhere in a sphere, the surface potential integral is zero. This fact is also valid for multilayer spherical model as the multipole expansion of neural electric sources in such a model is similar to the single sphere case (Chapter 4).

F.3 A DIPOLE IN A HALF-SPACE VOLUME CONDUCTOR

For a spherical volume conductor, if the radius R tends to infinity, the local surface of the sphere will evolve to the boundary plane of a half-space volume conductor. For such a model, any a dipole moment may be decomposed into two components, one is oriented parallel to the boundary plane and the other perpendicular to the plane surface. Then, it is physically clear that the potential integral of the parallel one is zero as its positive and negative potential anti-symmetrically distributing on the plane. However, the potential integral of the perpendicular one will not be zero. The mathematical proof is shown in Yao (2017).

F.4 A DIPOLE IN A HOMOGENOUS VOLUME CONDUCTOR, EXCEPT A SPHERICAL CAVE

Start from a half-space volume conductor, if the surface further bends to the opposite direction of the dipole, a special case may appear that a dipole locates in a homogenous volume conductor, except a spherical cave. For such a case, if we assume a spherical coordinate system with origin at the center of the spherical cave with a dipole outside, the dipole moment may be decomposed into three components, one radial component perpendicular to the spherical surface and the other two oriented tangentially to the spherical surface. Similar to the half-space volume conductor model, the potential integrals of the two tangential components are zero, while that of the radial one is not zero. The mathematical proof is shown in Yao (2017), where the zero potential integral of a dipole in a spherical volume conductor is also proved passingly.

F.5 A DIPOLE IN A HOMOGENEOUS SEMI-SPHERE VOLUME CONDUCTOR

The above two models, a half-space and a homogeneous volume conductor with a spherical cave, look not so like a head as the head would have a vivid almost closed surface. Here, we further consider a homogeneous semi-sphere volume conductor.

First, we need the potential solution of a dipole in a homogeneous semi-sphere volume conductor, which must satisfies the Poisson equation and the Neumann boundary condition. Suppose an actual dipole in the upper hemi-sphere of a spherical volume conductor, we may set a mirror dipole, both position and orientation mirrored, in the lower semi-sphere, and take the potential of the primary dipole in the sphere as Φ_1, and that of the mirror dipole in the same sphere as Φ_2, then we have both Φ_1 and $\Phi = \Phi_1 + \Phi_2$ satisfy the Poisson's equation in the upper semi-sphere and the Neumann boundary condition on the upper semi-spherical surface, both Φ_2 and $\Phi = \Phi_1 + \Phi_2$ satisfy the Poisson's equation in the lower semi-sphere and the Neumann boundary condition on the lower semi-spherical surface. For the circular plane, which divides the whole sphere into two semi-spheres, due to the orientation of the virtual dipole in the lower semi-sphere is mirror to the primary one, the Neumann boundary condition is also satisfied. Here the orientation mirror means that the two components parallel to the plane of the two dipoles are along the same direction, but the two components of them perpendicular to the plane are along opposite direction. These facts mean that $\Phi = \Phi_1 + \Phi_2$ is the potential solution of a dipole in a semi-sphere volume conductor as it satisfies the Poisson's equation inside the semi-sphere conductor and the Neumann boundary condition on its whole surface. As we have the closed form of Φ_1 and Φ_2 (Yao, 2000), we actually have the closed solution of a dipole in such a hemi-sphere volume conductor.

Now, let us check the potential integral over the whole surface of the semi-sphere conductor, we may consider the integrals over the hemi-spherical surface and the circular plane, separately. It is clear that the potential Φ_2 at a point on the upper semi-spherical surface is the same as the potential Φ_1 at a mirror point on the lower semi-spherical surface, thus the integral of $\Phi = \Phi_1 + \Phi_2$ over the upper semi-spherical surface is equal to the integral of Φ_1 over the whole sphere surface. And it is well-known that this integral is always zero (Bertrand et al., 1985; Yao, 2017). Now, for the potential integral of $\Phi = \Phi_1 + \Phi_2$ over the circular plane, as the two dipoles mirror to each other about the circular plane, the potential Φ_2 at a point on the circular plane is the same as the potential Φ_1 at the same point, the sum $\Phi = \Phi_1 + \Phi_2$ over the circular plane is the double of Φ_1 and the integral problem actually reduces to the potential integral of the potential Φ_1 over the circular plane. The potential integral of a tangential dipole at the Cartesian coordinates $(0, 0, z_0)$ with the origin of the coordinate system at the center of the sphere is always zero as its potential distribution over the plane is anti-symmetrical; however, the mathematical deduction shown in Yao (2017) indicates that the potential integral of a radial dipole at the Cartesian coordinates $(0, 0, z_0)$ is not zero but dependent on the radius R of the sphere and the value of z_0. Specifically, if the radius R tends to infinity, the integral reduces to the same as the half-space model. The proof is as follows.

According to bio-electromagnetic theory, biological electric source is current source density, and for any current source density distribution $s(\vec{r})$, its potential satisfies Poisson's equation (Stratton, 1941):

$$\nabla^2\Phi(\vec{r}) = -s(\vec{r})/\sigma \tag{F.2}$$

where $\vec{r}$ is the field point and σ is the conductivity of the conductor (Geselowitz, 1960). This equation is known to have a solution of the following form:

$$\Phi(\vec{r}) = \frac{1}{4\pi\sigma}\int_v \frac{s(\vec{r}')}{|\vec{r}-\vec{r}'|}d^3\vec{r}' \tag{F.3}$$

If the source distribution region v can be bounded by a closed surface S, then the potential outside of S satisfies Laplace's equation.

$$\nabla^2\Phi(\vec{r}) = 0,\ \vec{r} \notin v \tag{F.4}$$

As explained in the above, the potential Φ on the circular plane of a dipole in a homogeneous semi-sphere volume conductor is the double of the potential on the same plane of the same dipole in a homogeneous sphere volume conductor; thus, we may just work on the potential integral of the potential Φ_1 over the circular plane.

In general, for a dipole in a homogenous spherical volume conductor, the potential at anywhere in the sphere is shown by equation (4.15) (Yao, 2000).

Here, as an example, we specifically consider a dipole at Cartesian coordinates $(x_0, y_0, z_0) = (0, 0, z_0)$ and positively oriented along z–axis with $\vec{P} = P_x\vec{e}_x + P_y\vec{e}_y + P_z\vec{e}_z = P_z\vec{e}_z$, where (P_x, P_y, P_z) are the three components of the dipole moment $\vec{P}$ in accordance with the three unit vectors $(\vec{e}_x, \vec{e}_y, \vec{e}_z)$ of the Cartesian coordinate system with origin at the center of the sphere. For the potential on the circular plane ($z=0$), we have $\varphi = 90°$ and $\vec{r} = (x\vec{e}_x, y\vec{e}_y, z\vec{e}_z) = (x\vec{e}_x, y\vec{e}_y, 0\vec{e}_z)$.

For this special case, we get the following simplified formula (F.5) from equation (4.15).

$$\Phi = 2\Phi_1 = -2\frac{P_z}{4\pi\sigma}\left[\frac{z_0}{r_p^3} + \frac{r^2 z_0}{R^5 r_{pi}^3} + \frac{r^2 z_0}{R^5 r_{pi}(r_{pi}+1)}\right] \tag{F.5}$$

where Φ is the total potential of the dipole and its mirror, the double of Φ_1, $r_p = (r^2 + z_0^2)^{1/2}$, $r^2 = x^2 + y^2$, $r_{pi} = \left[1 + (r_0 r / R^2)^2\right]^{1/2} = 1/R^2\left(R^4 + (z_0 r)^2\right)^{1/2}$.

Then, the integral over the circular plane is

$$I = \int_0^{2\pi}\int_0^R \Phi r\, d\theta dr = -\frac{P_z}{\sigma}\int_0^R\left[\frac{z_0}{r_p^3} + \frac{r^2 z_0}{R^5 r_{pi}^3} + \frac{r^2 z_0}{R^5 r_{pi}(r_{pi}+1)}\right] r dr = -\frac{P_z}{\sigma}[I_1 + I_2 + I_3] \tag{F.6}$$

Specifically, if the dipole is just located on the circular plane $(z_0 = 0)$, the potential shown by equation (F.5) and the integral shown in equation (F.6) are all zero. However, this case is not the situation that we concerned that a dipole is in a semi-sphere volume conductor with $z_0 > 0$.

Now we check each item in equation (F.6) for $z_0 > 0$, we have

$$I_1 = z_0 \int_0^R \frac{1}{\left(r^2 + z_0^2\right)^{3/2}} r dr = \frac{z_0}{2} \int_0^R \frac{1}{\left(r^2 + z_0^2\right)^{3/2}} dr^2 = -\frac{z_0}{\left(R^2 + z_0^2\right)^{1/2}} + 1 \tag{F.7}$$

And

$$\begin{aligned} I_2 &= \int_0^R \frac{r^2 z_0}{R^3 \left(R^4 + (z_0 r)^2\right)^{3/2}} r dr = \frac{z_0}{2R^3} \int_0^R \frac{r^2}{\left(R^4 + z_0^2 r^2\right)^{3/2}} dr^2 \\ &= \frac{z_0}{2 z_0^2 R^3} \int_0^R \left[\frac{R^4 + z_0^2 r^2}{\left(R^4 + z_0^2 r^2\right)^{3/2}} - \frac{R^4}{\left(R^4 + z_0^2 r^2\right)^{3/2}} \right] dr^2 \\ &= \frac{2R^2 + z_0^2 - 2\left(R^4 + z_0^2 R^2\right)^{1/2}}{z_0^3 R \left(R^4 + z_0^2 R^2\right)^{1/2}} \end{aligned} \tag{F.8}$$

and

$$\begin{aligned} I_3 &= \frac{z_0}{2} \int_0^R \frac{r^2}{R^5 r_{pi} \left(r_{pi} + 1\right)} dr^2 \\ &= \frac{z_0}{2R} \int_0^R \frac{r^2}{\left(R^4 + (z_0 r)^2\right)^{1/2} \left[\left(R^4 + (z_0 r)^2\right)^{1/2} + R^2\right]} dr^2 \\ &= \frac{z_0}{2R} \frac{2}{z_0^2} \int_0^R \frac{r^2}{\left(R^4 + (z_0 r)^2\right)^{1/2} + R^2} d\left(R^4 + (z_0 r)^2\right)^{1/2} \\ &= \frac{z_0}{2R} \frac{2}{z_0^2} \frac{1}{z_0^2} \left[F_1 + F_2\right] \end{aligned} \tag{F.9}$$

where

$$F_1 = \int_0^R \frac{R^4 + (z_0 r)^2}{\left(R^4 + (z_0 r)^2\right)^{1/2} + R^2} d\left(R^4 + (z_0 r)^2\right)^{1/2} \tag{F.10}$$

Take $y=\left(R^4+(z_0r)^2\right)^{1/2}$, then y is $y_0=R^2$ and $y_R=\left(R^4+(z_0R)^2\right)^{1/2}$ for $r=0$ and $r=R$, respectively. Then, we have

$$F_1=\int_{y_0}^{y_R}\frac{y^2}{y+R^2}dy=\int_{y_0}^{y_R}\frac{y^2-R^4+R^4}{y+R^2}dy=\int_{y_0}^{y_R}\left[\frac{\left(y+R^2\right)\left(y-R^2\right)}{y+R^2}+\frac{R^4}{y+R^2}\right]dy$$

$$=\int_{y_0}^{y_R}\left[y-R^2+\frac{R^4}{y+R^2}\right]dy$$

$$=\frac{1}{2}(Rz_0)^2-R^2\left[\left(R^4+(Rz_0)^2\right)^{1/2}-R^2\right]+R^4\left[\ln\left|\left(R^4+(Rz_0)^2\right)^{1/2}+R^2\right|-\ln\left(2R^2\right)\right] \tag{F.11}$$

And

$$\begin{aligned}F_2&=-\int_0^R\frac{R^4}{\left(R^4+(z_0r)^2\right)^{1/2}+R^2}d\left(R^4+(z_0r)^2\right)^{1/2}\\&=-R^4\left[\ln\left|\left(R^4+(Rz_0)^2\right)^{1/2}+R^2\right|-\ln\left(2R^2\right)\right]\end{aligned} \tag{F.12}$$

Take equations (F.11) and (F.12) into (F.9), we have

$$\begin{aligned}I_3&=\frac{z_0}{2R}\frac{2}{z_0^2}\frac{1}{z_0^2}[F_1+F_2]\\&=\frac{1}{2Rz_0^3}\left[(Rz_0)^2-R^2\left(R^4+(Rz_0)^2\right)^{1/2}+R^4\right]\end{aligned} \tag{F.13}$$

Finally, we have

$$I=-\frac{P_z}{\sigma}[I_1+I_2+I_3]=-\frac{P_z}{\sigma}\left\{-\frac{z_0}{\left(R^2+z_0^2\right)^{1/2}}+1+\frac{2R^2+z_0^2-2\left(R^4+z_0^2R^2\right)^{1/2}}{z_0^3R\left(R^4+z_0^2R^2\right)^{1/2}}\right.$$

$$\left.+\frac{1}{2Rz_0^3}\left[(Rz_0)^2-R^2\left(R^4+(Rz_0)^2\right)^{1/2}+R^4\right]\right\}$$

$$=-\frac{P_z}{\sigma}\frac{-2z_0^4R^2+\left(-4R+2z_0^3R^2+z_0^2R^3+R^5\right)\left(R^2+z_0^2\right)^{1/2}+2z_0^2+4R^2-z_0^2R^4-R^6}{2z_0^3R^2\left(R^2+z_0^2\right)^{1/2}} \tag{F.14}$$

Specifically, let $\sigma = 1$, $P_z = 1$, $R = 1$, we have

$$I = -\frac{-2z_0^4 + \left(-4 + 2z_0^3 + z_0^2 + 1\right)\left(1 + z_0^2\right)^{1/2} + 2z_0^2 + 4 - z_0^2 - 1}{2z_0^3\left(1 + z_0^2\right)^{1/2}}$$
$$= -\frac{-2z_0^4 + z_0^2 + \left(-3 + 2z_0^3 + z_0^2\right)\left(1 + z_0^2\right)^{1/2} + 3}{2z_0^3\left(1 + z_0^2\right)^{1/2}} \tag{F.15}$$

For equation (F.15), it is easy to test by MATLAB that its value is always smaller than 0 and z_0 value dependent. For example, for $Z_0 = 0.001$, 0.01, 0.05, 0.1, 0.2, 0.3, 0.4, 0.5, 0.6, 0.7, 0.8, 0.9, 0.99, the values of I are –250.9993, –25.9931, –5.9657, –3.4315, –2.1142, –1.6327, –1.3629, –1.1803, –1.0438, –0.9355, –0.8464, –0.7713, –0.7131, respectively. In fact, here the dipole is oriented along the positive z-axis in the upper semi-sphere, the potential on the circular plane would be negative.

Specifically, if R tends to infinity, then according to equation (F.6), we have

$$I = \int_0^{2\pi}\int_0^{R} \Phi r\, d\theta dr = -\frac{P_z}{\sigma}\int_0^{R}\left[\frac{z_0}{r_p^3} + \frac{r^2 z_0}{R^5 r_{pi}^3} + \frac{r^2 z_0}{R^5 r_{pi}\left(r_{pi} + 1\right)}\right] r dr = -\frac{P_z}{\sigma} I_1 = -\frac{P_z}{\sigma} \tag{F.16}$$

It is the case of a dipole in a half-space volume conductor (Yao, 2017) except a sign because the dipole here is located upper the plane, and it is below the plane in the half-space volume model.

REFERENCES

Bertrand, O., F. Perrin, J. Pernier. 1985. A theoretical justification of the average reference in topographic evoked-potential studies. *EEG Clin Neurophysiol* 62:462–464.

Geselowitz, D.B. 1960. Multiple representation for an equivalent cardiac generator. *Proc IRE* 48(1):75–79.

Geselowitz, D.B. 1998. The zero of potential. *IEEE Eng Med Bio* 1:128–132.

Goldman, D. 1950. The clinical use of the "average" reference electrode in monopolar recording. *EEG Clin Neurophysiol* 2:209–212.

Nunez, P.L., R. Srinivasan. 2006. *Electric Fields of the Brain: The Neurophysics of EEG*. New York: Oxford University Press.

Offner, F.F. 1950. The EEG as potential mapping: The value of the average monopolar reference. *EEG Clin Neurophysiol* 2:213.

Stratton, J.A. 1941. *Electromagnetic Theory*. New York: McGraw Hill Education.

Yao, D. 2000. Electric potential produced by a dipole in a homogeneous conducting sphere. *IEEE Trans Biomed Eng* 47(7):964–966.

Yao, D. 2017. Is the surface potential integral of a dipole in a volume conductor always zero? A cloud over the average reference of EEG and ERP. *Brain Topogr* 30:161–171. doi: 10.1007/s10548-016-0543-x.

Appendix G
Reference Electrode Problem in Mathematics

The EEG reference problem has been discussed in terms of algebra (Section 13.2.2.6.1) and physics (Appendix F). In Chapter 14, matrix and statistic theory are introduced to understand the intrinsic relations and differences among the known main references. The content of this appendix is mainly from Hu et al. (2018a,b, 2019), and the mathematical tools involved are from Magnus and Neudecker (2007), Mardia et al. (1979), Baksalary et al. (2003) and Trenkler (2000).

G.1 BAYESIAN ESTIMATION OF ZERO REFERENCE POTENTIAL

Following previous studies (Hu et al., 2018a, 2019), we have

$$\Phi_r = T_r\Phi + e, e = T_r\varepsilon \tag{G.1}$$

Φ is the potential with zero reference, ε is the sensor noise, the reference transform T_r is a matrix of the rank as $N-1$, N is the number of channels, r denotes a specific reference, and due to the "no memory" property of T_r (Hu et al., 2019), r may be the on-line unipolar reference (UR) or the later offline unipolar re-reference. Thus, the estimate of Φ may be taken as an undetermined generalized linear inverse problem.

For the estimate problem (G.1), three different approaches may lead to the same results (Appendix E). Here, we take the maximum likelihood approach:

$$P(\Phi \mid \Phi_r, T_r, \varepsilon) \propto P(\Phi_r \mid \Phi, T_r, \varepsilon) P(\Phi) P(\varepsilon) \tag{G.1a}$$

$P(\Phi \mid \Phi_r, T_r, \varepsilon)$ is the posterior given the likelihood $P(\Phi_r \mid \Phi, T_r, \varepsilon)$ and priors $P(\Phi)$, $P(\varepsilon)$. Based on Hu et al. (2019), we take it as maximum a posteriori (MAP) estimate by maximizing the likelihood of the joint Gaussian distribution of data and a priori model parameters, and the final objective function

$$\mathrm{O} = (\Phi_r - T_r\Phi)^H \Sigma_{ee}^{+} (\Phi_r - T_r\Phi) + \Phi^H \Sigma_{\Phi\Phi}^{+} \Phi \tag{G.2}$$

Here, H denotes matrix transpose. This function can also be explicitly explained as minimizing a weighted sum of L2 prediction error and L2 solution simplicity

(Appendix E) with the corresponding covariances as the weights. After finding the partial derivative of equation (G.2) with respect to Φ and reorganizing the equation, we get equation (G.3):

$$\hat{\Phi} = \Sigma_{\Phi\Phi} T_r^{\ H} \left(T_r \Sigma_{\Phi\Phi} T_r^{\ H} + \Sigma_{ee} \right)^{+} \Phi_r \tag{G.3}$$

which is taken as the unified Bayesian estimator in reconstructing EEG potentials at infinity.

To derive the explicit expression of equation (G.3), in addition to assuming $\Sigma_{ee} = \sigma^2 T_r T_r^{\ H}$, $\Sigma_{\Phi\Phi}$ is assumed to have one of the following two different forms.

G.1.1 Uncorrelated Prior

$$\sum\nolimits_{\Phi\Phi} = \alpha^2 I_N \tag{G.4}$$

which means that the EEG potentials Φ have independent priors across all the channels and α^2 is the mean of variances of the potentials over each electrode.

Substituting (G.4), $\Phi_r = T_r \Phi$, and $\sum_{ee} = \sigma^2 T_r T_r^{\ H}$ into (G.3), it becomes

$$\hat{\Phi} = T_r^{\ +} T_r \Phi / \left(1 + \sigma^2/\alpha^2\right) \tag{G.5}$$

As $T_r^{\ +} T_r = I_N - 11^H/N$, which is the average reference transforming matrix (Appendix in Hu et al., 2018a). If we take sensor noise-to-the scalp EEG signal ratio as $nsr_1 = \sigma^2/\alpha^2$ and $T_{ar} = I_N - 11^T/N$, (G.5) is rewritten as

$$\hat{\Phi} = T_{ar} \Phi / (1 + nsr_1) \tag{G.6}$$

which we called the regularized average reference (rAR) (Hu et al., 2018a). It is obvious that the usual AR is the special case of rAR when $nsr_1 = 0$.

G.1.2 Correlated Prior

$$\sum\nolimits_{\Phi\Phi} = K_\infty \sum\nolimits_{jj} K_\infty^T \tag{G.7}$$

which models the EEG potentials across all the channels as correlated due to the effect of volume conduction on neural current sources, that is, we assume that $\Phi = K_\infty \boldsymbol{j}$; K_∞ is the leadfield matrix with infinity reference, $\boldsymbol{j}$ is the primal current density of the neural current sources with $\boldsymbol{j} \sim N\left(0, \beta^2 I_P\right)$, P is the number of neural current sources, and β^2 is the variance of the multivariate Gaussian signal $\boldsymbol{j}$.

Equation (G.3) is transformed by substituting (G.7) and defining $Kr = T_r K_\infty$ as

$$\hat{\Phi} = K_\infty \sum_{jj} K_r^H \left(K_r \sum_{jj} K_r^H + \sum_{ee} \right)^+ \Phi_r \tag{G.8}$$

which is the estimator for reconstructing the EEG potentials with reference at infinity named as the regularized reference electrode standardization technique (rREST) (Hu et al., 2018a). This process can be interpreted as processing in two stages,

$$\textit{Stage 1: } \hat{\boldsymbol{j}} = \sum_{jj} K_r^H \left(K_r \sum_{jj} K_r^H + \sum_{ee} \right)^+ \Phi_r$$

$$\textit{Stage 2: } \hat{\Phi} = K_\infty \hat{\boldsymbol{j}}$$

the first stage is solving the inverse problem with lead field K_r that has the same reference as the EEG potentials Φ_r and the second one is taking the forward calculation to reconstruct the EEG potentials with the theoretical neutral infinity reference.

Defining the sensor noise-to-the brain source signal ratio as $nsr_2 = \sigma^2/\beta^2$, and plugging, $\sum_{jj} = \beta^2 I_P$, $\sum_{ee} = \sigma^2 T_r T_r^H$ into (G.8), it becomes

$$\hat{\Phi} = K_\infty K_r^H \left(K_r K_r^H + nsr_2 T_r T_r^H \right)^+ \Phi_r \tag{G.9}$$

which is the solution to reconstruct the EEG potential at infinity through solving the inverse solution by incorporating the identity diagonal structure of Σ_{jj}. Apparently, REST (Yao, 2001) $\hat{\Phi} = K_\infty K_r^+ \Phi_r$ is the special case of rREST when $nsr_2 = 0$ in equation (G.9). For clarity, we summarize the general reference model and unified reference estimator in Table G.1.

G.2 WEIGHTED MINIMUM NORM SOLUTION FOR ZERO REFERENCE POTENTIAL

Linking to the consistence between maximum likelihood estimate and L2 least squares solution for Gaussian data (Section 12.1 and Appendix E), the reference problem can also be investigated in terms of weighted minimum norm solution. The following materials are mainly modified from Salido-Ruiz et al. (2019).

Consider the classical EEG linear model given as

$$\Phi = K_\infty j \tag{G.10}$$

where $\Phi \in R^{N\times 1}$ is a vector with unknown real potentials under each electrode with respect to infinity, from here defined as the true potentials for the N electrodes referenced to zero, $K_\infty \in R^{N\times P}$ is the leadfield matrix and $j \in R^{P\times 1}$ is the source vector.

TABLE G.1
EEG Reference Model, Unified Estimator, and Schemes

<table>
<tr><td>General reference model</td><td colspan="5">$\Phi_r = T_r\Phi + e, \quad e = T_r\varepsilon$</td></tr>
<tr><td>Unified reference estimator</td><td colspan="5">$\hat{\Phi} = \sum_{\Phi\Phi} T_r^H \left(T_r \sum_{\Phi\Phi} T_r^H + \sum_{ee}\right)^+ \Phi_r$</td></tr>
<tr><td>Prior of Φ</td><td colspan="2">$\sum_{\Phi\Phi} = \alpha^2 I_N$</td><td colspan="3">$\sum_{\Phi\Phi} = K_\infty \sum_{jj} K_\infty^H$</td></tr>
<tr><td>Solutions</td><td colspan="2">$\hat{\Phi} = T_{ar}\Phi/(1+nsr_1)$</td><td colspan="3">$\hat{\Phi} = K_\infty \sum_{jj} K_r^H \left(K_r \sum_{jj} K_r^H + \sum_{ee}\right)^+ \Phi_r$</td></tr>
<tr><td>Prior of j</td><td colspan="2"></td><td colspan="2">$\sum_{jj} = \beta^2 I_P$</td><td>$\sum_{jj} \neq \beta^2 I_P$</td></tr>
<tr><td>Sensor noise</td><td>Zero</td><td>Nonzero</td><td>Zero</td><td>Nonzero</td><td>Nonzero</td></tr>
<tr><td>Reference schemes</td><td>AR</td><td>rAR</td><td>REST</td><td>rREST</td><td></td></tr>
</table>

The actually measured potentials are given by the common UR, with potentials Φ_r modeled by subtracting the potential of the chosen reference electrode from the other electrodes. This can be seen as a matrix transform of the true potentials from equation (G.10):

$$\Phi_r = T_r\Phi = T_r K_\infty j = K_r j \tag{G.11}$$

with T_r the $N-1$ rank matrix:

$$T_r = \left[I_{N-1} - 1_{N-1}\right] \tag{G.12}$$

where I_{N-1} is the $(N-1)\times(N-1)$ identity matrix and $1_{N-1} \in R^{N-1\times 1}$ is a vector of 1's. As mentioned above and with no loss of generality, we assume the reference electrode potential as the true potential in EEG linear model (G.10). Note that unlike Φ in equation (G.2), the dimension of Φ_r is $N-1\times N_t$ (number of available signals).

This Φ_r did not include the on-line unipolar reference channel, so it is $N-1$.

G.2.1 AR Reference

In practice, as the only available signals are Φ_r, AR signals are obtained by calculating the sum of the $N-1$ signals of Φ_r, dividing it by the total number of electrodes N and subtracting it from each channel. More formally, the AR is obtained by averaging over all electrodes (reference electrode included with a null potential (i.e., its potential with respect to itself)):

$$\Phi_{\text{AR}} = \left(I_N - \frac{1}{N}1_N 1_N^H\right)\begin{bmatrix}\Phi_r \\ 0\end{bmatrix} = T_{\text{AR}}\Phi_r \tag{G.13}$$

with

$$T_{\text{AR}} = \begin{bmatrix} I_{N-1} - \dfrac{1}{N} 1_{N-1} 1_{N-1}^{H} \\ -\dfrac{1}{N} 1_{N-1}^{H} \end{bmatrix} \quad \text{(G.14)}$$

As we can see, Φ_{AR} has a dimension of $N \times N_t$ and rank of $N-1$, i.e., it preserves the rank of Φ_r (one less than Φ).

Here, N_t is the number of time samples. Φ_r is a vector of $N-1$ dimensions, and by taking equations (G.13) and (G.14), you get a vector of N dimensions Φ_{AR}. Here, the derivation reminding us the channel number of EEG/ERP recordings may be a confusion factor in practice. One says N channels which include the reference channel, while the other says N channels which may not contain the reference channel. It is due to the reference channel numerically being zero, and thus often be directly ignored. But when we perform reference transform, the number of channels we refer to must include the reference channel.

G.2.2 REST Reference

According to the strategy of REST (Yao, 2001), the chosen dipolar layer will yield a specific forward model between the dipoles situated on this layer and the actual electrodes placed on the scalp surface. Let this model be K_∞. The equivalent dipolar sources on this layer are estimated from unipolar scalp EEG recordings by using a simple inverse problem. Here, the given is the UR-based approach:

$$\hat{J} = K_r^{\ +}\Phi_r \quad \text{(G.15)}$$

with + designating the classical Moore–Penrose pseudo-inverse. Of course, K_r depends on a mixing model equation (G.10). With the estimated equivalent sources $\hat{J}$, the REST estimation of the EEG absolute potential:

$$\Phi_{\text{REST}} = K_\infty \hat{J} = K_\infty K_r^{\ +}\Phi_r = T_{\text{REST}}\Phi_r \quad \text{(G.16)}$$

It is important to recall that different REST solutions can be obtained by choosing different equivalent source configurations and thus different K_∞ models (Chapters 6 and 9).

G.2.3 REST and AR

According to the weighted length approach or the vector space approach (Menke, 2012), we may further deepen and unify the mathematical representation of the reference problem.

Consider the case of an EEG recording with a cephalic reference as given in equation (G.11). The estimation of true potentials Φ from the measured Φ_r and matrix transformation K_r is an ill-posed inverse problem somehow similar to the classical EEG source estimation with a known mixing model K_r.

Thus, the true potentials inverse problem is given as follows:

$$\hat{\Phi} = T\Phi_r = TT_r\Phi \tag{G.17}$$

where the unknown matrix $T \in R^{N \times N-1}$ is a generalized inverse of the common reference transformation matrix T_r:

$$T = W^{-1}T_r^H\left[T_rW^{-1}T_r^H\right]^{-1} \tag{G.18}$$

With W a weighting matrix, the form is similar to equations (12.42) or (E.5) where the weight is a priori of the source space, here W is about the true potential, and if no special priors, it should be a unit matrix. Actually, equation (G.18) is the weighted minimum length solution of an underdetermined problem (equation (3.43) in Menke (2012)).

According to equation (G.17), where we assume a T fit $\Phi = T\Phi_r$, then, as $\Phi = K_\infty \boldsymbol{j}$, $\Phi_r = T_rK_\infty \boldsymbol{j}$, we have

$$T = K_\infty \sum\nolimits_{jj} K_\infty{}^H T_r^H \left[T_rK_\infty \sum\nolimits_{jj} K_\infty{}^H T_r^H\right]^{-1} \tag{G.19}$$

where K_∞ the head model and $\sum_j = jj^H$ the source scatter matrix. This optimal solution is equivalent to equation (G.18) for $W^{-1} = \sum_{\Phi\Phi} = K_\infty \sum_{jj} K_\infty{}^H$ with $\sum_{\Phi\Phi} = \Phi\Phi^H$ being the scatter matrix of the true potentials. Of course, it remains theoretical because neither the true potentials scatter matrix nor, equivalently, the propagation coefficients between the actual source J and the sensors (depending on the sources positions and orientations and the head model) and the source scatter matrix, are unknown in practice.

In practice, Σ_{jj} is not known and prior covariance matrices $\Sigma_{\overline{jj}}$ are difficult to construct, we have to ignore the effect of the source amplitudes (or their covariance matrix) and take the covariance matrix being a unit matrix. Making $\Sigma_{jj} = I_p$ but assuming a known K_∞, one obtains:

$$T = K_\infty K_\infty{}^H T_r^H \left[T_rK_\infty K_\infty{}^T T_r^H\right]^{-1} = K_\infty\left[T_rK_\infty\right]^+ = K_\infty K_r^+ \tag{G.20}$$

This is the actual REST algorithm (Yao, 2001). This derivation just shows us that if we know something of the sources, the generalized transform operator (G.19) can be adopted, otherwise, we may take the simplest one (G.20) for realizing zero reference. Actually the assumption of j in deriving (G.20) is the iid assumption of the sources as in equations (14.12) and (G.9). And if we take the equivalent source principle, we may assume the positions and orientations of the equivalent sources then we have the K_∞. Interestingly, equation (G.20) enlightens the fact that any a full-rank matrix K_∞ can be used to construct a generalized inverse of T_r. Indeed,

$$T_r K_\infty K_\infty{}^H T_r^H \left[T_r K_\infty K_\infty{}^H T_r^H \right]^{-1} = I_{N-1} \qquad \text{(G.20a)}$$

Regardless of the accuracy and realness of the model K_∞. In this sense, a completely false or random model K_∞ will yield false estimates of the true potentials $\tilde{\Phi}$, but they will still verify the measured common reference signals $\Phi_r = T_r \tilde{\Phi}$. This fact shows that only mathematical skills are not enough in solving practical problem. Mathematical variables must correspond to real physical quantities in order to truly solve practical physical problems. The previous simulation studies show that the head model accuracy has a small but real impact on REST, which is the explanation to this equation (G.20a).

In fact, the source distribution assumption in REST is very important. It should have the ability to fully express the function of the real sources, so that it can be the equivalent source of the real source. In general, it can be a closely distributed source layer enclosing all latent real sources inside (Chapter 6), or a 3D distributed source overlap the whole space of the latent source space or other sources combination which are of quite similar spherical spectra (Chapter 9). In terms of operation, if the assumed equivalent layer is too close to the scalp surface, we may need a very dense discrete distributed source and need it to almost cover the whole head to make it able to equivalently represent the contribution of the all latent actual sources inside. On the other hand, if the layer is too deep (below the brain surface), it may not able to represent the cortical sources as it may not able to have enough high spatial spectra to represent the shallow sources.

If we have no any *a priori* information about the sources (i.e., neither on K_∞ nor on the source covariance), then a minimum norm solution (MN) obtained when $W = I_N$ from equation (G.18) will deduce an estimation of the true potential:

$$\hat{\Phi} = T_r^H \left(T_r T_r^H \right)^{-1} \Phi_r = T_r^+ \Phi_r \qquad \text{(G.21)}$$

with T_r^+ the Moore–Penrose pseudo–inverse of the UR transformation matrix T_r.

Proposition G.1: The Minimum Norm Solution to the Inverse Unipolar Reference Problem Is the AR Solution

$$T_r^+ = T_{\text{AR}} \qquad \text{(G.21a)}$$

Proof. By Sherman–Morrison formula and using the definition of the Moore–Penrose pseudo-inverse and the expression of T_r (G.12), one can write:

$$T_r^+ = T_r^H \left(T_r T_r^H\right)^{-1} = \begin{bmatrix} I_{N-1} \\ -1_{N-1}^T \end{bmatrix} \left(\begin{bmatrix} I_{N-1} - 1_{N-1} \end{bmatrix} \begin{bmatrix} I_{N-1} \\ -1_{N-1}^T \end{bmatrix} \right)^{-1}$$

$$= \begin{bmatrix} I_{N-1} \\ -1_{N-1}^T \end{bmatrix} \left[I_{N-1} + 1_{N-1} 1_{N-1}^T \right]^{-1} = \begin{bmatrix} I_{N-1} \\ -1_{N-1}^T \end{bmatrix} \left(I_{N-1} - 1_{N-1} \frac{1_{N-1}^T}{1 + 1_{N-1}^T 1_{N-1}} \right) \quad \text{(G.22)}$$

$$= \begin{bmatrix} I_{N-1} \\ -1_{N-1}^T \end{bmatrix} \left(I_{N-1} - 1_{N-1} \frac{1}{N} 1_{N-1}^T \right) = T_{\text{AR}}$$

Equation (G.21a) was repeatedly confirmed in Hu et al. (2018a) and Salido-Ruiz et al. (2019). It is also called "Orthogonal projector centering property" of UR as shown in following equation (G.34). Comparing equation (G.21a) with (G.20), the AR montage obtained by pseudo-inverting T_r can be seen as another particular case of REST with $K=I$. Apparently, equation (G.21a) just like a placebo, because T_r is an underdetermined matrix, its ordinary inverse does not exist, and the real potential cannot be restored through its generalized inverse, but it can induce the AR-based potential. Meanwhile, if $K_\infty = I$ and sources are iid, the scalp recordings are iid too, and that is the primary requirement of the AR to be zero reference as noted in equation (14.4) and (Hu et al., 2019).

The above two approaches, MAP estimate and weighted minimum norm solution are similar in taking the zero reference problem as a neutral reference potential estimate but different in the object function (Menke, 2012). The former one takes strength on the statistic properties, when the scalp channel signals are independent to each other (iid). The estimate reduces to AR. The latter one focuses on the physical scene, when the leadfield matrix K_∞ is a unit matrix, REST reduces to AR, too. Apparently, $K_\infty = I$ means that each equivalent source only contributes to one channel, and if we assume that the equivalent sources are mutually independent, the channel recordings are mutually independent, too. It can be seen that the theoretical basis of these two approaches is the same thing. However, in practice, each source will contribute to the all channels; thus, it is impossible to have the channel signals being mutually independent, and the potential referenced to a neutral reference is not able to be reconstructed from a direct generalized inverse from AR recordings. In physics, if the equivalent sources are very close to the scalp surface, the mixture effect of volume conductor becomes neglected; each equivalent source approximately contributes to the closest point on the scalp and thus make the scalp channel signals looks independent. For this case, we may need a very dense source distribution, and it cover the whole head so that it is able to represent the contribution of the all latent actual sources in the brain. For this case, it is similar to the whole brain integral, and if we know the integral value (a prior), the inverse is unique. For example, if the brain is a sphere, the integral is zero, otherwise, we need to seek the integral additionally (Appendix F).

G.3 STATISTICAL ANALYSIS OF UNIPOLAR REFERENCE

In Chapter 14 and related literature, various references such as the online single recording reference electrode Cz, Fz, Oz, FCz, etc. and the offline re-references such as the REST zero reference linked mastoids (LM), average reference (AR), and are comparatively investigated. All these are unipolar references (URs), meaning all the active electrodes are referenced to a unique reference electrode. However, there is still something vague and may be the core behind the phenomenal differences. For instances, what are the common properties of the UR?

Here presented are the general forms of the EEG UR, the demonstration of REST as a UR, and the notable properties of URs. These following materials are modified from Hu et al. (2019).

G.3.1 The Family of Unipolar References

This part takes the UR operator as T_r; thus, the UR transforming is

$$\Phi_r = T_r\Phi + e_r \tag{G.23}$$

UR is the overwhelming body of the reference issue, as its goal is to approach the ideal potentials with infinity reference.

The physical UR is usually the electrode (e.g., Cz, Fz, Oz, T3, T4, and FCz) placed on the scalp or the body surface during online recording. The virtual reference is the linearly combined signal of the recordings from all or part of the electrodes, during offline processing after the EEG data acquisition. Typical examples of virtual references are the LM, AR, and REST.

The reference operator in equation (G.23) has a common structure (Hu et al., 2018b, 2019) for the family of URs as,

$$T_r = I_N - 1f_r^T \tag{G.24}$$

where f_r consists of the linear combination weights of all the electrodes shown in Table G.2, where N including the reference channel.

AR is one of the most widely used methods to estimate the potentials Φ with infinity reference as

$$T_{\text{AR}} = I_N - 1f_{\text{AR}}^T, \quad f_{\text{AR}} = 1/N \tag{G.25}$$

It is justified that for a perfect layered spherical head, with neural currents spreading in an isotropic way, the integral of the potential over the scalp surface is zero (Appendix F). Thus, the averaged potential over all electrodes might tend to zero and would be suitable as the reference signal. However, these assumptions cannot fitted in general.

REST employs the equivalent source technique to transform one reference recording to another as

$$\hat{\Phi}_{\text{REST}} = K_\infty \left(K_r^+ \Phi_r\right) = \left(K_\infty K_r^+\right)\Phi_r = R_r \Phi_r \tag{G.26}$$

where $R_r = K_\infty K_r^+$ is the reference standardization matrix depending on the reference transform T_r in the EEG data Φ_r, and the equivalent source is approximately estimated as $\hat{j} = K_r^+ \Phi_r$. Note that R_r is transforming the referenced data Φ_r, thus, REST was previously taken as a re-reference. LM and AR are also re-references, both transform the ideal potentials Φ with the infinity reference. To evaluate the performance of REST, one needs to check how it transform the ideal potentials Φ with the infinity reference that is the root of the reference issue. For this reason, the REST operator is further defined as

$$T_{\text{REST}} = K_\infty \left(T_r K_\infty\right)^+ T_r = K^\infty \left(K_r\right)^+ T_r \tag{G.27}$$

by post-multiplying R_r with the reference T_r hidden in the data Φ_r (Hu et al., 2018b). The unipolar form of T_{REST} is derived next.

G.3.2 Demonstration of REST as a UR

The lead field referenced to the same UR as that in equation (G.24) is

$$K_r = T_r K_\infty = K_\infty + 1\left(-K_\infty^T f_r\right)^T \tag{G.28}$$

Since the number of distributed neural sources is much larger than the number of electrodes and because of the volume conductivities, K_∞ has all independent rows, namely, full row rank, leading $K_\infty K_\infty^+ = I_N$, and rank $rk\left(K_r\right) = rk\left(T_r\right)$. Noting that T_r is with full rank deficient by 1; thus, $rk\left(K_r\right) = rk\left(K_\infty\right) - 1$, which is the case (↓) of Theorem 1.1 in Baksalary et al. (2003). Define $d = -K_\infty^+ 1$ as the formula (1.3) in Baksalary et al. (2003), we have

$$K_r^+ K_r = K_\infty K_\infty^+ - \frac{dd^T}{d^T d} = I_N - \frac{K_\infty^+ 1 1^T K_\infty^{+T}}{1^T K_\infty^{+T} K_\infty^+ 1} \tag{G.29}$$

according to the case (↓) in the list 2.2 of Theorem 2.1 in Baksalary et al. (2003).

Post-multiplying $K_\infty K_\infty^+ = I_N$, the REST operator in equation (G.27) is equivalent to

$$T_{\text{REST}} = K_\infty K_r^+ K_r K_\infty^+ = I_N - 1\frac{1^T K_\infty^{+T} K_\infty^+}{1^T K_\infty^{+T} K_\infty^+ 1} \tag{G.30}$$

TABLE G.2
The Family of Unipolar References

Unipolar references	$\Phi_r = T_r\Phi + \varepsilon_r,\ T_r = I_{N_c} - 1f_r^T,\ f_r \in \{f_{\text{RR}}, f_{\text{LM}}, f_{\text{AR}}, f_{\text{REST}}\}$	
Online recording references	Cz, Fz, Oz, FCz, etc.	$f_{\text{RR}} = [0,\ldots,0,1,0,\ldots,0]^T$
Offline re-references	LM	$f_{\text{LM}} = [0,\ldots,0.5,\ldots,0.5,\ldots,0]^T$
	AR	$f_{\text{AR}} = 1/N$
	REST	$f_{\text{REST}} = K_\infty^{+T} K_\infty^{+} 1 / \left[1^T K_\infty^{+T} K_\infty^{+} 1\right]$

LMs: linked mastoids; AR, average reference.

Obviously, the REST operator belongs to the family of URs. Written as $T_{\text{REST}} = I_N - 1f_{\text{REST}}^T$, the linear combination weights for REST are

$$f_{\text{REST}} = K_\infty^{+T} K_\infty^{+} 1 / \left[1^T K_\infty^{+T} K_\infty^{+} 1\right] \tag{G.31}$$

Therefore, REST operator admits the same form of URs defined in equation (G.24). Different from the reference standardization matrix R_r, the REST operator T_{REST} is independent to the specific UR in the EEG data, and will be identical whichever T_r is adopted in equation (G.27), noting that f_r disappears in equation (G.31). The demonstration of REST as a UR enable absorbing REST into the family of URs.

G.3.3 Properties of URs: No Memory

Supposing $T_{r1} = I_N - 1f_{r1}^T$ is the latest reference one intends to apply, and T_{r2} is the previous reference already applied in the recorded data, as long as $f_{r1}^T 1 = 1$, one will have

$$T_{r1} = T_{r1} T_{r2} \tag{G.32}$$

where T_{r2} could be any UR operator.

Note that $f_r^T 1 = 1$ for $f_r \in \{f_{\text{RR}}, f_{\text{LM}}, f_{\text{AR}}, f_{\text{REST}}\}$, this no memory property holds true for the family of URs including both online recording references, e.g., Cz, Fz, Oz, and FCz, and the offline references such as LM, AR, and REST.

REFERENCES

Baksalary, J.K., O.M. Baksalary, G. Trenkler. 2003. A revisitation of formulae for the Moore-Penrose inverse of modified matrices. *Linear Algebra Appl* 372:207–224. doi: 10.1016/S0024-3795(03)00508-1.

Hu, S., D. Yao, M.L. Bringas-Vega, et al. 2019. The statistical estimation of unipolar EEG references: Derivations and properties. *Brain Topogr* 32:696–703.

Hu, S., D. Yao, P.A. Valdes-Sosa. 2018a. Unified Bayesian estimator of EEG reference at infinity: rREST (Regularized Reference Electrode Standardization Technique). *Front Neurosci* 3(12):297. doi: 10.3389/fnins.2018.00297.

Hu, S., Y. Lai, P.A. Valdes-Sosa, et al. 2018b. How do reference montage and electrodes setup affect the measured scalp EEG potentials. *J Neural Eng* 15(2):026013. doi: 10.1088/1741-2552/aaa13f.

Magnus, J.R., H. Neudecker. 2007. *Matrix Differential Calculus with Applications in Statistics and Econometrics*, 3rd edn. New York: John Wiley & Sons Press.

Mardia, K.V., J.T. Kent, J.M. Bibby. 1979. *Multivariate Analysis*. New York: Academic Press.

Menke, W. 2012. *Geophysical Data Analysis: Discrete Inverse Theory*. London: Academic Press.

Salido-Ruiz, R.A., R. Ranta, G. Korats, et al. 2019. A unified weighted minimum norm solution for the reference inverse problem in EEG. *Comput Biol Med* 115:103510.

Trenkler, G. 2000. *On a Generalisation of the Covariance Matrix of the Multinomial Distribution*. Boston, MA: Springer.

Yao, D. 2001. A method to standardize a reference of scalp EEG recordings to a point at infinity. *Physiol Meas* 22(4):693–711.

Appendix H
Vector, Tensor, and Matrix

H.1 UNIT VECTOR TRANSFORM BETWEEN SPHERICAL POLAR COORDINATES AND CARTESIAN COORDINATES

$$\vec{r} = \sin\theta\cos\varphi\vec{x}_1 + \sin\theta\sin\varphi\vec{x}_2 + \cos\theta\vec{x}_3 \tag{H.1a}$$

$$\vec{\theta} = \cos\theta\cos\varphi\vec{x}_1 + \cos\theta\sin\varphi\vec{x}_2 - \sin\theta\vec{x}_3 \tag{H.1b}$$

$$\vec{\varphi} = -\sin\varphi\vec{x}_1 + \cos\varphi\vec{x}_2 \tag{H.1c}$$

$$\vec{x}_1 = \sin\theta\cos\varphi\vec{r} + \cos\theta\cos\varphi\vec{\theta} - \sin\varphi\vec{\varphi} \tag{H.2a}$$

$$\vec{x}_2 = \sin\theta\sin\varphi\vec{r} + \cos\theta\sin\varphi\vec{\theta} + \cos\varphi\vec{\varphi} \tag{H.2b}$$

$$\vec{x}_3 = \cos\theta\vec{r} - \sin\theta\vec{\theta} \tag{H.2c}$$

Directed line element

$$d\vec{ii} = dr\vec{r} + rd\theta\vec{\theta} + r\sin\theta d\varphi\vec{\varphi} \tag{H.3}$$

Solid angle element

$$d\Omega = \sin\theta d\theta d\varphi \tag{H.4}$$

Directed area element

$$d\boldsymbol{S} = dS\vec{r} = r^2 d\Omega\vec{r} \tag{H.5}$$

Volume element

$$dV = drdS = r^2 drd\Omega \tag{H.6}$$

H.2 VECTOR FORMULAE

Vector (bold) algebraic identities (with definition of δ_{ij} and ϵ_{ijk} in equations (H.36) and (H.38))

$$\boldsymbol{a}\cdot\boldsymbol{b}=\boldsymbol{b}\cdot\boldsymbol{a}=\delta_{ij}\boldsymbol{a}_i\boldsymbol{b}_j=\boldsymbol{ab}\cos\theta \tag{H.7}$$

θ the angle between the two vectors

$$\boldsymbol{a}\times\boldsymbol{b}=-\boldsymbol{b}\times\boldsymbol{a}=\epsilon_{ijk}\boldsymbol{a}_i\boldsymbol{b}_k\vec{x}_i \tag{H.8}$$

$$\boldsymbol{a}\cdot(\boldsymbol{b}\times\boldsymbol{c})=(\boldsymbol{a}\times\boldsymbol{b})\cdot\boldsymbol{c} \tag{H.9}$$

$$\boldsymbol{a}\times(\boldsymbol{b}\times\boldsymbol{c})=\boldsymbol{b}(\boldsymbol{a}\cdot\boldsymbol{c})-\boldsymbol{c}(\boldsymbol{a}\cdot\boldsymbol{b}) \tag{H.10}$$

$$\boldsymbol{a}\times(\boldsymbol{b}\times\boldsymbol{c})+\boldsymbol{b}\times(\boldsymbol{c}\times\boldsymbol{a})+\boldsymbol{c}\times(\boldsymbol{a}\times\boldsymbol{b})=0 \tag{H.11}$$

$$(\boldsymbol{a}\times\boldsymbol{b})\cdot(\boldsymbol{c}\times\boldsymbol{d})=\boldsymbol{a}\cdot\left[\boldsymbol{b}\times(\boldsymbol{c}\times\boldsymbol{d})\right]=(\boldsymbol{a}\cdot\boldsymbol{c})(\boldsymbol{b}\cdot\boldsymbol{d})-(\boldsymbol{a}\cdot\boldsymbol{d})(\boldsymbol{b}\cdot\boldsymbol{c}) \tag{H.12}$$

$$(\boldsymbol{a}\times\boldsymbol{b})\times(\boldsymbol{c}\times\boldsymbol{d})=(\boldsymbol{a}\times\boldsymbol{b}\cdot\boldsymbol{d})\boldsymbol{c}-(\boldsymbol{a}\times\boldsymbol{b}\cdot\boldsymbol{c})\boldsymbol{d} \tag{H.13}$$

Vector analytic identities

$$\nabla(\alpha\beta)=\alpha\nabla\beta+\beta\nabla\alpha \tag{H.14}$$

$$\nabla\cdot(\alpha\boldsymbol{a})=\boldsymbol{a}\cdot\nabla\alpha+\alpha\nabla\cdot\boldsymbol{a} \tag{H.15}$$

$$\nabla\times(\alpha\boldsymbol{a})=\alpha\nabla\times\boldsymbol{a}-\boldsymbol{a}\times\nabla\alpha \tag{H.16}$$

$$\nabla\cdot(\boldsymbol{a}\times\boldsymbol{b})=\boldsymbol{b}\cdot(\nabla\times\boldsymbol{a})-\boldsymbol{a}\cdot(\nabla\times\boldsymbol{b}) \tag{H.17}$$

$$\nabla\times(\boldsymbol{a}\times\boldsymbol{b})=\boldsymbol{a}(\nabla\cdot\boldsymbol{b})-\boldsymbol{b}(\nabla\cdot\boldsymbol{a})+(\boldsymbol{b}\cdot\nabla)\boldsymbol{a}-(\boldsymbol{a}\cdot\nabla)\boldsymbol{b} \tag{H.18}$$

$$\nabla(\boldsymbol{a}\cdot\boldsymbol{b})=\boldsymbol{a}\times(\nabla\times\boldsymbol{b})+\boldsymbol{b}\times(\nabla\times\boldsymbol{a})+(\boldsymbol{b}\cdot\nabla)\boldsymbol{a}+(\boldsymbol{a}\cdot\nabla)\boldsymbol{b} \tag{H.19}$$

$$\nabla\cdot\nabla\alpha=\nabla^2\alpha \tag{H.20}$$

$$\nabla\times\nabla\alpha=0 \tag{H.21}$$

$$\nabla\cdot(\nabla\times\boldsymbol{a})=0 \tag{H.22}$$

$$\nabla\times(\nabla\times\boldsymbol{a})=\nabla(\nabla\cdot\boldsymbol{a})-\nabla^2\boldsymbol{a} \tag{H.23}$$

Some special identities

assuming $x' = \sum_{i=1}^{3} x_i' \vec{x}_i$, $\boldsymbol{k}$ an arbitrary *constant* vector, $\boldsymbol{a} = \boldsymbol{a}(\boldsymbol{x})$ an arbitrary vector field

$$\nabla \cdot \boldsymbol{x} = 3 \tag{H.24}$$

$$\nabla \times \boldsymbol{x} = 0 \tag{H.25}$$

$$\nabla(\boldsymbol{k} \cdot \boldsymbol{x}) = \boldsymbol{k} \tag{H.26}$$

$$\nabla|\boldsymbol{x}| = \frac{\boldsymbol{x}}{|\boldsymbol{x}|} \tag{H.27}$$

$$\nabla(|\boldsymbol{x} - \boldsymbol{x}'|) = \frac{\boldsymbol{x} - \boldsymbol{x}'}{|\boldsymbol{x} - \boldsymbol{x}'|} = -\nabla'(|\boldsymbol{x} - \boldsymbol{x}'|) \tag{H.28}$$

$$\nabla\left(\frac{1}{|\boldsymbol{x}|}\right) = -\frac{\boldsymbol{x}}{|\boldsymbol{x}|^3} \tag{H.29}$$

$$\nabla\left(\frac{1}{|\boldsymbol{x} - \boldsymbol{x}'|}\right) = -\frac{\boldsymbol{x} - \boldsymbol{x}'}{|\boldsymbol{x} - \boldsymbol{x}'|^3} = -\nabla'\left(\frac{1}{|\boldsymbol{x} - \boldsymbol{x}'|}\right) \tag{H.30}$$

$$\nabla \cdot \left(\frac{\boldsymbol{x} - \boldsymbol{x}'}{|\boldsymbol{x} - \boldsymbol{x}'|^3}\right) = -\nabla^2\left(\frac{1}{|\boldsymbol{x} - \boldsymbol{x}'|}\right) = 4\pi\delta(\boldsymbol{x} - \boldsymbol{x}') \tag{H.31}$$

H.3 INTEGRAL RELATIONS

Assume $V(S)$ be a volume with a closed surface $S(V)$,and denote the 3D volume element as $dV = d^3x$ and the surface element, $dS\left(\equiv d^2x\vec{n}\right)$with surface normal unit vector $\vec{n}$ directed along the outward direction. Then

$$\int_V (\nabla \cdot \boldsymbol{a}) d^3x = \oint_s d\boldsymbol{S} \cdot \boldsymbol{a} \tag{H.32}$$

$$\int_V (\nabla \alpha) d^3x = \oint_s d\boldsymbol{S}\alpha \tag{H.33}$$

$$\int_V (\nabla \times \boldsymbol{a}) d^3x = \oint_s d\boldsymbol{S} \times \boldsymbol{a} \tag{H.34}$$

H.4 TENSOR FIELDS

An arbitrary *tensor field* $\boldsymbol{A}(\boldsymbol{x})$ in R^3 can be represented in the following *matrix form*:

$$\left(A_{ij}\left(x_k\right)\right)\underline{\text{def}}\begin{pmatrix} A_{11}(\boldsymbol{x}) & A_{12}(\boldsymbol{x}) & A_{13}(\boldsymbol{x}) \\ A_{21}(\boldsymbol{x}) & A_{22}(\boldsymbol{x}) & A_{23}(\boldsymbol{x}) \\ A_{31}(\boldsymbol{x}) & A_{32}(\boldsymbol{x}) & A_{33}(\boldsymbol{x}) \end{pmatrix} \tag{H.35}$$

The 3D *Kronecker delta* with symbol δ_{ij} is a particularly tensor of rank 2 in R^3:

$$\delta_{ij} = \begin{cases} 0 & \text{if } i \neq j \\ 1 & \text{if } i = j \end{cases} \tag{H.36}$$

And it has the following matrix representation

$$\left(\delta_{ij}\right) = \begin{pmatrix} 1 & 0 & 0 \\ 0 & 1 & 0 \\ 0 & 0 & 1 \end{pmatrix} \tag{H.37}$$

The *Levi-Civita tensor,* a common and useful tensor of rank 3, is the fully antisymmetric tensor:

$$\varepsilon_{ijk} = \begin{cases} 1, & \text{if } i,j,k \text{ is an even permutation of } 1,2,3 \\ 0, & \text{if at least two of } i,j,k \text{ are equal} \\ -1, & \text{if } i,j,k \text{ is an odd permutation of } 1,2,3 \end{cases} \tag{H.38}$$

With the following further property

$$\varepsilon_{ijk}\varepsilon_{ilm} = \delta_{jl}\delta_{km} - \delta_{jm}\delta_{kl} \tag{H.39}$$

Tensors may have any rank n. A scalar is considered to be a tensor of rank $n=0$ and a vector a tensor of rank $n=1$. Consequently, the notation where a vector (tensor) is represented in its component form is called the *tensor notation.* A tensor of rank $n=2$ may be represented by a two-dimensional matrix, whereas higher rank tensor are best represented in their component forms (tensor notation).

H.5 MATRIX

A matrix is defined as a square or rectangular array of numbers or functions, a matrix is not a determinant. It is an ordered array of numbers, not a single number.

In general, matrix multiplication is not commutative:

$$AB \neq BA \tag{H.40}$$

But an associative law holds, $(AB)C = A(BC)$, distributive law holds, $A(B+C) = AB + AC$

If A is an $n \times n$ matrix with determinant $|A| \neq 0$, then it has a unique inverse A^{-1} satisfying $AA^{-1} = A^{-1}A = 1$. If B is also an $n \times n$ matrix with inverse B^{-1}, then the product AB has unique inverse:

$$(AB)^{-1} = B^{-1}A^{-1} \tag{H.41}$$

If A is an $m \times m$ matrix and B is an $n \times n$ matrix, then the Kronecker product is $C = A \otimes B$ which is a mn×mn matrix with elements.

$$C_{\alpha\beta} = A_{ij}B_{kl} \quad \text{with } \alpha = m(i-1)+k, \ \beta = n(j-1)+1 \tag{H.42}$$

If A and B are diagonal, then

$$AB = BA \tag{H.43}$$

In any a square matrix the sum of the diagonal elements is called the **trace**, and

$$\text{trace}(A - B) = \text{trace}(A) - \text{trace}(B) \tag{H.44}$$

If

$$A^H = A \tag{H.45}$$

the matrix is called **symmetric**, whereas if

$$A = -A^H \tag{H.46}$$

it is called **antisymmetric**. If $A^H A = I$ or $A^H = A^{-1}$, A is orthogonal.

For more tools, details, and more formulae, refers to Thide (2004), Magnus and Neudecker (2007), Baksalary et al. (2003), Hager (1989), and Rakha (2004).

REFERENCES

Baksalary, J.K., O.M. Baksalary, G. Trenkler. 2003. A revisitation of formulae for the Moore-Penrose inverse of modified matrices. *Linear Algebra Appl* 372:207–224. doi: 10.1016/S0024-3795(03)00508-1.

Hager, W.W. 1989. Updating the inverse of a matrix. *SIAM Rev* 31(2):221–239.

Magnus, J.R., H. Neudecker. 2007. *Matrix Differential Calculus with Applications in Statistics and Econometrics*, 3rd edn. New York: John Wiley & Sons Press.

Rakha, M.A. 2004. On the Moore-Penrose generalized inverse matrix. *App Math Comput* 158:185–200.

Thide, B. 2004. *Electromagnetic Field Theory*. Braşov: Transilvania University Press.

Index

For Product Safety Concerns and Information please contact our EU representative GPSR@taylorandfrancis.com
Taylor & Francis Verlag GmbH, Kaufingerstraße 24, 80331 München, Germany

www.ingramcontent.com/pod-product-compliance
Lightning Source LLC
LaVergne TN
LVHW020550110826
845149LV00002B/221

* 9 7 8 1 0 3 2 6 3 9 2 4 6 *